W0114723

Static & Dynamic Game Theory: Foundations & Applications

Series Editor
Tamer Başar, University of Illinois, Urbana-Champaign, IL, USA

Editorial Advisory Board
Daron Acemoglu, MIT, Cambridge, MA, USA
Pierre Bernhard, INRIA, Sophia-Antipolis, France
Maurizio Falcone, Università degli Studi di Roma "La Sapienza", Rome, Italy
Alexander Kurzhanski, University of California, Berkeley, CA, USA
Ariel Rubinstein, Tel Aviv University, Ramat Aviv, Israel; New York University, New York, NY, USA
William H. Sandholm, University of Wisconsin, Madison, WI, USA
Yoav Shoham, Stanford University, Palo Alto, CA, USA
Georges Zaccour, GERAD, HEC Montréal, Canada

For further volumes:
www.springer.com/series/10200

David W.K. Yeung · Leon A. Petrosyan

Subgame Consistent Economic Optimization

An Advanced Cooperative Dynamic Game Analysis

 Birkhäuser

David W.K. Yeung
SRS Consortium for Advanced Study
in Cooperative Dynamic Games
Hong Kong Shue Yan University
Hong Kong
People's Republic of China
dwkyeung@hksyu.edu

Center of Game Theory
St. Petersburg State University
Saint Petersburg 198904
Russia

Leon A. Petrosyan
Faculty of Applied Mathematics
and Control Processes
St. Petersburg State University
Saint Petersburg 198904
Russia
spbuoasis7@peterlink.ru

ISBN 978-0-8176-8261-3   e-ISBN 978-0-8176-8262-0
DOI 10.1007/978-0-8176-8262-0
Springer New York Dordrecht Heidelberg London

Library of Congress Control Number: 2011942872

Mathematics Subject Classification (2010): 91A12, 92A25, 91B62, 78M50, 91B70

© Springer Science+Business Media, LLC 2012
All rights reserved. This work may not be translated or copied in whole or in part without the written permission of the publisher (Springer Science+Business Media, LLC, 233 Spring Street, New York, NY 10013, USA), except for brief excerpts in connection with reviews or scholarly analysis. Use in connection with any form of information storage and retrieval, electronic adaptation, computer software, or by similar or dissimilar methodology now known or hereafter developed is forbidden.
The use in this publication of trade names, trademarks, service marks, and similar terms, even if they are not identified as such, is not to be taken as an expression of opinion as to whether or not they are subject to proprietary rights.

Printed on acid-free paper

Springer is part of Springer Science+Business Media (www.birkhauser-science.com)

To Stella and Nina

Foreword

In the middle of the twentieth century, the seminal work of Rufus Isaacs on differential games established the foundation for the analysis of interactive behavior over time. Since then, dynamic game theory has been applied in many disciplines; in the past three and a half decades, its applications in economics and business have been growing rapidly. Market failures give rise to the need for cooperation in conducting economic activities. Limited success from dynamic cooperation can be expected if there is no guarantee that the participants' agreed-upon scheme will always be maintained amid changes over time within the entire duration of the cooperation.

The solution mechanism for obtaining time consistent and subgame consistent cooperative schemes developed by Leon Petrosyan and David Yeung—the authors of this text—provides an effective prescription to this "classic" game-theoretic problem. Practical policy menus can be formulated with this mechanism to tackle some of the serious problems facing the global market economy. This text covers the main development of subgame consistent economic optimization and includes most of the existing economic studies involving subgame consistent solutions. Illuminating applications are presented to illustrate the detailed workings of the fundamental theorems established.

Atypical of current mainstream economic studies, which adopted existing mathematics in their analysis, this text uses novel game-theoretic mathematical techniques developed by its authors.

The text is truly a world-leading treatise in the field of dynamically consistent economic optimization and a Russian classic in mathematics and economics. It is a timely publication to tackle the increasingly crucial issues of consistency and dynamic stability in collaborative activities in the economic arena. The elegant mathematics developed by the authors and their practical applications in economics are prevalent in the analysis. The text significantly expands L.V. Kantorovich's award-winning work in economic optimization in the new directions of game-theoretic interaction, dynamic evolution, stochasticity, and subgame consistency.

Subgame Consistent Economic Optimization is undoubtedly a needed and important addition to the field of dynamic interactive optimization in economics.

Karelian Institute of   Vladimir Mazalov
Applied Mathematical Research,
Russian Academy of Sciences, Russia

Preface

The postulation that individually rational, self-maximizing behaviors bring about group (Pareto) optimality constitutes one of the most appealing characteristics of the perfectly competitive market. The market is often regarded as an effective means to allocate economic resources efficiently. However, in the presence of an imperfect market structure, externalities, imperfect information, and public goods, the market fails to provide an effective mechanism for efficient resource use. Not only have inefficient outcomes appeared, but gravely detrimental events—such as the global financial crisis and catastrophe-bound industrial pollution problem—have also emerged under the current market system. With market failures prevailing, optimization in economic activities is one of the remedies available.

Strategic behaviors in the market are increasingly pervasive, and as a result, game theory has emerged as one of the fundamental tools in pure and applied research in economics. Because economic activities in the modern corporate world are dynamic processes, economic decisions are more appropriately analyzed in an intertemporal framework. Dynamic cooperation suggests the possibility of socially optimal and group efficient solutions to economic decision problems involving strategic action.

In dynamic cooperation, a stringent condition is required for a scheme to be dynamically stable. In particular, the optimality principle must remain optimal throughout the game, that is, at any instant of time along the optimal state trajectory determined at the outset. This condition is known as time consistency. In the presence of stochastic elements, a more stringent condition—that of subgame consistency—is required for a credible cooperative solution. In particular, a cooperative solution is subgame consistent if an extension of the solution policy to a situation with a later starting time, and any realizable state brought about by prior optimal behavior, would remain optimal. The notion of subgame consistency originated in Yeung and Petrosyan (2004), which develops a generalized theorem for the derivation of an analytically tractable "payoff distribution procedure" leading to subgame consistent solutions. Time consistency for the economic optimization problem requires dynamical consistency for all subgames along the group optimal trajectory; then time consistency in this context reflects optimal-trajectory-subgame consistency.

This book provides a treatise on subgame consistent economic optimization. In particular, dynamically stable game-theoretic optimization techniques are developed to establish the foundation for an effective policy menu to tackle suboptimal problems that the conventional market mechanism fails to resolve. The book is expected to be used as an analytical tool for advanced graduate students, game theorists, economists, mathematicians, and researchers in this field.

We are very grateful to our esteemed friends George Leitmann and John Nash for inspiration from their classic work in game theory, on which many of the results in the book are based. Our families have been an enormous and continuing source of encouragement throughout our careers. We thank Stella and Patricia (DY) and Nina and Ovanes (LP), for their love and patience during this and other projects, which, on occasion, may have diverted our attention away from them. We thank Cynthia Yingxuan Zhang for her outstanding research assistance and manuscript formatting. Financial support from the Research Grants Council of the HKSAR (Grant Number 32-07-028), the European Union Research Commission (Contract Number 044287), and HKSYU is gratefully acknowledged.

Finally, we would like to dedicate this text to honor the memory of a pioneering researcher and Nobel Laureate in the field of economic optimization—our late Saint Petersburg colleague Leonid Vitalyevich Kantorovich—in his 100th birthday tribute.

Saint Petersburg, Russia   D.W.K. Yeung
                                                                                  L.A. Petrosyan

Contents

Chapter 1
Introduction

> Emile de Laveleye (1882): "Political economy may be defined as the science which determines what laws men ought to adopt in order that they may, with the least possible exertion, procure the greatest abundance of things useful for the satisfaction of their wants, may distribute them justly and consume them rationally."

The most appealing characteristic of the perfectly competitive market is perhaps the postulation that individually rational self-maximizing behaviors bring about group (Pareto) optimality. Hence the market is regarded as an effective means to allocate economic resources efficiently. However, a competitive market will fail to provide an efficient allocation mechanism if there exists an imperfect market structure, externalities, imperfect information, or public goods. These phenomena are prevalent in the current global economy. As a result, though the market is perceived to be the most effective instrument in conducting economic activities, it fails to guarantee its efficiency under many current conditions. Not only have inefficient market outcomes appeared, but gravely detrimental events—such as the worldwide financial crisis and catastrophe-bound industrial pollution problem—have also emerged under the conventional market system.

The 2008 worldwide financial tsunami has revealed, perhaps, the gravest situation generated by the global market economy. The event led to serious challenges on the performance of existing market systems. In one of the most thought-provoking publications about financial markets in the last century, Soros (1998) presented the thesis that the global capitalist system is in crisis. In particular, he pointed out that financial markets are inherently unstable. Market fundamentalism is defined as the widespread belief that markets are self-correcting, that a global economy can flourish under the belief that the common interest is served by allowing everyone to look out for his or her interests, and that attempts to protect the common interest by collective decision making distort the market mechanism. "This belief is false" said Soros. "Instead of acting like a pendulum, financial markets have recently acted more like a wrecking ball, knocking over one economy after another."

D.W.K. Yeung, L.A. Petrosyan, *Subgame Consistent Economic Optimization*, Static & Dynamic Game Theory: Foundations & Applications, DOI 10.1007/978-0-8176-8262-0_1, © Springer Science+Business Media, LLC 2012

Iceland, Greece, Italy, Spain, and Portugal are current examples of knocked-over economies in the midst of the 2008 financial tsunami.

In the presence of market failures, the optimization of economic activities provides an effective remedial measure. One of the most successful pioneers in economic optimization was Leonid Vitalyevich Kantorovich. His approach to optimizing resource use was of crucial importance in the economy of the former Soviet Union. Moreover, the work of Kantorovich played a significant role in introducing mathematics into economic optimization. One of the main features of Kantorovich's economic optimization paradigm was the development of novel mathematical tools to handle real-life economic problems rather than just borrowing the techniques from existing mathematics.

Economic analysis no longer treats the economic system as a given since the appearance of Hurwicz's (1973) pioneering work on mechanism design. The term "design" stresses that the structure of the economic system is to be regarded as an unknown. The design point of view enlarges our vision and helps economics avoid a narrow focus on existing institutions. After the 2008 financial tsunami, Soros (2009) stressed that the global market needs new international rules if another global financial collapse is to be avoided. Like Soros, many believe that international cooperation is the way out of the current global financial crisis. International endeavors such as the Kyoto Protocol, Copenhagen Accord, Organization for Economic Cooperation & Development (OECD), Asia-Pacific Economic Cooperation (APEC), and G-20 summits are vivid examples of joint optimization initiatives to seek remedies for the failed market mechanism. In a broad sense, economic optimization can be applied in the micro-framework involving a particular industry or a group of firms, in the macro-national level, and in a global international framework.

Hurwicz (1973) also pointed out that in economics, one deals with goal conflicts due to the multiplicity of consumers and one-objective-function problems, which fail to address the crucial issue of goal conflict. The role of strategic interactions in the current market system is increasingly recognized in theory and practice. As a result, game theory has emerged as one of the fundamental tools in pure and applied research in economics. The discipline of game theory studies decision making in an interactive environment. In canonical form, an economics game arises when an economic agent pursues an objective(s) in a situation in which other agents concurrently pursue other (possibly conflicting, possibly overlapping) objectives: The problem is then to determine each agent's optimal decision, how these decisions interact to produce equilibria, and the properties of such outcomes. The foundations of game theory were established some 60 years ago by von Neumann and Morgenstern (1944).

Advances in technology, communications, industrial organization, regulation methodology, international trade, economic integration, and political reform have created rapidly expanding social and economic networks incorporating cross-personal and cross-country interactions. From a decision- and policy-maker's perspective, it has become increasingly important to recognize and accommodate the interdependencies and interactions of human decisions under such circumstances. The strategic aspects of decision making are often crucial in areas as diverse as trade negotiation, foreign and domestic investment, multinational pollution planning, market development and integration, joint venture, technological research and development (R&D), resource extraction, competitive marketing, and regional cooperation.

Game theory is perhaps one of the most sophisticated and fertile paradigms that applied mathematics can offer to study and analyze decision making under real-world conditions. Since economic activities in the modern corporate world are dynamic processes, economic decisions would be more appropriately analyzed in an inter-temporal framework. One particularly complex and fruitful branch of game theory is dynamic or differential games, which investigate interactive decision making over time.

Nash's (1950) noncooperative equilibrium theory clearly demonstrated that individually rational behaviors by players would seem to deviate from a jointly optimal outcome when strategic interactions are present. Even worse, there is no guarantee such equilibrium self-maximizing behaviors will not bring about highly undesirable outcomes as those that can be illustrated in the Prisoner's Dilemma paradigm. Economic optimization continues to be a much needed remedy to many of the current economic activities in which strategic interactions are significant.

Cooperative optimization points to the possibility of socially optimal and individually rational solutions to decision making problems involving strategic action over time. However, one may find it hard to be convinced that dynamic cooperation can offer a long-term solution unless there is a guarantee that participants will always be better off throughout the entire cooperation period, and the agreed-upon optimality principle be maintained from beginning to end. Many cooperation schemes become unstable and may fail any time within the cooperation period because of the lack of this kind of guarantee.

To guarantee that cooperation will last throughout the agreement period, a stringent condition is required: The optimality principle agreed upon at the outset must remain effective throughout the game, at any instant of time along the optimal state trajectory. This condition is known as time consistency. In other words, the dynamic stability of solutions in cooperative differential games involves the property that, as the game proceeds along an optimal trajectory, players are guided by the same optimality principle at each instant of time, and hence do not possess incentives to deviate from the previously adopted optimal behavior throughout the game.

In the presence of stochastic elements, a more stringent condition—that of subgame consistency—is required for a credible cooperative solution. In particular, the optimality principle agreed upon at the outset must remain effective in any subgame starting at a later time with a realizable state brought about by prior optimal behavior. The notion of subgame consistency originated in our 2004 article (Yeung and Petrosyan 2004), in which a generalized theorem for the derivation of an analytically tractable "payoff distribution procedure" leading to subgame consistent solutions is developed. A series of further developments and extensions can be found in Petrosyan and Yeung (2006, 2007), Yeung (2005, 2006, 2007, 2008, 2011a), Yeung and Petrosyan (2005, 2006a, 2006b, 2007a, 2007b, 2007c, 2008, 2010), and Yeung et al. (2007).

Because time consistency for the economic optimization problem requires dynamical consistency for all subgames along the group optimal trajectory, time consistency in this context reflects an optimal-trajectory-subgame consistency.

This book presents a treatise on subgame consistent economic optimization. In particular, game-theoretic optimization is developed to establish the foundation for an effective policy menu to tackle suboptimal problems that the conventional market mechanism fails to resolve. The book expands Kantorovich's single-agent economic optimization paradigm to a multiple-agent framework. Moreover, novel mathematics developed by the authors is provided for dealing with game-theoretic economic optimization problems.

The text is organized as follows. Chapter 2 examines the dynamic strategic interactions in the economic system and typical dynamic economic game paradigms. Market equilibrium outcomes and the characterization of the solutions in open-loop and feedback strategies are provided. The dynamic stochastic interactive economic system and characterization of corresponding market outcomes are also presented.

Chapter 3 examines two fundamental elements in dynamic economic optimization—group optimality and individual rationality. Group optimal cooperative strategies and cooperative state trajectories are characterized. The conditions under which individual rationality are maintained throughout the cooperation period are identified.

In Chap. 4, time (optimal-trajectory-subgame) consistent economic optimization is analyzed. The principle of time consistency, time consistent cooperative strategies, imputation, and optimality principles are scrutinized. Payoff distribution procedures leading to time consistent solutions are derived. Infinite-horizon analysis is also analyzed.

Chapter 5 presents a dynamically stable joint venture. Time (optimal-trajectory-subgame) consistent profit sharing in joint ventures and instantaneous venture transfer payments are developed. Because the sizes and earning potentials of the firms in a corporate joint venture may vary significantly, the problem of profit sharing is inescapable in virtually every joint venture. The analysis first considers the case when the venture agrees to share the excess of the total cooperative payoff over the sum of individual noncooperative payoffs proportional to the firms' noncooperative payoffs. A Shapley value solution is also provided.

In Chap. 6, economic optimization is applied to collaborative environmental management. It is hard to be convinced that current multinational joint initiatives, such as the Kyoto Protocol or the Copenhagen Accord, can offer a long-term solution because there is no guarantee that the participants will always be better off within the entire duration of the agreement. Because of the lack of these kinds of incentives, current cooperative schemes fail to provide an effective means to solve the problem. This is a "classic" game-theoretic problem. A theoretical framework capturing the essence of a transboundary industrial pollution paradigm in the form of a differential game is adopted. A time (optimal-trajectory-subgame) consistent cooperative solution is illustrated in the chapter. Benefit distributions in collaborative environmental abatements fulfilling time consistency are obtained. Policy implications are also analyzed.

Chapter 7 looks into the special case of a cartel with dormant firms. Optimization in cartels, which restricts outputs to enhance their joint profits is examined. In particular, some firms have absolute cost disadvantages and this, therefore, forces them to become dormant partners. The Pareto optimal output path, time (optimal-trajectory-subgame) consistent imputation scheme, and cartel profit sharing are examined.

Chapter 8 considers dynamic economic optimization under uncertainty. The principles of subgame consistency, subgame consistent cooperative strategies, and imputation are scrutinized. Mechanisms for the derivation of a payoff distribution procedures leading to subgame consistent solutions and instantaneous transfer payments are presented.

Chapter 9 investigates a dynamic corporate joint venture under uncertainty and the corresponding subgame consistent solutions. Chapter 10 analyzes collaborative environmental management under uncertainty. Group optimality cooperative state trajectory, subgame consistent imputation, and benefit distribution are derived. Chapter 11 considers the dormant firm cartel under uncertainty and the corresponding subgame consistent solutions.

Chapter 12 considers the analysis in a discrete-time framework. In some economic situations, the economic process is in discrete time rather than in continuous time. The chapter presents a general formulation of dynamic economic games in discrete time and derives time (optimal-trajectory-subgame) consistent cooperative solutions with the corresponding payoff distribution procedures. An illustration in cooperative resource extraction is also provided.

Chapter 13 extends the analysis in discrete time to a stochastic framework. A general formulation of stochastic dynamic economic games in discrete time, subgame consistent cooperative solutions with corresponding payoff distribution procedures, and an illustration in stochastic cooperation are provided.

Additionally, dynamic optimization techniques are provided in the Technical Appendixes at the end of the book.

Chapter 2
Dynamic Strategic Interactions in Economic Systems

The recent globalization and emergence of multinational corporations turned many major economic activities into dynamic interactive endeavors. The number of decision makers involved is relatively small and it leads to significant strategic interdependence. With human life being lived over time, and institutions like markets, firms, and governments changing over time, the economic system is definitely a dynamic interactive entity. Section 2.1 provides a general overview of dynamic interactive economic systems. Market outcomes under open-loop equilibria are investigated in Sect. 2.2 and those under feedback equilibria are examined in Sect. 2.3. An extension of the analysis to a stochastic framework is provided in Sect. 2.4.

2.1 Dynamic Interactive Economic System

In this section, we provide the formulation of dynamic interactive economic systems, the typical dynamic game paradigm in economic analysis, and the characterization of market equilibria.

2.1.1 Basic Formulation

A fruitful way of modeling a dynamic interactive situation is by differential games. Differential games study a class of decision problems under which the evolution of the state is described by a differential equation and the players act throughout a time interval.

In economic analysis the general form of n-person differential games can be characterized as follows. Economic agent i seeks to maximize its objective

$$\int_{t_0}^{T} g^i \left[s, x(s), u_1(s), u_2(s), \ldots, u_n(s) \right] \exp \left[-\int_{t_0}^{s} r(y)\,dy \right] ds$$

$$+ \exp \left[-\int_{t_0}^{T} r(y)\,dy \right] q^i \left(x(T) \right), \quad \text{for } i \in N = \{1, 2, \ldots, n\}, \quad (2.1)$$

D.W.K. Yeung, L.A. Petrosyan, *Subgame Consistent Economic Optimization*,
Static & Dynamic Game Theory: Foundations & Applications,
DOI 10.1007/978-0-8176-8262-0_2, © Springer Science+Business Media, LLC 2012

where $r(y)$ is the discount rate, $x(s) \in X \subset R^m$ denotes the state variables of the game, $q^i(x(T))$ is agent i's valuation of the state at terminal time T, and $u_i \in U^i$ is the control of agent i, for $i \in N$.

The state variable evolves according to the dynamics

$$\dot{x}(s) = f[s, x(s), u_1(s), u_2(s), \ldots, u_n(s)], \quad x(t_0) = x_0, \tag{2.2}$$

where $x(s) \in X \subset R^m$ denotes the state variables of the game and $u_i \in U^i$ is the control of agent i, for $i \in N$. The functions $f[s, x, u_1, u_2, \ldots, u_n]$, $g^i[s, \cdot, u_1, u_2, \ldots, u_n]$, and $q^i(\cdot)$, for $i \in N$, and $s \in [t_0, T]$ are differentiable functions.

Examples of economic state variables include capital stock, resource biomass or deposits, the level of technology, market shares, economic assets, equity, prices, pollutants, and company goodwill. Examples of controls include investment, resource extraction rate, research and development (R&D) efforts, advertising rate, output produced, input used, taxes, subsidy, and expenditures.

In many economic situations, the terminal time of the game, T, is either very far in the future or unknown to the agents. For example, the value of a publicly listed firm is the present value of its discounted expected future earnings. Nobody knows when the firm will be out of business. As argued by Dockner et al. (2000), in this case setting $T = \infty$ may very well be the best approximation for the true game horizon. Even if the firm's management restricts itself to considering profit maximization over the next year, it should value its asset positions at the end of the year by the earning potential of these assets in the years to come.

In the case when the terminal horizon T approaches infinity, an autonomous game structure with constant discounting will replace (2.1) and (2.2). In particular, the game becomes

$$\max_{u_i} \int_{t_0}^{\infty} g^i[x(s), u_1(s), u_2(s), \ldots, u_n(s)] \exp[-r(s - t_0)] \, ds, \quad \text{for } i \in N, \tag{2.3}$$

subject to the state dynamics

$$\dot{x}(s) = f[x(s), u_1(s), u_2(s), \ldots, u_n(s)], \quad x(t_0) = x_0, \tag{2.4}$$

where r is a constant discount rate.

Since time s does not appear explicitly in the agent's payoff $g^i[x(s), u_1(s), u_2(s), \ldots, u_n(s)]$ and the state dynamics $f[x(s), u_1(s), u_2(s), \ldots, u_n(s)]$, the problem is an autonomous problem.

Theoretical research and the applications of differential games proceeded apace in the past in many areas of economics. An in-depth survey and analysis on differential games in economics and management science can be found in Dockner et al. (2000). A detailed account and a comprehensive list of differential games in marketing can be found in Zaccour (2003). A thorough survey of models of dynamic games in economics is given in Long (2010).

2.1.2 Typical Dynamic Economic Game Paradigms

In this section we present the general model structures and specific examples of some typical dynamic economic game paradigms.

2.1.2.1 Investment Games

A general structure of investment games can be characterized as follows. Let $K^i(s)$ be the physical capital stock of firm $i \in N$ at time s. Each firm in the industry accumulates capital according to the equation

$$\dot{K}^i(s) = I^i(s) - \delta^i K^i(s), \quad \text{for } i \in N, \tag{2.5}$$

where $I^i(s)$ is the gross investment of firm i at time s and $\delta^i \geq 0$ is the constant rate of depreciation. At initial time t_0, the capital stock $K^i(t_0) = K_0^i$, for $i \in N$, is given.

Denote the output of firm i by $q_i(s)$, the industry output by $Q(s) = \sum_{j=1}^{n} q_j(s)$, and the output price by $P[Q(s)]$.

The output of firm i is governed by the production function $q_i(s) = f^j[L^i(s), K^i(s)]$, where $L^i(s)$ is the quantity of noncapital input (like labor) employed at time s. The cost of production is $c_i\{f^i[K^i(s), L^i(s)]\}$.

At time instant s, the operating profit of firm i becomes

$$P\left(\sum_{j=1}^{n} f^j\left[L^j(s), K^j(s)\right]\right) f^i\left[L^i(s), K^i(s)\right] - c^i\{f[L^i(s), K^i(s)]\},$$

for $i \in N$. $\tag{2.6}$

The optimal choice of noncapital input by firm i satisfies

$$P'\left(\sum_{j=1}^{n} f^j\left[L^j(s), K^j(s)\right]\right) f_{L^i}^i\left[L^i(s), K^i(s)\right] f^i\left[L^i(s), K^i(s)\right]$$

$$+ P\left(\sum_{j=1}^{n} f^j\left[L^j(s), K^j(s)\right]\right) f_{L^i}^i\left[L^i(s), K^i(s)\right]$$

$$- c_{L^i}^i\{f[L^i(s), K^i(s)]\} = 0, \tag{2.7}$$

for $i \in N$.

If an instantaneous industry equilibrium exists, the optimal choice of noncapital inputs by these n firms can be found by solving (2.7) and be expressed as

$$L^{i*}(s) = \ell^i\left[K^1(s), K^2(s), \ldots, K^n(s)\right] = \ell^i\left[K(s)\right], \quad \text{for } i \in N. \tag{2.8}$$

Upon substituting $L^{i*}(s)$ in (2.8) into the firm's instantaneous operating profit in (2.6) yields

$$\pi^i[K(s)] = P\left(\sum_{j=1}^{n} f^j[\ell^j(K(s)), K^j(s)]\right) f^i[\ell^i(K(s)), K^i(s)]$$

$$- c^i\{f[\ell^i(K(s)), K^i(s)]\}, \quad \text{for } i \in N. \tag{2.9}$$

The cost of investment is $m^i[I^i(s)]$. Firms will choose an investment path over the time period $[t_0, T]$ to maximize their future streams of profits. The present value of future profits of firm i can then be expressed as

$$\int_{t_0}^{T} \left(\pi^i[K(s)] - m^i[I(s)]\right) \exp\left[-\int_{t_0}^{s} r(y)\,dy\right] ds, \quad \text{for } i \in N. \tag{2.10}$$

The maximization of (2.10) by firm $i \in N$ subject to (2.5) forms a differential game.

If the time horizon approaches infinity, that is, $T = \infty$, an infinite-horizon version of the game in (2.5) and (2.10) can be set up as

$$\int_{t_0}^{\infty} \left(\pi^i[K(s)] - m^i[I(s)]\right) \exp(-rs)\,ds, \quad \text{for } i \in N, \tag{2.11}$$

subject to (2.15).

Example 2.1 Consider a specific example of investment games in which the demand function is given as

$$P[Q(s)] = a - Q(s),$$

$$m^i[I^i(s)] = a_i[I(s)]^2,$$

$$c^i\{f[\ell^i(K(s)), K^i(s)]\} = b_i K^i(s) + \hat{b}_i[K^i(s)]^2.$$

The interest rate is r.

With these specifications, different investment games with a linear quadratic type can be formulated.

Example 2.2 A knowledge investment game, with knowledge being a public good, can be formulated as follows. The level of knowledge $K(s)$ will change according the accumulation equation

$$\dot{K}(s) = \sum_{j=1}^{n} I^j(s) - \delta K(s), \quad K(t_0) = K_0.$$

Economic agent i's cost of investment in the public knowledge capital is

$$m^i[I^i(s)] = \rho I^i(s) + \frac{1}{2}[I^i(S)]^2, \quad \text{for } i \in N.$$

Economic agent i's instantaneous operating net revenue is

$$\pi^i[K(s)] = K(s)[a^i - K(s)].$$

Once again the interest rate is r.

Dockner (1992), Fershtman and Muller (1984, 1986), Fudenberg and Tirole (1983, 1986, 1991), Reynolds (1987, 1991), and Spence (1979) presented various examples of this class of investment games.

2.1.2.2 Renewable Resource Extraction Games

A general structure of renewable resource extraction games can be characterized as follows. Consider an economy endowed with a single renewable resource, with $n \geq 2$ resource extractors (firms). Let $u_i(s)$ denote the quantity of the resource extracted by firm i at time s, for $i \in N$, where each firm controls its rate of extraction. Let U^i be the set of admissible extraction rates and $x(s)$ the size of the resource stock at time s. In particular, we have $U^i \in R^+, x(s) > 0$, and $U^i = \{0\}$ for $x(s) = 0$. The growth dynamics of the renewable resource stock becomes

$$\dot{x}(s) = f[s, x(s)] - \sum_{j=1}^{n} u_j(s), \quad x(t_0) = x_0 > 0, \tag{2.12}$$

where $f[s, x(s)]$ is the natural rate of evolution of the resource.

The extraction cost for firm $i \in N$ depends on the quantity of the resource extracted $u_i(s)$, the resource stock size $x(s)$, and some other input parameters. In particular, the extraction cost can be specified as

$$C^i[u_i(s), x(s)]. \tag{2.13}$$

The cost per unit of the resource extracted by firm i is negatively related to the size of the resource stock.

The market price of the resource depends on the total amount of the resource extracted and supplied to the market. The price-output relationship at time s is given by the following downward-sloping demand curve:

$$p = P[s, Q(s)], \tag{2.14}$$

where p is the market price of the resource and $Q(s) = \sum_{j=1}^{n} u_j(s)$ is the total amount of the resource extracted and marketed at time s. The firm's horizon is $[t_0, T]$, and at time T a terminal payment $q^i[x(T)]$ will be given to firm i.

Firm i seeks to maximize the present value of its profits

$$\int_{t_0}^{T} \left(P[s, Q(s)] u_i(s) - C^i[u_i(s), x(s)] \right) \exp\left[-\int_{t_0}^{s} r(y)\, dy \right] ds$$

$$+ \exp\left[-\int_{t_0}^{T} r(y)\, dy \right] q^i[x(T)], \quad \text{for } i \in N, \tag{2.15}$$

subject to (2.12).

If the time horizon approaches infinity, that is, $T = \infty$, an infinite-horizon version of the game in (2.12) and (2.15) can be set up as

$$\int_{t_0}^{\infty} \left(P[Q(s)] u_i(s) - C^i[u_i(s), x(s)] \right) \exp[-r(s - t_0)]\, ds, \quad \text{for } i \in N, \tag{2.16}$$

subject to

$$\dot{x}(s) = f[x(s), u_1(s), u_2(s), \ldots, u_n(s)], \quad x(t_0) = x_0 > 0. \tag{2.17}$$

Example 2.3 Consider the deterministic version of the Jørgensen and Yeung (1996) renewable resource game in which the growth dynamics is governed by

$$\dot{x}(s) = ax(s)^{1/2} - bx(s) - \sum_{j=1}^{n} u_j(s) \quad \text{and} \quad x(t_0) = x_0 > 0. \tag{2.18}$$

The natural growth function is $ax^{1/2} - bx = x[ax^{-1/2} - b]$. This function represents that pure compensation, viz., the proportional growth rate $ax^{-1/2}$ is a decreasing function of x.

The extraction cost for firm $i \in N$ depends on the quantity of the resource extracted $u_i(s)$, the resource stock size $x(s)$, and a parameter c. In particular, the extraction cost can be specified as follows:

$$C^i[u_i(s), x(s)] = \frac{c}{x(s)^{1/2}} u_i(s).$$

This specification implies that the cost per unit of the resource extracted by firm $icx(s)^{-1/2}$ decreases when $x(s)$ increases. The above cost structure was also adopted by Jørgensen and Yeung (1996). A decreasing unit cost follows from two assumptions: (i) The cost of extraction is proportional to the extraction effort and (ii) the amount of the resource extracted, seen as the output of a production function of two inputs (effort and stock level), is increasing in both inputs (cf. Clark 1990).

The market price of the resource depends on the total amount extracted and supplied to the market. The price-output relationship at time s is given by the following downward-sloping inverse demand curve:

$$P(s) = Q(s)^{-1/2},$$

where $Q(s) = \sum_{i \in N} u_i(s)$ is the total amount of the resource extracted and marketed at time s.

The objective of extractor $i \in N$ is to maximize the present value of the stream of future profits

$$\int_{t_0}^{T} \left[\left(\sum_{j=1}^{n} u_j(s) \right)^{-1/2} u_i(s) - \frac{c}{x(s)^{1/2}} u_i(s) \right] e^{-r(s-t_0)} \, ds + e^{-r(T-t_0)} x(T)^{1/2},$$

$$\text{for } i \in N, \tag{2.19}$$

subject to the stock dynamics of (2.18).

Chiarella et al. (1984), Reinganum and Stokey (1985), Clemhout and Wan (1985a, 1994), Dockner and Kaitala (1989), Plourde and Yeung (1989), Jørgensen and Sorger (1990), Fischer and Mirman (1992), Kaitala (1993), and Dockner et al. (1989) presented specific dynamic resource extraction games.

2.1.2.3 Marketing Games

Three major types of marketing games are presented below.

(i) Market Share Models

Market share models derive their name from the fact that the state variables of the game are the firm's market shares. In an n-firm oligopoly, let $x_i(s)$ denote the market share of firm $i \in N$. The state space X is represented by

$$X = \left\{ x^i(s) \in R \, \middle| \, x^i(s) \in [0, 1], i \in N, \sum_{j=1}^{n} x^j(s) = 1 \right\}. \tag{2.20}$$

Let $u_i(s) \in R^m$ denote the advertising efforts of firm i at time s; a general version of the market shares dynamics can be expressed as

$$\dot{x}^i(s) = \left[1 - x^i(s) \right] f^i \left[u_i(s) \right] - x^i(s) \sum_{\substack{j=1 \\ j \neq i}}^{n} f^j \left[u_j(s) \right]$$

$$= f^i \left[u_i(s) \right] - x^i(s) \sum_{j=1}^{n} f^j \left[u_j(s) \right], \quad \text{for } i \in N. \tag{2.21}$$

The advertising response function $f^i[u_i(s)]$ is positive for positive advertising efforts. A diminishing (or nonincreasing) marginal product of advertising efforts is assumed, leading to the second-order derivative of $f^i[u_i(s)]$ being nonpositive.

Instead of market shares, the state variable may also represent the sales rates in a market where sales are fixed, say at level $\bar{m}$, and therefore

$$\sum_{j=1}^{n} \varpi^i(s) = \bar{m}. \tag{2.22}$$

Equation (2.22) reflects a market at its maturity stage with a stationary total sales volume of $\bar{m}$. The market of firm i is then $\varpi^i(s)/\bar{m} = x^i(s)$. The dynamics of the change in sales rates can be formulated as

$$\dot{\varpi}^i(s) = \left[\bar{m} - \varpi^i(s)\right] f^i\left[u_i(s)\right] - \varpi^i(s) \sum_{\substack{j=1 \\ j \neq i}}^{n} f^j\left[u_j(s)\right]$$

$$= \bar{m} f^i\left[u_i(s)\right] - \varpi^i(s) \sum_{j=1}^{n} f^j\left[u_j(s)\right], \quad \text{for } i \in N. \tag{2.23}$$

Firm i's cost of advertising efforts is $c^i[u_i(s)]$ and the gross profit of a unit of sales is P_i. The terminal valuation of the sales (or market shares) yields firm i a value $q^i[\varpi^i(T)]$.

The profit to firm i can be expressed as

$$\int_{t_0}^{T} \left(P_i \varpi^i(s) - c^i\left[u_i(s)\right]\right) \exp\left[-\int_{t_0}^{s} r(y)\,dy\right] ds$$

$$+ \exp\left[-\int_{t_0}^{T} r(y)\,dy\right] q^i\left[\varpi^i(T)\right]. \tag{2.24}$$

If the time horizon approaches infinity, that is, $T = \infty$, an infinite-horizon version of the game can be set up as

$$\max_{u^i} \int_{t_0}^{\infty} \left(P_i \varpi^i - c^i\left[u_i(s)\right]\right) \exp\left[-r(s - t_0)\right] ds, \quad \text{for } i \in N, \tag{2.25}$$

subject to (2.23).

Example 2.4 A popular specification of the response function is

$$f^i\left[u_i(s)\right] = \beta_i\left[u_i(s)\right]^{\alpha_i},$$

with $B_i > 0$ and $\alpha_i \in (0, 1]$.

The cost of advertising

$$c^i\left[u_i(s)\right] = c_i\left[u_i(s)\right]^2,$$

where c_i is a positive constant.

Case (1979), Chintagunta and Jain (1995), Chintagunta and Vilcassim (1994), Erickson (1985, 1993, 1992, 1997), Fruchter (1999a, 1999b, 2001), Fruchter and Kalish (1998), Fruchter et al. (2001), Mesak and Calloway (1995), Mesak and Darrat (1993), Olsder (2001), and Sorger (1989) developed and analyzed market share models along this line.

A model closely related to the market shares model is the sales response model. A sales response game model specifies the rate of change of a firm's sales rate $\varpi^i(s)$ as a function of the marketing instruments of all the firms in the market. Let $u_i(s) \in R^m$ be the marketing instruments of firm i; a general specification of the sales dynamics is

$$\dot{\varpi}^i(s) = f^i\big[s, \varpi^1(s), \varpi^2(s), \ldots, \varpi^n(s), u_1(s), u_2(s), \ldots, u_n(s)\big],$$
$$\varpi^i(t_0) = \varpi_0^i, \quad \text{for } i \in N. \tag{2.26}$$

Example 2.5 Mukundan and Elsner (1975) presented a model with sales dynamics

$$\dot{\varpi}^i(s) = \gamma_i u_i(s)\left[1 - \frac{\varpi^i(s)}{\varpi^1(s) + \varpi^1(s)}\right] - \delta_i \varpi^i(s), \quad \text{for } i \in \{1, 2\}.$$

Erickson (1995) presented a model with sales dynamics

$$\dot{\varpi}^i(s) = \gamma_i \sqrt{u_i(s)}\left[\bar{m}(s) - \sum_{j=1}^n \varpi^j(s)\right] - \delta_i \varpi^i(s), \quad \text{for } i \in N,$$

where $\bar{m}(s)$ is the time-varying market potential.

Deal (1979), Feichtinger and Dockner (1984), Jørgensen (1982), Little (1979), Sethi (1973), and Wang and Wu (2001) presented various sales response models.

(ii) New Product Diffusion Models

New product diffusion models are paradigms in which new products or services are introduced and their reputations built up in the market. The cumulative sales affect the current instantaneous sales as the market becomes more mature and the knowledge of the products becomes more available. Using $x^i(s)$ to denote the cumulative sales of product i at time s, the time derivative of $x^i(s)$ then represents the sales rate at time s. A general diffusion process governing the sales dynamics can be expressed as

$$\dot{x}^i(s) = f^i\big[s, u_1(s), u_2(s), \ldots, u_n(s), x^1(s), x^2(s), \ldots, x^n(s)\big],$$
$$x^i(t_0) = x_0^i, \tag{2.27}$$

for $i \in N$, where $u_j(s)$ are the advertising strategies of firm i at time s.

The function f^i is assumed to satisfy the conditions $f_{u_i}^i > 0$, $f_{u_i u_j}^i < 0$. In a market with all products being substitutes of each other, $f_{u_j}^i < 0$ for $i \neq j$.

The instantaneous profit to firm i is

$$\pi^i\big[\dot{x}^i(s), u_i(s), x^i(s)\big].$$

In particular, $\pi^i[\dot{x}^i(s), u_i(s), x^i(s)]$ can take on a formulation like $R^i[\dot{x}^i(s), x^i(s)] - c^i[u_i(s)]$, where $R^i[\dot{x}^i(s), x^i(s)]$ is the instantaneous net revenue from sales $\dot{x}^i(s)$ and $c^i[u_i(s)]$ is the cost of advertising. The cumulative sales may affect the cost of production if experience counts.

A general dynamic game model can be formulated as

$$\max_{u^i} \int_{t_0}^{T} \big\{\pi^i\big[\dot{x}^i(s), u_i(s), x^i(s)\big]\big\} \exp\left[-\int_{t_0}^{s} r(y)\,dy\right] ds, \quad \text{for } i \in N, \quad (2.28)$$

subject to the dynamics in (2.27).

Example 2.6 Consider an oligopolistic extension of the Horsky and Simon (1983) model in which firm i seeks to maximize

$$\int_{t_0}^{T} \big\{\pi_i\big[\dot{x}^i(s)\big] - u_i(s)\big\} \exp\big[-r(s - t_0)\big]\,ds, \quad \text{for } i \in N,$$

where π_i is the nonnegative unit margin of firm i's product.

The sales dynamics is

$$\dot{x}^i(s) = \left[\alpha + \beta \ln\big(u_i(s)\big) + \gamma \sum_{j=1}^{n} x^j(s)\right]\left[\bar{m} - \sum_{j=1}^{n} x^j(s)\right], \quad \text{for } i \in N.$$

Industry-wide positive effects are realized as the new products' cumulative sales increase.

For other new product diffusion models one can see Dockner and Jørgensen (1988, 1992).

(iii) Goodwill Models

Another class of advertising games is one that deals with the accumulation of a stock of goodwill or brand image. Let $G^i(s)$ denote the stock of goodwill of firm i at time s. A general form of the dynamics of the goodwill of firm i is

$$\dot{G}^i(s) = h^i\big[s, u_i(s), G^i(s), x^i(s)\big], \quad \text{for } i \in N, \quad (2.29)$$

where $u_i(s)$ is the effort on the creation of goodwill and $x^i(s)$ is the market share or sales rate of firm i.

The market share (sales rate) of firm i may be affected by all the firms' goodwill stocks and market shares. The dynamics of the market share or sales rate of firm i

yields the relationships

$$\dot{x}^i(s) = f^i\left[s, u_1(s), u_2(s), \ldots, u_n(s), x^1(s), x^2(s), \ldots, x^n(s), G^1(s),\right.$$
$$\left. G^2(s), \ldots, G^n(s)\right], \tag{2.30}$$

for $i \in N$.

Firm i seeks to maximize

$$\int_{t_0}^T \left\{\pi^i\left[x^i(s), u_i(s), G^1(s), G^2(s), \ldots, G^n(s)\right]\right\} \exp\left[-\int_{t_0}^s r(y)\,dy\right] ds,$$
$$\text{for } i \in N, \tag{2.31}$$

subject to (2.29) and (2.30).

The term $\pi^i[x^i(s), u_i(s), G^1(s), G^2(s), \ldots, G^n(s)]$ represents the instantaneous net revenue of firm i.

An infinite-horizon game problem can be formulated with $T = \infty$, a constant discount rate, autonomous versions of the goodwill dynamics in (2.29), and of the market share dynamics in (2.30).

Example 2.7 Fornell et al. (1985) exploited the concept of consumption as a form of production and assumed that production learning took place. This resulted in consumption experience. In an oligopolistic market, brand-specific consumption experience stocks are denoted by $G^1, G^2, \ldots, G^n$. The dynamics of these experience stocks is

$$\dot{G}^i(s) = x^i(s) - \delta G^i(s), \qquad G^i(t_0) = G_0^i, \quad \text{for } i \in N,$$

where $x^i(s)$ is the market share of firm i.

Firm i controls the ratio of its advertising expenditure to unit sales $a^i(s)$ and the ration of its promotion expense to unit sales $b^i(s)$. The market shares of firm i evolve according to

$$\dot{x}^i(s) = \sum_{j=1}^n x^i(s)x^j(s)\left[f\left[a^i(s), x(s), G^i(s)\right]\right.$$
$$\left. - f\left[a^j(s), x(s), G^j(s)\right] + g\left[b^i(s)\right] - g\left[b^j(s)\right]\right],$$
$$x^i(t_0) = x_0^i, \quad \text{for } i \in N,$$

where $x(s) = \{x^1(s), x^2(s), \ldots, x^n(s)\}$.

Example 2.8 Consider a duopoly in which the goodwill dynamics is

$$\dot{G}^i(s) = \sqrt{u_i(s)} - \delta G^i(s), \qquad G^i(t_0) = G_0^i, \quad \text{for } i \in \{1, 2\}.$$

The sales rate of firm i at time s is

$$\varpi^i\big[G^1(s), G^1(s)\big] = \alpha_i G^i(s) - \beta_i G^j(s) - \gamma_i\big[G^i(s)\big]^2 + \theta_i\big[G^j(s)\big]^2$$
$$+ \varsigma_i G^i(s)G^j(s),$$

for $i, j \in \{1, 2\}$ and $i \neq j$, where $\alpha_i, \beta_i, \gamma_i, \theta_i$, and ς_i are positive constants.

The objectives of the duopolists are

$$\int_{t_0}^{\infty} \big(\pi_i \varpi^i\big[G^1(s), G^2(s)\big] - u_i(s)\big) \exp\big[-r(s - t_0)\big]\,ds, \quad \text{for } i, j \in \{1, 2\},$$

where π_i is the constant unit margin of firm i.

Chintagunta (1993), Feichtinger et al. (1994), Fershtman (1984), Sethi and Thompson (2000), and Tapiero (1979) considered games involving goodwill.

2.1.3 Market Equilibrium

The outcome in the economic system (often known as market outcome when the system is driven by markets) is characterized by an equilibrium in which each participant is maximizing its objective given the other participants' optimal choices of controls/strategies.

A set of strategies $\{v_1^*(s), v_2^*(s), \ldots, v_n^*(s)\}$ is said to constitute a *noncooperative Nash equilibrium solution* for the n-person differential game equations (2.1) and (2.2), if the following inequalities are satisfied for all $v_i(s) \in U^i, i \in N$:

$$\int_{t_0}^{T} g^i\big[s, x^*(s), v_1^*(s), v_2^*(s), \ldots, v_n^*(s)\big] \exp\bigg[-\int_{t_0}^{s} r(y)\,dy\bigg]\,ds$$

$$+ \exp\bigg[-\int_{t_0}^{T} r(y)\,dy\bigg] q^i\big(x^*(T)\big)$$

$$\geq \int_{t_0}^{T} g^i\big[s, \hat{x}^i(s), v_1^*(s), v_2^*(s), \ldots, v_{i-1}^*(s), v_i(s), v_{i+1}^*(s), \ldots, v_n^*(s)\big]$$

$$\times \exp\bigg[-\int_{t_0}^{s} r(y)\,dy\bigg]\,ds + \exp\bigg[-\int_{t_0}^{T} r(y)\,dy\bigg] q^i\big(\hat{x}^i(T)\big), \tag{2.32}$$

where, on the time interval $[t_0, T]$,

$$\dot{x}^*(s) = f\big[s, x^*(s), v_1^*(s), v_2^*(s), \ldots, v_n^*(s)\big], \quad x^*(t_0) = x_0, \quad \text{and}$$

$$\dot{\hat{x}}^i(s) = f\big[s, \hat{x}^i(s), v_1^*(s), v_2^*(s), \ldots, v_{i-1}^*(s), v_i(s), v_{i+1}^*(s), \ldots, v_n^*(s)\big],$$

$$\hat{x}(t_0) = x_0.$$

Similarly, a set of strategies $\{v_1^*(s), v_2^*(s), \ldots, v_n^*(s)\}$ constitutes a noncooperative Nash equilibrium solution for the infinite-horizon n-person differential game in (2.3) and (2.4), if there exists a set of inequalities similar to (2.32) with $T = \infty$, discount factor $\exp[-r(s - t_0)]$, objective functions g^i, and state growth f as in (2.3) and (2.4), and the omission of the terminal condition q^i.

Since the game is being played over time, the conditions on the commitment of the agents' strategies at the beginning of the game duration has to be specified. If economic agents choose to commit their strategies from the outset, they are using *open-loop* strategies. If economic agents can revise their strategies contingent upon the state variables, they are using feedback strategies.

2.2 Market Outcomes Under Open-Loop Nash Equilibria

If the agents have to commit their strategies from the outset, the agents' information structure can be seen as an *open-loop* pattern in which $\eta^i(s) = \{x_0\}, s \in [t_0, T]$. Their strategies become functions of the initial state x_0 and time s and can be expressed as $\{u_i(s) = \vartheta_i(s, x_0), \text{ for } i \in N\}$.

2.2.1 Characterization of Open-Loop Equilibria

An open-loop Nash equilibrium for the game in (2.1) and (2.2) is characterized as follows.

Theorem 2.1 *If a set of strategies $\{u_i^*(s) = \zeta_i^*(s, x_0), \text{for } i \in N\}$ provides an open-loop Nash equilibrium solution to the game in (2.1) and (2.2) , and $\{x^*(s), t_0 \leq s \leq T\}$ is the corresponding optimal state trajectory, then there exist m costate functions $\Lambda^i(s) : [t_0, T] \to R^m, \text{for } i \in N$, such that the following relations are satisfied:*

$$
\zeta_i^*(s, x_0)
$$

$$
\equiv u_i^*(s)
$$

$$
= \underset{u_i \in U^i}{\arg\max} \left\{ g^i\left[s, x^*(s), u_1^*(s), u_2^*(s), \ldots, u_{i-1}^*(s), u_i(s), u_{i+1}^*(s), \ldots, u_n^*(s)\right] \right.
$$

$$
\times \exp\left[-\int_{t_0}^{s} r(y)\,dy\right]
$$

$$
\left. + \Lambda^i(s) f\left[s, x^*(s), u_1^*(s), u_2^*(s), \ldots, u_{i-1}^*(s), u_i(s), u_{i+1}^*(s), \ldots, u_n^*(s)\right] \right\},
$$

$$
\dot{x}^*(s) = f\left[s, x^*(s), u_1^*(s), u_2^*(s), \ldots, u_n^*(s)\right], \quad x^*(t_0) = x_0, \tag{2.33}
$$

$$\Lambda^i(s) = -\frac{\partial}{\partial x^*} \left\{ g^i \left[s, x^*(s), u_1^*(s), u_2^*(s), \ldots, u_n^*(s) \right] \exp \left[-\int_{t_0}^s r(y) \, dy \right] \right.$$

$$\left. + \Lambda^i(s) f \left[s, x^*(s), u_1^*(s), u_2^*(s), \ldots, u_n^*(s) \right] \right\},$$

$$\Lambda^i(T) = \frac{\partial}{\partial x^*} q^i \left(x^*(T) \right) \exp \left[-\int_{t_0}^T r(y) \, dy \right];$$

for $i \in N$.

Proof Consider the problem of choosing a control path $\upsilon_i^*(s) = u_i^*(s) = \zeta_i^*(s, x_0)$ that maximizes

$$\int_{t_0}^T g^i \left[s, x(s), u_1^*(s), u_2^*(s), \ldots, u_{i-1}^*(s), u_i(s), u_{i+1}^*(s), \ldots, u_n^*(s) \right]$$

$$\times \exp \left[-\int_{t_0}^s r(y) \, dy \right] ds + \exp \left[-\int_{t_0}^T r(y) \, dy \right] q^i \left(x(T) \right),$$

subject to the state dynamics

$$\dot{x}(s) = f \left[s, x(s), u_1^*(s), u_2^*(s), \ldots, u_{i-1}^*(s), u_i(s), u_{i+1}^*(s), \ldots, u_n^*(s) \right],$$

$$x(t_0) = x_0,$$

for $i \in N$.

This is a standard optimal control problem for agent i, treating $u_j^*(s)$ for $j \in N$ and $j \neq i$ as time paths given at the beginning of the game.

Invoking Theorem A.3 in the Technical Appendixes, the conditions for a maximum for agent i's problem is characterized by the ith set of equalities in Theorem 2.1. Since the set of equalities for all n agents holds, a Nash equilibrium as in (2.32) will arise.   □

There may be multiple Nash equilibria. We assume that the agents will choose an equilibrium at time t_0 and stick with the corresponding strategies for the entire game interval.

The derivation of open-loop equilibria in nonzero-sum deterministic differential games first appeared in Berkovitz (1964) and Ho et al. (1965), with open-loop and feedback Nash equilibria in nonzero-sum deterministic differential games being presented in Case (1967, 1969) and Starr and Ho (1969a, 1969b).

In the case when the game horizon approaches infinity, we can characterize an open-loop equilibrium solution to the infinite-horizon game in (2.3) and (2.4) as follows.

Theorem 2.2 *If a set of strategies $\{u_i^*(s) = \zeta_i^*(s, x_t), for \, i \in N\}$ provides an open-loop Nash equilibrium solution to the infinite-horizon game in (2.3) and (2.4) and $\{x^*(s), t \leq s \leq T\}$ is the corresponding optimal state trajectory, then there exist m* *costate functions* $\Lambda^i(s) : [t, T] \to R^m$, *for* $i \in N$, *such that the following relations are satisfied:*

$$\zeta_i^*(s, x) \equiv u_i^*(s)$$
$$= \arg\max_{u_i \in U^i}\{g^i[x^*(s), u_1^*(s), u_2^*(s), \ldots, u_{i-1}^*(s), u_i(s), u_{i+1}^*(s), \ldots, u_n^*(s)]$$
$$+ \lambda^i(s) f[x^*(s), u_1^*(s), u_2^*(s), \ldots, u_{i-1}^*(s), u_i(s),$$
$$u_{i+1}^*(s), \ldots, u_n^*(s)]\}, \tag{2.34}$$
$$\dot{x}^*(s) = f[x^*(s), u_1^*(s), u_2^*(s), \ldots, u_n^*(s)], \quad x^*(t) = x_t,$$
$$\dot{\lambda}^i(s) = r\lambda(s) - \frac{\partial}{\partial x^*}\{g^i[x^*(s), u_1^*(s), u_2^*(s), \ldots, u_n^*(s)]$$
$$+ \lambda^i(s) f[x^*(s), u_1^*(s), u_2^*(s), \ldots, u_n^*(s)]\}, \quad for\ i \in N.$$

Proof Consider the problem of choosing a control path $v_i^*(s) = u_i^*(s)$ that maximizes

$$\int_t^\infty g^i[x(s), u_1^*(s), u_2^*(s), \ldots, u_{i-1}^*(s), u_i(s), u_{i+1}^*(s), \ldots, u_n^*(s)]$$
$$\times \exp[-r(s - t)]\,ds$$

subject to the state dynamics

$$\dot{x}(s) = f[x(s), u_1^*(s), u_2^*(s), \ldots, u_{i-1}^*(s), u_i(s), u_{i+1}^*(s), \ldots, u_n^*(s)], \quad x(t) = x,$$

for $i \in N$.

This is an infinite-horizon optimal control problem for agent i, treating $u_j^*(s)$, for $j \in N$ and $j \neq i$ as time paths given at the beginning of the game.

Invoking Theorem A.4 in the Technical Appendixes, the conditions for a maximum for agent i's problem is characterized by the ith set of equalities in Theorem 2.2. Since the set of equalities for all n agents hold, a Nash equilibrium as in (2.32) will arise. $\qquad\square$

A detailed account of the applications of open-loop equilibria in marketing, economics, and management science can be found in Zaccour (2003) and Dockner et al. (2000).

2.2.2 Open-Loop Solution in Competitive Advertising

Consider the competitive dynamic advertising game in Sorger (1989). There are two firms in a market and the profits of firm 1 and that of firm 2 are, respectively,

$$\int_0^T \left[q_1 x(s) - \frac{c_1}{2}u_1(s)^2\right]\exp(-rs)\,ds + \exp(-rT)S_1 x(T),$$

and

$$\int_0^T \left[q_2\big(1 - x(s)\big) - \frac{c_2}{2} u_2(s)^2 \right] \exp(-rs)\, ds + \exp(-rT) S_2 \big[1 - x(T)\big], \quad (2.35)$$

where r, q_i, c_i, S_i, for $i \in \{1, 2\}$, are positive constants, $x(s)$ is the market share of firm 1 at time s, $[1 - x(s)]$ is that of firm 2's, and $u_i(s)$ is the advertising rate for firm $i \in \{1, 2\}$.

It is assumed that market potential is constant over time. The only marketing instrument used by the firms is advertising. Advertising has diminishing returns since there are increasing marginal costs of advertising as reflected through the quadratic cost function. The dynamics of firm 1's market share is governed by

$$\dot{x}(s) = u_1(s)\big[1 - x(s)\big]^{1/2} - u_2(s)x(s)^{1/2}, \quad x(0) = x_0. \quad (2.36)$$

Consider that the firms would like to seek an open-loop solution. Using open-loop strategies requires the firms to determine their action's path at the outset. This is realistic only if there are restrictive commitments concerning advertising. Invoking Theorem 2.1, an open-loop solution to the game in (2.35) and (2.36) has to satisfy the following conditions:

$$u_1^*(s) = \arg\max_{u_1} \left\{ \left[q_1 x^*(s) - \frac{c_1}{2} u_1(s)^2 \right] \exp(-rs) \right.$$

$$\left. + \Lambda^1(s)\big(u_1(s)\big[1 - x^*(s)\big]^{1/2} - u_2(s)x^*(s)^{1/2}\big) \right\},$$

$$u_2^*(s) = \arg\max_{u_2} \left\{ \left[q_2\big(1 - x^*(s)\big) - \frac{c_2}{2} u_2(s)^2 \right] \exp(-rs) \right.$$

$$\left. + \Lambda^2(s)\big(u_1(s)\big[1 - x^*(s)\big]^{1/2} - u_2(s)x^*(s)^{1/2}\big) \right\},$$

$$\dot{x}^*(s) = u_1^*(s)\big[1 - x^*(s)\big]^{1/2} - u_2^*(s)x^*(s)^{1/2}, \quad x^*(0) = x_0, \quad (2.37)$$

$$\dot{\Lambda}^1(s) = \left\{ -q_1 \exp(-rs) + \Lambda^1(s)\left(\frac{1}{2} u_1^*(s)\big[1 - x^*(s)\big]^{-1/2} + \frac{1}{2} u_2^*(s)x^*(s)^{-1/2} \right) \right\},$$

$$\dot{\Lambda}^2(s) = \left\{ q_2 \exp(-rs) + \Lambda^2(s)\left(\frac{1}{2} u_1^*(s)\big[1 - x^*(s)\big]^{-1/2} + \frac{1}{2} u_2^*(s)x^*(s)^{-1/2} \right) \right\},$$

$$\Lambda^1(T) = \exp(-rT) S_1,$$

$$\Lambda^2(T) = -\exp(-rT) S_2.$$

Using (2.37), we obtain

$$u_1^*(s) = \frac{\Lambda^1(s)}{c_1}\left[1 - x^*(s)\right]^{1/2}\exp(rs) \quad \text{and} \quad u_2^*(s) = \frac{\Lambda^2(s)}{c_2}\left[x^*(s)\right]^{1/2}\exp(rs).$$

Substituting $u_1^*(s)$ and $u_2^*(s)$ into (2.37) yields

$$\dot{\Lambda}^1(s) = \left\{-q_1\exp(-rs) + \left(\frac{[\Lambda^1(s)]^2}{2c_1} + \frac{\Lambda^1(s)\Lambda^2(s)}{2c_2}\right)\right\},$$

$$\dot{\Lambda}^2(s) = \left\{q_2\exp(-rs) + \left(\frac{[\Lambda^2(s)]^2}{2c_2} + \frac{\Lambda^1(s)\Lambda^2(s)}{2c_1}\right)\right\},$$

(2.38)

with boundary conditions $\Lambda^1(T) = \exp(-rT)S_1$ and $\Lambda^2(T) = -\exp(-rT)S_2$.
   The game equilibrium state dynamics becomes

$$\dot{x}^*(s) = \frac{\Lambda^1(s)\exp(rs)}{c_1}\left[1 - x^*(s)\right] - \frac{\Lambda^2(s)\exp(rs)}{c_2}x^*(s), \quad x^*(0) = x_0. \quad (2.39)$$

Solving the system of differential equations in (2.38) gives the solution time paths
of $\Lambda^1(s)$ and $\Lambda^2(s)$. Using these time paths in (2.39), a solution time path for $x^*(s)$
can be derived. Substituting these solution paths into $u_1^*(s)$ and $u_2^*(s)$ yields the
open-loop game equilibrium strategies.

2.3   Market Outcomes Under Feedback Equilibria

In many economic analyses we could not assume that agents would commit to fixed
control paths at the outset of the game, as in the case of the open-loop solution.
In particular, there are hardly any means that can prevent the agents from revising
their strategies during duration of the game. Instead, agents would consider adopting
feedback strategies, which are decision rules that are dependent upon the current
state $x(t)$ and current time t for $t_0 \le t \le s$.

2.3.1   Characterization of Feedback Equilibria

For the n-person differential game of (2.1) and (2.2), an n-tuple of feedback strategies $\{u_i^*(s) = \phi_i^*(s, x) \in U^i, \text{for } i \in N\}$ constitutes a Nash equilibrium solution if
the following relations for each $i \in N$ are satisfied:

$$\int_t^T g^i\left[s, x^*(s), \phi_1^*\left(s, x^*(s)\right), \phi_2^*\left(s, x^*(s)\right), \dots, \phi_n^*\left(s, x^*(s)\right)\right]$$

$$\times \exp\left[-\int_{t_0}^s r(y)\,dy\right]ds + q^i\left(x^*(T)\right)\exp\left[-\int_{t_0}^T r(y)\,dy\right]$$

$$\geq \int_t^T g^i\left[s, x^i(s), \phi_1^*\left(s, x^i(s)\right), \phi_2^*\left(s, x^i(s)\right), \ldots, \phi_{i-1}^*\left(s, x^i(s)\right), \phi_i\left(s, x^i(s)\right),\right.$$

$$\left.\phi_{i+1}^*\left(s, x^i(s)\right), \ldots, \phi_n^*\left(s, x^i(s)\right)\right]$$

$$\times \exp\left[-\int_{t_0}^s r(y)\,dy\right] ds + q^i\left(x^i(T)\right) \exp\left[-\int_{t_0}^T r(y)\,dy\right],$$

$$\forall \phi_i^*(s, x) \in U^i, x \in R^m, \tag{2.40}$$

where, on the interval $[t_0, T]$,

$$\dot{x}^*(s) = f\left[s, x^*(s), \phi_1^*\left(s, x^*(s)\right), \phi_2^*\left(s, x^*(s)\right), \ldots, \phi_n^*\left(s, x^*(s)\right)\right], \quad x^*(t) = x;$$

and

$$\dot{x}^i(s) = f\left[s, x^i(s), \phi_1^*\left(s, x^i(s)\right), \phi_2^*\left(s, x^i(s)\right), \ldots, \phi_{i-1}^*\left(s, x^i(s)\right), \phi_i\left(s, x^i(s)\right),\right.$$

$$\left.\phi_{i+1}^*\left(s, x^i(s)\right), \ldots, \phi_n^*\left(s, x^i(s)\right)\right], \quad x^i(t) = x, \text{ for } i \in N.$$

One salient feature of the concept introduced above is that if an n-tuple $\{\phi_i^*; i \in N\}$ provides a feedback Nash equilibrium solution (FNES) to an N-person differential game with duration $[t_0, T]$, its restriction to the time interval $[t, T]$ provides an FNES to the same differential game defined on the shorter time interval $[t, T]$, with the initial state taken as $x(t)$, and this being so for all $t_0 \leq t \leq T$. An immediate consequence of this observation is that feedback Nash equilibrium strategies will depend only on the time variable and the current value of the state, but not on memory (including the initial state x_0). Therefore the agents' strategies can be expressed as $\{u_i(s) = \phi_i(s, x), \text{ for } i \in N\}$. The following theorem provides a set of conditions characterizing a feedback Nash equilibrium solution for the game in (2.1) and (2.2) and is characterized as follows.

Theorem 2.3 *An n-tuple of strategies $\{u_i^*(s) = \phi_i^*(t, x) \in U^i, \text{for } i \in N\}$ provides a feedback Nash equilibrium solution to the game in (2.1) and (2.2) if there exist continuously differentiable functions $V^{(t_0)i}(t, x) : [t_0, T] \times R^m \to R, i \in N$, satisfying the following set of partial differential equations*:

$$-V_t^{(t_0)i}(t, x) = \max_{u_i}\left\{g^i\left[t, x, \phi_1^*(t, x), \phi_2^*(t, x), \ldots, \phi_{i-1}^*(t, x), u_i(t, x),\right.\right.$$

$$\left.\phi_{i+1}^*(t, x), \ldots, \phi_n^*(t, x)\right] \exp\left[-\int_{t_0}^t r(y)\,dy\right]$$

$$+ V_x^{(t_0)i}(t, x) f\left[t, x, \phi_1^*(t, x), \phi_2^*(t, x), \ldots, \phi_{i-1}^*(t, x), u_i(t, x),\right.$$

$$\left.\left.\phi_{i+1}^*(t, x), \ldots, \phi_n^*(t, x)\right]\right\}$$

$$= g^i\left[t, x, \phi_1^*(t,x), \phi_2^*(t,x), \ldots, \phi_n^*(t,x)\right]\exp\left[-\int_{t_0}^t r(y)\,dy\right]$$

$$+ V_x^{(t_0)i}(t,x)f\left[t, x, \phi_1^*(t,x), \phi_2^*(t,x), \ldots, \phi_n^*(t,x)\right],$$

$$V^{(t_0)i}(T,x) = q^i(x)\exp\left[-\int_{t_0}^T r(y)\,dy\right], \quad i \in N.$$

Proof Invoking Theorem A.1 in the Technical Appendixes, $V^{(t_0)i}(t,x)$ is the maximized payoff associated with the optimal control problem of agent i for given strategies $\{u_j^*(s) = \phi_j^*(t,x) \in U^j, \text{ for } j \in N \text{ and } j \neq i\}$ of the other $n-1$ agents. The conditions in Theorem 2.3 imply the expressions in (2.40), and hence yield a Nash equilibrium.   □

Again, there may be multiple Nash equilibria; the agents are assumed to choose an equilibrium at time t_0 and stick with the corresponding strategies for the entire game interval. Moreover, $V^{(t_0)i}(t,x)$ is the game equilibrium payoff of agent i at time $t \in [t_0, T]$ with the state being x, that is,

$$V^{(t_0)i}(t,x) = \int_t^T g^i\left[s, x^*(s), \phi_1^*(s, x^*(s)), \phi_2^*(s, x^*(s)), \ldots, \phi_n^*(s, x^*(s))\right]$$

$$\times \exp\left[-\int_{t_0}^s r(y)\,dy\right]ds + q^i\left(x^*(T)\right)\exp\left[-\int_{t_0}^T r(y)\,dy\right].$$

We also call it the value function of agent i in the game.

A remark that will be utilized in the subsequent analysis is given below.

Remark 2.1 Let $V^{(\tau)i}(t,x)$ denote the value function of agent i in a game with the payoffs in (2.1) and dynamics in (2.2), which starts at time τ for $\tau \in [t_0, T)$. Note that the equilibrium feedback strategies are Markovian in the sense that they depend on the current time and current state. One can readily verify that

$$\exp\left[\int_{t_0}^\tau r(y)\,dy\right]V^{(t_0)i}(t,x)$$

$$= \exp\left[\int_{t_0}^\tau r(y)\,dy\right]$$

$$\times \int_t^T g^i\left[s, x^*(s), \phi_1^*(s, x^*(s)), \phi_2^*(s, x^*(s)), \ldots, \phi_n^*(s, x^*(s))\right]$$

$$\times \exp\left[-\int_{t_0}^s r(y)\,dy\right]ds$$

$$= \int_t^T g^i\left[s, x^*(s), \phi_1^*(s, x^*(s)), \phi_2^*(s, x^*(s)), \ldots, \phi_n^*(s, x^*(s))\right]$$

$$\times \exp\left[-\int_\tau^s r(y)\,\mathrm{d}y\right]\mathrm{d}s$$

$$= V^{(\tau)i}(t, x),$$

for $\tau \in [t_0, T)$.

We now turn to the infinite-horizon autonomous game in (2.3) and (2.4). First, consider the infinite-horizon subgame that starts at time $\tau \in [t_0, \infty)$ with initial state $x(\tau) = x$

$$\max_{u_i} \int_\tau^\infty g^i\big[x(s), u_1(s), u_2(s), \ldots, u_n(s)\big]\exp\big[-r(s-\tau)\big]\mathrm{d}s, \quad \text{for } i \in N, \quad (2.41)$$

subject to the dynamics

$$\dot{x}(s) = f\big[x(s), u_1(s), u_2(s), \ldots, u_n(s)\big], \quad x(\tau) = x. \qquad (2.42)$$

The infinite-horizon autonomous game in (2.41) and (2.42) is independent of the choice of τ and dependent only upon the state at the starting time, that is, x.

In the infinite-horizon optimization problem in Sect. A.1 in the Technical Appendixes, the feedback control is shown to be a function the state variable x only. With the validity of the game equilibrium $\{u_i^*(s) = \phi_i^*(x) \in U^i, \text{for } i \in N\}$ to be verified later, we first define the following.

Definition 2.1 For the n-person differential game in (2.41) and (2.42), an n-tuple of feedback strategies $\{u_i^*(s) = \phi_i^*(x) \in U^i, \text{for } i \in N\}$ constitutes a *feedback Nash equilibrium solution* if the following relations for each $i \in N$ are satisfied:

$$\int_t^\infty g^i\big[x^*(s), \phi_1^*(x^*(s)), \phi_2^*(x^*(s)), \ldots, \phi_n^*(x^*(s))\big]\exp\big[-r(s-\tau)\big]\mathrm{d}s$$

$$\geq \int_t^\infty g^i\big[x^i(s), \phi_1^*(x^i(s)), \phi_2^*(x^i(s)), \ldots, \phi_{i-1}^*(x^i(s)), \phi_i(x^i(s)),$$

$$\phi_{i+1}^*(x^i(s)), \ldots, \phi_n^*(x^i(s))\big]\exp\big[-r(s-\tau)\big]\mathrm{d}s,$$

$$\forall \phi_i(\cdot) \in \Gamma^i, x \in R^m, \qquad (2.43)$$

where on the interval $[\tau, \infty)$,

$$\dot{x}^*(s) = f\big[x^*(s), \phi_1^*(x^*(s)), \phi_2^*(x^*(s)), \ldots, \phi_n^*(x^*(s))\big], \quad x^*(s) = x;$$

$$\dot{x}^i(s) = f\big[x^i(s), \phi_1^*(x^i(s)), \phi_2^*(x^i(s)), \ldots, \phi_{i-1}^*(x^i(s)), \phi_i(x^i(s)),$$

$$\phi_{i+1}^*(x^i(s)), \ldots, \phi_n^*(x^i(s))\big], \quad x^i(t) = x.$$

We can express the value function of agent i as

$$V^{(\tau)i}(t, x) = \exp\left[-r(t - \tau)\right] \int_t^\infty g^i\left[x^*(s), \phi_1^*\left(x^*(s)\right), \phi_2^*\left(x^*(s)\right), \ldots, \phi_n^*\left(x^*(s)\right)\right]$$
$$\times \exp\left[-r(s - t)\right] ds,$$

for $x(t) = x^*(t) = x$.

Since $\int_t^\infty g^i[x^*(s), \phi_1^*(x^*(s)), \phi_2^*(x^*(s)), \ldots, \phi_n^*(x^*(s))] \exp[-r(s - t)] ds$ is independent of the choice of t and dependent only upon the state at the starting time x, we can write

$$\hat{V}^i(x) = \int_t^\infty g^i\left[x^*(s), \phi_1^*\left(x^*(s)\right), \phi_2^*\left(x^*(s)\right), \ldots, \phi_n^*\left(x^*(s)\right)\right] \exp\left[-r(s - t)\right] ds.$$

It follows that

$$V^{(\tau)i}(t, x) = \exp\left[-r(t - \tau)\right] \hat{V}^i(x),$$

$$V_t^{(\tau)i}(t, x) = -r \exp\left[-r(t - \tau)\right] \hat{V}^i(x), \quad \text{and} \qquad (2.44)$$

$$V_x^{(\tau)i}(t, x) = \exp\left[-r(t - \tau)\right] \hat{V}_x^i(x), \quad \text{for } i \in N.$$

A feedback Nash equilibrium solution for the infinite-horizon autonomous game in (2.41) and (2.42) can be characterized as follows.

Theorem 2.4 *An n-tuple of strategies $\{u_i^* = \phi_i^*(\cdot) \in U^i, for\ i \in N\}$, provides a feedback Nash equilibrium solution to the infinite-horizon game in (2.3) and (2.4) if there exist continuously differentiable functions $\hat{V}^i(x) : R^m \to R, i \in N$, satisfying the following set of partial differential equations:*

$$r\hat{V}^i(x) = \max_{u_i}\left\{g^i\left[x, \phi_1^*(x), \phi_2^*(x), \ldots, \phi_{i-1}^*(x), u_i, \phi_{i+1}^*(x), \ldots, \phi_n^*(x)\right]\right.$$

$$\left. + \hat{V}_x^i(x) f\left[x, \phi_1^*(x), \phi_2^*(x), \ldots, \phi_{i-1}^*(x), u_i, \phi_{i+1}^*(x), \ldots, \phi_n^*(x)\right]\right\}$$

$$= \left\{g^i\left[x, \phi_1^*(x), \phi_2^*(x), \ldots, \phi_n^*(x)\right]\right.$$

$$\left. + \hat{V}_x^i(x) f\left[x, \phi_1^*(x), \phi_2^*(x), \ldots, \phi_n^*(x)\right]\right\},$$

for $i \in N$.

Proof By Theorem A.2 in the Technical Appendixes, $\hat{V}^i(x)$ is the value function associated with the optimal control problem of agent $i, i \in N$. Together with the expressions in Definition 2.1, the conditions in Theorem 2.4 imply a Nash equilibrium. $\square$

Since time t is not explicitly involved in the partial differential equations in Theorem 2.4, the validity that the feedback Nash equilibrium $\{u_i^* = \phi_i^*(x), \text{for } i \in N\}$, are functions independent of time is obtained.

Substituting the game equilibrium strategies in Theorem 2.4 into (2.4) yields the game equilibrium dynamics of the state path as

$$\dot{x}(s) = f[x(s), \phi_1^*(x(s)), \phi_2^*(x(s)), \ldots, \phi_n^*(x(s))], \quad x(t_0) = x_0.$$

Solving the above dynamics yields the optimal state trajectory $\{x^*(t)\}_{t \geq t_0}$ as

$$x^*(t) = x_0 + \int_{t_0}^{t} f[x^*(s), \phi_1^*(x^*(s)), \phi_2^*(x^*(s)), \ldots, \phi_n^*(x^*(s))] \, ds,$$

for $t \geq t_0$.   (2.45)

We denote term $x^*(t)$ by x_t^*. The feedback Nash equilibrium strategies for the infinite-horizon game in (2.3) and (2.4) can be obtained as

$$[\phi_1^*(x_t^*), \phi_2^*(x_t^*), \ldots, \phi_n^*(x_t^*)], \quad \text{for } t \geq t_0.$$

2.3.2 Feedback Equilibria in Resource Extraction

Consider an economy endowed with a renewable resource and with $n \geq 2$ resource extractors (firms). The lease for resource extraction begins at time t_0 and ends at time T. Let $u_i(s)$ denote the rate of resource extraction of firm i at time s, $i \in N = \{1, 2, \ldots, n\}$, where each extractor controls its rate of extraction. Let U^i be the set of admissible extraction rates and $x(s)$ the size of the resource stock at time s. In particular, we have $U^i \in R^+$ for $x > 0$ and $= \{0\}$ for $x = 0$. The extraction cost for firm $i \in N$ depends on the quantity of the resource extracted $u_i(s)$, the resource stock size $x(s)$, and a parameter c.

In particular, the extraction cost can be specified as $C^i = cu_i(s)/x(s)^{1/2}$. The market price of the resource depends on the total amount extracted and supplied to the market. The price-output relationship at time s is given by the following downward-sloping inverse demand curve $P(s) = Q(s)^{-1/2}$, where $Q(s) = \sum_{j=1}^{n} u_j(s)$ is the total amount of the resource extracted and marketed at time s. A terminal bonus $wx(T)^{1/2}$ is offered to each extractor and r is a discount rate that is common to all extractors. Extractor i seeks to maximize the present value of the profits

$$\int_{t_0}^{T} \left[\left(\sum_{j=1}^{n} u_j(s) \right)^{-1/2} u_i(s) - \frac{c}{x(s)^{1/2}} u_i(s) \right] \exp[-r(s - t_0)] \, ds$$

$$+ \exp[-r(T - t_0)] wx(T)^{1/2}, \quad \text{for } i \in N,$$ (2.46)

subject to the resource dynamics

$$\dot{x}(s) = ax(s)^{1/2} - bx(s) - \sum_{j=1}^{n} u_j(s), \quad x(t_0) = x_0 \in X.$$ (2.47)

The model is a deterministic version of the Jørgensen and Yeung (1996) fishery game model. Invoking Theorem 2.3, a set of feedback strategies $\{u_i^*(t) = \phi_i^*(t, x); i \in N\}$ constitutes a feedback Nash equilibrium solution for the game in (2.46) and (2.47), if there exist functions $V^{(t_0)i}(t, x) : [t_0, T] \times R \to R$ for $i \in N$, which satisfy the following set of partial differential equations:

$$
-V_t^{(t_0)i}(t, x) = \max_{u_i \in U^i} \left\{ \left[u_i \left(\sum_{\substack{j=1 \\ j \neq i}}^n \phi_j^*(t, x) + u_i \right)^{-1/2} - \frac{c}{x^{1/2}} u_i(t) \right] \exp[-r(t - t_0)] \right.
$$

$$
\left. + V_x^{(t_0)i} \left[ax^{1/2} - bx - \sum_{\substack{j=1 \\ j \neq i}}^n \phi_j^*(t, x) - u_i \right] \right\}, \quad \text{and} \quad (2.48)
$$

$$
V^{(t_0)i}(T, x) = \exp[-r(T - t_0)] w x^{1/2}.
$$

Applying the maximization operator on the right-hand side of the first equation in (2.49) for agent i yields the condition for a maximum as

$$
\left[\left(\sum_{\substack{j=1 \\ j \neq i}}^n \phi_j^*(t, x) + \frac{1}{2} \phi_i^*(t, x) \right) \left(\sum_{j=1}^n \phi_j^*(t, x) \right)^{-3/2} - \frac{c}{x^{1/2}} \right] \exp[-r(t - t_0)]
$$

$$
- V_x^{(t_0)i} = 0, \quad (2.49)
$$

for $i \in N$.

Summing over $i = 1, 2, \ldots, n$ in (2.49) yields

$$
\left(\sum_{j=1}^n \phi_j^*(t, x) \right)^{1/2} = \left(n - \frac{1}{2} \right) \left(\sum_{j=1}^n \left[\frac{c}{x^{1/2}} + \exp[r(t - t_0)] V_x^{(t_0)j} \right] \right)^{-1}. \quad (2.50)
$$

Substituting (2.50) into (2.49) produces

$$
\left(\sum_{\substack{j=1 \\ j \neq i}}^n \phi_j^*(t, x) + \frac{1}{2} \phi_i^*(t, x) \right) \left(n - \frac{1}{2} \right)^{-3} \left(\sum_{j=1}^n \left[\frac{c}{x^{1/2}} + \exp[r(t - t_0)] V_x^{(t_0)j} \right] \right)^3
$$

$$
- \frac{c}{x^{1/2}} - \exp[r(t - t_0)] V_x^{(t_0)i} = 0, \quad \text{for } i \in N. \quad (2.51)
$$

Rearranging the terms in (2.51) yields

$$
\left(\sum_{\substack{j=1 \\ j \neq i}}^n \phi_j^*(t, x) + \frac{1}{2} \phi_i^*(t, x) \right)
$$

$$
= \left(n - \frac{1}{2} \right)^3 \frac{[c + \exp[r(t - t_0)] V_x^{(t_0)i} x^{1/2}] x}{(\sum_{j=1}^n [c + \exp[r(t - t_0)] V_x^{(t_0)j} x^{1/2}])^3}, \quad (2.52)
$$

for $i \in N$.

Condition (2.52) represents a system of equations that is linear in $\{\phi_1^*(t, x),$ $\phi_2^*(t, x), \ldots, \phi_n^*(t, x)\}$. Solving (2.52) yields

$$\phi_i^*(t, x) = \frac{x(2n-1)^2}{2[\sum_{j=1}^{n}[c+\exp[r(t-t_0)]V_x^{(t_0)j}x^{1/2}]]^3} \left\{ \sum_{\substack{j=1 \\ j \neq i}}^{n} \left[c + \frac{V_x^{(t_0)j}x^{1/2}}{\exp[-r(t-t_0)]} \right] \right.$$

$$\left. - \left(n - \frac{3}{2}\right)\left[c + \frac{V_x^{(t_0)i}x^{1/2}}{\exp[-r(t-t_0)]} \right] \right\}, \quad \text{for } i \in N. \tag{2.53}$$

Substituting $\phi_i^*(t, x)$ in (2.53) into (2.49); upon solving it yields the following.

Proposition 2.1 *The system in* (2.49) *admits a solution*

$$V^{(t_0)i}(t, x) = \exp[-r(t-t_0)][A(t)x^{1/2} + B(t)], \quad \text{for } i \in N, \tag{2.54}$$

where $A(t)$ *and* $B(t)$ *satisfy*

$$\dot{A}(t) = \left[r + \frac{b}{2}\right]A(t) - \frac{(2n-1)}{2n^2}\left(c + \frac{A(t)}{2}\right)^{-1}$$

$$+ \frac{c(2n-1)^2}{4n^3}\left(c + \frac{A(t)}{2}\right)^{-2} + \frac{(2n-1)^2 A(t)}{8n^2(c + \frac{A(t)}{2})^2}, \tag{2.55}$$

$$\dot{B}(t) = rB(t) - \frac{a}{2}A(t),$$

$$A(T) = w \quad \text{and} \quad B(T) = 0.$$

Proof Substituting $V^i(t, x)$ and the relevant derivatives $V_t^i(t, x)$ and $V_x^i(t, x)$ into (2.53) and (2.49) yields the results in Proposition 2.1.   □

The first equation in (2.55) can be further reduced to

$$\dot{A}(t) = \left\{ \left(r + \frac{1}{8}\sigma^2 + \frac{b}{2}\right)\frac{[A(t)]^3}{4} + \left(r + \frac{1}{8}\sigma^2 + \frac{b}{2}\right)c[A(t)]^2 \right.$$

$$\left. + \left[\left(r + \frac{1}{8}\sigma^2 + \frac{b}{2}\right)c^2 + \frac{(4n^2 - 8n + 3)}{8n^2}\right]A(t) - \frac{(2n-1)c}{4n^3} \right\}$$

$$\bigg/ \left(c + \frac{A(t)}{2}\right)^2. \tag{2.56}$$

The denominator of the right-hand side of (2.56) is always positive. Denote the numerator of the right-hand side of (2.56) by

$$F[A(t)] - \frac{(2n-1)c}{4n^3}. \tag{2.57}$$

Fig. 2.1 Phase diagram for $\dot{A}(t)$ and $A(t)$

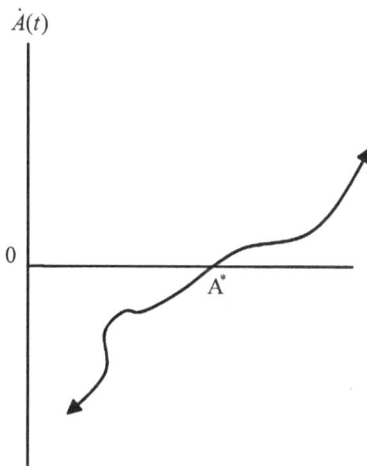

In particular, $F[A(t)]$ is a polynomial function in $A(t)$ of degree 3. Moreover, $F[A(t)] = 0$ for $A(t) = 0$, and for any $A(t) \in (0, \infty)$,

$$\frac{\mathrm{d}F[A(t)]}{\mathrm{d}A(t)} = \left(r + \frac{1}{8}\sigma^2 + \frac{b}{2}\right)\frac{3[A(t)]^2}{4} + 2\left(r + \frac{1}{8}\sigma^2 + \frac{b}{2}\right)c[A(t)]$$

$$+ \left[\left(r + \frac{1}{8}\sigma^2 + \frac{b}{2}\right)c^2 + \frac{(4n^2 - 8n + 3)}{8n^2}\right] > 0. \qquad (2.58)$$

Therefore, there exists a unique level of $A(t)$, denoted by A^*, at which

$$F[A^*] - \frac{(2n-1)c}{4n^3} = 0. \qquad (2.59)$$

If $A(t)$ equals A^*, $\dot{A}(t) = 0$. For values of $A(t)$ less than A^*, $\dot{A}(t)$ is negative. For values of $A(t)$ greater than A^*, $\dot{A}(t)$ is positive. A phase diagram depicting the relationship between $\dot{A}(t)$ and $A(t)$ is provided in Fig. 2.1, while the time paths of $A(t)$ in relation to A^* are illustrated in Fig. 2.2.

For a given value of w that is less than A^*, the time path $\{A(t)\}_{t=t_0}^T$ will start at a value $A(t_0)$, which is greater than w and less than A^*. The value of $A(t)$ will decrease over time and reach w at time T. On the other hand, for a given value of w that is greater than A^*, the time path $\{A(t)\}_{t=t_0}^T$ will start at a value $A(t_0)$, which is less than w and greater than A^*. The value of $A(t)$ will increase over time and reach w at time T. Therefore $A(t)$ is a monotonic function and $A(t) > 0$, for $t \in [t_\tau, T]$.

Using $A(t)$, the solution to $B(t)$ can be readily obtained as

$$B(t) = \exp(rt)\left(K - \int_{t_0}^t \frac{a}{2}A(s)\exp(-rs)\,\mathrm{d}s\right), \qquad (2.60)$$

where $K = \int_{t_0}^T \frac{a}{2}A(s)\exp(-rs)\,\mathrm{d}s$.

Fig. 2.2 Time paths of $A(t)$

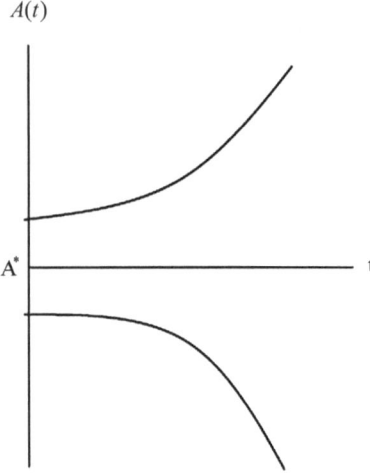

Substituting the relevant derivatives of the value functions in Proposition 2.1 into the game equilibrium strategies of (2.53) gives the feedback Nash equilibrium of the resource extraction game of (2.46) and (2.47).

2.3.3 Feedback Solution in Competitive Advertising

Consider the competitive advertising game in Sect. 2.2.2. Instead of an open-loop solution we seek a feedback Nash equilibrium. Invoking Theorem 2.3, a set of feedback strategies $\{u_i^*(t) = \phi_i^*(t, x); i \in N\}$ constitutes a feedback Nash equilibrium solution for the game in (2.35) and (2.36), if there exist functions $V^{(t_0)i}(t, x) :$ $[t_0, T] \times R \to R$ for $i \in \{1, 2\}$, which satisfy the following set of partial differential equations:

$$-V_t^{(t_0)1}(t, x) = \max_{u_1} \left\{ \left(q_1 x - \frac{c_1}{2} u_1^2 \right) \exp(-rt) \right.$$
$$\left. + V_x^1(t, x)\left[u_1(1 - x)^{1/2} - \phi_2^*(t, x)x^{1/2} \right] \right\},$$

$$V^{(t_0)1}(T, x) = \exp(-rT)S_1 x$$

and

$$-V_t^{(t_0)2}(t, x) = \max_{u_1} \left\{ \left[q_2(1 - x) - \frac{c_2}{2} u_2^2 \right] \exp(-rt) \right.$$
$$\left. + V_x^2(t, x)\left[\phi_1^*(t, x)(1 - x)^{1/2} - u_2 x^{1/2} \right] \right\}, \qquad (2.61)$$

$$V^{(t_0)2}(T, x) = \exp(-rT)S_2(1 - x).$$

Performing the indicated maximization in (2.62) yields the condition for a maximum as

$$\phi_1^*(t, x) = \frac{\exp rt}{c_1} V_x^{(t_0)1}(t, x)(1 - x)^{1/2} \quad \text{and}$$

$$\phi_2^*(t, x) = \frac{\exp rt}{c_2} V_x^{(t_0)2}(t, x)x^{1/2}. \tag{2.62}$$

Substituting $\phi_1^*(t, x)$ and $\phi_2^*(t, x)$ into (2.62) and solving it yields

$$- V_t^{(t_0)1}(t, x) = \left(q_1 x - \frac{(\exp rt)^2}{2c_1} \left[V_x^{(t_0)1}(t, x) \right]^2 (1 - x) \right) \exp(-rt)$$

$$+ V_x^{(t_0)1}(t, x) \left[\frac{\exp rt}{c_1} V_x^{(t_0)1}(t, x)(1 - x) - \frac{\exp rt}{c_2} V_x^{(t_0)2}(t, x)x \right],$$

$$V^{(t_0)1}(T, x) = \exp(-rT)S_1 x;$$

$$- V_t^{(t_0)2}(t, x) = \left(q_2(1 - x) - \frac{(\exp rt)^2}{2c_2} \left[V_x^{(t_0)2}(t, x) \right]^2 x \right) \exp(-rt) \tag{2.63}$$

$$+ V_x^{(t_0)2}(t, x) \left[\frac{\exp rt}{c_1} V_x^{(t_0)1}(t, x)(1 - x) - \frac{\exp rt}{c_2} V_x^{(t_0)2}(t, x)x \right],$$

$$V^{(t_0)2}(T, x) = \exp(-rT)S_2(1 - x).$$

Proposition 2.2 *The system in (2.63) admits a solution*

$$V^{(t_0)1}(t, x) = \exp\left[-r(t)\right]\left[A_1(t)x + B_1(t)\right],$$

$$V^{(t_0)2}(t, x) = \exp\left[-r(t)\right]\left[A_2(t)x + B_2(t)\right],$$

where $A(t)$ and $B(t)$ satisfy

$$\dot{A}_1(t) = r A_1(t) - q_1 + \frac{[A_1(t)]^2}{2c_1} + \frac{A_1(t)A_2(t)}{c_2}, \quad A_1(T) = S_1,$$

$$\dot{B}_1(t) = r B_1(t) - \frac{[A_1(t)]^2}{2c_1}, \quad B_1(T) = 0;$$

$$\dot{A}_2(t) = r A_2(t) - q_2 + \frac{[A_2(t)]^2}{2c_2} + \frac{A_1(t)A_2(t)}{c_1}, \quad A_2(T) = S_2,$$

$$\dot{B}_2(t) = r B_2(t) - \frac{[A_2(t)]^2}{2c_2}, \quad B_2(T) = 0.$$

Proof Substituting $V^i(t, x)$ and the relevant derivatives $V_t^i(t, x)$ and $V_x^i(t, x)$ into (2.63) yields the results in Proposition 2.2. □

With the value functions in Proposition 3.2, one can characterize the game equilibrium strategies in (2.62) over the game interval $[t_0, T]$, the equilibrium state path, and the profits of the firms over time.

2.3.4 Duopolistic Competition in Infinite Horizon

Consider a dynamic duopoly in which there are two publicly listed firms selling a homogeneous good. Since the value of a publicly listed firm is the present value of its discounted expected future earnings. The terminal time of the game T may be very far in the future and nobody knows when the firms will be out of business. Therefore, setting $T = \infty$ may very well be the best approximation for the true game horizon. Even if the firm's management restricts itself to considering profit maximization over the next year, it should value its asset positions at the end of the year by the earning potential of these assets in the years to come. There is a lag in price adjustment so the evolution of market price over time is assumed to be a function of the current market price and the price specified by the current demand condition. In particular, we follow Tsutsui and Mino (1990) and assume that

$$\dot{P}(s) = k[a - u_1(s) - u_2(s) - P(s)], \quad P(t_0) = P_0, \tag{2.64}$$

where $P(s)$ is the market price at time s, $u_i(s)$ is the output supplied firm $i \in \{1, 2\}$, the current demand condition is specified by the instantaneous inverse demand function $P(s) = [a - u_1(s) - u_2(s)]$ and $k > 0$ represents the price adjustment velocity.

The payoff of firm i is given as the present value of the stream of discounted profits

$$\int_{t_0}^{\infty} \left\{ P(s)u_i(s) - cu_i(s) - (1/2)[u_i(s)]^2 \right\} \exp[-r(s - t_0)] \, ds,$$

$$\text{for } i \in \{1, 2\}, \tag{2.65}$$

where $cu_i(s) + (1/2)[u_i(s)]^2$ is the cost of producing output $u_i(s)$ and r is the interest rate.

Once again, we consider the infinite-horizon game that starts at time $t \in [t_0, \infty)$ with initial state $P(t) = P$

$$\max_{u_i} \int_{t}^{\infty} \left\{ P(s)u_i(s) - cu_i(s) - (1/2)[u_i(s)]^2 \right\} \exp[-r(s - t)] \, ds,$$

$$\text{for } i \in \{1, 2\}, \tag{2.66}$$

subject to

$$\dot{P}(s) = k[a - u_1(s) - u_2(s) - P(s)], \quad P(t) = P. \tag{2.67}$$

The infinite-horizon game in (2.66) and (2.67) has autonomous structures and a constant rate. Therefore, we can apply Theorem 3.2 to characterize a feedback Nash equilibrium solution as

$$r\hat{V}^i(P) = \max_{u_i}\left\{\left[Pu_i - cu_i - (1/2)(u_i)^2\right]\right.$$
$$\left. + \hat{V}_P^i\left[k\left(a - u_i - \phi_j^*(P) - P\right)\right]\right\}, \quad \text{for } i \in \{1, 2\}. \quad (2.68)$$

Performing the indicated maximization in (2.68), we obtain

$$\phi_i^*(P) = P - c - k\hat{V}_P^i(P), \quad \text{for } i \in \{1, 2\}. \quad (2.69)$$

Substituting the results from (2.69) into (2.68), and upon solving (2.68), yields

$$\hat{V}^i(P) = \frac{1}{2}AP^2 - BP + C, \quad (2.70)$$

where

$$A = \frac{r + 6k - \sqrt{(r + 6k)^2 - 12k^2}}{6k^2},$$

$$B = \frac{-akA + c - 2kcA}{r - 3k^2A + 3k}, \quad \text{and}$$

$$C = \frac{c^2 + 3k^2B^2 - 2kB(2c + a)}{2r}.$$

Again, one can readily verify that $\hat{V}^i(P)$ in (2.70) indeed solves (2.68) by substituting $\hat{V}^i(P)$ and its derivative into (2.68) and (2.69).

The game equilibrium strategy can then be expressed as

$$\phi_i^*(P) = P - c - k(AP - B), \quad \text{for } i \in \{1, 2\}.$$

Substituting the game equilibrium strategies above into (2.64) yields the game equilibrium state dynamics of the game in (2.64) and (2.65) as

$$\dot{P}(s) = k\left[a - 2(c + kB) - (3 - kA)P(s)\right], \quad P(t_0) = P_0.$$

Solving the above dynamics yields the optimal state trajectory as

$$P^*(t) = \left[P_0 - \frac{k[a - 2(c + kB)]}{k(3 - kA)}\right]\exp\left[-k(3 - kA)t\right] + \frac{k[a - 2(c + kB)]}{k(3 - kA)}.$$

We denote term $P^*(t)$ by P_t^*. The feedback Nash equilibrium strategies for the infinite-horizon game in (2.64) and (2.65) can be obtained as

$$\phi_i^*(P_t^*) = P_t^* - c - k(AP_t^* - B), \quad \text{for } i \in \{1, 2\}.$$

2.4 Dynamic Stochastic Interactive Economic System

One way to incorporate stochastic elements in dynamic interactive economic systems is to introduce stochastic dynamics. Uncertainties in the evolution of economic state variables are prevalent. For instance, the natural growth rate of renewable resources, the development of technology, capital accumulation, the build-up of goodwill, and special skills are often subject to stochastic impacts.

2.4.1 Game Formulation and Solution Characterization

A stochastic formulation of state dynamics is by adopting a vector-valued stochastic differential equation

$$dx(s) = f\big[s, x(s), u_1(s), u_2(s), \ldots, u_n(s)\big]\,ds + \sigma\big[s, x(s)\big]\,dz(s),$$
$$x(t_0) = x_0, \tag{2.71}$$

where $\sigma[s, x(s)]$ is a $m \times \Theta$ matrix, $z(s)$ is a Θ-dimensional Wiener process, and the initial state x_0 is given. Let $\Omega[s, x(s)] = \sigma[s, x(s)], \sigma[s, x(s)]$ denote the covariance matrix with its element in row h and column ζ denoted by $\Omega^{h\zeta}[s, x(s)]$. Moreover, $E[dz_\varpi] = 0$, $E[dz_\varpi\,dt] = 0$ and $E[(dz_\varpi)^2] = dt$ for $\varpi \in [1, 2, \ldots, \Theta]$; $E[dz_\varpi\,dz_\omega] = 0$ for $\varpi \in [1, 2, \ldots, \Theta]$, $\varpi \in [1, 2, \ldots, \Theta]$, and $\varpi \neq \omega$.

Given the stochastic nature of the state dynamics, the economic agent i's objective becomes

$$E_{t_0}\Bigg\{\int_{t_0}^{T} g^i\big[s, x(s), u_1(s), u_2(s), \ldots, u_n(s)\big]\exp\Big[-\int_{t_0}^{s} r(y)\,dy\Big]\,ds$$
$$+ \exp\Big[-\int_{t_0}^{T} r(y)\,dy\Big]q^i\big(x(T)\big)\Bigg\}, \quad \text{for } i \in N, \tag{2.72}$$

with $E_{t_0}\{\cdot\}$ denoting the expectation operation taken at time t_0.

The system in (2.71) and (2.72) is a stochastic differential game. Basar (1977a, 1977b, 1980) was the first to derive explicit results for stochastic linear quadratic differential games. Examples of solvable stochastic differential games in economics include Clemhout and Wan (1985b), Kaitala (1993), Jørgensen and Yeung (1996, 1999), and Yeung (1998, 1999, 2001).

A Nash equilibrium of the stochastic game in (2.71) and (2.72) can be characterized as follows.

Theorem 2.5 *An N-tuple of feedback strategies $\{\phi_i^*(t, x) \in U^i; i \in N\}$ provides a Nash equilibrium solution to the game in (2.71) and (2.72) if there exist suitably* smooth functions $V^{(t_0)i}(t, x) : [t_0, T] \times R^m \to R, i \in N$, satisfying the partial differential equations

$$-V_t^{(t_0)i}(t, x) - \frac{1}{2} \sum_{h,\zeta=1}^{m} \Omega^{h\zeta}(t, x) V_{x^h x^\zeta}^{(t_0)i}(t, x)$$

$$= \max_{u_i} \left\{ g^i \left[t, x, \phi_1^*(t, x), \phi_2^*(t, x), \ldots, \phi_{i-1}^*(t, x), u_i(t), \phi_{i+1}^*(t, x), \ldots, \phi_n^*(t, x) \right] \right.$$

$$\times \exp \left[-\int_{t_0}^{s} r(y) \, dy \right] + V_x^{(t_0)i}(t, x)$$

$$\left. \times f \left[t, x, \phi_1^*(t, x), \phi_2^*(t, x), \ldots, \phi_{i-1}^*(t, x), u_i(t), \phi_{i+1}^*(t, x), \ldots, \phi_n^*(t, x) \right] \right\},$$

$$V^{(t_0)i}(T, x) = q^i(x) \exp \left[-\int_{t_0}^{T} r(y) \, dy \right], \quad i \in N.$$

Proof This result follows readily from the definition of the Nash equilibrium and from the stochastic control result in Theorem A.5 of the Technical Appendixes. □

In particular, $V^{(t_0)i}(t, x)$ represents the expected game equilibrium payoff of agent i at time $t \in [t_0, T]$ with the state being x, that is,

$$E_{t_0} \left\{ V^{(t_0)i}(t, x) = \int_t^T g^i \left[s, x^*(s), \phi_1^*(s, x^*(s)), \phi_2^*(s, x^*(s)), \ldots, \phi_n^*(s, x^*(s)) \right] \right.$$

$$\left. \times \exp \left[-\int_{t_0}^{s} r(y) \, dy \right] ds + q^i(x^*(T)) \exp \left[-\int_{t_0}^{T} r(y) \, dy \right] \right\}.$$

A remark that will be utilized in the subsequent analysis is given below.

Remark 2.2 Let $V^{(\tau)i}(t, x)$ denote the value function of nation i in a game with stochastic dynamics found in (2.71) and expected payoffs in (2.72), which starts at time τ for $\tau \in [t_0, T)$. Note that the equilibrium feedback strategies are Markovian in the sense that they depend on the current time and the current state. One can readily verify that

$$\exp \left[\int_{t_0}^{\tau} r(y) \, dy \right] V^{(t_0)i}(t, x)$$

$$= \exp \left[\int_{t_0}^{\tau} r(y) \, dy \right]$$

$$\times E_{t_0} \left\{ \int_t^T g^i \left[s, x^*(s), \phi_1^*(s, x^*(s)), \phi_2^*(s, x^*(s)), \ldots, \phi_n^*(s, x^*(s)) \right] \right.$$

$$\times \exp\left[-\int_{t_0}^{s} r(y)\,dy\right]ds\Bigg\}$$

$$= E_t\Bigg\{\int_t^T g^i\left[s, x^*(s), \phi_1^*(s, x^*(s)), \phi_2^*(s, x^*(s)), \ldots, \phi_n^*(s, x^*(s))\right]$$

$$\times \exp\left[-\int_{\tau}^{s} r(y)\,dy\right]ds\Bigg\} = V^{(\tau)i}(t, x), \quad \text{for } \tau \in [t_0, T).$$

In the case when the terminal horizon T approaches infinity, an autonomous game structure with constant discounting will replace (2.71) and (2.72). In particular, the game becomes

$$\max_{u_i} E_{t_0}\Bigg\{\int_{t_0}^{\infty} g^i\left[x(s), u_1(s), u_2(s), \ldots, u_n(s)\right]\exp\left[-r(s - t_0)\right]ds\Bigg\},$$

$$\text{for } i \in N, \tag{2.73}$$

subject to the stochastic dynamics

$$dx(s) = f\left[x(s), u_1(s), u_2(s), \ldots, u_n(s)\right]ds + \sigma\left[x(s)\right]dz(s), \quad x(t_0) = x_0. \tag{2.74}$$

Consider the alternative infinite-horizon game that starts at time $t \in [t_0, \infty)$ with initial state $x(t) = x$

$$\max_{u_i} E_t\Bigg\{\int_t^{\infty} g^i\left[x(s), u_1(s), u_2(s), \ldots, u_n(s)\right]\exp\left[-r(s - t)\right]ds\Bigg\}, \tag{2.75}$$

for $i \in N$, subject to the stochastic dynamics

$$dx(s) = f\left[x(s), u_1(s), u_2(s), \ldots, u_n(s)\right]ds + \sigma\left[x(s)\right]dz(s), \quad x(t) = x_t. \tag{2.76}$$

Let $\Omega[x(s)] = \sigma[x(s)]\sigma[x(s)]^T$ denote the covariance matrix with its element in row h and column ζ denoted by $\Omega^{h\zeta}[x(s)]$.

The infinite-horizon autonomous game in (2.75) and (2.76) is independent of the choice of t and dependent only upon the state at the starting time, that is, x.

A Nash equilibrium solution for the infinite-horizon stochastic differential game in (2.75) and (2.76) can be characterized as follows.

Theorem 2.6 *An n-tuple of strategies $\{u_i^* = \phi_i^*(\cdot) \in U^i, \text{for } i \in N\}$, provides a Nash equilibrium solution to the game in (2.75) and (2.76) if there exist continuously twice differentiable functions $\hat{V}^i(x) : R^m \to R, i \in N$, satisfying the following set of partial differential equations:*

$$r\hat{V}^i(x) - \frac{1}{2}\sum_{h,\zeta=1}^{m} \Omega^{h\zeta}(x)\hat{V}^i_{x^h x^\zeta}(x)$$

$$= \max_{u_i}\{g^i\left[x, \phi_1^*(x), \phi_2^*(x), \ldots, \phi_{i-1}^*(x), u_i(x), \phi_{i+1}^*(x), \ldots, \phi_n^*(x)\right]$$

$$+ \hat{V}_x^i(x) f\left[x, \phi_1^*(x), \phi_2^*(x), \ldots, \phi_{i-1}^*(x), u_i(x), \phi_{i+1}^*(x), \ldots, \phi_n^*(x)\right]\}$$
$$= \{g^i\left[x, \phi_1^*(x), \phi_2^*(x), \ldots, \phi_n^*(x)\right] + \hat{V}_x^i(x) f\left[x, \phi_1^*(x), \phi_2^*(x), \ldots, \phi_n^*(x)\right]\},$$

for i ∈ N.

Proof This result follows readily from the definition of a Nash equilibrium and from the infinite-horizon stochastic control Theorem A.6 in the Technical Appendixes. □

2.4.2 An Application of Stochastic Differential Games in Resource Extraction

Consider the resource extraction game in Sect. 2.3.2. To present a stochastic model we replace the deterministic state dynamics with a stochastic dynamics

$$dx(s) = \left[ax(s)^{1/2} - bx(s) - \sum_{j=1}^{n} u_j(s)\right] ds + \sigma x(s) dz(s),$$

$$x(t_0) = x_0 \in X. \tag{2.77}$$

In the absence of human harvesting, the resource stock will grow according to the dynamics

$$dx(s) = \left[ax(s)^{1/2} - bx(s)\right] ds + \sigma x(s) dz(s).$$

The deterministic part of the natural growth function is $G(x) = ax^{1/2} - bx = x[ax^{-1/2} - b]$. This function represents pure compensation, viz., the proportional growth rate $G(x)/x$ is a decreasing function of x. The stochastic term reflects the randomness in the deathrate b. The resource stock has a nondegenerate stationary equilibrium level, which is characterized by the stationary density function $\varphi(x)$ (see Jørgensen and Yeung 1996)

$$\varphi(x) = \left\{K/\left[\sigma^2 x^{2(1+b/\sigma^2)}\right]\right\} \exp\left[-(4a/\sigma^2)x^{-1/2}\right],$$

where K is a normalization factor such that $\int_0^\infty \varphi(x) \, dx = 1$.

Extractor i seeks to maximize the expected payoff

$$E_{t_0} \left\{ \int_{t_0}^{T} \left[\left(\sum_{j=1}^{n} u_j(s)\right)^{-1/2} u_i(s) - \frac{c}{x(s)^{1/2}} u_i(s)\right] \exp\left[-r(t - t_0)\right] ds \right.$$

$$\left. + \exp\left[-r(T - t_0)\right] w x(T)^{1/2} \right\}, \quad \text{for } i \in N, \tag{2.78}$$

subject to the resource dynamics in (2.77).

Invoking Theorem A.5 in the Technical Appendixes, a set of feedback strategies $\{u_i^*(t) = \phi_i^*(t, x); i \in N\}$ constitutes a Nash equilibrium solution for the game in (2.77) and (2.78), if there exist functions $V^{(t_0)i}(t, x) : [t_0, T] \times R \to R$, for $i \in N$, which satisfy the following set of partial differential equations:

$$
- V_t^{(t_0)i}(t, x) - \frac{1}{2}\sigma^2 x^2 V_{xx}^{(t_0)i}(t, x)
$$

$$
= \max_{u_1 \in U^i} \left\{ \left[u_i \left(\sum_{\substack{j=1 \\ j \neq i}}^{n} \phi_j^*(t, x) + u_i \right)^{-1/2} - \frac{c}{x^{1/2}} u_i(t) \right] \exp[-r(t - t_0)] \right.
$$

$$
\left. + V_x^{(t_0)i}\left[ax^{1/2} - bx - \sum_{\substack{j=1 \\ j \neq i}}^{n} \phi_j^*(t, x) - u_i \right] \right\}, \quad \text{and}
$$

$$
V^{(t_0)i}(T, x) = \exp[-r(T - t_0)]wx^{1/2}.
$$

Applying the maximization operator on the right-hand side of the first equation in (2.79) for agent i yields the condition for a maximum as

$$
\left[\left(\sum_{\substack{j=1 \\ j \neq i}}^{n} \phi_j^*(t, x) + \frac{1}{2}\phi_i^*(t, x) \right) \left(\sum_{j=1}^{n} \phi_j^*(t, x) \right)^{-3/2} - \frac{c}{x^{1/2}} \right] \exp[-r(t - t_0)]
$$

$$
- V_x^{(t_0)i} = 0, \tag{2.80}
$$

for $i \in N$.

Following the analysis in Sect. 2.3.2, we obtain

$$
\phi_i^*(t, x) = \frac{x(2n - 1)^2}{2[\sum_{j=1}^{n}[c + \exp[r(t - t_0)]V_x^{(t_0)j} x^{1/2}]]^3} \left\{ \sum_{\substack{j=1 \\ j \neq i}}^{n} \left[c + \frac{V_x^{(t_0)j} x^{1/2}}{\exp[-r(t - t_0)]} \right] \right.
$$

$$
\left. - \left(n - \frac{3}{2} \right) \left[c + \frac{V_x^{(t_0)i} x^{1/2}}{\exp[-r(t - t_0)]} \right] \right\}, \quad \text{for } i \in N. \tag{2.81}
$$

Substituting $\phi_i^*(t, x)$ in (2.81) into (2.79), and upon solving it, yields the following.

Proposition 2.3 *The system in* (2.79) *admits a solution*

$$
V^{(t_0)i}(t, x) = \exp[-r(t - t_0)][A(t)x^{1/2} + B(t)], \quad \text{for } i \in N, \tag{2.82}
$$

where $A(t)$ and $B(t)$ satisfy

$$\dot{A}(t) = \left[r + \frac{1}{8}\sigma^2 + \frac{b}{2} \right] A(t) - \frac{(2n-1)}{2n^2} \left(c + \frac{A(t)}{2} \right)^{-1}$$
$$+ \frac{c(2n-1)^2}{4n^3} \left(c + \frac{A(t)}{2} \right)^{-2} + \frac{(2n-1)^2 A(t)}{8n^2(c + \frac{A(t)}{2})^2},$$

$$\dot{B}(t) = r B(t) - \frac{a}{2} A(t),$$

$$A(T) = w, \quad and$$

$$B(T) = 0.$$

(2.83)

Proof Substituting $V^{(t_0)i}(t, x)$ and the relevant derivatives $V_t^{(t_0)i}(t, x)$, $V_x^{(t_0)i}(t, x)$, and $V_{xx}^{(t_0)i}(t, x)$ into (2.81) and (2.79) yields the results in Proposition 2.3. $\quad\square$

Substituting the relevant derivatives of the value functions in Proposition 2.3 into the game equilibrium strategies in (2.81) gives a Nash equilibrium of the stochastic resource extraction game in (2.77) and (2.78).

2.4.3 Infinite-Horizon Resource Extraction

Consider the infinite-horizon game in which extractor i seeks to maximize the expected payoff

$$E_{t_0} \left\{ \int_{t_0}^{\infty} \left[\left(\sum_{j=1}^{n} u_j(s) \right)^{-1/2} u_i(s) - \frac{c}{x(s)^{1/2}} u_i(s) \right] \exp[-r(s - t_0)] \, ds \right\},$$

for $i \in N$,

(2.84)

subject to the resource dynamics

$$dx(s) = \left[ax(s)^{1/2} - bx(s) - \sum_{j=1}^{n} u_j(s) \right] ds + \sigma x(s) \, dz(s),$$

$$x(t_0) = x_0 \in X.$$

(2.85)

Consider the alternative problem that starts at time $t \in [t_0, \infty)$ with initial state $x(t) = x$

$$E_t \left\{ \int_t^{\infty} \left[\left(\sum_{j=1}^{n} u_j(s) \right)^{-1/2} u_i(s) - \frac{c}{x(s)^{1/2}} u_i(s) \right] \exp[-r(s - t)] \, ds \right\},$$

for $i \in N$,

(2.86)

subject to the resource dynamics

$$dx(s) = \left[ax(s)^{1/2} - bx(s) - \sum_{j=1}^{n} u_j(s)\right] ds + \sigma x(s) dz(s),$$

$$x(t) = x \in X. \tag{2.87}$$

Invoking Theorem 2.6, we obtain a set of feedback strategies $\{\phi_i^*(x), i \in N\}$ constituting of a Nash equilibrium solution for the game in (2.86) and (2.87) if there exist functions $\hat{V}^i(x) : R \to R$ for $i \in N$ that satisfy the following set of partial differential equations:

$$r\hat{V}^i(x) - \frac{1}{2}\sigma^2 x^2 W_{xx}^i(x)$$

$$= \max_{u_i \in U^i} \left\{ \left[u_i \left(\sum_{\substack{j=1 \\ j \neq i}}^{n} \phi_j^*(x) + u_i \right)^{-1/2} - \frac{c}{x^{1/2}} u_i \right] \right.$$

$$\left. + \hat{V}_x^i \left[ax^{1/2} - bx - \sum_{\substack{j=1 \\ j \neq i}}^{n} \phi_j^*(x) - u_i \right] \right\}, \quad \text{for } i \in N. \tag{2.88}$$

Applying the maximization operator in (2.88) for agent i yields the condition for a maximum as

$$\left[\left(\sum_{\substack{j=1 \\ j \neq i}}^{n} \phi_j^*(x) + \frac{1}{2}\phi_i^*(x) \right) \left(\sum_{j=1}^{n} \phi_j^*(x) \right)^{-3/2} - \frac{c}{x^{1/2}} \right] - \hat{V}_x^i = 0,$$

$$\text{for } i \in N. \tag{2.89}$$

Summing over $i = 1, 2, \ldots, n$ in (2.89) yields

$$\left(\sum_{j=1}^{n} \phi_j^*(x) \right)^{1/2} = \left(n - \frac{1}{2} \right) \left(\sum_{j=1}^{n} \left[\frac{c}{x^{1/2}} + \hat{V}_x^j \right] \right)^{-1}. \tag{2.90}$$

Substituting (2.90) into (2.89) produces

$$\left(\sum_{\substack{j=1 \\ j \neq i}}^{n} \phi_j^*(x) + \frac{1}{2}\phi_i^*(x) \right) \left(n - \frac{1}{2} \right)^{-3} \left(\sum_{j=1}^{n} \left[\frac{c}{x^{1/2}} + \hat{V}_x^j \right] \right)^{3} - \frac{c}{x^{1/2}} - \hat{V}_x^i = 0,$$

$$\text{for } i \in N. \tag{2.91}$$

Rearranging the terms in (2.91) yields

$$\left(\sum_{\substack{j=1 \\ j\neq i}}^{n}\phi_j^*(x)+\frac{1}{2}\phi_i^*(x)\right)=\left(n-\frac{1}{2}\right)^3\frac{[c+\hat{V}_x^i x^{1/2}]x}{(\sum_{j=1}^{n}[c+\hat{V}_x^j x^{1/2}])^3},$$

for $i\in N$.   (2.92)

The condition in (2.92) represents a system of equations that is linear in $\{\phi_1^*(x),\phi_2^*(x),\ldots,\phi_n^*(x)\}$. Solving (2.92) yields the game equilibrium strategies

$$\phi_i^*(x)=\frac{x(2n-1)^2}{2[\sum_{j=1}^{n}[c+\hat{V}_x^j x^{1/2}]]^3}\left\{\sum_{\substack{j=1 \\ j\neq i}}^{n}[c+\hat{V}_x^j x^{1/2}]-\left(n-\frac{3}{2}\right)[c+\hat{V}_x^i x^{1/2}]\right\},$$

for $i\in N$.   (2.93)

Substituting $\phi_i^*(t,x)$ in (2.93) into (2.88), and upon solving it, yields the following.

Proposition 2.4 *The system in* (2.88) *admits a solution*

$$\hat{V}^i(x)=\left[Ax^{1/2}+B\right],\quad\text{for } i\in N,$$ (2.94)

where A and B satisfy

$$0=\left[r+\frac{1}{8}\sigma^2+\frac{b}{2}\right]-\frac{(2n-1)}{2n^2}\left(c+\frac{A}{2}\right)^{-1}+\frac{c(2n-1)^2}{4n^3}\left(c+\frac{A}{2}\right)^{-2}$$

$$+\frac{(2n-1)^2A}{8n^2(c+\frac{A}{2})^2},$$ (2.95)

$$B=\frac{a}{2r}A.$$

Proof Substituting $\hat{V}^i(x)$ and the relevant derivatives $\hat{V}_x^i(x)$ and $\hat{V}_{xx}^i(x)$ into (2.93) and (2.88) yields the results in Proposition 2.4.   □

A feedback Nash equilibrium can be readily obtained by substituting the relevant derivatives of the value functions in Proposition 2.4 into the game equilibrium strategies of (2.93).

2.5 Exercises

2.1 Consider the competitive dynamic advertising game in which there are two firms in a market. The profits of firm 1 and that of firm 2 are, respectively,

$$\int_0^5\left[10x(s)-2u_1(s)^2\right]\exp(-0.05s)\,ds+\exp\left[(-0.05)5\right]12x(5),$$

and

$$\int_0^5 \left[8\bigl(1 - x(s)\bigr) - u_2(s)^2\right] \exp(-0.05s)\, ds + \exp\bigl[(-0.05)5\bigr]9\bigl[1 - x(5)\bigr],$$

where $x(s)$ is the market share of firm 1 at time s, $[1 - x(s)]$ is that of firm 2, and $u_i(s)$ is the advertising rate for firm $i \in \{1, 2\}$.

It is assumed that market potential is constant over time. The only marketing instrument used by the firms is advertising. Advertising has diminishing returns since there are increasing marginal costs of advertising as reflected through the quadratic cost function. The dynamics of firm 1's market share is governed by

$$\dot{x}(s) = u_1(s)\bigl[1 - x(s)\bigr]^{1/2} - u_2(s)x(s)^{1/2}, \quad x(0) = 0.6.$$

Derive an open-loop solution for the market equilibrium.

2.2 Consider an economy endowed with a renewable resource and with $n \geq 2$ resource extractors (firms). The lease for resource extraction begins at time 0 and ends at time 10. Let $u_i(s)$ denote the rate of resource extraction of firm i at time s, $i \in \{1, 2\}$, and $x(s)$ is the size of the resource stock. The extraction cost is $C^i = 2u_i(s)/x(s)^{1/2}$ for firm $i \in \{1, 2\}$. The demand for the resource is $P(s) = Q(s)^{-1/2}$, where $Q(s) = \sum_{j=1}^2 u_j(s)$. A terminal bonus $4x(T)^{1/2}$ is offered to each extractor and the discount rate is 0.1. Extractor i seeks to maximize the present value of profits

$$\int_0^4 \left[\left(\sum_{j=1}^2 u_j(s)\right)^{-1/2} u_i(s) - \frac{2}{x(s)^{1/2}}u_i(s)\right] \exp(-0.1s)\, ds$$

$$+ \exp\bigl[-0.1(10)\bigr]4x(T)^{1/2},$$

for $i \in \{1, 2\}$, subject to the resource dynamics

$$\dot{x}(s) = 5x(s)^{1/2} - x(s) - \sum_{j=1}^2 u_j(s), \quad x(0) = 100.$$

Derive a feedback Nash equilibrium solution.

2.3 Consider a dynamic duopoly in which there are two publicly listed firms selling a homogeneous good. The payoff of firm i is given as the present value of the stream of discounted profits

$$\int_0^\infty \left\{P(s)u_i(s) - 2u_i(s) - (1/2)\bigl[u_i(s)\bigr]^2\right\} \exp[-0.05s]\, ds, \quad \text{for } i \in \{1, 2\},$$

where $2u_i(s) + (1/2)[u_i(s)]^2$ is the cost of the producing output $u_i(s)$ and the interest rate is 0.05.

There is a lag in price adjustment so the evolution of the market price over time is assumed to be a function of the current market price and the price specified by the current demand condition. In particular, the price dynamics follows

$$\dot{P}(s) = 0.5\big[50 - u_1(s) - u_2(s) - P(s)\big], \quad P(0) = 5.$$

Characterize a feedback equilibrium for the duopoly.

2.4 Consider an economy endowed with a renewable resource and with two resource extraction firms. The lease for resource extraction begins at time 0 and ends at time 3. Let $u_i(s)$ denote the rate of resource extraction of firm i at time $s \in [0, 3]$, $i \in \{1, 2\}$, where each extractor controls its rate of extraction. Let $x(s)$ denote the size of the resource stock at time s; the resource growth dynamics is stochastic. The extraction cost depends on the quantity of the resource extracted $u_i(s)$ and the resource stock size $x(s)$. In particular, the extraction cost can be specified as $C^i = u_i(s)/x(s)^{1/2}$ for firm $i \in \{1, 2\}$. The market price of the resource depends on the total amount extracted and supplied to the market. The price-output relationship at time s is given by the following downward-sloping inverse demand curve $P(s) = 0.5Q(s)^{-1/2}$, where $Q(s) = \sum_{j=1}^{2} u_j(s)$ is the total amount of the resource extracted and marketed at time s. A terminal bonus $2x(T)^{1/2}$ is offered to each extractor and the discount rate is 0.1. Extractor $i \in \{1, 2\}$ seeks to maximize the present value of the expected profits

$$E_0\left\{ \int_0^3 \left[0.5\left(\sum_{j=1}^{2} u_j(s)\right)^{-1/2} u_i(s) - \frac{u_i(s)}{x(s)^{1/2}} \right] \exp(-rs)\, ds \right.$$

$$\left. + \exp\big[-3(0.1)\big] 2x(T)^{1/2} \right\}$$

subject to the stochastic resource dynamics

$$dx(s) = \left[10x(s)^{1/2} - x(s) - \sum_{j=1}^{2} u_j(s) \right] ds + 0.4x(s)\, dz(s), \quad x(0) = 120.$$

Derive a feedback equilibrium for the above stochastic dynamic economy.

Chapter 3
Dynamic Economic Optimization: Group Optimality and Individual Rationality

The most appealing characteristic of perfectly competitive markets is that individually rational behaviors bring about group (Pareto) optimality in economic resource allocation. However, the market fails to provide an effective mechanism for optimal resource use because of the prevalence of imperfect market structure, externalities, imperfect information, and public goods in the current global economy. As a result, though the market is one of the most effective instruments in conducting economic activities, it fails to guarantee its efficiency under the current arrangement. The noncooperative outcomes characterized in Chap. 2 vividly demonstrate that Pareto optimality could not be achieved by markets. Removing market suboptimality is not just a task of achieving a better alternative, but sometimes it can be an absolute necessity. For instance, efforts to alleviate the worldwide financial tsunami and catastrophe-bound industrial pollution are currently pressing issues.

Cooperative games suggest the possibility of socially optimal and group efficient solutions to decision problems involving strategic action. The formulation of optimal behavior for players (economic agents) is a fundamental element in this theory. Two essential factors for economic optimization are group optimality and individual rationality. Group optimality ensures that all potential gains from cooperation are captured. The failure to fulfill group optimality leads to the condition where the participants prefer to deviate from the agreed-upon solution plan to extract the unexploited gains. Individual rationality is required to hold so that the payoff allocated to an economic agent under cooperation will be no less than its noncooperative payoff. The failure to guarantee individual rationality leads to the condition where the concerned participants will reject the agreed upon solution plan and play noncooperatively. For the optimization scheme to be upheld throughout the game horizon both group rationality and individual rationality are required to be satisfied at any time.

Section 3.1 examines the notion of group optimality and shows the derivation of group optimal strategies and the cooperative state trajectory. Individual rationality and transfer payments leading to the satisfaction of individual rationality are given in Sect. 3.2. The analysis is extended to the infinite-horizon scenario in Sect. 3.3 and Sect. 3.4 presents cooperative economic games satisfying group optimality and individual rationality.

D.W.K. Yeung, L.A. Petrosyan, *Subgame Consistent Economic Optimization*,
Static & Dynamic Game Theory: Foundations & Applications,
DOI 10.1007/978-0-8176-8262-0_3, © Springer Science+Business Media, LLC 2012

3.1 Group Optimality

Consider the general form of n-person differential games in the economics characterized in Chap. 2. Economic agent $i \in N$ seeks to maximize its objective

$$\int_{t_0}^{T} g^i\big[s, x(s), u_1(s), u_2(s), \ldots, u_n(s)\big] \exp\left[-\int_{t_0}^{s} r(y)\, dy\right] ds$$

$$+ \exp\left[-\int_{t_0}^{T} r(y)\, dy\right] q^i\big(x(T)\big), \tag{3.1}$$

subject to the state dynamics

$$\dot{x}(s) = f\big[s, x(s), u_1(s), u_2(s), \ldots, u_n(s)\big], \quad x(t_0) = x_0. \tag{3.2}$$

Now consider the case when the agents agree to act cooperatively. The agents agree to act according to an agreed-upon optimality principle. The agreement on how to act cooperatively and allocate cooperative payoff constitutes the solution optimality principle of a cooperative scheme. In particular, the solution optimality principle includes (i) an agreement on a set of cooperative strategies/controls and (ii) a mechanism to distribute the total payoff among agents.

3.1.1 Optimal Strategies and Cooperative State Trajectories

Since payoffs are transferable, group optimality requires the agents to maximize their joint payoff. The agents must then solve the following optimal control problem:

$$\max_{u_1, u_2, \ldots, u_n} \left\{ \int_{t_0}^{T} \sum_{j=1}^{n} g^j\big[s, x(s), u_1(s), u_2(s), \ldots, u_n(s)\big] \exp\left[-\int_{t_0}^{s} r(y)\, dy\right] ds \right.$$

$$\left. + \exp\left[-\int_{t_0}^{T} r(y)\, dy\right] \sum_{j=1}^{n} q^j\big(x(T)\big) \right\} \tag{3.3}$$

subject to (3.2).

Both optimal control and dynamic programming can be used to solve the problem in (3.2) and (3.3). The technique of optimal control is given in Sect. A.2 of the Technical Appendixes. For the sake of comparison with other derived results, and for expositional convenience, a dynamic programming technique is adopted. The set of group optimal control strategies can be characterized as follows.

Theorem 3.1 *A set of controls* $\{\psi_i^*(t, x), \text{ for } i \in N \text{ and } t \in [t_0, T]\}$ *provides an optimal solution to the control problem in (3.2) and (3.3) if there exists a continuously* *differentiable function* $W^{(t_0)}(t, x) : [t_0, T] \times R^m \to R$ *satisfying the following Bellman equation*:

$$-W_t^{(t_0)}(t, x) = \max_{u_1, u_2, \ldots, u_n} \left\{ \sum_{j=1}^n g^j[t, x, u_1, u_2, \ldots, u_n] \exp\left[-\int_{t_0}^t r(y) \, dy \right] \right.$$

$$\left. + W_x^{(t_0)} f[t, x, u_1, u_2, \ldots, u_n] \right\},$$

$$W^{(t_0)}(T, x) = \exp\left[-\int_{t_0}^T r(y) \, dy \right] \sum_{j=1}^n q^j(x).$$

Proof Follow the proof of Theorem A.1 in the Technical Appendixes. □

Hence the agents will adopt the cooperative control $\{\psi_i^*(t, x)$, for $i \in N$ and $t \in [t_0, T]\}$, to obtain the maximized level of joint profit. In a cooperative framework, the issue of the nonuniqueness of the optimal controls can be resolved by the agreement between the agents on a particular set of controls. Substituting this set of controls into (3.2) yields the dynamics of the optimal (cooperative) trajectory as

$$\dot{x}(s) = f\left[s, x(s), \psi_1^*\left(s, x(s)\right), \psi_2^*\left(s, x(s)\right), \ldots, \psi_n^*\left(s, x(s)\right)\right],$$
$$x(t_0) = x_0. \tag{3.4}$$

Let $x^*(t)$ denote the solution to (3.4). The optimal trajectory $\{x^*(t)\}_{t=t_0}^T$ can be expressed as

$$x^*(t) = x_0 + \int_{t_0}^t f\left[s, x^*(s), \psi_1^*\left(s, x^*(s)\right), \psi_2^*\left(s, x^*(s)\right), \ldots, \psi_n^*\left(s, x^*(s)\right)\right] ds.$$

For notational convenience, we use the terms $x^*(t)$ and x_t^* interchangeably.

The cooperative control for the game in (3.1) and (3.2) over the time interval $[t_0, T]$ can be expressed more precisely as

$$\left\{ \psi_i^*\left(t, x^*(t)\right), \text{ for } i \in N \text{ and } t \in [t_0, T] \right\}. \tag{3.5}$$

Note that, for group optimality to be achievable, the cooperative controls $\{\psi_i^*(t, x^*(t))$, for $i \in N$ and $t \in [t_0, T]\}$, must be exercised throughout time interval $[t_0, T]$.

The cooperative payoff over the interval $[t, T]$, for $t \in [t_0, T)$, can be expressed as

$$W^{(t_0)}\left(t, x_t^*\right) = \int_t^T \sum_{j=1}^n g^j\left[s, x^*(s), \psi_1^*\left(s, x^*(s)\right), \psi_2^*\left(s, x^*(s)\right), \ldots, \psi_n^*\left(s, x^*(s)\right)\right]$$

$$\times \exp\left[-\int_{t_0}^s r(y) \, dy \right] ds + \exp\left[-\int_{t_0}^T r(y) \, dy \right] \sum_{j=1}^n q^j\left(x^*(T)\right).$$

$$\tag{3.6}$$

To verify whether the agent would find it optimal to adopt the cooperative controls in (3.5) throughout the cooperative duration, we consider an optimal control problem with the dynamics in (3.2) and payoff in (3.3), which begins at time $\tau \in [t_0, T]$ with initial state x_τ^*. At time τ, the optimality principle ensuring group rationality requires the agents to solve the problem

$$
\max_{u_1, u_2, \ldots, u_n} \left\{ \int_\tau^T \sum_{j=1}^n g^j \left[s, x(s), u_1(s), u_2(s), \ldots, u_n(s) \right] \exp\left[-\int_\tau^s r(y)\,dy \right] ds \right.
$$

$$
+ \exp\left[-\int_\tau^T r(y)\,dy \right] \sum_{j=1}^n q^j \left(x(T) \right) \Bigg\}, \tag{3.7}
$$

subject to

$$
\dot{x}(s) = f\left[s, x(s), u_1(s), u_2(s), \ldots, u_n(s) \right], \quad x(\tau) = x_\tau^*. \tag{3.8}
$$

The problem in (3.7) can alternatively be written as

$$
\max_{u_1, u_2, \ldots, u_n} \left\{ \exp\left[\int_{t_0}^\tau r(y)\,dy \right] \left(\int_\tau^T \sum_{j=1}^n g^j \left[s, x(s), u_1(s), u_2(s), \ldots, u_n(s) \right] \right. \right.
$$

$$
\times \exp\left[-\int_{t_0}^s r(y)\,dy \right] ds + \exp\left[-\int_{t_0}^T r(y)\,dy \right] \sum_{j=1}^n q^j \left(x(T) \right) \Bigg) \Bigg\}
$$

$$
= \exp\left[\int_{t_0}^\tau r(y)\,dy \right] \max_{u_1, u_2, \ldots, u_n} \left\{ \left(\int_\tau^T \sum_{j=1}^n g^j \left[s, x(s), u_1(s), u_2(s), \ldots, u_n(s) \right] \right. \right.
$$

$$
\times \exp\left[-\int_{t_0}^s r(y)\,dy \right] ds + \exp\left[-\int_{t_0}^T r(y)\,dy \right] \sum_{j=1}^n q^j \left(x(T) \right) \Bigg) \Bigg\}. \tag{3.9}
$$

Invoking the backward induction property of the principle of optimality one can readily verify that the optimal controls strategies for the problems in (3.8) and (3.9) are analogous to the optimal controls strategies for the problems in (3.1) and (3.2) in the time interval $[t, T]$.

A remark that will be utilized in the subsequent analysis is given below.

Remark 3.1 Let $W^{(\tau)}(t, x_t^*)$ denote the total cooperative payoff function of the control problem of (3.8) and (3.9). One can readily verify that

$$
\exp\left[\int_{t_0}^\tau r(y)\,dy \right] W^{(t_0)}(t, x_t^*)
$$

$$= \exp\left[\int_{t_0}^{\tau} r(y)\,dy\right]$$

$$\times \left\{\int_t^T \sum_{j=1}^n g^j\left[s, x^*(s), \psi_1^*\left(s, x^*(s)\right), \psi_2^*\left(s, x^*(s)\right), \ldots, \psi_n^*\left(s, x^*(s)\right)\right]\right.$$

$$\times \exp\left[-\int_{t_0}^s r(y)\,dy\right]ds + \exp\left[-\int_{t_0}^T r(y)\,dy\right]\sum_{j=1}^n q^j\left(x^*(T)\right)\right\}$$

$$= \int_t^T \sum_{j=1}^n g^j\left[s, x^*(s), \psi_1^*\left(s, x^*(s)\right), \psi_2^*\left(s, x^*(s)\right), \ldots, \psi_n^*\left(s, x^*(s)\right)\right]$$

$$\times \exp\left[-\int_{\tau}^s r(y)\,dy\right]ds + \exp\left[-\int_{\tau}^T r(y)\,dy\right]\sum_{j=1}^n q^j\left(x^*(T)\right)$$

$$= W^{(\tau)}\left(t, x_t^*\right),$$

for $\tau \in [t_0, T]$ and $t \in [\tau, T]$.

Next, we present a cooperative economic game yielding group optimality.

3.1.2 Group Optimality in Resource Extraction

Consider the two firms' version of the resource extraction game in Sect. 2.3.2 in Chap. 2.

The resource stock $x(s) \in X \subset R$ follows the dynamics

$$\dot{x}(s) = ax(s)^{1/2} - bx(s) - u_1(s) - u_2(s), \quad x(t_0) = x_0 \in X, \tag{3.10}$$

where $u_1(s)$ is the harvest rate of economic agent 1 and $u_2(s)$ is the harvest rate of economic agent 2. The instantaneous payoffs at time $s \in [t_0, T]$ for agents 1 and 2 are, respectively,

$$\left[u_1(s)^{1/2} - \frac{c_1}{x(s)^{1/2}}u_1(s)\right] \quad \text{and} \quad \left[u_2(s)^{1/2} - \frac{c_2}{x(s)^{1/2}}u_2(s)\right],$$

where c_1 and c_2 are constants and $c_1 \neq c_2$.

At time T, each agent will receive a termination bonus

$$qx(T)^{\frac{1}{2}},$$

which depends on the resource remaining at the terminal time.

Payoffs are transferable between agents 1 and 2 over time. Given the discount rate r, the values received t after time t_0 have to be discounted by the factor $\exp[-r(t - t_0)]$.

Consider the case when these two economic agents agree to cooperate and maximize the sum of their payoffs

$$\int_{t_0}^T \left(\left[u_1(s)^{1/2} - \frac{c_1}{x(s)^{1/2}} u_1(s) \right] + \left[u_2(s)^{1/2} - \frac{c_2}{x(s)^{1/2}} u_2(s) \right] \right)$$

$$\times \exp[-r(s - t_0)] \, ds + 2 \exp[-r(T - t_0)] q x(T)^{\frac{1}{2}}, \qquad (3.11)$$

subject to (3.10).

Let $[\psi_1^*(t, x), \psi_2^*(t, x)]$ denote a set of controls that provide a solution to the optimal control problem in (3.10) and (3.11), and $W^{(t_0)}(t, x) : [t_0, T] \times R \to R$ denotes the maximized joint payoff function that satisfies the equations (see Theorem 3.1)

$$-W_t^{(t_0)}(t, x) = \max_{u_1, u_2} \left\{ \left(\left[u_1^{1/2} - \frac{c_1}{x^{1/2}} u_1 \right] + \left[u_2^{1/2} - \frac{c_2}{x^{1/2}} u_2 \right] \right) \exp[-r(t - t_0)] \right.$$

$$\left. + W_x^{(t_0)}(t, x) \left[a x^{1/2} - b x - u_1 - u_2 \right] \right\}, \quad \text{and} \qquad (3.12)$$

$$W^{(t_0)}(T, x) = 2 \exp[-r(T - t_0)] q x^{\frac{1}{2}}.$$

Performing the indicated maximization we obtain

$$\psi_1^*(t, x) = \frac{x}{4[c_1 + W_x^{(t_0)} \exp[r(t - t_0)] x^{1/2}]^2}, \quad \text{and}$$

$$\psi_2^*(t, x) = \frac{x}{4[c_2 + W_x^{(t_0)} \exp[r(t - t_0)] x^{1/2}]^2}.$$

Substituting $\psi_1^*(t, x)$ and $\psi_2^*(t, x)$ above into (3.13) yields the value function

$$W^{(t_0)}(t, x) = \exp[-r(t - t_0)] [\hat{A}(t) x^{1/2} + \hat{B}(t)], \qquad (3.13)$$

where

$$\dot{\hat{A}}(t) = \left[r + \frac{b}{2} \right] \hat{A}(t) - \frac{1}{2[c_1 + \hat{A}(t)/2]} - \frac{1}{2[c_2 + \hat{A}(t)/2]} + \frac{c_1}{4[c_1 + \hat{A}(t)/2]^2}$$

$$+ \frac{c_2}{4[c_2 + \hat{A}(t)/2]^2} + \frac{\hat{A}(t)}{8[c_1 + \hat{A}(t)/2]^2} + \frac{\hat{A}(t)}{8[c_2 + \hat{A}(t)/2]^2},$$

$$\dot{\hat{B}}(t) = r \hat{B}(t) - \frac{a}{2} \hat{A}(t), \qquad \hat{A}(T) = 2q, \quad \text{and} \quad \hat{B}(T) = 0.$$

The optimal cooperative controls can then be obtained as

$$\psi_1^*(t, x) = \frac{x}{4[c_1 + \hat{A}(t)/2]^2}, \quad \text{and} \quad \psi_2^*(t, x) = \frac{x}{4[c_2 + \hat{A}(t)/2]^2}.$$

Substituting these control strategies into (3.10) yields the dynamics of the state trajectory under cooperation

$$\dot{x}(s) = ax(s)^{1/2} - bx(s) - \frac{x(s)}{4[c_1 + \hat{A}(s)/2]^2} - \frac{x(s)}{4[c_2 + \hat{A}(s)/2]^2}, \quad x(t_0) = x_0. \tag{3.14}$$

Solving (3.14) yields the optimal cooperative state trajectory as

$$x^*(s) = \varpi(t_0, s)^2 \left[x_0^{1/2} + \int_{t_0}^{s} \varpi^{-1}(t_0, t) H_1 \, dt \right]^2, \quad \text{for } s \in [t_0, T], \tag{3.15}$$

where $\varpi(t_0, s) = \exp[\int_{t_0}^{s} H_2(\tau) \, d\tau]$, $H_1 = \frac{1}{2}a$, and

$$H_2(s) = -\left[\frac{1}{2}b + \frac{1}{8[c_1 + \hat{A}(s)/2]^2} + \frac{1}{8[c_2 + \hat{A}(s)/2]^2} \right].$$

The outcome of cooperation in the game in (3.10) and (3.11) is completely characterized by its optimal cooperative state trajectory, cooperative strategies, and joint payoff above.

3.1.3 Group Optimality in Infinite-Horizon Problems

Consider the n-person infinite-horizon general economic problem in which agent i's payoff is

$$\int_{t_0}^{\infty} g^i [x(s), u_1(s), u_2(s), \ldots, u_n(s)] \exp[-r(s - t_0)] \, ds, \quad \text{for } i \in N. \tag{3.16}$$

The state dynamics is

$$\dot{x}(s) = f[x(s), u_1(s), u_2(s), \ldots, u_n(s)], \quad x(t_0) = x_0. \tag{3.17}$$

In the case where the agents agree to cooperate, group optimality can be achieved if the agents agree to maximize the sum of their payoffs, that is,

$$\max_{u_1, u_2, \ldots, u_n} \left\{ \int_{t_0}^{\infty} \sum_{j=1}^{N} g^j [x(s), u_1(s), u_2(s), \ldots, u_n(s)] \exp[-r(s - t_0)] \, ds \right\}, \tag{3.18}$$

subject to (3.17).

Now consider the alternative infinite problem that starts at time $t \in [t_0, \infty)$ with initial state $x(t) = x$:

$$\max_{u_1, u_2, \ldots, u_n} \left\{ \int_t^\infty \sum_{j=1}^N g^j [x(s), u_1(s), u_2(s), \ldots, u_n(s)] \exp[-r(s-t)] ds \right\},$$

subject to

$$\dot{x}(s) = f[x(s), u_1(s), u_2(s), \ldots, u_n(s)], \quad x(t) = x. \tag{3.19}$$

The infinite-horizon autonomous problem in (3.19) is independent of the choice of t and dependent only upon the state x. We define

$$W(x) = \max_{u_1, u_2, \ldots, u_n} \left\{ \int_t^\infty \sum_{j=1}^N g^j [x(s), u_1(s), u_2(s), \ldots, u_n(s)] \right.$$

$$\left. \times \exp[-r(s-t)] ds \,\middle|\, x(t) = x \right\}.$$

Again, for the sake of comparison in later analysis in the book we characterize the group optimal control strategies as follows.

Theorem 3.2 *A set of controls $\{\psi_i^*(x), \text{for } i \in N\}$ provides a solution to the optimal control problem in (3.19) if there exists a continuously differentiable function $W(x)$: $R^m \to R$ satisfying the infinite-horizon Bellman equation*

$$r W(x) = \max_{u_1, u_2, \ldots, u_n} \left\{ \sum_{j=1}^n g^j [x, u_1, u_2, \ldots, u_n] + W_x f[x, u_1, u_2, \ldots, u_n] \right\}.$$

Proof Follow the proof of Theorem A.2 in the Technical Appendixes.   □

Hence the agents will adopt the cooperative control $\{\psi_i^*(x), \text{for } i \in N\}$ characterized in Theorem 3.2. Note that these controls are functions of the current state x only. Substituting this set of controls into (3.17) yields the dynamics of the optimal (cooperative) trajectory as

$$\dot{x}(s) = f[x(s), \psi_1^*(x(s)), \psi_2^*(x(s)), \ldots, \psi_n^*(x(s))], \quad x(t_0) = x_0. \tag{3.20}$$

Let $x^*(t)$ denote the solution to (3.20). The optimal trajectory $\{x^*(t)\}_{t=t_0}^\infty$ can be expressed as

$$x^*(t) = x_0 + \int_{t_0}^t f[x^*(s), \psi_1^*(x^*(s)), \psi_2^*(x^*(s)), \ldots, \psi_n^*(x^*(s))] ds.$$

Once again, for notational convenience, we use the terms $x^*(t)$ and x_t^* interchangeably.

The cooperative control for the game in (3.6) and (3.7) can be expressed more precisely as

$$\{\psi_i^*(x_t^*), \text{ for } i \in N \text{ and } t \in [t_0, \infty)\}.$$

Note that these controls are functions of the current state x_t^* only. The current-value maximized cooperative payoff at current time $t \in [t_0, \infty)$, given that the state is x_t^* at t, can be expressed as

$$W(x_t^*) = \int_t^\infty \sum_{j=1}^n g^j \left[x^*(s), \psi_1^*(x^*(s)), \psi_2^*(x^*(s)), \ldots, \psi_n^*(x^*(s)) \right]$$

$$\times \exp\left[-r(s-t)\right] ds. \tag{3.21}$$

An illustration of an infinite-horizon cooperative economic game achieving group optimality is provided in the following example.

Example 3.1 Consider an infinite-horizon version of the resource extraction example in Sect. 3.1.2. At time t_0, the payoff functions of agent 1 and agent 2, are respectively,

$$\int_{t_0}^\infty \left[u_1(s)^{1/2} - \frac{c_1}{x(s)^{1/2}} u_1(s) \right] \exp\left[-r(t-t_0)\right] ds,$$

and

$$\int_{t_0}^\infty \left[u_2(s)^{1/2} - \frac{c_2}{x(s)^{1/2}} u_2(s) \right] \exp\left[-r(t-t_0)\right] ds. \tag{3.22}$$

The resource stock $x(s) \in X \subset R$ follows the dynamics in (3.10).

These two extractors agree to cooperate and maximize the sum of their payoffs. The agents have to solve the control problem of maximizing

$$\int_{t_0}^\infty \left(\left[u_1(s)^{1/2} - \frac{c_1}{x(s)^{1/2}} u_1(s) \right] + \left[u_2(s)^{1/2} - \frac{c_2}{x(s)^{1/2}} u_2(s) \right] \right)$$

$$\times \exp\left[-r(t-t_0)\right] ds, \tag{3.23}$$

subject to (3.10).

Consider the alternative problem of maximizing

$$\int_t^\infty \left(\left[u_1(s)^{1/2} - \frac{c_1}{x(s)^{1/2}} u_1(s) \right] + \left[u_2(s)^{1/2} - \frac{c_2}{x(s)^{1/2}} u_2(s) \right] \right)$$

$$\times \exp\left[-r(t-t)\right] ds, \tag{3.24}$$

subject to

$$\dot{x}(s) = ax(s)^{1/2} - bx(s) - u_1(s) - u_2(s), \quad x(t) = x_t.$$

Invoking Theorem 3.2 we obtain

$$rW(x) = \max_{u_1,u_2}\left\{\left(\left[u_1^{1/2} - \frac{c_1}{x^{1/2}}u_1\right] + \left[u_2^{1/2} - \frac{c_2}{x^{1/2}}u_2\right]\right)\right.$$
$$\left. + W_x(x)\left[ax^{1/2} - bx - u_1 - u_2\right]\right\}. \quad (3.25)$$

Performing the indicated maximization we obtain

$$\psi_1^*(x) = \frac{x}{4[c_1 + W_x(x)x^{1/2}]^2}, \quad \text{and}$$

$$\psi_2^*(x) = \frac{x}{4[c_2 + W_x(x)x^{1/2}]^2}.$$

Substituting $\psi_1^*(x)$ and $\psi_2^*(x)$ above into (3.25) yields the value function

$$W(x) = \left[\hat{A}x^{1/2} + \hat{B}\right], \quad (3.26)$$

where

$$\left[r + \frac{b}{2}\right]\hat{A} - \frac{1}{2[c_1 + \hat{A}/2]} - \frac{1}{2[c_2 + \hat{A}/2]}$$

$$+ \frac{c_1}{4[c_1 + \hat{A}/2]^2} + \frac{c_2}{4[c_2 + \hat{A}/2]^2} + \frac{\hat{A}}{8[c_1 + \hat{A}/2]^2} + \frac{\hat{A}}{8[c_2 + \hat{A}/2]^2} = 0, \quad \text{and}$$

$$\hat{B} = \frac{a}{2r}\hat{A}.$$

The optimal cooperative controls can then be obtained as

$$\psi_1^*(x) = \frac{x}{4[c_1 + \hat{A}/2]^2} \quad \text{and} \quad \psi_2^*(x) = \frac{x}{4[c_2 + \hat{A}/2]^2}. \quad (3.27)$$

Substituting these control strategies into (3.10) yields the dynamics of the state trajectory under cooperation

$$\dot{x}(s) = ax(s)^{1/2} - bx(s) - \frac{x(s)}{4[c_1 + \hat{A}/2]^2} - \frac{x(s)}{4[c_2 + \hat{A}/2]^2}, \quad x(t_0) = x_0. \quad (3.28)$$

Solving (3.28) yields the optimal cooperative state trajectory for the infinite-horizon control problem in (3.10) and (3.24) as

$$x^*(s) = \left[\frac{a}{2H} + \left(x_0^{1/2} - \frac{a}{2H}\right)\exp\left[-H(s - t_0)\right]\right]^2,$$

where

$$H = -\left[\frac{b}{2} + \frac{1}{8[c_1 + \hat{A}/2]^2} + \frac{1}{8[c_2 + \hat{A}/2]^2}\right]. \quad (3.29)$$

In Sect. 3.1, the notion of group optimality and the derivation of group optimal strategies and the cooperative state trajectory for finite and infinite horizons are presented. The issue of individual rationality will be examined in the following section.

3.2 Individual Rationality

After the economic agents agree to cooperate and maximize their joint payoff, they have to distribute the cooperative payoff among themselves. At time t_0, with the state being x_0, the term $\xi^{(t_0)i}(t_0, x_0)$ is used to denote the imputation of the payoff (received over the time interval $[t_0, T]$) to agent i. A necessary condition for group optimality and individual rationality to be upheld is

$$\text{(i)} \quad \sum_{j=1}^{n} \xi^{(t_0)j}(t_0, x_0) = W^{(t_0)}(t_0, x_0), \quad \text{and}$$

$$\text{(ii)} \quad \xi^{(t_0)i}(t_0, x_0) \geq V^{(t_0)i}(t_0, x_0), \quad \text{for } i \in N. \tag{3.30}$$

Condition (i) of (3.30) ensures group optimality and condition (ii) guarantees individual rationality at time t_0. Failure to ensure group optimality leads to disputes over the unexploited gains. If individual rationality is not guaranteed, economic agents whose cooperative payoff is less than their noncooperative payoff will not participate in the cooperative plan at the outset.

For the optimization scheme to be upheld throughout the game horizon, both group rationality and individual rationality are required to be satisfied throughout the cooperation period $[t_0, T]$. At time $\tau \in [t_0, T]$, let $\xi^{(\tau)i}(\tau, x_\tau^*)$ denote the imputation of the payoff to agent i over the time interval $[\tau, T]$. Therefore, the conditions

$$\text{(i)} \quad \sum_{j=1}^{n} \xi^{(\tau)j}(\tau, x_\tau^*) = W^{(\tau)}(\tau, x_\tau^*), \quad \text{and}$$

$$\text{(ii)} \quad \xi^{(\tau)i}(\tau, x_\tau^*) \geq V^{(\tau)i}(\tau, x_\tau^*), \quad \text{for } i \in N \text{ and } \tau \in [t_0, T], \tag{3.31}$$

have to be fulfilled.

In particular, condition (i) ensures Pareto optimality and condition (ii) guarantees individual rationality throughout the cooperation period $[t_0, T]$. Failure to guarantee individual rationality leads to the condition where the concerned participants would reject the agreed-upon solution plan and play noncooperatively.

3.2.1 Lump-Sum and Continuous Transfer Payments

To guarantee individual rationality transfer payments have to be arranged. First consider the case when individual rationality is required to hold only at t_0. In the case where part (ii) of (3.30) is required to be satisfied lump-sum transfers can be used. With agents using the cooperative strategies $\{\psi_i^*(s, x_s^*)$, for $s \in [t_0, T]$ and $i \in N\}$, agent i would derive a payoff

$$
\begin{aligned}
W^{(t_0)i}(t_0, x_0) = \int_{t_0}^{T} & g^i \left[s, x^*(s), \psi_1^*(s, x^*(s)), \psi_2^*(s, x^*(s)), \dots, \psi_n^*(s, x^*(s))\right] \\
& \times \exp\left[-\int_{t_0}^{s} r(y)\,dy\right] ds + \exp\left[-\int_{t_0}^{T} r(y)\,dy\right] q^i\left(x^*(T)\right) \\
& + \exp\left[-\int_{t_0}^{T} r(y)\,dy\right] q^i\left(x^*(T)\right),
\end{aligned}
\tag{3.32}
$$

for $i \in N$.

Given that the agreed-upon imputation to agent i is $\xi^{(t_0)i}(t_0, x_0)$, a lump-sum transfer $\bar{\chi}^i$ has to be incurred to agent i. In particular,

$$
\bar{\chi}^i = \xi^{(t_0)i}(t_0, x_0) - W^{(t_0)i}(t_0, x_0), \quad \text{for } i \in N; \text{ and } \sum_{j=1}^{n} \bar{\chi}^i = 0.
\tag{3.33}
$$

Individual rationality would hold at time t_0 if $\xi^{(t_0)i}(t_0, x_0) \geq V^{(t_0)i}(t_0, x_0)$.

On the other hand, transfer payments can be paid continuously to satisfy part (ii) of (3.30). Let $\chi^i(s)$ denote the instantaneous transfer payment allocated to agent i at time $s \in [t_0, T]$. With agents using the cooperative strategies $\{\psi_i^*(s, x_s^*)$, for $s \in [t_0, T]$ and $i \in N\}$, for a given agreed-upon imputation $\xi^{(t_0)i}(t_0, x_0)$, we can express agent i's cooperative payoff as

$$
\begin{aligned}
\xi^{(t_0)i} & (t_0, x_0) \\
= \int_{t_0}^{T} & \left\{g^i \left[s, x^*(s), \psi_1^*(s, x^*(s)), \psi_2^*(s, x^*(s)), \dots, \psi_n^*(s, x^*(s))\right] + \chi^i(s)\right\} \\
& \times \exp\left[-\int_{t_0}^{s} r(y)\,dy\right] ds + \exp\left[-\int_{t_0}^{T} r(y)\,dy\right] q^i\left(x^*(T)\right) \\
& + \exp\left[-\int_{t_0}^{T} r(y)\,dy\right] q^i\left(x^*(T)\right),
\end{aligned}
$$

for $i \in N$; and

$$
\sum_{j=1}^{n} \int_{t_0}^{T} \chi^j(s) \exp\left[-\int_{t_0}^{s} r(y)\,dy\right] ds = 0.
\tag{3.34}
$$

Once again, individual rationality would hold at time t_0 if $\xi^{(t_0)i}(t_0, x_0) \geq V^{(t_0)i}(t_0, x_0)$.

However, requiring individual rationality to hold only at t_0 does not guarantee that individual rationality will hold for the rest of the cooperation duration. Credible threats must be created to deter agents from abandoning the cooperative strategies at a later time in the cooperation duration.

Now we consider the case when individual rationality is required to be satisfied throughout the cooperation period $[t_0, T]$. In general, only continuous instantaneous transfer payments can guarantee the satisfaction of (3.31). To uphold part (ii) of (3.31) one has to devise a set of instantaneous transfer payments $\chi^i(s)$ for $s \in [t_0, T]$ satisfying

$$\int_\tau^T \left\{ g^i \left[s, x^*(s), \psi_1^*(s, x^*(s)), \psi_2^*(s, x^*(s)), \ldots, \psi_n^*(s, x^*(s)) \right] + \chi^i(s) \right\}$$

$$\times \exp\left[-\int_\tau^s r(y)\, dy \right] ds + \exp\left[-\int_\tau^T r(y)\, dy \right] q^i\left(x^*(T) \right) \geq V^{(\tau)i}\left(\tau, x_\tau^* \right),$$

for $i \in N$;   (3.35)

$$\text{and} \quad \sum_{j=1}^n \int_\tau^T \chi^j(s) \exp\left[-\int_\tau^s r(y)\, dy \right] ds = 0, \quad \text{for } \tau \in [t_0, T]. \quad (3.36)$$

Remark 3.2 In the special case when the economic agents are identical, transfer payments may not be necessary for the maintenance of individual rationality because

$$W^{(\tau)}\left(\tau, x_\tau^* \right) \geq \sum_{j=1}^n V^{(\tau)j}\left(\tau, x_\tau^* \right) = n V^{(\tau)i}\left(\tau, x_\tau^* \right), \quad \text{and}$$

$$\int_\tau^T \left\{ g^i \left[s, x^*(s), \psi_1^*(s, x^*(s)), \psi_2^*(s, x^*(s)), \ldots, \psi_n^*(s, x^*(s)) \right] \right\}$$

$$\times \exp\left[-\int_\tau^s r(y)\, dy \right] ds + \exp\left[-\int_\tau^T r(y)\, dy \right] q^i\left(x^*(T) \right)$$

$$= \frac{1}{n} W^{(\tau)}\left(\tau, x_\tau^* \right) V^{(\tau)i}\left(\tau, x_\tau^* \right) \geq V^{(\tau)i}\left(\tau, x_\tau^* \right),$$

for $i \in N$ and $\tau \in [t_0, T]$.

An illustration of a cooperative economic game involving individually rational imputation is provided in the next section.

3.2.2 Individually Rational Imputation in Cooperative Resource Extraction

Consider the two firms' resource extraction game in Sect. 3.1.2. Let $[\phi_1^*(t, x), \phi_2^*(t, x)]$, for $t \in [t_0, T]$, denote a set of strategies that provide a feedback Nash equilibrium solution to the game in (3.10) and (3.36). Invoking Theorem 2.3 in Chap. 2, the value function $V^{(t_0)i}(t, x) : [t_0, T] \times R^m \to R$, for $i \in \{1, 2\}$, satisfies the equations

$$-V_t^{(t_0)i}(t, x) = \max_{u_i}\left\{\left[u_i(t)^{1/2} - \frac{c_i}{x^{1/2}}u_i(t)\right]\exp[-r(t - t_0)]\right.$$

$$\left. + V_x^{(t_0)i}(t, x)\left[ax^{1/2} - bx - u_i(t) - \phi_j^*(t, x)\right]\right\}, \quad \text{and} \qquad (3.37)$$

$$V^{(t_0)i}(T, x) = \exp[-r(T - t_0)]qx(T)^{\frac{1}{2}}, \quad \text{for } i \in \{1, 2\} \text{ and } j \in \{1, 2\}, \text{ and } j \neq i.$$

Performing the indicated maximization in (3.38) yields

$$\phi_i^*(t, x) = \frac{x}{4[c_i + V_x^{(t_0)i}\exp[r(t - t_0)]x^{1/2}]^2}, \quad \text{for } i \in \{1, 2\}. \qquad (3.38)$$

Proposition 3.1 *The value function of agent $i \in \{1, 2\}$ satisfying (3.38) is*

$$V^{(t_0)i}(t, x) = \exp[-r(t - t_0)][A_i(t)x^{1/2} + B_i(t)], \qquad (3.39)$$

where for $i, j \in \{1, 2\}$ and $i \neq j$, $A_i(t)$, $B_i(t)$, $A_j(t)$, and $B_j(t)$ satisfy

$$\dot{A}_i(t) = \left[r + \frac{b}{2}\right]A_i(t) - \frac{1}{2[c_i + A_i(t)/2]} + \frac{c_i}{4[c_i + A_i(t)/2]^2}$$

$$+ \frac{A_i(t)}{8[c_i + A_i(t)/2]^2} + \frac{A_i(t)}{8[c_j + A_j(t)/2]^2}, \qquad (3.40)$$

$$\dot{B}_i(t) = rB_i(t) - \frac{a}{2}A_i(t), \quad A_i(T) = q, \quad \text{and} \quad B_i(T) = 0.$$

Proof Substituting $\phi_1^*(t, x)$ and $\phi_2^*(t, x)$ into (3.38) and upon solving (3.38) one obtains Proposition 3.1. □

Now consider the situation when the firms agree to cooperate and want to share their profits with individual rationality being upheld. As mentioned in Sect. 3.2.1, individual rationality may just hold at the outset of the game or be maintained throughout the game. We shall consider transfer schemes leading to the former case first and then schemes leading to the latter case.

(i) Transfers Satisfying Individual Rationality at the Outset

We begin with transfers that satisfy individual rationality at initial time t_0 and consider the case when lump-sum transfers are given out so that (3.30) is fulfilled. The imputation to agent i satisfying individual rationality and group optimality as in (3.30) requires

$$\xi^{(t_0)i}(t_0, x_0) = V^{(t_0)i}(t_0, x_0) + \varpi_i, \quad \text{for } i \in \{1, 2\} \quad \text{and} \quad \varpi_i \geq 0, \qquad (3.41)$$

where

$$\sum_{j=1}^{2} \varpi_j = W^{(t_0)}(t_0, x_0) - \sum_{j=1}^{2} V^{(t_0)j}(t_0, x_0)$$

$$= \left[\hat{A}(t_0)x_0^{1/2} + \hat{B}(t_0)\right] - \sum_{j=1}^{2}\left[A_j(t_0)x_0^{1/2} + B_j(t_0)\right].$$

Under cooperation, agent i would derive a payoff

$$W^{(t_0)i}(t_0, x_0) = \int_{t_0}^{T}\left[\left[\psi_i^*\big(s, x^*(s)\big)\right]^{1/2} - \frac{c_i}{x^*(s)^{1/2}}\psi_i^*\big(s, x^*(s)\big)\right]$$

$$\times \exp\left[-r(s - t_0)\right]ds$$

$$+ \exp\left[-r(T - t_0)\right]qx^*(T)^{\frac{1}{2}}, \quad \text{for } i \in \{1, 2\}, \qquad (3.42)$$

where $\psi_i^*(s, x^*(s)) = \frac{x^*(s)}{4[c_i+\hat{A}(s)/2]^2}$ and $x^*(s)$ as in (3.15).

Given that the agreed-upon imputation to agent i is $\xi^{(t_0)i}(t_0, x_0) = V^{(t_0)i}(t_0, x_0) + \varpi_i$, a lump-sum transfer $\bar{\chi}^i$ has to be incurred to agent i. In particular,

$$\bar{\chi}^i = \xi^{(t_0)i}(t_0, x_0) - W^{(t_0)i}(t_0, x_0)$$

$$= V^{(t_0)i}(t_0, x_0) + \varpi_i - W^{(t_0)i}(t_0, x_0), \quad \text{for } i \in N; \text{ and } \sum_{j=1}^{2}\bar{\chi}^i = 0.$$

Now we consider the case of continuous instantaneous transfer payments satisfying (3.30). Let $\chi^i(s)$ denote the instantaneous transfer payment allocated to agent i at time $s \in [t_0, T]$. To maintain part (ii) of (3.30) the chosen $\chi^i(s)$ must satisfy

$$\xi^{(t_0)i}(t_0, x_0) = V^{(t_0)i}(t_0, x_0) + \varpi_i = \left[A_i(t_0)x_0^{1/2} + B_i(t_0)\right] + \varpi_i$$

$$= \int_{t_0}^{T}\left[\left[\psi_i^*\big(s, x^*(s)\big)\right]^{1/2} - \frac{c_i}{x^*(s)^{1/2}}\psi_i^*\big(s, x^*(s)\big) + \chi(s)\right]$$

$$\times \exp\left[-r(s - t_0)\right]ds \qquad (3.43)$$

$$+ \exp\left[-r(T - t_0)\right]qx^*(T)^{\frac{1}{2}} \geq V^{(t_0)i}(t_0, x_0), \quad \text{for } i \in \{1, 2\},$$

and $$\sum_{j=1}^{2}\int_{t_0}^{T}\chi^j(s)\exp\left[-\int_{t_0}^{s}r(y)\,dy\right]ds = 0.$$

Then we proceed to consider the case with transfer payments satisfying individual rationality throughout the cooperative period.

(ii) Transfers Satisfying Individual Rationality Throughout

For individual rationality and group optimality to be satisfied throughout the cooperation period, (3.31) has to be maintained. The use of a lump sum cannot secure an outcome fulfilling (3.31). Hence continuous instantaneous transfer payments will be considered. Given that an instantaneous transfer payment $\chi^i(s)$ allocated to agent i at time $s \in [t_0, T]$, the imputation to him over the period $[\tau, T]$ as viewed at time $\tau \in [t_0, T]$ can be expressed as

$$\xi^{(\tau)i}(\tau, x_\tau^*) = \int_\tau^T \left[\left[\psi_i^*(s, x^*(s)) \right]^{1/2} - \frac{c_i}{x^*(s)^{1/2}} \psi_i^*(s, x^*(s)) + \chi(s) \right]$$

$$\times \exp\left[-r(s - \tau) \right] ds$$

$$+ \exp\left[-r(T - \tau) \right] q x^*(T)^{\frac{1}{2}}, \quad \text{for } i \in \{1, 2\} \text{ and } \tau \in [t_0, T], \tag{3.44}$$

$$\text{and} \quad \sum_{j=1}^2 \int_{t_0}^T \chi^j(s) \exp\left[-\int_{t_0}^s r(y)\, dy \right] ds = 0.$$

For individual rationality to be satisfied throughout the cooperation period, it is required that condition $\xi^{(\tau)i}(\tau, x_\tau^*) \geq V^{(\tau)i}(\tau, x_\tau^*)$, for $i \in \{1, 2\}$ and $\tau \in [t_0, T]$.

Invoking Remark 2.1 in Chap. 2 and (3.39), we can obtain the value functions $V^{(\tau)i}(\tau, x_\tau^*)$, for $i \in \{1, 2\}$, as

$$V^{(\tau)i}(\tau, x_\tau^*) = \left[A_i(\tau)(x_\tau^*)^{1/2} + B_i(\tau) \right]. \tag{3.45}$$

Therefore any set of chosen instantaneous transfer payment $\chi^i(s)$, for $s \in [t_0, T]$, satisfying $\xi^{(\tau)i}(\tau, x_\tau^*) \geq V^{(\tau)i}(\tau, x_\tau^*)$, for $i \in \{1, 2\}$ and $\tau \in [t_0, T]$, will ensure individual rationality throughout the cooperation duration.

3.3 Individual Rationality Under Infinite Horizon

In many economic situations, the terminal time of the game T is either very far in the future or unknown to the agents. Consider the case in Sect. 3.1.3 where the agents agree to cooperate and maximize the sum of their payoffs in (3.18) subject to (3.17). They have to distribute the cooperative payoff among themselves.

3.3.1 Individually Rational at the Outset

At time t_0, with the state being x_0, the term $\xi^{(t_0)i}(x_0) = \xi^i(x_0)$ is used to denote the imputation of a payoff (over the time interval $[t_0, \infty)$) to agent i.

A necessary condition for group optimality and individual rationality to be upheld is

$$\text{(i)} \quad \sum_{j=1}^{n} \xi^{(t_0)j}(x_0) = W(x_0), \quad \text{and}$$

$$\text{(ii)} \quad \xi^{(t_0)i}(x_0) \geq \hat{V}^{(t_0)i}(x_0) = \hat{V}^i(x_0), \quad \text{for } i \in N,$$

(3.46)

where $\hat{V}^{(t_0)i}(x_0) = \hat{V}^i(x_0)$ is the payoff value function of agent i in the n-person infinite-horizon general economic game problem

$$\max_{u_i} \int_{t_0}^{\infty} g^i\big[x(s), u_1(s), u_2(s), \ldots, u_n(s)\big] \exp\big[-r(s - t_0)\big] ds,$$

for $i \in N$.

Subject to the state dynamics

$$\dot{x}(s) = f\big[x(s), u_1(s), u_2(s), \ldots, u_n(s)\big], \quad x(t_0) = x_0.$$

Condition (i) of (3.46) ensures group optimality and condition (ii) guarantees individual rationality at time t_0. Consider first the case where lump-sum transfer payments are given out at time t_0. With agents using the cooperative strategies $\{\psi_i^*(x_s^*), \text{for } s \in [t_0, \infty) \text{ and } i \in N\}$, agent i would derive a payoff

$$W^{(t_0)i}(x_0) = \int_{t_0}^{\infty} g^i\big[x^*(s), \psi_1^*\big(x^*(s)\big), \psi_2^*\big(x^*(s)\big), \ldots, \psi_n^*\big(x^*(s)\big)\big]$$

$$\times \exp\big[-r(s - t_0)\big] ds$$

$$= W^i(x_0),$$

(3.47)

for $i \in N$.

Given that the agreed-upon imputation to agent i is $\xi^{(t_0)i}(x_0)$, a lump-sum transfer $\bar{\chi}^i$ has to be incurred to agent i. In particular,

$$\bar{\chi}^i = \xi^{(t_0)i}(x_0) - W^i(x_0), \quad \text{for } i \in N; \text{ and } \sum_{j=1}^{n} \bar{\chi}^i = 0.$$

(3.48)

Individual rationality would hold at time t_0 if $\xi^{(t_0)i}(x_0) \geq \hat{V}^i(x_0)$.

On the other hand, transfer payments can be paid continuously to satisfy (3.30). Let $\chi^i(s)$ denote the instantaneous transfer payment allocated to agent i at time $s \in [t_0, \infty)$. With agents using the cooperative strategies $\{\psi_i^*(x_s^*), \text{for } s \in [t_0, \infty) \text{ and } i \in N\}$ and given an agreed-upon imputation $\xi^{(t_0)i}(x_0)$, we can express agent i's cooperative payoff as

$$\xi^{(t_0)i}(x_0) = \int_{t_0}^{\infty} \left\{ g^i\left[x^*(s), \psi_1^*(x^*(s)), \psi_2^*(x^*(s)), \dots, \psi_n^*(x^*(s))\right] + \chi^i(s) \right\}$$

$$\times \exp\left[-r(s - t_0)\right] ds, \quad \text{for } i \in N; \tag{3.49}$$

$$\text{and} \quad \sum_{j=1}^{n} \int_{t_0}^{\infty} \chi^j(s) \exp\left[-r(s - t)\right] ds = 0.$$

Once again, individual rationality would hold at time t_0 with $\xi^{(t_0)i}(x_0) \geq V^i(x_0)$.

Note that requiring individual rationality to hold only at t_0 does not guarantee that individual rationality will hold for the rest of the cooperation period. Credible threats must be created to deter agents from abandoning the cooperative strategies at a later time in the cooperation period.

3.3.2 Individually Rational Throughout the Cooperative Duration

For the optimization scheme to be upheld throughout the game horizon, both group rationality and individual rationality are required to be satisfied throughout the cooperation period $[t_0, \infty)$. As mentioned earlier, only continuous instantaneous transfer payments can guarantee the satisfaction of individual rationality throughout the duration of cooperation. Therefore along the optimal cooperative path $\{x_\tau^*\}_{\tau \in [t_0, \infty)}$, one has to devise a set of instantaneous transfer payments $\chi^i(s)$ for $s \in [t_0, \infty)$ satisfying

$$\xi^{(\tau)i}\left(x_\tau^*\right) = \xi^i\left(x_\tau^*\right)$$

$$= \int_{\tau}^{\infty} \left\{ g^i\left[x^*(s), \psi_1^*(x^*(s)), \psi_2^*(x^*(s)), \dots, \psi_n^*(x^*(s))\right] + \chi^i(s) \right\}$$

$$\times \exp\left[-r(s - \tau)\right] ds \tag{3.50}$$

$$\geq \hat{V}^i\left(x_\tau^*\right), \quad \text{for } i \in N;$$

$$\text{and} \quad \sum_{j=1}^{n} \int_{\tau}^{\infty} \chi^j(s) \exp\left[-r(s - \tau)\right] ds = 0, \quad \text{for } \tau \in [t_0, \infty).$$

A set of instantaneous transfer payments $\chi^i(s)$ for $s \in [t_0, \infty)$ satisfying (3.51) satisfies

(i) $\displaystyle\sum_{j=1}^{n} \xi^j\left(x_\tau^*\right) = W\left(x_\tau^*\right), \quad$ and

$$\tag{3.51}$$

(ii) $\xi^i\left(x_\tau^*\right) \geq \hat{V}^i\left(x_\tau^*\right); \quad$ for $i \in N$ and $\tau \in [t_0, \infty)$ along the path $\{x_\tau^*\}_{\tau \in [t_0, \infty)}$.

In particular, condition (i) ensures Pareto optimality and condition (ii) guarantees individual rationality throughout the cooperation period $[t_0, \infty)$. The failure to guarantee individual rationality leads to the condition where the concerned participants will reject the agreed-upon solution plan and play noncooperatively.

In the steady state of the infinite-horizon problem, the state dynamics under cooperation become

$$\dot{x}^*(s) = f\left[x^*(s), \psi_1^*(x^*(s)), \psi_2^*(x^*(s)), \ldots, \psi_n^*(x^*(s))\right] = 0. \qquad (3.52)$$

Let x_∞^* denote the steady-state level of x^* that satisfies (3.52). In the steady state, the joint cooperative payoffs can be expressed as $W(x_\infty^*)$ and the noncooperative payoff value function of agent i as $\hat{V}^i(x_\infty^*)$.

At any time $t \in [t_0, \infty)$ at which a steady state has been attained, agents will use the cooperative strategies $\{\psi_i^*(x_\infty^*)$, for $i \in N\}$, and agent i will derive a payoff

$$W^i(x_\infty^*) = \int_t^\infty g^i\left[x_\infty^*(s), \psi_1^*(x_\infty^*), \psi_2^*(x_\infty^*), \ldots, \psi_n^*(x_\infty^*)\right] \exp\left[-r(s-t)\right] ds, \qquad (3.53)$$

for $i \in N$.

To fulfill individual rationality in a steady state an infinite stream of constant instantaneous transfer payments χ_∞^i satisfying

$$\xi^i(x_\infty^*) = \int_\tau^\infty \left\{ g^i\left[x^*(s), \psi_1^*(x^*(s)), \psi_2^*(x^*(s)), \ldots, \psi_n^*(x^*(s))\right] + \chi_\infty^i \right\}$$
$$\times \exp\left[-r(s-\tau)\right] ds,$$
$$\sum_{j=1}^n \int_\tau^\infty \chi_\infty^j \exp\left[-r(s-\tau)\right] ds = 0, \quad \text{and} \qquad (3.54)$$
$$\xi^i(x_\infty^*) \geq \hat{V}^i(x_\infty^*), \quad \text{for } i \in N,$$

will be given out.

An illustration with an economic cooperation involving the satisfaction of individual rationality is shown below.

3.3.3 Individuall Rationality in Resource Extraction

Consider the infinite-horizon resource extraction game in (3.10) and (3.22) in Sect. 3.1. Again, we first examine the alternative game problem that starts at time $t \in [t_0, \infty)$ with initial state $x(t) = x$

$$\max_{u_1} \int_t^\infty \left[u_1(s)^{1/2} - \frac{c_1}{x(s)^{1/2}} u_1(s) \right] \exp\left[-r(t-t)\right] ds,$$

and

$$\max_{u_2} \int_t^\infty \left[u_2(s)^{1/2} - \frac{c_2}{x(s)^{1/2}} u_2(s) \right] \exp[-r(t-t)] \, ds, \quad \text{where } t \in [t_0, \infty),$$

(3.55)

subject to

$$\dot{x}(s) = ax(s)^{1/2} - bx(s) - u_1(s) - u_2(s), \quad x(t) = x_t.$$

(3.56)

Let $[\phi_1^*(x), \phi_2^*(x)]$, for $t \in [t_0, \infty)$, denote a set of strategies that provides a feedback Nash equilibrium solution to the game in (3.55) and (3.56). Invoking Theorem 2.4 in Chap. 2, the value function $\hat{V}^i(x) : R \to R$, for $i \in \{1, 2\}$, satisfies the Isaacs–Bellman equations

$$r\hat{V}_t^i(x) = \max_{u_i} \left\{ \left[u_i(t)^{1/2} - \frac{c_i}{x^{1/2}} u_i(t) \right] + \hat{V}_x^i(x) \left[ax^{1/2} - bx - u_i(t) - \phi_j^*(x) \right] \right\},$$

(3.57)

for $i \in \{1, 2\}$ and $j \in \{1, 2\}$, and $j \neq i$.

Performing the indicated maximization in (3.57) yields

$$\phi_i^*(x) = \frac{x}{4[c_i + \hat{V}_x^i(x)x^{1/2}]^2}, \quad \text{for } i \in \{1, 2\}.$$

(3.58)

Proposition 3.2 *The value function of agent $i \in \{1, 2\}$ satisfying (3.57) is*

$$\hat{V}^i(x) = [\hat{A}_i x^{1/2} + \hat{B}_i],$$

(3.59)

where for $i, j \in \{1, 2\}$ and $i \neq j$, $\hat{A}_i$, $\hat{B}_i$, $\hat{A}_j$, and $\hat{B}_j$ satisfy

$$0 = \left[r + \frac{b}{2} \right] \hat{A}_i - \frac{1}{2[c_i + \hat{A}_i/2]} + \frac{c_i}{4[c_i + \hat{A}_i/2]^2} + \frac{\hat{A}_i}{8[c_i + \hat{A}_i/2]^2} + \frac{A_i(t)}{8[c_j + \hat{A}_j/2]^2},$$

and

$$\hat{B}_i(t) = \frac{a}{2r} \hat{A}_i.$$

Proof Substituting $\phi_1^*(x)$ and $\phi_2^*(x)$ into (3.57) and upon solving (3.57) one obtains Proposition 3.2. □

The joint payoff of the firms can be obtained as in (3.26) and the cooperative strategies are given in (3.27). Again, we consider both the case where individual rationality holds at the outset of the game and the case where it is maintained throughout the game.

(i) Transfers Satisfying Individual Rationality at the Outset

We then consider transfers that satisfy individual rationality at initial time t_0. First we present the cases when lump-sum transfers are given out so that (3.46) is fulfilled. The imputation to agents i satisfying individual rationality and group optimality as, in (3.46), requires

$$\xi^i(x_0) = \hat{V}^i(x_0) + \varpi_i, \quad \text{for } i \in \{1, 2\} \text{ and } \varpi_i \geq 0,$$

where

$$\sum_{j=1}^{2} \varpi_j = W(x_0) - \sum_{j=1}^{2} \hat{V}^j(x_0) = \left[\hat{A}x_0^{1/2} + \hat{B}\right] - \sum_{j=1}^{2}\left[\hat{A}_j x_0^{1/2} + \hat{B}_j\right].$$

Under cooperation, agent i would derive a payoff

$$W^i(x_0) = \int_{t_0}^{\infty}\left[\left[\psi_i^*(x^*(s))\right]^{1/2} - \frac{c_i}{x^*(s)^{1/2}}\psi_i^*(x^*(s))\right]\exp[-r(s - t_0)]\,ds, \tag{3.60}$$

for $i \in \{1, 2\}$, where

$$\psi_1^*(x^*(s)) = \frac{x^*(s)}{4[c_1 + \hat{A}/2]^2} \quad \text{and} \quad \psi_2^*(x^*) = \frac{x^*(s)}{4[c_2 + \hat{A}/2]^2},$$

as given in (3.26), and $x^*(s)$ is given in (3.28).

With the agreed-upon imputation to agent i being $\xi^i(x_0) = \hat{V}^i(x_0) + \varpi_i$, a lump-sum transfer $\bar{\chi}^i$ has to be incurred to agent i. In particular,

$$\bar{\chi}^i = \xi^i(x_0) - W^i(x_0) = \hat{V}^i(x_0) + \varpi_i - W^i(x_0),$$

for $i \in N$; and $\sum_{j=1}^{2} \bar{\chi}^i = 0$.

Now we consider the case of continuous instantaneous transfer payments satisfying (3.46). Let $\chi^i(s)$ denote the instantaneous transfer payment allocated to agent i at time $s \in [t_0, \infty)$. To fulfill (3.46) the chosen $\chi^i(s)$ must satisfy

$$\begin{aligned}\xi^i(x_0) = \hat{V}^i(x_0) + \varpi_i &= \left[\hat{A}_i x_0^{1/2} + \hat{B}_i\right] + \varpi_i \\ &= \int_{t_0}^{\infty}\left[\left[\psi_i^*(x^*(s))\right]^{1/2} - \frac{c_i}{x^*(s)^{1/2}}\psi_i^*(x^*(s)) + \chi(s)\right] \\ &\quad \times \exp[-r(s - t_0)]\,ds, \quad \text{for } i \in \{1, 2\},\end{aligned} \tag{3.61}$$

$$\text{and} \quad \sum_{j=1}^{2}\int_{t_0}^{\infty}\chi^j(s)\exp[-r(s - t_0)]\,ds = 0.$$

Next, we proceed to consider the case with transfer payments satisfying individual rationality throughout the cooperative period.

(ii) Transfers Satisfying Individual Rationality Throughout

For individual rationality and group optimality to be satisfied throughout the co-operation period, (3.51) has to be maintained. The use of a lump sum cannot secure an outcome fulfilling (3.51). Hence continuous instantaneous transfer payments will be considered. Given that an instantaneous transfer payment $\chi^i(s)$ is allocated to agent i at time $s \in [t_0, \infty)$, the imputation to him over the period $[\tau, T]$ as viewed at time $\tau \in [t_0, \infty)$ can be expressed as

$$\xi^i(x_\tau^*) = \int_\tau^\infty \left[[\psi_i^*(x^*(s))]^{1/2} - \frac{c_i}{x^*(s)^{1/2}} \psi_i^*(x^*(s)) + \chi(s) \right] \exp[-r(s - \tau)] \, ds,$$

(3.62)

for $i \in \{1, 2\}$ and $\tau \in [t_0, \infty)$.

For individual rationality to be satisfied throughout the cooperation period, it is required that condition $\xi^i(x_\tau^*) \geq V^i(x_\tau^*)$, for $i \in \{1, 2\}$ and $\tau \in [t_0, \infty)$. To fulfill (3.51) the chosen $\chi^i(s)$ must satisfy

$$\xi^i(x_\tau^*) = \int_{t_0}^\infty \left[[\psi_i^*(x^*(s))]^{1/2} - \frac{c_i}{x^*(s)^{1/2}} \psi_i^*(x^*(s)) + \chi(s) \right] \exp[-r(s - t_0)] \, ds$$

$$\geq \hat{V}^i(x_\tau^*) = \left[\hat{A}_i(x_\tau^*)^{1/2} + \hat{B}_i \right], \quad \text{for } i \in \{1, 2\} \text{ and } \tau \in [t_0, \infty), \quad (3.63)$$

$$\text{and} \quad \sum_{j=1}^2 \int_\tau^\infty \chi^j(s) \exp\left[-\int_\tau^s r(y) \, dy \right] ds = 0.$$

Now consider the case in a steady state. The state dynamics under cooperation become

$$\dot{x}^*(s) = a[x^*(s)]^{1/2} - bx^*(s) - \frac{x^*(s)}{4[c_1 + \hat{A}/2]^2} - \frac{x^*(s)}{4[c_2 + \hat{A}/2]^2} = 0. \quad (3.64)$$

Solving (3.64) yields the steady-state level of the state variable as

$$x_\infty^* = a^2 \Big/ \left[b + \frac{1}{4[c_1 + \hat{A}/2]^2} + \frac{1}{4[c_2 + \hat{A}/2]^2} \right]^2. \quad (3.65)$$

The joint cooperative payoffs in a steady state are

$$W(x_\infty^*) = \left[\hat{A}(x_\infty^*)^{1/2} + \hat{B} \right],$$

and the payoff value function of agent i is

$$\hat{V}^i(x_\infty^*) = \left[\hat{A}_i(x_\infty^*)^{1/2} + \hat{B}_i \right], \quad \text{for } i \in \{1, 2\}.$$

At any time $t \in [t_0, \infty)$ at which a steady state has been attained, agents will use the cooperative strategies $\{\psi_1^*(x_\infty^*), \psi_2^*(x_\infty^*)\}$, and agent i will derive a payoff

$$W^i(x_\infty^*) = \int_t^\infty \left[[\psi_i^*(x^*(s))]^{1/2} - \frac{c_i}{x^*(s)^{1/2}} \psi_i^*(x^*(s)) \right] \exp[-r(s - t)] \, ds,$$

(3.66)

for $i \in \{1, 2\}$.

To fulfill individual rationality in a steady state an infinite stream of constant instantaneous transfer payments χ_∞^i must satisfy

$$\xi^i\left(x_\infty^*\right) = \int_t^\infty \left[\left[\psi_i^*\left(x^*(s)\right)\right]^{1/2} - \frac{c_i}{x^*(s)^{1/2}}\psi_i^*\left(x^*(s)\right) + \chi_\infty^i\right]\exp\left[-r(s-t)\right]ds$$

$$\geq \hat{V}^i\left(x_\infty^*\right) = \left[\hat{A}_i\left(x_\infty^*\right)^{1/2} + \hat{B}_i\right], \quad \text{for } i \in \{1,2\},$$

$$\sum_{j=1}^2 \int_t^\infty \chi_\infty^j \exp\left[-r(s-t)\right]ds = 0.$$

3.4 Cooperative Economic Games Satisfying Individual Rationality and Group Optimality

Cooperative games suggest the possibility of socially optimal and group efficient solutions to decision problems involving strategic action. As discussed above, individual rationality and group optimality are essential elements of a cooperative game solution. Dockner and Jørgensen (1984), Dockner and Long (1993), Tahvonen (1994), Mäler and de Zeeuw (1998), and Rubio and Casino (2002) presented cooperative solutions satisfying group optimality in differential games. The majority of cooperative differential games adopt solutions satisfying the essential criteria for dynamic stability—group optimality and individual rationality. Haurie and Zaccour (1986, 1991), Kaitala and Pohjola (1988, 1990, 1995), Kaitala et al. (1995), and Jørgensen and Zaccour (2001) presented classes of transferable-payoff cooperative differential games with solutions that satisfy group optimality and individual rationality.

In the following sections, cooperative economic games satisfying individual rationality and group optimality are presented.

3.4.1 Cooperative Resource Extraction Game

Consider the resource extraction game presented in Sects. 3.2 and 3.3. The growth rate of the fish biomass is characterized by the differential equations

$$\dot{x}(s) = ax(s)^{1/2} - bx(s) - u_1(s) - u_2(s), \quad x(t_0) = x_0 \in X, \tag{3.67}$$

where $u_i \in U_i$ is the (nonnegative) amount of fish harvested by nation i, for $i \in \{1,2\}$; a and b are positive constants.

The harvesting cost for firm $i \in \{1,2\}$ depends on the quantity of resource extracted $u_i(s)$, the resource stock size $x(s)$, and a parameter c_i. In particular, firm i's extraction cost can be specified as $c_i u^i(s)x(s)^{-1/2}$. The fish harvested by nation i at time s will generate a net benefit of the amount $[u_i(s)]^{1/2}$. The horizon of concern is $[t_0, T]$. At time T, nation i will receive a termination bonus $q_i x(T)^{1/2}$, where q_i is nonnegative. There exists a discount rate r, and profits received at time t have to be discounted by the factor $\exp[-r(t - t_0)]$.

At time t_0 the payoff of nation $i \in \{1, 2\}$ is

$$
\int_{t_0}^{T} \left[[u_i(s)]^{1/2} - \frac{c_i}{x(s)^{1/2}} u_i(s) \right] \exp[-r(s - t_0)] \, ds
$$

$$
+ \exp[-r(T - t_0)] q_i x(T)^{\frac{1}{2}}. \tag{3.68}
$$

The game structure is a deterministic version of an example in Yeung and Petrosyan (2004). According to Proposition 3.1, the value function of agent $i \in \{1, 2\}$ is

$$
V^{(t_0)i}(t, x) = \exp[-r(t - t_0)] \left[A_i(t) x^{1/2} + B_i(t) \right], \tag{3.69}
$$

where for $i, j \in \{1, 2\}$ and $i \neq j$, $A_i(t)$, $B_i(t)$, $A_j(t)$, and $B_j(t)$ satisfy (3.41).

According to (3.45), the value functions $V^{(\tau)i}(\tau, x_\tau^*)$, for $i \in \{1, 2\}$, are

$$
V^{(\tau)i}(\tau, x_\tau^*) = \left[A_i(\tau)(x_\tau^*)^{1/2} + B_i(\tau) \right]. \tag{3.70}
$$

The firms agree to cooperate and seek to solve the following joint profit maximization problem to achieve a Pareto optimum by

$$
\max_{u_1, u_2} \int_{t_0}^{T} \left(\left[u_1(s)^{1/2} - \frac{c_1}{x(s)^{1/2}} u_1(s) \right] + \left[u_2(s)^{1/2} - \frac{c_2}{x(s)^{1/2}} u_2(s) \right] \right)
$$

$$
\times \exp[-r(t - t_0)] \, ds + 2 \exp[-r(T - t_0)] q x(T)^{\frac{1}{2}}, \tag{3.71}
$$

subject to (3.66).

According to (3.13) the maximized value function under cooperation is obtained as

$$
W^{(t_0)}(t, x) = \exp[-r(t - t_0)] \left[\hat{A}(t) x^{1/2} + \hat{B}(t) \right], \tag{3.72}
$$

with $\hat{A}(t)$ and $\hat{B}(t)$ satisfying the corresponding differential equations in (3.13).

The optimal cooperative controls can then be obtained as

$$
\psi_1^*(t, x) = \frac{x}{4[c_1 + \hat{A}(t)/2]^2}, \quad \text{and} \quad \psi_2^*(t, x) = \frac{x}{4[c_2 + \hat{A}(t)/2]^2},
$$

with the optimal cooperative state trajectory $\{x^*(s)\}_{s=t_0}^{T}$ given in (3.15).

Under cooperation, firm i would derive a payoff

$$
W^{(t_0)i}(t_0, x_0) = \int_{t_0}^{T} \left[[\psi_i^*(s, x^*(s))]^{1/2} - \frac{c_i}{x^*(s)^{1/2}} \psi_i^*(s, x^*(s)) \right]
$$

$$
\times \exp[-r(s - t_0)] \, ds + \exp[-r(T - t_0)] q_i x^*(T)^{\frac{1}{2}}, \tag{3.73}
$$

where

$$\psi_i^*\left(s, x^*(s)\right) = \frac{x^*(s)}{4[c_i + \hat{A}(s)/2]^2}, \quad \text{for } i \in \{1, 2\}.$$

The firms decide to share the excess gain from cooperation equally. There can be different methods of payment to achieve this.

(i) A Lump-sum Transfer at the Outset

First consider the case when a lump-sum transfer is arranged at the outset of the game. Given that the firms agree to share the excess gain from cooperation equally, therefore,

$$\xi^{(t_0)i}(t_0, x_0) = V^{(t_0)i}(t_0, x_0) + \frac{1}{2}\left[W^{(t_0)}(t_0, x_0) - \sum_{j=1}^{2} V^{(t_0)j}(t_0, x_0) \right]$$

$$= \frac{1}{2}\{[\hat{A}(t_0)x^{1/2} + \hat{B}(t_0)] + [A_i(t_0)x^{1/2} + B_i(t_0)]$$

$$- [A_j(t_0)x^{1/2} + B_j(t_0)]\}, \tag{3.74}$$

for $i \in \{1, 2\}$.

Since firm i's receipt under cooperation is $W^{(t_0)i}(t_0, x_0)$, a lump-sum transfer $\bar{\chi}^i$ has to be incurred to agent i to achieve $\xi^{(t_0)i}(t_0, x_0)$. In particular,

$$\bar{\chi}^i = \xi^{(t_0)i}(t_0, x_0) - W^{(t_0)i}(t_0, x_0), \quad \text{for } i \in \{1, 2\}.$$

Note that $\sum_{j=1}^{n} \bar{\chi}^i = 0$.

Group optimality is satisfied and individual rationality holds at time t_0.

(ii) Continuous Instantaneous Transfer Satisfying Individual Rationality at the Outset

Now we consider the case of continuous instantaneous transfer payments satisfying individual rationality at the outset. Let $\chi^i(s)$ denote the instantaneous transfer payment allocated to agent i at time $s \in [t_0, T]$. The chosen $\chi^i(s)$ must satisfy

$$\xi^{(t_0)i}(t_0, x_0) = \frac{1}{2}\{[\hat{A}(t_0)x^{1/2} + \hat{B}(t_0)] + [A_i(t_0)x^{1/2} + B_i(t_0)]$$

$$- [A_j(t_0)x^{1/2} + B_j(t_0)]\}$$

$$= \int_{t_0}^{T} \left[[\psi_i^*(s, x^*(s))]^{1/2} - \frac{c_i}{x^*(s)^{1/2}} \psi_i^*(s, x^*(s)) - \chi^i(s) \right] \tag{3.75}$$

$$\times \exp[-r(s - t_0)]\, ds + \exp[-r(T - t_0)]q_i x^*(T)^{\frac{1}{2}}, \quad \text{for } i \in \{1, 2\};$$

and $\displaystyle\sum_{j=1}^{n} \int_{t_0}^{T} \chi^j(s) \exp\left[-\int_{t_0}^{s} r(y)\, dy \right] ds = 0.$

Once again, group optimality is satisfied and individual rationality holds at time t_0.

(iii) Transfers Satisfying Individual Rationality Throughout

Now we consider the case of continuous instantaneous transfer payments satisfying individual rationality throughout the cooperative period. Given that an instantaneous transfer payment $\chi^i(s)$, allocated to agent i at time $s \in [t_0, T]$, the imputation to him over the period $[\tau, T]$ as viewed at time $\tau \in [t_0, T]$ can be expressed as

$$\xi^{(\tau)i}(\tau, x_\tau^*) = \int_\tau^T \left[\left[\psi_i^*(s, x^*(s)) \right]^{1/2} - \frac{c_i}{x^*(s)^{1/2}} \psi_i^*(s, x^*(s)) - \chi(s) \right]$$

$$\times \exp\left[-r(s - \tau) \right] ds + \exp\left[-r(T - \tau) \right] q_i x^*(T)^{\frac{1}{2}} \geq V^{(\tau)i}(\tau, x_\tau^*)$$

$$= A_i(\tau)(x_\tau^*)^{1/2} + B_i(\tau),$$

(3.76)

for $i \in \{1, 2\}$, and $\displaystyle\sum_{j=1}^n \int_{t_0}^T \chi^j(s) \exp\left[-\int_{t_0}^s r(y)\, dy \right] ds = 0.$

Any set of chosen instantaneous transfer payments $\chi^i(s)$, for $s \in [t_0, T]$, satisfying (3.76) will ensure group optimality and individual rationality throughout the cooperation period.

3.4.2 Fully Coordinated Pollution Control

Dockner and Long (1993) presented a differential game of international pollution control. There are two nations and each nation produces a single consumption good with a given fixed factor of production and a given technology. Let the quantity of the good produced at time s be denoted by $Q_i(s)$, for $i \in \{1, 2\}$. The production of a unit of the consumption good results in $\varepsilon_i(s)$ amount of pollutants. An emission consumption trade-off function (see Forster 1973, 1975) states that

$$Q_i(s) = f^i[\varepsilon_i(s)], \quad \text{for } i \in \{1, 2\}. \tag{3.77}$$

The function $f^i[\varepsilon_i(s)]$ is strictly concave in $\varepsilon_i(s)$ and satisfies $f^i[0] = 0$. Let $x(s)$ denote the level of pollution stock at time s. The pollution accumulation dynamics is governed by the kinematic equation

$$\dot{x}(s) = \varepsilon_1(s) + \varepsilon_2(s) - \delta x(s), \quad x(0) = x_0, \tag{3.78}$$

where δ is the rate of natural purification.

In each nation there are n identical consumers. The representative consumer in nation i derives utility from consuming $q_i(s) = Q_i(s)/n$ and faces the costs of the polluted environment $c^i[x(s)]$. Consumer preference $U^i[Q_i(s)/n]$ is assumed to be strictly concave and the cost function $c^i[x(s)]$ is strictly convex. The net benefits of the representative consumer in nation i are given by

$$U^i[Q_i(s)/n] - c^i[x(s)] \equiv U^i[f^i(\varepsilon_i(s))/n] - c^i[x(s)]. \qquad (3.79)$$

The objective of government i is to choose a pollution control strategy $\varepsilon_i(s)$, or equivalently, an output strategy that maximizes the discounted stream of net benefits from consumption of a representative consumer, that is,

$$\max_{\varepsilon_i} \int_0^\infty \{U^i[f^i(\varepsilon_i(s))/n] - c^i[x(s)]\} \exp(-rs) \, ds, \quad \text{for } i \in \{1, 2\}, \qquad (3.80)$$

subject to the pollutant accumulation dynamics in (3.78).

The outcome of a cooperative game is interpreted as the scenario in which the nations are able to reach a pollution control agreement (they coordinate their own control efforts) leading to a Pareto optimum. It is used as a reference scenario yielding a first-best solution.

An explicit first-best solution is characterized with the normalization of n to unity and the specification of the functional forms of preferences and technologies as

$$c^i[x(s)] = \frac{c}{2}[x(s)]^2 \quad \text{and} \quad U^i[f^i(\varepsilon_i(s))/n] = A\varepsilon_i(s) - \frac{1}{2}[\varepsilon_i(s)]^2,$$

where $A > 0$.

In particular, a first-best solution can be obtained by solving the optimization problem

$$\max_{\varepsilon_1, \varepsilon_2} \int_0^\infty \sum_{j=1}^2 \left\{ A\varepsilon_j(s) - \frac{1}{2}[\varepsilon_j(s)]^2 - \frac{c}{2}[x(s)]^2 \right\} \exp(-rs) \, ds, \qquad (3.81)$$

subject to (3.78).

To solve the optimization problem in (3.78) and (3.81) we invoke Theorem 3.2 to characterize the solution as follows.

A set of controls $\{\psi_1^*(x), \psi_2^*(x)\}$ provide a solution to the optimal control problem in (3.78) and (3.81) if there exists a continuously differentiable function $W(x) : R \to R$ satisfying the infinite-horizon Bellman equation

$$rW(x) = \max_{\varepsilon_1, \varepsilon_2} \left\{ \sum_{j=1}^2 \left[A\varepsilon_j - \frac{1}{2}\varepsilon_j^2 - \frac{c}{2}x^2 \right] + W_x(x)[\varepsilon_1 + \varepsilon_2 - \delta x] \right\}. \qquad (3.82)$$

Performing the maximization operation in (3.82) yields

$$\psi_1^*(x) = A + W_x(x) \quad \text{and} \quad \psi_2^*(x) = A + W_x(x). \qquad (3.83)$$

Substituting (3.83) into the Bellman equation yields

$$rW(x) = 2A[A + W_x(x)] - [A + W_x(x)]^2 - cx^2 + W_x(x)\{2[A + W_x(x)] - \delta x\}.$$

Upon the cancellation of terms we have

$$r W(x) = A^2 + 2A W_x(x) + \left[W_x(x)\right]^2 - cx^2 - \delta x W_x(x). \tag{3.84}$$

Proposition 3.3

$$W(x) = \frac{1}{2}\alpha x^2 + \beta x + \gamma, \tag{3.85}$$

where

$$\alpha = -\frac{1}{2}\left[\sqrt{\left(\delta + \frac{r}{2}\right)^2 + 4c} - \left(\delta + \frac{r}{2}\right)\right] < 0,$$

$$\beta = \frac{2A\alpha}{r + \delta - 2\alpha} < 0, \quad and \tag{3.86}$$

$$\gamma = \frac{(-\beta - A)^2}{r} > 0.$$

Proof Substituting $W(x)$ and $W_x(x)$ from (3.85) into (3.84) allows the Bellman equation (3.82) to be expressed as

$$r\gamma + r\beta x + \frac{r\alpha}{2}x^2$$

$$= A^2 + 2A\alpha x + 2A\beta + \alpha^2 x^2 + 2\alpha\beta x + \beta^2 - cx^2 - \alpha\delta x^2 - \beta\delta x.$$

Grouping terms together, one obtains

$$\left(\frac{1}{2}r\alpha - \alpha^2 + c + \alpha\delta\right)x^2 + (r\beta - 2A\alpha - 2\alpha\beta + \beta\gamma)x + r\gamma - A^2 - 2A\beta - \beta^2 = 0. \tag{3.87}$$

For (3.87) to hold, it is required that

$$\frac{1}{2}r\alpha - \alpha^2 + c + \alpha\delta = 0,$$

$$r\beta - 2A\alpha - 2\alpha\beta + \beta\gamma = 0, \quad and \tag{3.88}$$

$$r\gamma - A^2 - 2A\beta - \beta^2 = 0.$$

Solving (3.89) yields (3.86). Hence Proposition 3.3 follows.   □

Using (3.85) and (3.83) the cooperative emission controls can be obtained as

$$\psi_1^*(x) = A + \alpha x + \beta \quad and \quad \psi_2^*(x) = A + \alpha x + \beta. \tag{3.89}$$

Substituting these controls into (3.78) yields the optimal pollution accumulation dynamics under cooperation

$$\dot{x}(s) = 2\left[A + \alpha x(s) + \beta\right] - \delta x(s), \quad x(0) = x_0. \tag{3.90}$$

Let $x^*(s) = x_s^*$, for $s \in [t_0, \infty)$, denote the solution to (3.90). The net benefits to nation i under cooperation can be obtained as

$$W^i(x_0) = \int_0^\infty \left\{ A[A + \alpha x^*(s) + \beta] - \frac{1}{2}[A + \alpha x^*(s) + \beta]^2 - \frac{c}{2}[x^*(s)]^2 \right\}$$

$$\times \exp(-rs)\,ds$$

$$= \frac{1}{2}W(x_0) = \frac{1}{4}\alpha x_0^2 + \frac{1}{2}\beta x_0 + \frac{1}{2}\gamma, \quad \text{for } i \in \{1, 2\}. \tag{3.91}$$

Moreover, along the cooperative trajectory $\{x_t^*\}_{t \in [0, \infty)}$, the net benefits to nation i under cooperation can be obtained as

$$W^i(x_t^*) = \int_t^\infty \left\{ A[A + \alpha x^*(s) + \beta] - \frac{1}{2}[A + \alpha x^*(s) + \beta]^2 - \frac{c}{2}[x^*(s)]^2 \right\}$$

$$\times \exp(-rs)\,ds$$

$$= \frac{1}{2}W(x_t^*) = \frac{1}{4}\alpha(x_t^*)^2 + \frac{1}{2}\beta x_t^* + \frac{1}{2}\gamma, \quad \text{for } i \in \{1, 2\}. \tag{3.92}$$

Given that nations are symmetrical, splitting the cooperative gains would guarantee individual rationality throughout the cooperation period because

$$W(x_t^*) \geq \hat{V}^1(x_t^*) + \hat{V}^2(x_t^*) \quad \text{and}$$

$$W^i(x_t^*) = \frac{1}{2}W(x_t^*) \geq \hat{V}^i(x_t^*), \quad \text{for } i \in \{1, 2\} \text{ and } t \in [0, \infty).$$

As mentioned in Remark 3.2, even without transfer payments, the individual rationality of identical agents can be maintained.

3.5 Exercises

3.1 Consider the resource stock $x(s) \in X \subset R$, which follows the dynamics

$$\dot{x}(s) = 40x(s)^{1/2} - 2x(s) - u_1(s) - u_2(s), \quad x(0) = 50,$$

where $u_1(s)$ is the harvest rate of economic agent 1 and $u_2(s)$ is the harvest rate of economic agent 2. The extractors are entitled to harvest the resource in the period $[0, 4]$. The instantaneous payoff at time $s \in [0, 4]$ for agents 1 and 2 are, respectively,

$$\left[u_1(s)^{1/2} - \frac{2}{x(s)^{1/2}}u_1(s) \right] \quad \text{and} \quad \left[u_2(s)^{1/2} - \frac{1}{x(s)^{1/2}}u_2(s) \right].$$

At terminal time 4, each agent will receive a termination bonus equaling

$$6x(4)^{\frac{1}{2}},$$

which depends on the resource remaining at the terminal time.

Payoffs are transferable between agents 1 and 2 and over time; the discount rate is 0.05.

Consider the case when these two firms agree to cooperate and maximize the sum of their payoffs

$$\int_0^4 \left(\left[u_1(s)^{1/2} - \frac{2}{x(s)^{1/2}} u_1(s) \right] + \left[u_2(s)^{1/2} - \frac{1}{x(s)^{1/2}} u_2(s) \right] \right) \exp(-0.05s)\, ds$$
$$+ 2\exp\left[-0.05(4) \right] 6x(4)^{\frac{1}{2}},$$

subject to the resource dynamics above.

Derive the optimal cooperative strategies and the optimal trajectory path of the resource stock.

3.2 Solve a feedback equilibrium solution for the resource extraction game in Exercise 3.1.

3.3 The agents agree to cooperate and share the excess gain over the noncooperative profits equally. They also agree to distribute the excess gain at the end of the game. Compute the transfers.

Chapter 4
Time Consistency and Optimal-Trajectory-Subgame Consistent Economic Optimization

The noncooperative games discussed in Chap. 2 fail to reflect all the facets of optimal behavior in n-person market games. In particular, equilibria in noncooperative games do not take into consideration Pareto efficiency or group optimality. Chapter 3 considers cooperation in economic optimization and it is shown that group optimality and individual rationality are two essential properties for cooperation. However, merely satisfying group optimality and individual rationality does not necessarily bring about a dynamically stable solution in economic cooperation because there is no guarantee that the agreed-upon optimality principle is fulfilled throughout the cooperative period. In this chapter we consider dynamically stable economic optimization.

The formulation of a solution for dynamic economic optimization is given in Sect. 4.1. The principle of time consistency and the characterization of time consistent solutions are provided in Sect. 4.2. The derivation of payoff distribution procedures leading to a time consistent solution is investigated in Sect. 4.3. Section 4.4 depicts solutions from specific optimality principles and Sect. 4.5 presents an illustration in the cooperative harvesting of a fishery. The analysis is extended to an infinite-horizon framework in Sect. 4.6 and an example of optimizing infinite-horizon resource extraction is presented in Sect. 4.7.

4.1 Solution in Dynamic Economic Optimization

The formulation of optimal behaviors for participating agents is a fundamental element in the theory of cooperative games. The agents' behaviors satisfying the agreed-upon optimality principles constitute a solution of the game. In other words, the solution of a cooperative economic game is generated by a set of optimality principles.

Consider again the situation when economic agents agree to optimize cooperatively in a dynamic context. Let $\Gamma_c(x_0, T - t_0)$ denote a cooperative game in which

D.W.K. Yeung, L.A. Petrosyan, *Subgame Consistent Economic Optimization*,
Static & Dynamic Game Theory: Foundations & Applications,
DOI 10.1007/978-0-8176-8262-0_4, © Springer Science+Business Media, LLC 2012

agent i's payoff is

$$\int_{t_0}^{T} g^i\left[s, x(s), u_1(s), u_2(s), \ldots, u_n(s)\right] \exp\left[-\int_{t_0}^{s} r(y)\,dy\right] ds$$

$$+ \exp\left[-\int_{t_0}^{T} r(y)\,dy\right] q^i\left(x(T)\right), \quad \text{for } i \in N,   \tag{4.1}$$

and the dynamics of the state is

$$\dot{x}(s) = f\left[s, x(s), u_1(s), u_2(s), \ldots, u_n(s)\right], \quad x(t_0) = x_0.   \tag{4.2}$$

The participating agents agree to act according to an agreed-upon optimality principle. The solution generated by the agreed-upon optimality principle includes agreements on how to act cooperatively and allocate the cooperative payoff.

Let there be an optimality principle agreed upon by all agents in the cooperative game $\Gamma_c(x_0, T - t_0)$. Based on this optimality principle, the solution $P(x_0, T - t_0)$ of the game $\Gamma_c(x_0, T - t_0)$ at time t_0 includes the following.

(i) A set of cooperative strategies $u^{(t_0)*}(s) = [u_1^{(t_0)*}(s), u_2^{(t_0)*}(s), \ldots, u_n^{(t_0)*}(s)]$, for $s \in [t_0, T]$.

(ii) An imputation vector $\xi^{(t_0)}(t_0, x_0) = [\xi^{(t_0)1}(t_0, x_0), \xi^{(t_0)2}(t_0, x_0), \ldots, \xi^{(t_0)n}(t_0, x_0)]$ to allocate the cooperative payoff to the agents.

(iii) A payoff distribution procedure $B^{t_0}(s) = [B_1^{t_0}(s), B_2^{t_0}(s), \ldots, B_n^{t_0}(s)]$ for $s \in [t_0, T]$, where $B_i^{t_0}(s)$ is the instantaneous payments for agent i at time s. In particular,

$$\xi^{(t_0)i}(t_0, x_0) = \int_{t_0}^{T} B_i^{t_0}(s) \exp\left[-\int_{t_0}^{s} r(y)\,dy\right] ds$$

$$+ q^i(x_T) \exp\left[-\int_{t_0}^{T} r(y)\,dy\right],   \tag{4.3}$$

for $i \in N$.

This means that the agents agree at the outset on a set of cooperative strategies $u^{(t_0)*}(s)$, an imputation $\xi^{(t_0)i}(t_0, x_0)$ of the gains to the ith agent covering the time interval $[t_0, T]$, and a payoff distribution procedure $\{B^{t_0}(s)\}_{s=t_0}^{T}$ to allocate payments to the agents over the game interval.

Using the agreed-upon cooperative strategies the state evolves according to the state dynamics

$$\dot{x}(s) = f\left[s, x(s), u_1^{(t_0)*}(s), u_2^{(t_0)*}(s), \ldots, u_n^{(t_0)*}(s)\right], \quad x(t_0) = x_0.   \tag{4.4}$$

The solution to (4.4) yields the conditional optimal trajectory, which is denoted by $\{x^c(s)\}_{s=t_0}^{T}$. For notational convenience we use $x^c(s)$ and x_s^c interchangeably.

When time $t \in (t_0, T]$ has arrived, the situation becomes a cooperative game in which economic agent i's payoff is

$$\int_t^T g^i \big[s, x(s), u_1(s), u_2(s), \ldots, u_n(s)\big] \exp\bigg[-\int_t^s r(y)\,dy\bigg]\,ds$$

$$+ \exp\bigg[-\int_t^T r(y)\,dy\bigg] q^i\big(x(T)\big), \quad \text{for } i \in N, \tag{4.5}$$

and the evolutionary dynamics of the state is

$$\dot{x}(s) = f\big[s, x(s), u_1(s), u_2(s), \ldots, u_n(s)\big], \quad x(t) = x_t^c. \tag{4.6}$$

We use $\Gamma_c(x_t^c, T - t)$ to denote a cooperative game in which economic agent i's objective is (4.5) with the state dynamics in (4.6).

At time $t \in (t_0, T]$ when the state is x_t^c, according to the agreed-upon optimality principle, the solution $P(x_t^c, T - t)$ of the game $\Gamma_c(x_t^c, T - t)$ at time t includes the following.

(i) A set of cooperative strategies $u^{(t)*}(s) = [u_1^{(t)*}(s), u_2^{(t)*}(s), \ldots, u_n^{(t)*}(s)]$, for $s \in [t, T]$.

(ii) An imputation vector $\xi^{(t)}(t, x_t^c) = [\xi^{(t)1}(t, x_t^c), \xi^{(t)2}(t, x_t^c), \ldots, \xi^{(t)n}(t, x_t^c)]$ to allocate the cooperative payoff to the agents.

(iii) A payoff distribution procedure $B^t(s) = [B_1^t(s), B_2^t(s), \ldots, B_n^t(s)]$ for $s \in [t, T]$, where $B_i^t(s)$ is the instantaneous payments for agent i at time s. In particular,

$$\xi^{(t)i}\big(t, x_t^c\big) = \int_t^T B_i^t(s) \exp\bigg[-\int_t^s r(y)\,dy\bigg]\,ds$$

$$+ q^i\big(x_T^c\big) \exp\bigg[-\int_t^T r(y)\,dy\bigg], \tag{4.7}$$

for $i \in N$ and $t \in [t_0, T]$.

This means that under the agreed-upon optimality principle, the agents agree on a set of cooperative strategies $u^{(t)*}(s)$, an imputation of the gains in such a way that the gain under cooperation of the ith agent over the time interval $[t, T]$ is equal to $\xi^{(t)i}(t, x_t^c)$, and a payoff distribution procedure $\{B^t(s)\}_{s=t}^T$ to allocate payments to the agents over the game interval $[t, T]$.

Let there exist solutions $P(x_t^c, T - t) \neq \emptyset, t_0 \leq t \leq T$ along the conditionally optimal trajectory $\{x^c(t)\}_{t=t_0}^T$. If this condition is not satisfied it is impossible for the agents to adhere to the chosen principle of optimality since at the very first instant t, when $P(x_t^c, T - t) = \emptyset$, the agents cannot follow this optimality principle.

For $\xi^{(t)}(t, x_t^c), t \in [t_0, T]$, to be valid imputations, it is required that both group optimality and individual rationality have to be satisfied. Hence a valid optimality principle would yield a solution $P(x_t^c, T - t)$, which contains

(i) $\sum_{j=1}^{n} \xi^{(t)j}(t, x_t^c) = W^{(t)}(t, x_t^c)$, for $t \in [t_0, T]$, and

(ii) $\xi^{(t)i}(t, x_t^c) \geq V^{(t)i}(t, x_t^c)$, for $i \in N$ and $t \in [t_0, T]$.

As discussed in Chap. 3, part (i) above guarantees group optimality, which yields the highest joint payoffs for the participating agents. Part (ii) yields individual rationality so that the payoff allocated to an economic agent under cooperation will be no less than its noncooperative payoff. The failure to guarantee group optimality and individual rationality would lead to the condition where participants will reject the agreed-upon optimality principle and play noncooperatively.

4.2 Principle of Time Consistency

To ensure stability in dynamic cooperation over time a stringent condition is required: the specific agreed-upon optimality principle must be maintained at any instant of time throughout the game along the optimal state trajectory. This condition is known as *time consistency*. Assume that at the start of the game the agents execute the solution $P(x_0, T - t_0)$ generated by an agreed-upon optimality principle (which includes a set of cooperative strategies, an imputation to distribute the cooperative payoff, and a payoff distribution procedure). When the game proceeds to time t, the continuation of the scheme in $P(x_0, T - t_0)$ has to be consistent with the solution $P(x_t^c, T - t)$ to the game $\Gamma_c(x_t^c, T - t)$ under the same optimality principle. If this consistency condition is violated, some of the agents will have an incentive to deviate from the initially chosen trajectory. If this happens, instability arises. In particular, the dynamic stability of a solution of a cooperative differential game is the property that, when the game proceeds along the cooperative state trajectory, at each instant of time the agents are guided by the same optimality principle; therefore, they do not have any incentive to deviate from the previously adopted optimal behavior.

The question of time consistency or dynamic stability in differential games has been explored rigorously in the past three decades. Haurie (1976) discussed the problem of dynamic instability in extending the Nash bargaining solution to differential games. Petrosyan (1977) formalized mathematically the notion of dynamic stability in solutions of differential games. Petrosyan and Danilov (1979, 1982) introduced the notion of "imputation distribution procedure" for a cooperative solution. Tolwinski et al. (1986) considered cooperative equilibria in differential games in which memory-dependent strategies and threats are introduced to maintain the agreed-upon control path. Petrosyan and Zenkevich (1996) and Petrosyan (1997) provided a detailed analysis of dynamic stability in cooperative differential games. In particular, the method of regularization was introduced to construct time consistent solutions. Yeung and Petrosyan (2001) designed a time consistent solution in differential games and characterized the conditions that the allocation distribution procedure must satisfy. Petrosyan (1995, 2003) employed the regularization method to construct time consistent bargaining procedures. Yeung and Petrosyan (2006a) developed a generalized method for the derivation of analytically tractable time consistent solutions for games with transferable payoffs.

4.2.1 Characterization of Time Consistent Solution

Let there be an optimality principle agreed upon by all agents in the cooperative game $\Gamma_c(x_0, T - t_0)$. At time t_0, the solution generated by this optimality principle is $P(x_0, T - t_0)$. At time $t \in (t_0, T]$ when the state is x_t^c, according to the agreed-upon optimality, the solution of the game at time t with state x_t^c is $P(x_t^c, T - t)$.

A cooperative game $\Gamma_c(x_0, T - t_0)$ has a time consistent solution $P(x_0, T - t_0)$ if the continuation of the scheme from the solution $P(x_0, T - t_0) = \{u^{(t_0)*}(s)$ and $B^{t_0}(s)$ for $s \in [t_0, T]; \xi^{(t_0)}(t_0, x_0)\}$ over the time period $[t, T]$ coincides with the solution $P(x_t^c, T - t) = \{u^{(t)*}(s)$ and $B^t(s)$ for $s \in [t, T]; \xi^{(t)}(t, x_t^c)\}$ generated by the same agreed-upon optimality principle at any time instant $t \in [t_0, T]$ along the conditional optimal trajectory $\{x_s^c\}_{s=t_0}^T$.

If this coincidence does not appear, there is no guarantee that the agents will not abandon the solution $P(x_0, T - t_0)$ and switch to $P(x_t^c, T - t)$. Dynamical instability would arise as participants found that their agreed-upon optimality principle could not be maintained after cooperation has gone on for some time.

To verify whether the solution $P(x_0, T - t_0) = \{u^{(t_0)*}(s)$ and $B^{t_0}(s)$ for $s \in [t_0, T]; \xi^{(t_0)}(t_0, x_0)\}$ is indeed time consistent, one has to verify whether the agreed-upon cooperative strategies, payoff distribution procedures, and imputations are all time consistent.

4.2.2 Time Consistent Cooperative Strategies

First, we consider the cooperative strategies adopted under the solution $P(x_0, T - t_0)$ generated by the agreed-upon optimality principle. At time t_0 when the initial state is x_0, the set of cooperative strategies according to $P(x_0, T - t_0)$ is

$$u^{(t_0)*}(s) = \left[u_1^{(t_0)*}(s), u_2^{(t_0)*}(s), \ldots, u_n^{(t_0)*}(s)\right], \quad \text{for } s \in [t_0, T].$$

Consider the case when the game has proceeded to time t and the state variable becomes x_t^c. Then one has a cooperative game $\Gamma_c(x_t^c, T - t)$, which starts at time t with initial state x_t^c. According to the solution $P(x_t^c, T - t)$ generated by the adopted optimality principle a set of cooperative strategies

$$u^{(t)*}(s) = \left[u_1^{(t)*}(s), u_2^{(t)*}(s), \ldots, u_n^{(t)*}(s)\right], \quad \text{for } s \in [t, T],$$

will be adopted.

Definition 4.1 The set of cooperative strategies $u^{(t_0)*}(s) = [u_1^{(t_0)*}(s), u_2^{(t_0)*}(s), \ldots,$ $u_n^{(t_0)*}(s)] \in P(x_0, T - t_0)$ is time consistent if, for $s \in [t, T]$ and $t \in [t_0, T]$,

$$\left[u_1^{(t_0)*}(s), u_2^{(t_0)*}(s), \ldots, u_n^{(t_0)*}(s)\right]$$
$$= \left[u_1^{(t)*}(s), u_2^{(t)*}(s), \ldots, u_n^{(t)*}(s)\right] \in P\left(x_t^c, T - t\right).$$

If the condition in Definition 4.1 is satisfied at each instant of time $t \in [t_0, T]$ along the conditional optimal trajectory $\{x^c(t)\}_{t=t_0}^{T}$, the continuation of the original cooperative strategies $u^{(t_0)*}(s)$ coincides with the cooperative strategies $u^{(t)*}(s)$ in the cooperative game $\Gamma_c(x_t^c, T - t)$. Hence the set of cooperative strategies $u^{(t_0)*}(s) \in P(x_0, T - t_0)$ is time consistent.

Recall that to ensure group optimality the agents have to maximize the agents' joint payoffs. An optimality principle that requires group optimality will yield a solution $P(x_0, T - t_0)$, which includes the set of cooperative controls that solves the problem

$$
\max_{u_1, u_2, \ldots, u_n} \left\{ \int_{t_0}^{T} \sum_{j=1}^{n} g^j \left[s, x(s), u_1(s), u_2(s), \ldots, u_n(s) \right] \exp \left[- \int_{t_0}^{s} r(y) \, dy \right] ds \right.
$$

$$
\left. + \exp \left[- \int_{t_0}^{T} r(y) \, dy \right] \sum_{j=1}^{n} q^j \left(x(T) \right) \right\}, \tag{4.8}
$$

subject to

$$
\dot{x}(s) = f \left[s, x(s), u_1(s), u_2(s), \ldots, u_n(s) \right], \quad x(t_0) = x_0. \tag{4.9}
$$

A set of group optimal cooperative strategies $\{\psi_i^*(s, x^*(s)), \text{ for } i \in N \text{ and } s \in [t_0, T]\}$, which solves the problem in (4.8) and (4.9), can be characterized by Theorem 3.1 in Chap. 3. In particular, $\{x^*(t)\}_{t=t_0}^{T}$ is the solution path of the optimal cooperative trajectory

$$
\dot{x}(s) = f \left[s, x(s), \psi_1^*(s, x(s)), \psi_2^*(s, x(s)), \ldots, \psi_n^*(s, x(s)) \right], \quad x(t_0) = x_0.
$$

Invoking Remark 3.1 in Chap. 3, one can show that the joint payoff maximizing controls for the cooperative game $\Gamma_c(x_t^*, T - t)$ over the time interval $[t, T]$ is identical to the joint payoff maximizing controls for the cooperative game $\Gamma_c(x_0, T - t_0)$ over the same time interval.

Therefore, the solution to an optimality principle that requires group optimality yields a system of time consistent cooperative strategies. Given that group optimality is an essential element in dynamic cooperation, a valid optimality principle will require the maximization of joint payoff and the cooperative strategies $u^{(t_0)*}(s) = u_1^{(t)*}(s) = \psi_i^*(s, x^*(s))$, for $s \in [t, T]$ and $t \in [t_0, T]$. Hence the conditional optimal trajectory $\{x^c(t)\}_{t=t_0}^{T}$ coincides with $\{x^*(t)\}_{t=t_0}^{T}$ in games where the optimality principle requires group optimality.

4.2.3 Time Consistency in Imputation and Payoff Distribution Procedure

Now we consider time consistency in imputation and the payoff distribution procedure. At time t_0 when the initial state is x_0, according to the solution $P(x_0, T - t_0)$

generated by the agreed-upon optimality principle, the economic agents will use the payoff distribution procedure $\{B^{t_0}(s)\}_{s=t_0}^{T}$ to bring about an imputation to agent i as

$$\xi^{(t_0)i}(t_0, x_0) = \int_{t_0}^{T} B_i^{t_0}(s) \exp\left[-\int_{t_0}^{s} r(y)\,dy\right] ds + q^i(x_T) \exp\left[-\int_{t_0}^{T} r(y)\,dy\right],$$

(4.10)

for $i \in N$.

When the game proceeds to time $t \in (t_0, T]$, the current state is x_t^c. According to the solution $P(x_0, T - t_0)$, agent i will receive an imputation (in the present value viewed at time t_0) equaling

$$\xi^{(t_0)i}\left(t, x_t^c\right) = \int_{t}^{T} B_i^{t_0}(s) \exp\left[-\int_{t_0}^{s} r(y)\,dy\right] ds + q^i\left(x_T^c\right) \exp\left[-\int_{t_0}^{T} r(y)\,dy\right],$$

(4.11)

over the time interval $[t, T]$.

At time $t \in (t_0, T]$ when the current state is x_t^c, we have a cooperative game $\Gamma_c(x_t^c, T - t)$. According to the solution $P(x_t^c, T - t)$ generated by the agreed-upon optimality principle, the economic agents will use the payoff distribution procedure $\{B^t(s)\}_{s=t}^{T}$ to bring about an imputation to agent i as

$$\xi^{(t)i}\left(t, x_t^c\right) = \int_{t}^{T} B_i^t(s) \exp\left[-\int_{t}^{s} r(y)\,dy\right] ds + q^i\left(x_T^c\right) \exp\left[-\int_{t}^{T} r(y)\,dy\right],$$

(4.12)

for $i \in N$.

For the imputation and payoff distribution procedure from $P(x_0, T - t_0)$ to be consistent with those from $P(x_t^c, T - t)$, it is essential that

$$\exp\left[\int_{t_0}^{t} r(y)\,dy\right] \xi^{(t_0)}\left(t, x_t^c\right) = \xi^{(t)}\left(t, x_t^c\right) \in P\left(x_t^c, T - t\right), \quad \text{for } t \in [t_0, T].$$

In addition, at time t_0 when the initial state is x_0, according to the solution $P(x_0, T - t_0)$ generated by the agreed-upon optimality principle, the payoff distribution procedure is

$$B^{t_0}(s) = \left[B_1^{t_0}(s), B_2^{t_0}(s), \ldots, B_n^{t_0}(s)\right], \quad \text{for } s \in [t_0, T].$$

Consider the case when the game has proceeded to time t and the state variable became x_t^c. Then one has a cooperative game $\Gamma_c(x_t^c, T - t)$, which starts at time t with initial state x_t^c. According to the solution $P(x_t^c, T - t)$ generated by the agreed-upon optimality principle, the payoff distribution procedure

$$B^t(s) = \left[B_1^t(s), B_2^t(s), \ldots, B_n^t(s)\right], \quad \text{for } s \in [t, T],$$

will be adopted.

For the continuation of the payoff distribution procedure $B^{t_0}(s)$ under $P(x_0, T - t_0)$ to be consistent with $B^t(s) \in P(x_t^c, T - t)$, it is required that

$$B^{t_0}(s) = B^t(s), \quad \text{for } s \in [t, T] \quad \text{and} \quad t \in [t_0, T].$$

Therefore, a formal definition can be presented as follows.

Definition 4.2 The imputation and payoff distribution procedure $\{\xi^{(t_0)}(t_0, x_0)$ and $B^{t_0}(s)$, for $s \in [t_0, T]\} \in P(x_0, T - t_0)$, are time consistent if

(i)

$$\exp\left[\int_{t_0}^t r(y)\,dy\right]\xi^{(t_0)i}\left(t, x_t^c\right)$$

$$\equiv \exp\left[\int_{t_0}^t r(y)\,dy\right]\left\{\left[\int_t^T B_i^{t_0}(s)\exp\left[-\int_t^s r(y)\,dy\right]ds\right.\right.$$

$$\left.\left. + q^i\left(x_T^c\right)\exp\left[-\int_{t_0}^T r(y)\,dy\right]\right\}\right.$$

$$= \xi^{(t)i}\left(t, x_t^c\right)$$

$$\equiv \int_t^T B_i^t(s)\exp\left[-\int_t^s r(y)\,dy\right]ds + q^i\left(x_T^c\right)\exp\left[-\int_t^T r(y)\,dy\right]$$

$$\in P\left(x_t^c, T - t\right), \tag{4.13}$$

for $i \in N$ and $t \in [t_0, T]$, and
(ii) the payoff distribution procedure $B^{t_0}(s) = [B_1^{t_0}(s), B_2^{t_0}(s), \ldots, B_n^{t_0}(s)]$, for $s \in [t, T]$, is identical to

$$B^t(s) = \left[B_1^t(s), B_2^t(s), \ldots, B_n^t(s)\right] \in P\left(x_t^c, T - t\right). \tag{4.14}$$

Thus cooperative strategies, payoff distribution procedures, and imputations satisfying the conditions in Definitions 4.1 and 4.2 are time consistent.

4.3 Payoff Distribution Procedure Derivation and Time (Optimal-Trajectory-Subgame) Consistent Solutions

Crucial to obtaining a time consistent solution is the derivation of a payoff distribution procedure satisfying Definition 4.2 in Sect. 4.2.

4.3.1 Derivation of Payoff Distribution Procedures

Invoking part (ii) of Definition 4.2, we have $B^{t_0}(s) = B^t(s)$ for $t \in [t_0, T]$ and $s \in [t, T]$. We use $B(s) = \{B_1(s), B_2(s), \ldots, B_n(s)\}$ to denote $B^t(s)$ for all $t \in [t_0, T]$.

Along the conditional optimal trajectory $\{x^c(s)\}_{s=t_0}^T$ we then have

$$\xi^{(\tau)i}\left(\tau, x_\tau^c\right) = \int_\tau^T B_i(s) \exp\left[-\int_\tau^s r(y)\,dy\right] ds + q^i\left(x_T^c\right) \exp\left[-\int_\tau^T r(y)\,dy\right],$$

(4.15)

for $i \in N$ and $\tau \in [t_0, T]$; and

$$\sum_{j=1}^n B_j(s) = \sum_{j=1}^n g^j\left[s, x_s^c, u_1^{(\tau)*}(s), u_2^{(\tau)*}(s), \dots, u_n^{(\tau)*}(s)\right].$$

Moreover, for $t \in [\tau, T]$, we use the term

$$\xi^{(\tau)i}\left(t, x_t^c\right) = \int_t^T B_i(s) \exp\left[-\int_\tau^s r(y)\,dy\right] ds + q^i\left(x_T^c\right) \exp\left[-\int_\tau^T r(y)\,dy\right],$$

(4.16)

to denote the present value (with the initial time being τ) of agent i's payoff under cooperation over the time interval $[t, T]$ according to the solution $P(x_\tau^c, T - \tau)$ along the cooperative state trajectory.

Invoking (4.15) and (4.16) we have

$$\xi^{(\tau)i}\left(t, x_t^c\right) = \exp\left[-\int_\tau^t r(y)\,dy\right]\xi^{(t)i}\left(t, x_t^c\right),$$

(4.17)

for $i \in N$, and $\tau \in [t_0, T]$ and $t \in [\tau, T]$.

One can readily verify that a payoff distribution procedure $\{B(s)\}_{s=t_0}^T$ that satisfies (4.17) will give rise to time consistent imputations satisfying part (ii) of Definition 4.2. The next task is the derivation of a payoff distribution procedure $\{B(s)\}_{s=t_0}^T$ that leads to the realization of (4.15)–(4.17).

We first consider the following condition concerning the imputation $\xi^{(\tau)}(t, x_t^c)$, for $\tau \in [t_0, T]$ and $t \in [\tau, T]$.

Condition 4.1 For $i \in N, t \in [\tau, T]$, and $\tau \in [t_0, T]$, the imputation $\xi^{(\tau)i}(t, x_t^c)$, for $i \in N$, is a function that is continuously differentiable in t and x_t^c.

A theorem characterizing a formula for $B_i(s)$, for $s \in [t_0, T]$ and $i \in N$, which yields (4.15)–(4.17), can be provided as follows.

Theorem 4.1 *If Condition* 4.1 *is satisfied, a PDP with a terminal payment* $q^i(x_T^c)$ *at time* T, *and an instantaneous payment at time* $s \in [\tau, T]$

$$B_i(s) = -\left[\xi_t^{(s)i}\left(t, x_t^c\right)\big|_{t=\tau}\right]$$
$$\quad -\left[\xi_{x_\tau^c}^{(s)i}\left(\tau, x_\tau^c\right)\right]f\left[s, x_s^c, \psi_1^*(\tau, x_s^c), \psi_2^*(\tau, x_s^c), \dots, \psi_n^*(\tau, x_s^c)\right], \quad (4.18)$$

for $i \in N$, *yields the imputation vector* $\xi^{(\tau)}(\tau, x_\tau^c)$, *for* $\tau \in [t_0, T]$ *which satisfies* (4.15)–(4.17).

Proof Invoking (4.15)–(4.17), one can obtain

$$\xi^{(v)i}\left(v, x_v^c\right) = \int_v^{v+\Delta t} B_i(s) \exp\left[-\int_v^s r(y)\,dy\right] ds$$

$$+ \exp\left[-\int_v^{v+\Delta t} r(y)\,dy\right] \xi^{(v+\Delta t)i}\left(v + \Delta t, x_v^c + \Delta x_v^c\right), \quad (4.19)$$

for $v \in [\tau, T]$ and $i \in N$, where

$$\Delta x_v^c = f\left[v, x_v^c, \psi_1^*(v, x_v^c), \psi_2^*(v, x_v^c), \ldots, \psi_n^*(v, x_v^c)\right]\Delta t + o(\Delta t), \quad \text{and}$$
$$o(\Delta t)/\Delta t \to 0 \quad \text{as } \Delta t \to 0.$$

From (4.16) and (4.19), one obtains

$$\int_v^{v+\Delta t} B_i(s) \exp\left[-\int_v^s r(y)\,dy\right] ds$$

$$= \xi^{(v)i}\left(v, x_v^c\right) - \exp\left[-\int_v^{v+\Delta t} r(y)\,dy\right] \xi^{(v+\Delta t)i}\left(v + \Delta t, x_v^c + \Delta x_v^c\right)$$

$$= \xi^{(v)i}\left(v, x_v^c\right) - \xi^{(v)i}\left(v + \Delta t, x_v^c + \Delta x_v^c\right), \quad (4.20)$$

for all $v \in [t_0, T]$ and $i \in N$.

If the imputations $\xi^{(v)}(t, x_t^c)$, for $v \in [t_0, T]$, satisfy Condition 4.1, as $\Delta t \to 0$, one can express (4.20) as

$$B_i(v)\Delta t = -\left[\xi_t^{(v)i}\left(t, x_t^c\right)\big|_{t=v}\right]\Delta t - \left[\xi_{x_v^c}^{(v)i}\left(v, x_v^c\right)\right]$$
$$\times f\left[v, x_v^c, \psi_1^*(v, x_v^c), \psi_2^*(v, x_v^c), \ldots, \psi_n^*(v, x_v^c)\right]\Delta t - o(\Delta t). \quad (4.21)$$

Dividing (4.21) throughout by Δt, with $\Delta t \to 0$, yields (4.18). Thus the payoff distribution procedure in $B_i(s)$ in (4.18) will lead to the realization of $\xi^{(\tau)i}(\tau, x_\tau^c)$, for $\tau \in [t_0, T]$ which satisfies (4.15)–(4.17). □

Assigning the instantaneous payments according to the payoff distribution procedure in (4.18) leads to the realization of the imputation $\xi^{(\tau)}(\tau, x_\tau^c) \in P(x_\tau^c, T - \tau)$ for $\tau \in [t_0, T]$. Therefore, the payoff distribution procedure in $B_i(s)$ in (4.18) yields time consistent imputations.

4.3.2 Time (Optimal-Trajectory-Subgame) Consistent Solution

Given that group optimality has to be satisfied at every instant in dynamic cooperation we consider the following optimality principle.

Principle PI Principle PI is an optimality principle that entails (i) group optimality and individual rationality, and (ii) the distribution of the total cooperative payoff according to an imputation that equals $\xi^{(\tau)}(\tau, x_\tau^*)$ for the subgame in $[\tau, T]$ along the group optimal trajectory. Moreover, the function $\xi^{(\tau)}(\tau, x_\tau^*)$, for $\tau \in [t_0, T]$, is continuously differentiable in τ and x_τ^*.

The term "time consistency" has been applied in a wide range of problems, like dynamic optimization, noncooperative differential games, noncooperative dynamic games, and rational choice theory. However, time consistency for the economic optimization problem presented in this chapter requires dynamical consistency for all subgames along the group optimal trajectory. Hence time consistency in this context reflects optimal-trajectory-subgame consistency. Therefore, we use the term optimal-trajectory-subgame consistency as a qualifier to the general term of time consistency.

A theorem characterizing a time (optimal-trajectory-subgame) consistent solution for the cooperative game $\Gamma_c(x_0, T - t_0)$ under optimality Principle PI is presented below.

Theorem 4.2 *For the cooperative game* $\Gamma_c(x_0, T - t_0)$ *with optimality Principle* PI *the solution* $P(x_0, T - t_0) = \{u(s)$ *and* $B(s)$ *for* $s \in [t_0, T]$ *and* $\xi^{(t_0)}(t_0, x_0)\}$ *in which* (i) $u(s)$ *for* $s \in [t_0, T]$ *is the set of group optimal strategies* $\psi^*(s, x_s^*)$ *for the game* $\Gamma_c(x_0, T - t_0)$, *and* (ii) *the imputation distribution procedure* $B(s) = \{B_1(s), B_2(s), \ldots, B_n(s)\}$ *for* $s \in [t_0, T]$ *where*

$$B_i(s) = -\left[\xi_t^{(s)i}(t, x_t^*)\big|_{t=s}\right] - \left[\xi_{x_s^*}^{(s)i}(s, x_s^*)\right]$$
$$\times f\left[s, x_s^*, \psi_1^*(\tau, x_s^*), \psi_2^*(\tau, x_s^*), \ldots, \psi_n^*(\tau, x_s^*)\right], \quad \text{for } i \in N; \quad (4.22)$$

and

$$\xi^{(s)}(s, x_s^*) = \left[\xi^{(s)1}(s, x_s^*), \xi^{(s)2}(s, x_s^*), \ldots, \xi^{(s)n}(s, x_s^*)\right] \in P(x_s^*, T - s)$$

is the imputation at time $s \in [t_0, T]$ *with the state being* $x_s^* \in \{x^*(t)\}_{t=t_0}^T$ *according to optimality Principle* PI *and is time (optimal-trajectory-subgame) consistent.*

Proof Following the algorithm that specifies $P(x_0, T - t_0)$ as the solution to the game $\Gamma_c(x_0, T - t_0)$ one can readily obtain the solution of the cooperative game $\Gamma_c(x_\upsilon^*, T - \upsilon)$, for $\upsilon \in [t_0, T]$, as $P(x_\upsilon^*, T - \upsilon) = \{u(s)$ and $B(s)$ for $s \in [\upsilon, T]$ and $\xi^{(\upsilon)}(\upsilon, x_\upsilon^*)\}$ in which (i) $u(s)$ for $s \in [\upsilon, T]$ is the set of group optimal strategies $\psi^*(s, x_s^*)$ for the game $\Gamma_c(x_\upsilon^*, T - \upsilon)$, and (ii) $B(s) = \{B_1(s), B_2(s), \ldots, B_n(s)\}$ for $s \in [\upsilon, T]$ where

$$B_i(s) = -\left[\xi_t^{(s)i}(t, x_t^*)\big|_{t=s}\right] - \left[\xi_{x_s^*}^{(s)i}(s, x_s^*)\right]$$
$$\times f\left[s, x_s^*, \psi_1^*(\tau, x_s^*), \psi_2^*(\tau, x_s^*), \ldots, \psi_n^*(\tau, x_s^*)\right], \quad \text{for } i \in N; \quad (4.23)$$

and

$$\xi^{(s)}(s, x_s^*) = \left[\xi^{(s)1}(s, x_s^*), \xi^{(s)2}(s, x_s^*), \ldots, \xi^{(s)n}(s, x_s^*)\right] \in P(x_s^*, T - s)$$

is the imputation according to the agreed-upon optimality principle at time $s \in [\upsilon, T]$ with the state being $x_s^* \in \{x^*(t)\}_{t=\upsilon}^T$.

Invoking Remark 3.1 in Chap. 3 and Definition 4.1, one can show that the group optimal joint payoff maximizing strategies $\psi^*(s, x_s^*)$ for the cooperative game $\Gamma_c(x_0, T - t_0)$ over the time interval $[\upsilon, T]$ is identical to the joint payoff maximizing strategies controls for the cooperative game $\Gamma_c(x_\upsilon^*, T - \upsilon)$ over the same time interval.

Comparing (4.22) and (4.23) one can show that the payoff distribution procedure $B(s)$ for the cooperative game $\Gamma_c(x_0, T - t_0)$ over the time interval $[\upsilon, T]$ is identical to the payoff distribution procedure $B(s)$ for the cooperative game $\Gamma_c(x_\upsilon^*, T - \upsilon)$ over the same time interval.

Invoking Theorem 4.1 one can show that the payoff distribution procedure $B(s) = \{B_1(s), B_2(s), \ldots, B_n(s)\}$ in (4.22) will yield

$$\xi^{(\upsilon)i}\left(\upsilon, x_\upsilon^c\right) = \left\{\int_\tau^T B_i(s) \exp\left[-\int_\tau^s r(y)\,dy\right]ds\right.$$
$$\left. + q^i\left(x_T^*\right)\exp\left[-\int_\tau^T r(y)\,dy\right]\right\} \in P\left(x_\upsilon^*, T - \upsilon\right),$$

for $i \in N$, and $\upsilon \in [\tau, T]$.

Hence

$$\exp\left[\int_{t_0}^\upsilon r(y)\,dy\right]\xi^{(t_0)i}\left(\upsilon, x_\upsilon^*\right)$$
$$\equiv \exp\left[\int_{t_0}^\upsilon r(y)\,dy\right]\left\{\int_\upsilon^T B_i(s)\exp\left[-\int_{t_0}^s r(y)\,dy\right]ds\right.$$
$$\left. + q^i\left(x_T^*\right)\exp\left[-\int_{t_0}^T r(y)\,dy\right]\right\}$$
$$= \xi^{(\upsilon)i}\left(\upsilon, x_\upsilon^c\right) \in P\left(x_\upsilon^*, T - \upsilon\right).$$

In summary, the continuation of the solution $P(x_0, T - t_0)$ over the time interval $[\upsilon, T]$ coincides with the solution $P(x_\upsilon^*, T - \upsilon)$ of the game $\Gamma_c(x_\upsilon^*, T - \upsilon)$ under optimality Principle PI. Thus the solution $P(x_0, T - t_0)$ in Theorem 4.2 is indeed time consistent.   □

With agents using the cooperative strategies $\{\psi_i^*(\tau, x_\tau^*), \text{ for } \tau \in [t_0, T] \text{ and } i \in N\}$, the instantaneous payment received by agent i at time instant τ is

$$\zeta_i(\tau) = g^i\left[\tau, x_\tau^*, \psi_1^*(\tau, x_\tau^*), \psi_2^*(\tau, x_\tau^*), \ldots, \psi_n^*(\tau, x_\tau^*)\right], \qquad (4.24)$$

for $\tau \in [t_0, T]$ and $i \in N$.

According to Theorem 4.2, the instantaneous payment that agent i should receive under the agreed-upon optimality principle is $B_i(\tau)$, as stated in (4.22). Hence an instantaneous transfer payment

$$\chi^i(\tau) = B_i(\tau) - \zeta_i(\tau), \qquad (4.25)$$

has to be given to agent i at time τ, for $i \in N$ and $\tau \in [t_0, T]$.

Under an optimal-trajectory-subgame consistent solution, the agreed-upon optimality principle remains effective at any instant of time throughout the game along the optimal state trajectory. Moreover, group and individual rationality are satisfied throughout the entire game interval. Theorem 4.2 provides a handy tool to obtain optimal-trajectory-subgame consistent or time consistent cooperative solutions. Examples of cooperative differential games with solutions satisfying time (optimal-trajectory-subgame) consistency can be found in Petrosyan (1997), Jørgensen and Zaccour (2001), Yeung (2005, 2007), Yeung and Petrosyan (2004, 2006a, 2006b), and Filar and Petrosjan (2000). Moreover, Theorem 4.2 can be applied to obtain a time (optimal-trajectory-subgame) consistent cooperative solution for the existing differential games in economic analysis.

4.4 Solutions from Specific Optimality Principle

In this section we present examples of time consistent solutions from some optimality principles.

Case 1. Joint Payoff Maximization and Equal Sharing of Gains from Cooperation Consider the cooperative differential game $\Gamma_c(x_0, T - t_0)$. In particular, the agents agree with an optimality principle that entails (i) group optimality and (ii) the division of the excess of the total cooperative payoff over the sum of individual noncooperative payoffs equally.

We denote the above optimality principle as Principle PI. Recall in Chap. 3 that the total cooperative payoffs in the cooperative game $\Gamma_c(x_0, T - t_0)$ is $W^{(t_0)}(t_0, x_0)$, the noncooperative payoff for agent j is $V^{(t_0)j}(t_0, x_0)$ in the noncooperative game $\Gamma(x_0, T - t_0)$. According to optimality Principle PI the imputation to agent j in $\Gamma_c(x_0, T - t_0)$ is

$$\xi^{(t_0)i}(t_0, x_0) = V^{(t_0)i}(t_0, x_0) + \frac{1}{n}\left[W^{(t_0)}(t_0, x_0) - \sum_{j=1}^{n} V^{(t_0)j}(t_0, x_0)\right], \quad \text{for } i \in N.$$

(4.26)

As the game progresses along the conditional optimal cooperative path $\{x_s^c\}_{s=t_0}^{T}$, according to Principle PI the imputation to agent j in the cooperative game $\Gamma_c(x_\tau^c, T - \tau)$ is

$$\xi^{(\tau)i}\left(\tau, x_\tau^c\right) = V^{(\tau)i}\left(\tau, x_\tau^c\right) + \frac{1}{n}\left[W^{(\tau)}\left(\tau, x_\tau^c\right) - \sum_{j=1}^{n} V^{(\tau)j}\left(\tau, x_\tau^c\right)\right], \quad (4.27)$$

for $i \in N$ and $\tau \in (t_0, T]$.

The imputation in (4.26) and (4.27) yields

(i) $\xi^{(\tau)i}(\tau, x_\tau^c) \geq V^{(\tau)i}(\tau, x_\tau^c)$, for $i \in N$ and $\tau \in [t_0, T]$; and
(ii) $\sum_{j=1}^{n} \xi^{(\tau)j}(\tau, x_\tau^c) = W^{(\tau)}(\tau, x_\tau^c)$ for $\tau \in [t_0, T]$.

Hence the imputation vector $\xi^{(\tau)i}(\tau, x_\tau^*)$ satisfies individual rationality and group optimality.

Applying Theorem 4.2, a time (optimal-trajectory-subgame) consistent solution under the optimal Principle PI can be characterized as $P(x_0, T - t_0) = \{u(s)$ and $B(s)$ for $s \in [t_0, T]$ and $\xi^{(t_0)}(t_0, x_0)\}$ in which

(i) $u(s)$ for $s \in [t_0, T]$ is the set of group optimal strategies $\psi^*(s, x_s^*)$ in the game $\Gamma(x_0, T - t_0)$, and

(ii) the imputation distribution procedure $B(s) = \{B_1(s), B_2(s), \ldots, B_n(s)\}$ for $s \in [t_0, T]$, where

$$B_i(s) = -\frac{\partial}{\partial t}\left[V^{(s)i}\left(t, x_t^*\right) + \frac{1}{n}\left(W^{(s)}\left(t, x_t^*\right) - \sum_{j=1}^{n} V^{(s)j}\left(t, x_t^*\right)\right)\right]\Bigg|_{t=\tau}$$

$$-\frac{\partial}{\partial x_s^*}\left[V^{(s)i}\left(s, x_s^*\right) + \frac{1}{n}\left(W^{(s)}\left(s, x_s^*\right) - \sum_{j=1}^{n} V^{(s)j}\left(s, x_s^*\right)\right)\right]$$

$$\times f\left[s, x_s^*, \psi_1^*\left(s, x_s^*\right), \psi_2^*\left(s, x_s^*\right), \ldots, \psi_n^*\left(s, x_s^*\right)\right], \qquad (4.28)$$

for $i \in N$.

Case 2. Joint Payoff Maximization and Sharing Gains Proportional to Noncooperative Payoffs   Consider the cooperative differential game $\Gamma_c(x_0, T - t_0)$. In particular, the agents agree with an optimality principle that entails (i) group optimality and (ii) the sharing of the excess of the total cooperative payoff over the sum of the individual noncooperative payoffs proportional to the agents' noncooperative payoffs.

We denote the above optimality principle as Principle PII. According to optimality Principle PII the imputation to agent j in $\Gamma_c(x_0, T - t_0)$ is

$$\xi^{(t_0)i}(t_0, x_0) = V^{(t_0)i}(t_0, x_0) + \frac{V^{(t_0)i}(t_0, x_0)}{\sum_{j=1}^{n} V^{(t_0)j}(t_0, x_0)}$$

$$\times \left[W^{(t_0)}(t_0, x_0) - \sum_{j=1}^{n} V^{(\tau)j}\left(\tau, x_\tau^*\right)\right]$$

$$= \frac{V^{(t_0)i}(t_0, x_0)}{\sum_{j=1}^{n} V^{(t_0)j}(t_0, x_0)} W^{(t_0)}(t_0, x_0), \quad \text{for } i \in N. \qquad (4.29)$$

As the game progresses along the conditional optimal cooperative path $\{x_s^c\}_{s=t_0}^{T}$, according to Principle PII the imputation to agent j in the cooperative game $\Gamma_c(x_\tau^c, T - \tau)$ is

$$\xi^{(\tau)i}\left(\tau, x_\tau^c\right) = \frac{V^{(\tau)i}(\tau, x_\tau^c)}{\sum_{j=1}^{n} V^{(\tau)j}(\tau, x_\tau^c)} W^{(\tau)}\left(\tau, x_\tau^c\right), \quad \text{for } i \in N \quad \text{and} \quad \tau \in [t_0, T].$$

$$(4.30)$$

Again the imputation under optimality Principle PII satisfies individual rationality and group optimality.

Applying Theorem 4.2, a time (optimal-trajectory-subgame) consistent solution under the optimal Principle PII can be characterized as $P(x_0, T - t_0) =$
$\{u(s)$ and $B(s)$ for $s \in [t_0, T]$ and $\xi^{(t_0)}(t_0, x_0)\}$ in which (i) $u(s)$ for $s \in [t_0, T]$ is
the set of group optimal strategies $\psi^*(s, x_s^*)$ in the game $\Gamma(x_0, T - t_0)$, and (ii) the
imputation distribution procedure $B(s) = \{B_1(s), B_2(s), \ldots, B_n(s)\}$ for $s \in [t_0, T]$,
where

$$
\begin{aligned}
B_i(s) = & -\frac{\partial}{\partial t}\left[\frac{V^{(s)i}(t, x_t^*)}{\sum_{j=1}^n V^{(s)j}(t, x_t^*)} W^{(s)}(t, x_t^*)\Big|_{t=s}\right] \\
& -\frac{\partial}{\partial x_s^*}\left[\frac{V^{(s)i}(s, x_s^*)}{\sum_{j=1}^n V^{(s)j}(s, x_s^*)} W^{(s)}(s, x_s^*)\right] \\
& \times f\left[s, x_s^*, \psi_1^*(s, x_s^*), \psi_2^*(s, x_s^*), \ldots, \psi_n^*(s, x_s^*)\right], \quad \text{for } i \in N. \quad (4.31)
\end{aligned}
$$

**Case 3. Joint Payoff Maximization and Sharing Gains as a Combination of
the Imputations in Principles PI and PII**   Consider the cooperative differential
game $\Gamma_c(x_0, T - t_0)$. In particular, the agents agree with an optimality principle that
entails (i) group optimality and (ii) the sharing of the excess of the total cooperative
payoff over the sum of individual noncooperative payoffs as a linear combination of
the imputations in Principles PI and PII.

We denote the above optimality principle as PIII. According to optimality principle PIII the imputation to agent j in $\Gamma_c(x_0, T - t_0)$ is

$$
\begin{aligned}
\xi^{(t_0)i}(t_0, x_0) = & \alpha\left[V^{(t_0)i}(t_0, x_0) + \frac{1}{n}\left(W^{(t_0)}(t_0, x_0) - \sum_{j=1}^n V^{(t_0)j}(t_0, x_0)\right)\right] \\
& + (1 - \alpha)\frac{V^{(t_0)i}(t_0, x_0)}{\sum_{j=1}^n V^{(t_0)j}(t_0, x_0)} W^{(t_0)}(t_0, x_0), \quad \text{for } i \in N, \quad (4.32)
\end{aligned}
$$

where $\alpha \in (0, 1)$.

As the game progresses along the conditional optimal cooperative path $\{x_s^c\}_{s=t_0}^T$,
according to PIII the imputation to agent j in the cooperative game $\Gamma_c(x_\tau^c, T - \tau)$
is

$$
\begin{aligned}
& \xi^{(\tau)i}(\tau, x_\tau^c) \\
& = \alpha\left[V^{(\tau)i}(\tau, x_\tau^c) + \frac{1}{n}\left(W^{(\tau)}(\tau, x_\tau^c) - \sum_{j=1}^n V^{(\tau)j}(\tau, x_\tau^c)\right)\right] \\
& \quad + (1 - \alpha)\frac{V^{(\tau)i}(\tau, x_\tau^c)}{\sum_{j=1}^n V^{(\tau)j}(\tau, x_\tau^c)} W^{(\tau)}(\tau, x_\tau^c), \quad \text{for } i \in N \text{ and } \tau \in [t_0, T]. \quad (4.33)
\end{aligned}
$$

Again, the imputation under optimality Principle PIII satisfies individual rationality and group optimality.

Applying Theorem 4.2, a time (optimal-trajectory-subgame) consistent solution under the optimal Principle PIII can be characterized as $P(x_0, T - t_0) = \{u(s) \text{ and } B(s) \text{ for } s \in [t_0, T] \text{ and } \xi^{(t_0)}(t_0, x_0)\}$ in which (i) $u(s)$ for $s \in [t_0, T]$ is the set of group optimal strategies $\psi^*(s, x_s^*)$ in the game $\Gamma(x_0, T - t_0)$, and (ii) the imputation distribution procedure $B(s) = \{B_1(s), B_2(s), \ldots, B_n(s)\}$ for $s \in [t_0, T]$, where

$$
B_i(s) = -\frac{\partial}{\partial t}\left\{ \alpha\left[V^{(s)i}\left(t, x_t^*\right) + \frac{1}{n}\left(W^{(s)}\left(t, x_t^*\right) - \sum_{j=1}^{n} V^{(s)j}\left(t, x_t^*\right) \right) \right] \right.
$$
$$
\left. + (1 - \alpha)\frac{V^{(s)i}\left(t, x_t^*\right)}{\sum_{j=1}^{n} V^{(s)j}\left(t, x_t^*\right)} W^{(s)}\left(t, x_t^*\right)\Big|_{t=s} \right\}
$$
$$
- \frac{\partial}{\partial x_s^*}\left\{ \alpha\left[V^{(s)i}\left(s, x_s^*\right) + \frac{1}{n}\left(W^{(s)}\left(s, x_s^*\right) - \sum_{j=1}^{n} V^{(s)j}\left(s, x_s^*\right) \right) \right] \right.
$$
$$
\left. + (1 - \alpha)\frac{V^{(s)i}\left(s, x_s^*\right)}{\sum_{j=1}^{n} V^{(s)j}\left(s, x_s^*\right)} W^{(s)}\left(s, x_s^*\right) \right\}
$$
$$
\times f\left[s, x_s^*, \psi_1^*\left(s, x_s^*\right), \psi_2^*\left(s, x_s^*\right), \ldots, \psi_n^*\left(s, x_s^*\right)\right], \tag{4.34}
$$

for $i \in N$.

Case 4. Joint Payoff Maximization and Time Varying Sharing Weights Consider the cooperative differential game $\Gamma_c(x_0, T - t_0)$ with two agents. In particular, the agents agree with an optimality principle that entails (i) group optimality and (ii) the division of the excess of the total cooperative payoff over the sum of individual noncooperative payoffs by the time-varying weights—$\frac{\tau}{T+\alpha}$ for agent 1 and $\frac{T+\alpha-\tau}{T+\alpha}$ for agent 2 at time $\tau \in [t_0, T]$.

We denote the above optimality principle as PIV. According to optimality Principle PIV the imputations to agents 1 and 2 in $\Gamma_c(x_0, T - t_0)$ are

$$
\xi^{(t_0)1}(t_0, x_0) = V^{(t_0)1}(t_0, x_0) + \frac{t_0}{T + \alpha}\left(W^{(t_0)}(t_0, x_0) - \sum_{j=1}^{2} V^{(t_0)j}(t_0, x_0) \right),
$$

for agent 1, and

$$
\xi^{(t_0)2}(t_0, x_0) = V^{(t_0)2}(t_0, x_0) + \frac{T + \alpha - t_0}{T + \alpha}\left(W^{(t_0)}(t_0, x_0) - \sum_{j=1}^{2} V^{(t_0)j}(t_0, x_0) \right),
$$
$$\tag{4.35}$$

for agent 2.

As the game progresses along the conditional optimal cooperative path $\{x_s^c\}_{s=t_0}^{T}$, according to PIV the imputation to agent j in the cooperative game $\Gamma_c(x_\tau^c, T - \tau)$ is

$$
\xi^{(\tau)1}\left(\tau, x_\tau^c\right) = V^{(\tau)1}\left(\tau, x_\tau^c\right) + \frac{\tau}{T + \alpha}\left[W^{(\tau)}\left(\tau, x_\tau^c\right) - \sum_{j=1}^{2} V^{(\tau)j}\left(\tau, x_\tau^c\right) \right],
$$

for agent 1, and

$$\xi^{(\tau)2}\left(\tau,x_\tau^c\right) = V^{(\tau)2}\left(\tau,x_\tau^c\right) + \frac{T+\alpha-\tau}{T+\alpha}\left[W^{(\tau)}\left(\tau,x_\tau^c\right) - \sum_{j=1}^{2}V^{(\tau)j}\left(\tau,x_\tau^c\right)\right],$$

(4.36)

for agent 2; $\tau \in [t_0, T]$.

Again the imputation under optimality Principle PIV satisfies individual rationality and group optimality.

Applying Theorem 4.2, a time (optimal-trajectory-subgame) consistent solution under the optimal Principle PIV can be characterized as $P(x_0, T - t_0) = \{u(s) \text{ and } B(s) \text{ for } s \in [t_0, T] \text{ and } \xi^{(t_0)}(t_0, x_0)\}$ in which (i) $u(s)$ for $s \in [t_0, T]$ is the set of group optimal strategies $\psi^*(s, x_s^*)$ in the game $\Gamma(x_0, T - t_0)$, and (ii) the imputation distribution procedure $B(s) = \{B_1(s), B_2(s), \ldots, B_n(s)\}$ for $s \in [t_0, T]$, where

$$B_1(s) = -\frac{\partial}{\partial t}\left[V^{(s)1}\left(t,x_t^*\right) + \frac{t}{T+\alpha}\left(W^{(s)}\left(t,x_t^*\right) - \sum_{j=1}^{2}V^{(s)j}\left(t,x_t^*\right)\right)\Bigg|_{t=s}\right]$$

$$-\frac{\partial}{\partial x_s^*}\left[V^{(s)1}\left(s,x_s^*\right) + \frac{s}{T+\alpha}\left(W^{(s)}\left(s,x_s^*\right) - \sum_{j=1}^{2}V^{(s)j}\left(s,x_s^*\right)\right)\right]$$

$$\times f\left[s,x_s^*,\psi_1^*\left(s,x_s^*\right),\psi_2^*\left(s,x_s^*\right)\right],$$

(4.37)

$$B_2(s) = -\frac{\partial}{\partial t}\left[V^{(s)2}\left(t,x_t^*\right) + \frac{T-t+\alpha}{T+\alpha}\left(W^{(s)}\left(t,x_t^*\right) - \sum_{j=1}^{2}V^{(s)j}\left(t,x_t^*\right)\right)\Bigg|_{t=s}\right]$$

$$-\frac{\partial}{\partial x_s^*}\left[V^{(s)1}\left(s,x_s^*\right) + \frac{T-s+\varepsilon}{T+\alpha}\left(W^{(s)}\left(s,x_s^*\right) - \sum_{j=1}^{2}V^{(s)j}\left(s,x_s^*\right)\right)\right]$$

$$\times f\left[s,x_s^*,\psi_1^*\left(s,x_s^*\right),\psi_2^*\left(s,x_s^*\right)\right].$$

A variety of optimality principles with various imputation schemes like those in cases 1 to 4 can be constructed.

4.5   An Illustration in Cooperative Fishery

Consider the case of two nations harvesting fish in common waters. The growth rate of the fish biomass is characterized by the differential equation

$$\dot{x}(s) = ax(s)^{1/2} - bx(s) - u_1(s) - u_2(s), \quad x(t_0) = x_0 \in X, \qquad (4.38)$$

where $u_i \in U_i$ is the (nonnegative) amount of fish harvested by nation i, for $i \in \{1, 2\}$, a and b are positive constants.

The harvesting cost for nation $i \in \{1, 2\}$ depends on the quantity of the resource extracted $u_i(s)$, the resource stock size $x(s)$, and a parameter c_i. In particular, nation i's extraction cost can be specified as $c_i u_i(s) x(s)^{-1/2}$. The fish harvested by nation i at time s will generate a net benefit of the amount $[u_i(s)]^{1/2}$. The horizon of concern is $[t_0, T]$. At time T, nation i will receive a termination bonus $q_i x(T)^{1/2}$, where q_i is nonnegative. There exists a positive discount rate r.

At time t_0 the payoff of nation $i \in \{1, 2\}$ is

$$\int_{t_0}^{T} \left[[u_i(s)]^{1/2} - \frac{c_i}{x(s)^{1/2}} u_i(s) \right] \exp[-r(s - t_0)] \, ds + \exp[-r(T - t_0)] q_i x(T)^{\frac{1}{2}}.$$

(4.39)

The game is a deterministic version of an example in Yeung and Petrosyan (2004). A set of feedback strategies $\{u_i^*(t) = \phi_i^*(t, x), \text{for } i \in \{1, 2\}\}$, provides a feedback Nash equilibrium solution to the game in (4.38) and (4.39), if there exist continuously differentiable functions $V^{(t_0)i}(t, x) : [t_0, T] \times R \to R, i \in \{1, 2\}$, satisfying the following partial differential equations:

$$-V_t^{(t_0)i}(t, x) = \max_{u_i} \left\{ \left[u_i^{1/2} - \frac{c_i}{x^{1/2}} u_i \right] \exp[-r(t - t_0)] \right.$$

$$\left. + V_x^{(t_0)i}(t, x) \left[ax^{1/2} - bx - u_i - \phi_j^*(t, x) \right] \right\}, \quad \text{and} \quad (4.40)$$

$$V^{(t_0)i}(T, x) = q_i x^{1/2} \exp[-r(T - t_0)], \quad \text{for } i \in \{1, 2\}, j \in \{1, 2\} \text{ and } j \neq i.$$

Performing the indicated maximization yields the game equilibrium strategies

$$\phi_i^*(t, x) = \frac{x}{4[c_i + V_x^{(t_0)i} \exp[r(t - t_0)] x^{1/2}]^2}, \quad \text{for } i \in \{1, 2\}. \quad (4.41)$$

Proposition 4.1 *The value function of nation i in the game in* (4.38) *and* (4.39) *is*

$$V^{(t_0)i}(t, x) = \exp[-r(t - t_0)][A_i(t) x^{1/2} + C_i(t)], \quad \text{for } i \in \{1, 2\} \text{ and } t \in [t_0, T],$$

(4.42)

where $A_i(t), C_i(t), A_j(t)$, and $C_j(t)$, for $i \in \{1, 2\}$ and $j \in \{1, 2\}$, and $i \neq j$, satisfy

$$\dot{A}_i(t) = \left[r + \frac{b}{2} \right] A_i(t) - \frac{1}{2[c_i + A_i(t)/2]} + \frac{c_i}{4[c_i + A_i(t)/2]^2}$$

$$+ \frac{A_i(t)}{8[c_i + A_i(t)/2]^2} + \frac{A_i(t)}{8[c_j + A_j(t)/2]^2}, \quad (4.43)$$

$$\dot{C}_i(t) = rC_i(t) - \frac{a}{2} A_i(t), \quad \text{and} \quad A_i(T) = q, \quad \text{and} \quad C_i(T) = 0.$$

Proof By substituting $\phi_1^*(t, x)$ and $\phi_2^*(t, x)$ into (4.40) and upon solving (4.40) one can obtain the results in Proposition 4.1 (see also the proof of Proposition 3.1 in Chap. 3). ☐

Consider the alternative game $\Gamma(x_\tau, T - \tau)$ with the payoff structure in (4.39) and the dynamics in (4.38) starting at time $\tau \in [t_0, T]$ with initial state $x_\tau \in X$. Following the above analysis, the value function $V^{(\tau)i}(t, x) : [\tau, T] \times R \to R$, for $i \in \{1, 2\}$ and $\tau \in [t_0, T]$, for the game $\Gamma(x_\tau, T - \tau)$ can be obtained as follows.

Proposition 4.2 *The value function of nation $i \in \{1, 2\}$ in the game $\Gamma(x_\tau, T - \tau)$ is*

$$V^{(\tau)i}(t, x) = \exp[-r(t - \tau)][A_i(t)x^{1/2} + C_i(t)], \tag{4.44}$$

where for $i, j \in \{1, 2\}$ and $i \neq j$, $A_i(t), C_i(t), A_j(t)$, and $C_j(t)$ are the same as those in Proposition 4.1.

Proof Follow the proof of Proposition 4.1. ☐

Substituting the relevant derivatives of the value functions into the game equilibrium strategies of (4.41) yields a feedback Nash equilibrium for the game in (4.38) and (4.39).

Now consider the case when the nations agree to cooperate in harvesting the fishery. Let $\Gamma_c(x_0, T - t_0)$ denote a cooperative game with the game structure of $\Gamma(x_0, T - t_0)$ in which the agents agree to act according to the optimality principle that they would (i) maximize the sum of their payoffs and (ii) divide the excess of the total cooperative payoff over the sum of individual noncooperative payoffs equally.

To maximize the joint payoffs, the nations would consider the optimal control problem

$$\int_{t_0}^T \left(\left[u_1(s)^{1/2} - \frac{c_1}{x(s)^{1/2}} u_1(s) \right] + \left[u_2(s)^{1/2} - \frac{c_2}{x(s)^{1/2}} u_2(s) \right] \right)$$

$$\times \exp[-r(t - t_0)] \, ds + 2 \exp[-r(T - t_0)] q x(T)^{\frac{1}{2}}, \tag{4.45}$$

subject to (4.38).

Let $[\psi_1^*(t, x), \psi_2^*(t, x)]$ denote a set of controls that provides a solution to the optimal control problem in (4.38) and (4.45) and $W^{(t_0)}(t, x) : [t_0, T] \times R^n \to R$ denote the value function that satisfies the equations

$$-W_t^{(t_0)}(t, x) = \max_{u_1, u_2} \left\{ \left(\left[u_1^{1/2} - \frac{c_1}{x^{1/2}} u_1 \right] + \left[u_2^{1/2} - \frac{c_2}{x^{1/2}} u_2 \right] \right) \exp[-r(t - t_0)] \right.$$

$$\left. + W_x^{(t_0)}(t, x)[ax^{1/2} - bx - u_1 - u_2] \right\}, \quad \text{and} \tag{4.46}$$

$$W^{(t_0)}(T, x) = 2 \exp[-r(T - t_0)] q x^{\frac{1}{2}}.$$

Performing the indicated maximization we obtain

$$\psi_1^*(t, x) = \frac{x}{4[c_1 + W_x^{(t_0)} \exp[r(t - t_0)]x^{1/2}]^2}, \quad \text{and}$$

$$\psi_2^*(t, x) = \frac{x}{4[c_2 + W_x^{(t_0)} \exp[r(t - t_0)]x^{1/2}]^2}.$$

Substituting $\psi_1^*(t, x)$ and $\psi_2^*(t, x)$ above into (4.47) yields the value function

$$W^{(t_0)}(t, x) = \exp[-r(t - t_0)]\big[\hat{A}(t)x^{1/2} + \hat{C}(t)\big],$$

$$\text{where } \dot{\hat{A}}(t) = \left[r + \frac{b}{2}\right]\hat{A}(t) - \frac{1}{2[c_1 + \hat{A}(t)/2]} - \frac{1}{2[c_2 + \hat{A}(t)/2]}$$

$$+ \frac{c_1}{4[c_1 + \hat{A}(t)/2]^2} + \frac{c_2}{4[c_2 + \hat{A}(t)/2]^2} \qquad (4.47)$$

$$+ \frac{\hat{A}(t)}{8[c_1 + \hat{A}(t)/2]^2} + \frac{\hat{A}(t)}{8[c_2 + \hat{A}(t)/2]^2},$$

$$\dot{\hat{C}}(t) = r\hat{C}(t) - \frac{a}{2}\hat{A}(t), \; \hat{A}(T) = 2q, \; \text{and} \; \hat{B}(T) = 0.$$

The optimal cooperative controls can then be obtained as

$$\psi_1^*(t, x) = \frac{x}{4[c_1 + \hat{A}(t)/2]^2}, \quad \text{and} \quad \psi_2^*(t, x) = \frac{x}{4[c_2 + \hat{A}(t)/2]^2}. \qquad (4.48)$$

Substituting these control strategies into (4.38) yields the dynamics of the state trajectory under cooperation

$$\dot{x}(s) = ax(s)^{1/2} - bx(s) - \frac{x(s)}{4[c_1 + \hat{A}(s)/2]^2} - \frac{x(s)}{4[c_2 + \hat{A}(s)/2]^2},$$

$$x(t_0) = x_0. \qquad (4.49)$$

Solving (4.49) yields the optimal cooperative state trajectory for $\Gamma_c(x_0, T - t_0)$ as

$$x^*(s) = \varpi(t_0, s)^2 \left[x_0^{1/2} + \int_{t_0}^s \varpi^{-1}(t_0, t)H_1 \, dt\right]^2, \quad \text{for } s \in [t_0, T], \qquad (4.50)$$

where $\varpi(t_0, s) = \exp[\int_{t_0}^s H_2(\tau) \, d\tau]$, $H_1 = \frac{1}{2}a$, and

$$H_2(s) = -\left[\frac{1}{2}b + \frac{1}{8[c_1 + \hat{A}(s)/2]^2} + \frac{1}{8[c_2 + \hat{A}(s)/2]^2}\right].$$

The cooperative control for the game $\Gamma_c(x_0, T - t_0)$ over the time interval $[t_0, T]$ along the optimal trajectory $\{x^*(t)\}_{t=t_0}^{T}$ can be expressed precisely as

$$\psi_1^*(t, x_t^*) = \frac{x_t^*}{4[c_1 + \hat{A}(t)/2]^2}, \quad \text{and} \quad \psi_2^*(t, x_t^*) = \frac{x_t^*}{4[c_2 + \hat{A}(t)/2]^2}. \quad (4.51)$$

Following the above analysis, the value function of the optimal control problem with the dynamics structure of (4.38) and the payoff structure in (4.45) which starts at time τ with initial state x_τ^* can be obtained as $W^{(\tau)}(t, x) = \exp[-r(t - \tau)][\hat{A}(t)x^{1/2} + \hat{B}(t)]$, and the corresponding optimal controls as

$$\psi_1^*(t, x_t^*) = \frac{x_t^*}{4[c_1 + \hat{A}(t)/2]^2}, \quad \text{and} \quad \psi_2^*(t, x_t^*) = \frac{x_t^*}{4[c_2 + \hat{A}(t)/2]^2},$$

over the time interval $[\tau, T]$.

The agreed-upon optimality principle entails an imputation

$$\xi^{(\tau)i}(\tau, x_\tau^*) = V^{(\tau)i}(\tau, x_\tau^*) + \frac{1}{n}\left[W^{(\tau)}(\tau, x_\tau^*) - \sum_{j=1}^{n} V^{(\tau)j}(\tau, x_\tau^*)\right], \quad i \in \{1, 2\},$$

$$(4.52)$$

in the cooperative game $\Gamma_c(x_\tau^*, T - \tau)$ for $\tau \in \{t_0, T\}$.

Applying Theorem 4.2, a time (optimal-trajectory-subgame) consistent solution under the above optimal principle for the cooperative game $\Gamma_c(x_0, T - t_0)$ can be obtained as $P(x_0, T - t_0) = \{u(s) \text{ and } B(s) \text{ for } s \in [t_0, T] \text{ and } \xi^{(t_0)}(t_0, x_0)\}$ in which

(i) $u(s)$ for $s \in [t_0, T]$ is the set of group optimal strategies

$$\psi_1^*(s, x_s^*) = \frac{x_s^*}{4[c_1 + \hat{A}(s)/2]^2}, \quad \text{and} \quad \psi_2^*(s, x_s^*) = \frac{x_s^*}{4[c_2 + \hat{A}(s)/2]^2}.$$

(ii) The imputation distribution procedure $B(s) = \{B_1(s), B_2(s)\}$ for $s \in [t_0, T]$ where

$$\begin{aligned}
B_i(s) = \frac{-1}{2}\Bigg\{ &\left([\dot{A}_i(s)(x_s^*)^{1/2} + \dot{C}_i(s)] + r[A_i(s)(x_s^*)^{1/2} + C_i(s)]\right) \\
&+ \left[\frac{1}{2}A_i(s)(x_s^*)^{-1/2}\right] \\
&\times \left[a(x_s^*)^{1/2} - bx_s^* - \frac{x_s^*}{4[c_i + \hat{A}(s)/2]^2} - \frac{x_s^*}{4[c_j + \hat{A}(s)/2]^2}\right]\Bigg\} \\
&- \frac{1}{2}\Bigg\{\left([\dot{\hat{A}}(s)(x_s^*)^{1/2} + \dot{\hat{C}}(s)] + r[\hat{A}(s)(x_s^*)^{1/2} + \hat{C}(s)]\right) \\
&+ \left[\frac{1}{2}\hat{A}(s)(x_s^*)^{-1/2}\right]
\end{aligned}$$

$$\times \left[a \left(x_s^* \right)^{1/2} - b x_s^* - \frac{x_s^*}{4[c_i + \hat{A}(s)/2]^2} - \frac{x_s^*}{4[c_j + \hat{A}(s)/2]^2} \right] \right\}$$

$$+ \frac{1}{2} \left\{ \left([\dot{A}_j(s)\left(x_s^* \right)^{1/2} + \dot{C}_j(s)] + r[A_j(s)\left(x_s^* \right)^{1/2} + C_j(s)] \right) \right.$$

$$+ \left[\frac{1}{2} A_j(s)\left(x_s^* \right)^{-1/2} \right]$$

$$\times \left. \left[a \left(x_s^* \right)^{1/2} - b x_s^* - \frac{x_s^*}{4[c_i + \hat{A}(s)/2]^2} - \frac{x_s^*}{4[c_j + \hat{A}(s)/2]^2} \right] \right\}, \quad (4.53)$$

for $i, j \in \{1, 2\}$ and $i \neq j$, where $\dot{A}_i(s)$ and $\dot{C}_i(s)$ are given in (4.44) and $\hat{\dot{A}}(s)$ and $\hat{C}(s)$ are given in (4.47).

With agents using the cooperative strategies, the instantaneous receipt of agent i at time instant τ is

$$\zeta_i(\tau) = \frac{(x_\tau^*)^{1/2}}{2[c_i + A(\tau)/2]} - \frac{c_i(x_\tau^*)^{1/2}}{4[c_i + A(\tau)/2]^2}. \quad (4.54)$$

Under cooperation the instantaneous payment that agent i should receive is $B_i(\tau)$, as stated in (4.53). Hence an instantaneous transfer payment

$$\chi^i(\tau) = B_i(\tau) - \zeta_i(\tau) \quad (4.55)$$

has to be given to agent i at time τ, for $i \in \{1, 2\}$ and $\tau \in [t_0, T]$.

4.6 Consistent Economic Optimization Under Infinite Horizon

In many economic situations, the terminal time of the game T is either very far in the future or unknown to the agents. In this section, time consistent cooperation for games with infinite horizon are considered.

Consider the n-person infinite-horizon general economic problem in which economic agent i's payoff is

$$\int_\tau^\infty g^i \left[x(s), u_1(s), u_2(s), \ldots, u_n(s) \right] \exp\left[-r(s - \tau) \right] ds, \quad \text{for } i \in N. \ (4.56)$$

The state dynamics is

$$\dot{x}(s) = f \left[x(s), u_1(s), u_2(s), \ldots, u_n(s) \right], \quad x(\tau) = x_\tau. \quad (4.57)$$

Since s does not appear in $g^i[x(s), u_1(s), u_2(s)]$ or the state dynamics, the game in (4.56) and (4.57) is an autonomous problem. Consider the alternative game $\Gamma(x)$

that starts at time $t \in [t_0, \infty)$ with initial state $x(t) = x$

$$\max_{u_i} \int_t^\infty g^i\big[x(s), u_1(s), u_2(s), \ldots, u_n(s)\big] \exp\big[-r(s-t)\big] ds, \quad \text{for } i \in N,$$

(4.58)

subject to the state dynamics

$$\dot{x}(s) = f\big[x(s), u_1(s), u_2(s), \ldots, u_n(s)\big], \quad x(t) = x.$$
(4.59)

The infinite-horizon autonomous game $\Gamma(x)$ is independent of the choice of t and dependent only upon the state at the starting time, that is, x.

Now consider the case when the economic agents agree to act cooperatively. Let $\Gamma_c(\tau, x_\tau)$ denote a cooperative game in which agent i's payoff is (4.56) and the state dynamics is (4.57). The agents agree to act according to an agreed-upon optimality principle. As noted before, group optimality is an essential factor in cooperation and we let the agreed-upon optimality principle be as follows.

Principle PII It is an optimality principle that entails (i) group optimality and individual rationality, and (ii) the distribution of the total cooperative payoff according to an imputation that equals $\xi^{(\upsilon)}(\upsilon, x_\upsilon^*)$ for $\upsilon \in [\tau, \infty)$ over the game duration. Moreover, the function $\xi^{(\upsilon)i}(\upsilon, x_\upsilon^*) \in \xi^{(\upsilon)}(\upsilon, x_\upsilon^*)$, for $i \in N$, is continuously differentiable in υ and x_υ^*.

The solution $P(\tau, x_\tau)$ of the cooperative game $\Gamma_c(\tau, x_\tau)$ under optimality Principle PII includes the following.

(i) A set of group optimal cooperative strategies

$$u^{(\tau)*}(s) = \big[u_1^{(\tau)*}(s), u_2^{(\tau)*}(s), \ldots, u_n^{(\tau)*}(s)\big], \quad \text{for } s \in [\tau, \infty).$$

(ii) An imputation vector $\xi^{(\tau)}(\tau, x_\tau) = [\xi^{(\tau)1}(\tau, x_\tau), \xi^{(\tau)2}(\tau, x_\tau), \ldots, \xi^{(\tau)n}(\tau, x_\tau)]$ to allocate the cooperative payoff to the agents.

(iii) A payoff distribution procedure $B^\tau(s) = [B_1^\tau(s), B_2^\tau(s), \ldots, B_n^\tau(s)]$, for $s \in [\tau, \infty)$, where $B_i^\tau(s)$ is the instantaneous payment for agent i at time s. In particular,

$$\xi^{(\tau)i}(\tau, x_\tau) = \int_\tau^\infty B_i^\tau(s) \exp\big[-r(s-\tau)\big] ds, \quad \text{for } i \in N.$$
(4.60)

In the following sections, we explicitly characterize the solution $P(\tau, x_\tau)$ of the cooperative game $\Gamma_c(\tau, x_\tau)$ under the optimality principle.

4.6.1 Group Optimal Cooperative Strategies

To ensure group rationality, the agents maximize the sum of their payoffs, the agents solve the problem

$$\max_{u_1,u_2,\ldots,u_n} \left\{ \int_\tau^\infty \sum_{j=1}^n g^j \left[x(s), u_1(s), u_2(s), \ldots, u_n(s) \right] \exp\left[-r(s-\tau) \right] ds \right\}, \quad (4.61)$$

subject to (4.57).

Following Theorem 3.2 in Chap. 3, we note that a set of controls $\{\psi_i^*(x),$ for $i \in N\}$ provides a solution to the optimal control problem in (4.61) if there exists a continuously differentiable function $W(x) : R^m \to R$ satisfying the infinite-horizon Bellman equation

$$rW(x) = \max_{u_1,u_2,\ldots,u_n} \left\{ \sum_{j=1}^2 g^j [x, u_1, u_2, \ldots, u_n] + W_x f [x, u_1, u_2, \ldots, u_n] \right\}. \quad (4.62)$$

According to optimality Principle PII the agents will adopt the cooperative control $\{\psi_1^*(x),$ for $i \in N\}$ characterized in (4.62). Note that these controls are functions of the current state x only. Substituting this set of controls into the state dynamics yields the optimal (cooperative) trajectory as

$$\dot{x}(s) = f\left[x(s), \psi_1^*(x(s)), \psi_2^*(x(s)), \ldots, \psi_n^*(x(s)) \right], \quad x(\tau) = x_\tau. \quad (4.63)$$

Let $x^*(s)$ denote the solution to (4.63). The optimal trajectory $\{x^*(s)\}_{s=\tau}^\infty$ can be expressed as

$$x^*(s) = x_\tau + \int_\tau^s f\left[x^*(\upsilon), \psi_1^*(x^*(\upsilon)), \psi_2^*(x^*(\upsilon)), \ldots, \psi_n^*(x^*(\upsilon)) \right] d\upsilon.$$

For notational convenience, we use the terms $x^*(s)$ and x_s^* interchangeably.

The cooperative control for the game can be expressed more precisely as

$$\{\psi_i^*(x_s^*), \text{ for } i \in N \text{ and } s \in [\tau, \infty)\},$$

which are functions of the current state x_s^* only. The term

$$W(x_\tau^*) = \int_\tau^\infty \sum_{j=1}^n g^j \left[x^*(s), \psi_1^*(x^*(s)), \psi_2^*(x^*(s)), \ldots, \psi_n^*(x^*(s)) \right]$$

$$\times \exp\left[-r(s-\tau) \right] ds,$$

is the maximized cooperative payoff at current time τ, given that the state is x_τ^*.

Moreover, one can easily verify that the joint payoff maximizing controls for the cooperative game $\Gamma_c(\tau, x_\tau)$ over the time interval $[t, \infty)$ is identical to the joint payoff maximizing controls for the cooperative game $\Gamma_c(t, x_t^*)$ over the same time interval.

4.6.2 Consistent Imputation and Payoff Distribution Procedure

Let $P(\tau, x_\tau)$ denote the solution to the cooperative game $\Gamma_c(\tau, x_\tau)$ under the agreed-upon optimality Principle PII. According to $P(\tau, x_\tau)$, the economic agents would use the Payoff Distribution Procedure $\{B^\tau(s)\}_{s=\tau}^{\infty}$ to bring about an imputation to agent i as

$$\xi^{(\tau)i}(\tau, x_\tau) = \int_\tau^\infty B_i^\tau(s)\exp\big[-r(s-\tau)\big]\,ds, \quad \text{for } i \in N. \tag{4.64}$$

We define

$$\xi^{(\tau)i}(t, x_t^*) = \int_t^\infty B_i^\tau(s)\exp\big[-r(s-\tau)\big]\,ds, \quad \text{for } i \in N, \tag{4.65}$$

where $t > \tau$ and $x_t^* \in \{x^*(s)\}_{s=\tau}^{\infty}$.

According to $P(\tau, x_\tau)$, agent i is supposed to receive a payoff $\xi^{(\tau)i}(t, x_t^*)$ over the remaining time interval $[t, \infty)$.

Consider the case when the game has proceeded to time t and the state variable became x_t^*. Then one has a cooperative game $\Gamma_c(t, x_t^*)$ that starts at time t with initial state x_t^*. According to the solution $P(t, x_t^*)$, an imputation

$$\xi^{(t)i}(t, x_t^*) = \int_t^\infty B_i^t(s)\exp\big[-r(s-t)\big]\,ds,$$

will be allotted to agent i, for $i \in N$.

However, according to $P(\tau, x_\tau)$, the imputation (in the present value viewed at time τ) to agent i over the period $[t, \infty)$ is

$$\xi^{(\tau)i}(t, x_t^*) = \int_t^\infty B_i^\tau(s)\exp\big[-r(s-\tau)\big]\,ds, \quad \text{for } i \in N. \tag{4.66}$$

For the imputation from $P(\tau, x_\tau)$ to be consistent with those from $P(t, x_t^*)$, it is essential that

$$\exp\big[r(t-\tau)\big]\xi^{(\tau)i}(t, x_t^*) = \xi^{(t)i}(t, x_t^*) \in P(t, x_t^*), \quad \text{for } t \in (\tau, \infty).$$

In addition, at time τ when the initial state is x_τ, according to the solution $P(\tau, x_\tau)$ generated by optimality Principle PII, the payoff distribution procedure is

$$B^\tau(s) = \big[B_1^\tau(s), B_2^\tau(s), \ldots, B_n^\tau(s)\big], \quad \text{for } s \in [\tau, \infty).$$

When the game has proceeded to time t and the state variable has become x_t^*, according to the solution $P(t, x_t^*)$ generated by optimality Principle PII, the payoff distribution procedure

$$B^t(s) = \big[B_1^t(s), B_2^t(s), \ldots, B_n^t(s)\big], \quad \text{for } s \in [t, \infty),$$

will be adopted.

For the continuation of the payoff distribution procedure $B^\tau(s)$ under $P(\tau, x_\tau)$ to be consistent with $B^t(s) \in P(x_t^c, T - t)$, it is required that

$$B^{t_0}(s) = B^t(s), \quad \text{for } s \in [t, \infty) \text{ and } t \in [\tau, \infty).$$

Definition 4.3 The imputation and payoff distribution procedure $\{\xi^{(\tau)}(\tau, x_\tau)$ and $B^\tau(s)$ for $s \in [\tau, \infty)\} \in P(\tau, x_\tau)$ are time consistent if

(i)

$$\exp\big[r(t - \tau)\big]\xi^{(\tau)i}\big(t, x_t^*\big)$$

$$\equiv \exp\big[r(t - \tau)\big] \int_t^\infty B_i^\tau(s) \exp\big[-r(s - \tau)\big] ds$$

$$= \xi^{(t)i}\big(t, x_t^*\big) \in P\big(t, x_t^*\big), \quad \text{for } t \in (\tau, \infty) \text{ and } i \in N; \qquad (4.67)$$

and

(ii) the payoff distribution procedure $B^\tau(s) = [B_1^\tau(s), B_2^\tau(s), \ldots, B_n^\tau(s)]$ for $s \in [t, \infty)$ is identical to $B^t(s) = [B_1^t(s), B_2^t(s), \ldots, B_n^t(s)] \in P(t, x_t^*)$.

Definition 4.3 is the infinite-horizon counterpart of Definition 4.2 in characterizing the time consistent imputation and payoff distribution procedure.

4.6.3 Derivation of Consistent Payoff Distribution Procedure

A payoff distribution procedure leading to the time consistent imputation has to satisfy Definition 4.3. Invoking Definition 4.3, we have $B_i^\tau(s) = B_i^t(s) = B_i(s)$, for $s \in [\tau, \infty)$, $t \in [\tau, \infty)$, and $i \in N$.

Therefore, along the cooperative trajectory $\{x^*(t)\}_{t \geq t_0}$,

$$\xi^{(\tau)i}\big(\tau, x_\tau^*\big) = \int_\tau^\infty B_i(s) \exp\big[-r(s - \tau)\big] ds, \quad \text{for } i \in N,$$

$$\xi^{(\upsilon)i}\big(\upsilon, x_\upsilon^*\big) = \int_\upsilon^\infty B_i(s) \exp\big[-r(s - \upsilon)\big] ds, \quad \text{for } i \in N, \quad \text{and} \qquad (4.68)$$

$$\xi^{(t)i}\big(t, x_t^*\big) = \int_t^\infty B_i(s) \exp\big[-r(s - t)\big] ds, \quad \text{for } i \in N \text{ and } t \geq \upsilon \geq \tau.$$

Moreover, for $i \in N$ and $t \in [\tau, \infty)$, we define the term

$$\xi^{(\upsilon)i}\big(t, x_t^*\big) = \left\{ \left(\int_t^\infty B_i(s) \exp\big[-r(s - \upsilon)\big] ds \right) \bigg| x(t) = x_t^* \right\} \qquad (4.69)$$

to denote the present value of agent i's cooperative payoff over the time interval $[t, \infty)$, given that the state is x_t^* at time $t \in [\upsilon, \infty)$, under the solution $P(\upsilon, x_\upsilon^*)$.

Invoking (4.69) and (4.69), one can readily verify that $\exp[r(t-\tau)]\xi^{(\tau)i}(t,x_t^*) = \xi^{(t)i}(t,x_t^*)$, for $i \in N$, $\tau \in [t_0, T]$, and $t \in [\tau, T]$.

The next task is to derive $B_i(s)$, for $s \in [\tau, \infty)$ and $t \in [\tau, \infty)$ so that (4.69) can be realized. Consider again the following condition.

Condition 4.2 For $i \in N$, $t \geq v$, and $v \in [\tau, T]$, the term $\xi^{(v)i}(t, x_t^*)$ is a function that is continuously differentiable in t and x_t^*.

Lemma 4.1 *If Condition 4.2 is satisfied, a PDP with instantaneous payments at time s equaling*

$$B_i(s) = -\left[\xi_t^{(s)i}(t, x_t^*)\big|_{t=s}\right] - \xi_{x_s^*}^{(s)i}(s, x_s^*)f\left[x_s^*, \psi_1^*(x_s^*), \psi_2^*(x_s^*), \ldots, \psi_n^*(x_s^*)\right],$$

$$(4.70)$$

for $i \in N$ and $s \in [v, \infty)$, yields imputation $\xi^{(v)i}(v, x_v^c)$, for $v \in [\tau, \infty)$, which satisfies (4.69).

Proof Note that along the cooperative trajectory $\{x^*(t)\}_{t \geq \tau}$

$$\xi^{(v)i}(t, x_t^*) = \int_t^\infty B_i(s)\exp\left[-r(s-v)\right]ds = \exp\left[-r(t-v)\right]\xi^{(t)i}(t, x_t^*),$$

$$(4.71)$$

for $i \in N$ and $t \in [v, \infty)$.

For $\Delta t \to 0$, (4.69) can be expressed as

$$\xi^{(v)i}(\tau, x_\tau^*) = \int_v^\infty B_i(s)\exp\left[-r(s-v)\right]ds$$

$$= \int_v^{v+\Delta t} B_i(s)\exp\left[-r(s-v)\right]ds + \xi^{(v)i}(v+\Delta t, x_v^* + \Delta x_v^*),$$

$$(4.72)$$

where

$$\Delta x_v^* = f\left[x_v^*, \psi_1^*(x_v^*), \psi_2^*(x_v^*), \ldots, \psi_n^*(x_v^*)\right]\Delta t + o(\Delta t), \quad \text{and}$$
$$o(\Delta t)/\Delta t \to 0 \quad \text{as } \Delta t \to 0.$$

Replacing the term $x_v^* + \Delta x_v^*$ with $x_{v+\Delta t}^*$ and rearranging (4.72) yields

$$\int_v^{v+\Delta t} B_i(s)\exp\left[-r(s-v)\right]ds$$

$$= \xi^{(v)i}(v, x_v^*) - \xi^{(v)i}(v+\Delta t, x_{v+\Delta t}^*), \quad \text{for all } v \in [\tau, \infty) \text{ and } i \in N.$$

$$(4.73)$$

Consider the following condition concerning $\xi^{(\upsilon)i}(t, x_t^*)$, for $\upsilon \in [\tau, \infty)$ and $t \in [\upsilon, \infty)$.

With Condition 4.2 holding and $\Delta t \to 0$, (4.73) can be expressed as

$$B_i(\upsilon)\Delta t = -\left[\xi_t^{(\upsilon)i}(t, x_t^*)\big|_{t=\tau}\right]\Delta t$$
$$- \xi_{x_\upsilon^*}^{(\upsilon)i}(\upsilon, x_\upsilon^*) f\left[x_\upsilon^*, \psi_1^*(x_\upsilon^*), \psi_2^*(x_\upsilon^*), \ldots, \psi_n^*(x_\upsilon^*)\right]\Delta t - o(\Delta t).$$

$$(4.74)$$

Dividing (4.74) throughout by Δt, with $\Delta t \to 0$ yields (4.70). Thus the payoff distribution procedure in $B_i(\upsilon)$ in (4.70) will lead to the realization of the imputations that satisfy (4.70). $\qquad \square$

Since the payoff distribution procedure in $B_i(\tau)$ in (4.70) leads to the realization of (4.69), it will yield time consistent imputations satisfying Definition 4.3.

A more succinct form of Lemma 4.1 can be derived as follows.

If Condition 4.2 is satisfied, a PDP with instantaneous payments at time s equaling

$$B_i(s) = r\xi^{(s)i}(s, x_s^*) - \xi_{x_s^*}^{(s)i}(s, x_s^*) f\left[x_s^*, \psi_1^*(x_s^*), \psi_2^*(x_s^*), \ldots, \psi_n^*(x_s^*)\right], \quad (4.75)$$

for $i \in N$ and $s \in [\upsilon, \infty)$, yields imputation $\xi^{(\upsilon)i}(\upsilon, x_\upsilon^c)$, for $\upsilon \in [\tau, \infty)$, which satisfies (4.69).

To demonstrate that (4.75) is an alternative form for (4.70) in Lemma 4.1, we first define

$$\hat{\xi}^i(x_\upsilon^*) = \left\{\int_\upsilon^\infty B_i(s)\exp[-r(s - \upsilon)]\,ds \,\bigg|\, x(\upsilon) = x_\upsilon^*\right\} = \xi^{(\upsilon)i}(\tau, x_\upsilon^*), \quad \text{and}$$

$$\hat{\xi}^i(x_t^*) = \left\{\int_t^\infty B_i(s)\exp[-r(s - t)]\,ds \,\bigg|\, x(t) = x_t^*\right\} = \xi^{(t)i}(t, x_t^*),$$

for $i \in N$, $\upsilon \in [\tau, \infty)$, and $t \in [\upsilon, \infty)$ along the optimal cooperative trajectory $\{x_s^*\}_{s=\tau}^\infty$.

We then have

$$\xi^{(\upsilon)i}(t, x_t^*) = \exp[-r(t - \upsilon)]\hat{\xi}^i(x_t^*).$$

Differentiating $\xi^{(\upsilon)i}(t, x_t^*)$ with respect to t yields

$$\left[\xi_t^{(\upsilon)i}(t, x_t^*)\big|_{t=\upsilon}\right] = -r\exp[-r(t - \upsilon)]\hat{\xi}^i(x_t^*) = -r\xi^{(\upsilon)i}(t, x_t^*).$$

At $t = \upsilon$, $\xi^{(\upsilon)i}(t, x_t^*) = \xi^{(\upsilon)i}(\upsilon, x_\upsilon^*)$, therefore,

$$\left[\xi_t^{(\upsilon)i}(t, x_t^*)\big|_{t=\upsilon}\right] = r\xi^{(\upsilon)i}(t, x_t^*) = r\xi^{(\upsilon)i}(\upsilon, x_\upsilon^*). \quad (4.76)$$

Substituting (4.76) into (4.70) yields (4.75).

Using (4.75), a time (optimal-trajectory-subgame) consistent solution in an infinite-horizon framework is characterized in the next section.

4.6.4 Time (Optimal-Trajectory-Subgame) Consistent Solution

A theorem characterizing a time (optimal-trajectory-subgame) consistent solution $P(\tau, x_\tau)$ for the cooperative game $\Gamma_c(\tau, x_\tau)$ under optimality Principle PII is presented below.

Theorem 4.3 *For the cooperative game $\Gamma_c(\tau, x_\tau)$ with optimality Principle PII the solution $P(\tau, x_\tau) = \{u(s) \text{ and } B(s) \text{ for } s \in [\tau, \infty) \text{ and } \xi^{(\tau)}(\tau, x_\tau)\}$ in which*

(i) *$u(s)$ for $s \in [\tau, \infty)$ is the set of group optimal strategies $\psi^*(x_s^*)$ for the game $\Gamma_c(\tau, x_\tau)$, and*
(ii) *the imputation distribution procedure $B(s) = \{B_1(s), B_2(s), \ldots, B_n(s)\}$ for $s \in [\tau, \infty)$, where*

$$B_i(s) = r\xi^{(s)i}\left(s, x_s^*\right) - \xi_{x_s^*}^{(s)i}\left(s, x_s^*\right) f\left[x_s^*, \psi_1^*\left(x_s^*\right), \psi_2^*\left(x_s^*\right), \ldots, \psi_n^*\left(x_s^*\right)\right],$$

(4.77)

for $i \in N$, and

$$\xi^{(s)}\left(s, x_s^*\right) = \left[\xi^{(s)1}\left(s, x_s^*\right), \xi^{(s)2}\left(s, x_s^*\right), \ldots, \xi^{(s)n}\left(s, x_s^*\right)\right] \in P\left(s, x_s^*\right)$$

is the imputation at time $s \in [\tau, \infty)$ with the state being $x_s^* \in \{x^*(t)\}_{t \geq \tau}$ under optimality Principle PII and it is time (optimal-trajectory-subgame) consistent.

Proof Following the algorithm that specifies $P(\tau, x_\tau)$ as the solution to the game $\Gamma_c(\tau, x_\tau)$ one can readily obtain the solution of the cooperative game $\Gamma_c(\upsilon, x_\upsilon^*)$, for $\upsilon > \tau$, as $P(\upsilon, x_\upsilon^*) = \{u(s) \text{ and } B(s) \text{ for } s \in [\upsilon, \infty) \text{ and } \xi^{(\upsilon)}(\upsilon, x_\upsilon^*)\}$ in which

(i) $u(s)$ for $s \in [\upsilon, \infty)$ is the set of group optimal strategies $\psi^*(x_s^*)$ for the game $\Gamma_c(\upsilon, x_\upsilon^*)$, and
(ii) $B(s) = \{B_1(s), B_2(s), \ldots, B_n(s)\}$ for $s \in [\upsilon, \infty)$, where

$$B_i(s) = -\left[\xi_t^{(s)i}\left(t, x_t^*\right)\big|_{t=s}\right] - \left[\xi_{x_s^*}^{(s)i}\left(s, x_s^*\right)\right]$$

$$\times f\left[s, x_s^*, \psi_1^*\left(\tau, x_s^*\right), \psi_2^*\left(\tau, x_s^*\right), \ldots, \psi_n^*\left(\tau, x_s^*\right)\right],$$

(4.78)

for $i \in N$, and

$$\xi_{x_\tau^c}^{(s)}\left(s, x_s^*\right) = \left[\xi_{x_\tau^c}^{(s)1}\left(s, x_s^*\right), \xi_{x_\tau^c}^{(s)2}\left(s, x_s^*\right), \ldots, \xi_{x_\tau^c}^{(s)n}\left(s, x_s^*\right)\right] \in P\left(s, x_s^*\right)$$

is the imputation at time $s \in [\upsilon, \infty)$ with the state being $x_s^* \in \{x^*(t)\}_{t \geq \upsilon}$.

Using the characterization of optimal control strategies in (4.62), one can show that the group optimal joint payoff maximizing strategies $\psi^*(x_s^*)$ for the cooperative game $\Gamma_c(\tau, x_\tau)$ over the time interval $[\upsilon, \infty)$ is identical to the joint payoff maximizing strategies controls for the cooperative game $\Gamma_c(\upsilon, x_\upsilon^*)$ over the same time interval.

Comparing (4.77) and (4.78), one can show that the payoff distribution procedure $B(s)$ for the cooperative game $\Gamma_c(\tau, x_\tau)$ over the time interval $[\upsilon, \infty)$ is identical to the payoff distribution procedure $B(s)$ for the cooperative game $\Gamma_c(\upsilon, x_\upsilon^*)$ over the same time interval.

Invoking Lemma 4.1 and (4.75) one can show that the payoff distribution procedure $B(s) = \{B_1(s), B_2(s), \ldots, B_n(s)\}$ in (4.77) would yield

$$\xi^{(\upsilon)i}\left(\upsilon, x_\upsilon^*\right) = \left\{ \int_\upsilon^\infty B_i(s) \exp[-r(s - \upsilon)] \, ds \right\} \in P\left(\upsilon, x_\upsilon^*\right), \quad \text{for } i \in N,$$

and $\upsilon \in [\tau, \infty)$.

Hence

$$\exp[r(\upsilon - \tau)]\xi^{(\tau)i}\left(\upsilon, x_\upsilon^*\right) \equiv \exp[r(\upsilon - \tau)] \int_\upsilon^\infty B_i(s) \exp[-r(s - \tau)] \, ds$$

$$= \xi^{(\upsilon)i}\left(\upsilon, x_\upsilon^*\right) P\left(\upsilon, x_\upsilon^*\right), \quad \text{for } i \in N \text{ and } \upsilon \in [\tau, \infty).$$

In summary, the continuation of the solution $P(\tau, x_\tau)$ over the time interval $[\upsilon, \infty)$ is consistent with the solution $P(\upsilon, x_\upsilon^*)$ of the game $\Gamma_c(\upsilon, x_\upsilon^*)$ under optimality Principle PII. Thus the solution $P(\tau, x_\tau)$ in Theorem 4.3 is indeed time (optimal-subgame-consistent) consistent.   □

With agents using the cooperative strategies $\{\psi_i^*(x_\upsilon^*)$, for $i \in N$ and $\upsilon \in [\tau, \infty)\}$, the instantaneous receipt of agent i at time instant υ is

$$\zeta_i(\upsilon) = g^i\left[x_\upsilon^*, \psi_1^*(x_\upsilon^*), \psi_2^*(x_\upsilon^*), \ldots, \psi_n^*(x_\upsilon^*)\right], \quad \text{for } i \in N. \tag{4.79}$$

According to Theorem 4.3, the instantaneous payment that agent i should receive under the agreed-upon optimality principle is $B_i(\upsilon)$, as stated in (4.77). Hence an instantaneous transfer payment

$$\chi^i(\upsilon) = B_i(\upsilon) - \zeta_i(\upsilon) \tag{4.80}$$

has to be given to agent i at time υ, for $i \in N$.

4.7 Infinite-Horizon Resource Extraction Optimization

Consider an infinite-horizon version of the cooperative fishery game in Sect. 4.5. At initial time τ, the payoff function of nations 1 and 2 are, respectively,

$$\int_\tau^\infty \left[u_1(s)^{1/2} - \frac{c_1}{x(s)^{1/2}} u_1(s) \right] \exp[-r(t - \tau)] \, ds,$$

and

$$\int_\tau^\infty \left[u_2(s)^{1/2} - \frac{c_2}{x(s)^{1/2}} u_2(s) \right] \exp[-r(t - \tau)] \, ds. \tag{4.81}$$

The resource stock $x(s) \in X \subset R$ follows the dynamics

$$\dot{x}(s) = ax(s)^{1/2} - bx(s) - u_1(s) - u_2(s), \quad x(\tau) = x_\tau \in X. \quad (4.82)$$

Invoking Theorem 2.4 in Chap. 2, a noncooperative feedback Nash equilibrium solution of the game in (4.81) and (4.82) can be characterized as

$$r\hat{V}^i(x) = \max_{u_i}\left\{u_i^{1/2} - \frac{c_i}{x^{1/2}}u_i + \hat{V}_x^i(x)\left[ax^{1/2} - bx - u_i - \phi_j^*(x)\right]\right\}, \quad (4.83)$$

for $i, j \in \{1, 2\}$ and $i \neq j$.

Performing the indicated maximization in (4.83) yields

$$\phi_i^*(x) = \frac{x}{4[c_i + \hat{V}_x^i(x)x^{1/2}]^2}, \quad \text{for } i \in \{1, 2\}.$$

Substituting $\phi_1^*(x)$ and $\phi_2^*(x)$ above into (4.83) and upon solving (4.83) one obtains the value function of nation $i \in \{1, 2\}$ as

$$\hat{V}^i(t, x) = \left[A_i x^{1/2} + C_i\right], \quad (4.84)$$

where, for $i, j \in \{1, 2\}$ and $i \neq j$, A_i, C_i, A_j, and C_j satisfy

$$\left[r + \frac{b}{2}\right]A_i - \frac{1}{2[c_i + A_i/2]} + \frac{c_i}{4[c_i + A_i/2]^2}$$

$$+ \frac{A_i}{8[c_i + A_i/2]^2} + \frac{A_i}{8[c_j + A_j/2]^2} = 0, \quad \text{and}$$

$$C_i = \frac{a}{2}A_i.$$

The game equilibrium strategies can be obtained as

$$\phi_1^*(x) = \frac{x}{4[c_1 + A_1/2]^2}, \quad \text{and} \quad \phi_2^*(x) = \frac{x}{4[c_2 + A_2/2]^2}. \quad (4.85)$$

Consider the case when these two nations agree to act according to an agreed-upon optimality principle that entails (i) group optimality and (ii) the distribution of the cooperative payoff according to the imputation that equally divides the excess of the total cooperative payoff over the sum of individual noncooperative payoffs.

To maximize their joint payoff for group optimality, the nations have to solve the control problem of maximizing

$$\int_\tau^\infty \left(\left[u_1(s)^{1/2} - \frac{c_1}{x(s)^{1/2}}u_1(s)\right] + \left[u_2(s)^{1/2} - \frac{c_2}{x(s)^{1/2}}u_2(s)\right]\right)$$

$$\times \exp\left[-r(t - \tau)\right]ds, \quad (4.86)$$

subject to (4.82).

Invoking Theorem 3.2 in Chap. 3, we obtain

$$
rW(x) = \max_{u_1, u_2} \left\{ \left(\left[u_1^{1/2} - \frac{c_1}{x^{1/2}} u_1 \right] + \left[u_2^{1/2} - \frac{c_2}{x^{1/2}} u_2 \right] \right) \right.
$$
$$
\left. + W_x(x) \left[ax^{1/2} - bx - u_1 - u_2 \right] \right\}.
$$

Following similar procedures in previous analyses, one can obtain

$$
W(x) = \left[A x^{1/2} + C \right],
$$

where

$$
\left[r + \frac{b}{2} \right] A - \frac{1}{2[c_1 + A/2]} - \frac{1}{2[c_2 + A/2]} + \frac{c_1}{4[c_1 + A/2]^2} + \frac{c_2}{4[c_2 + A/2]^2}
$$
$$
+ \frac{A}{8[c_1 + A/2]^2} + \frac{A}{8[c_2 + A/2]^2} = 0, \quad \text{and}
$$
$$
C = \frac{a}{2r} A.
$$

The optimal cooperative controls can then be obtained as

$$
\psi_1^*(x) = \frac{x}{4[c_1 + A/2]^2} \quad \text{and} \quad \psi_2^*(x) = \frac{x}{4[c_2 + A/2]^2}. \tag{4.87}
$$

Substituting these control strategies into (4.82) yields the dynamics of the state trajectory under cooperation

$$
\dot{x}(s) = ax(s)^{1/2} - bx(s) - \frac{x(s)}{4[c_1 + A/2]^2} - \frac{x(s)}{4[c_2 + A/2]^2}, \quad x(\tau) = x_\tau. \tag{4.88}
$$

Solving (4.88) yields the optimal cooperative state trajectory $\{x^*(s)\}_{\tau=t_0}^{\infty}$ for the cooperative game in (4.81) and (4.82) as

$$
x^*(s) = \left[\frac{a}{2H} + \left((x_\tau)^{\frac{1}{2}} - \frac{a}{2H} \right) \exp\left[-H(s - \tau) \right] \right]^2, \tag{4.89}
$$

where

$$
H = - \left[\frac{b}{2} + \frac{1}{8[c_1 + A/2]^2} + \frac{1}{8[c_2 + A/2]^2} \right].
$$

According to the agreed-upon optimality principle these nations will distribute the cooperative payoff according to the imputation that equally divides the excess of the total cooperative payoff over the sum of individual noncooperative payoffs. Hence the imputation $\xi(v, x_v^*) = [\xi^1(v, x_v^*), \xi^2(v, x_v^*)]$ has to satisfy the following:

Condition 4.3

$$\xi^i\left(\upsilon, x_\upsilon^*\right) = \hat{V}^i\left(x_\upsilon^*\right) + \frac{1}{2}\left[W\left(x_\upsilon^*\right) - \sum_{j=1}^{2}\hat{V}^j\left(x_\upsilon^*\right)\right], \tag{4.90}$$

for $i \in \{1, 2\}$ and $\upsilon \in [\tau, \infty)$.

Applying Theorem 4.3 a time (optimal-trajectory-subgame) consistent solution for the cooperative game $\Gamma_c(\tau, x_\tau)$ can be obtained as $P(\tau, x_\tau) = \{u(s) \text{ and } B(s) \text{ for } s \in [\tau, \infty) \text{ and } \xi^{(\tau)}(\tau, x_\tau)\}$ in which

(i) $u(s)$ for $s \in [\tau, \infty)$ is the set of group optimal strategies

$$\psi_1^*\left(x_s^*\right) = \frac{x_s^*}{4[c_1 + A/2]^2} \quad \text{and} \quad \psi_2^*\left(x_s^*\right) = \frac{x_s^*}{4[c_2 + A/2]^2};$$

and
(ii) $B(s) = \{B_1(s), B_2(s), \ldots, B_n(s)\}$ for $s \in [\tau, \infty)$ where

$$B_i(s) = \frac{1}{2}\left\{r\left[A_i\left(x_s^*\right)^{1/2} + C_i\right] + r\left[A\left(x_s^*\right)^{1/2} + C\right] - r\left[A_j\left(x_s^*\right)^{1/2} + C_j\right]\right\}$$

$$- \frac{1}{4}\left\{A_i\left(x_s^*\right)^{-1/2} + A\left(x_s^*\right)^{-1/2} - A_j\left(x_s^*\right)^{-1/2}\right\}$$

$$\times \left[a\left(x_s^*\right)^{1/2} - bx_s^* - \frac{x_s^*}{4[c_1 + A/2]^2} - \frac{x_s^*}{4[c_2 + A/2]^2}\right], \tag{4.91}$$

for $i, j \in \{1, 2\}$ and $i \neq j$.

With agents using the cooperative strategies $\{\psi_i^*(x_s^*), i \in \{1, 2\}\}$ along the cooperative trajectory, the instantaneous receipt of agent i at time instant υ becomes

$$\zeta_i(\upsilon) = \frac{(x_\upsilon^*)^{1/2}}{2[c_i + A/2]} - \frac{c_i(x_\upsilon^*)^{1/2}}{4[c_i + A/2]^2}. \tag{4.92}$$

According to (4.91), the instantaneous payment that agent i should receive under the agreed-upon optimality principle is $B_i(\upsilon)$. Hence an instantaneous transfer payment

$$\chi^i(\upsilon) = B_i(\upsilon) - \zeta_i(\upsilon), \tag{4.93}$$

has to be given to agent i at time $\upsilon \in [\tau, \infty)$, for $i \in \{1, 2\}$.

4.8   Exercises

4.1 Consider the case of two nations harvesting fish in common waters. The growth rate of the fish biomass is characterized by the differential equation

$$\dot{x}(s) = 3x(s)^{1/2} - 0.5x(s) - u_1(s) - u_2(s), \quad x(0) = 50,$$

where $u_i \in U_i$ is the (nonnegative) amount of fish harvested by nation i, for $i \in \{1, 2\}$. The horizon of the game is $[0, 4]$.

The harvesting cost for nation $i \in \{1, 2\}$ depends on the quantity of resource extracted $u_i(s)$ and the resource stock size $x(s)$. In particular, nation 1's extraction cost is $2u_1(s)x(s)^{-1/2}$ and nation 2's is $u_2(s)x(s)^{-1/2}$. The fish harvested by nation i at time s will generate a net benefit of the amount $[u_i(s)]^{1/2}$. At terminal time 4, nations 1 and 2 will receive termination bonuses $7.5x(4)^{1/2}$ and $5x(4)^{1/2}$ while the interest rate is 0.05.

At time 0 the payoffs of nation 1 and nation 2 are, respectively,

$$\int_0^4 \left[[u_1(s)]^{1/2} - \frac{4}{x(s)^{1/2}} u_i(s) \right] \exp(-0.05s)\, ds + \exp[-r(4)]7.5x(4)^{\frac{1}{2}}, \quad \text{and}$$

$$\int_0^4 \left[[u_2(s)]^{1/2} - \frac{3}{x(s)^{1/2}} u_i(s) \right] \exp(-0.05s)\, ds + \exp[-r(4)]5x(4)^{\frac{1}{2}}.$$

Obtain a feedback Nash equilibrium solution for this transnational market activity.

4.2 If these nations agree to cooperate and maximize their joint payoff, compute the optimal cooperative strategies and optimal stock path of the fish biomass.

4.3 Furthermore, if these nations agree to share the excess of their gain equally along the optimal trajectory, obtain a time (optimal-trajectory-subgame) consistent solution.

4.4 Consider the case when the game horizon in exercise 1 is extended to infinity.

(i) Obtain a feedback Nash equilibrium solution for this transnational market activity.

(ii) If these nations agree to cooperate and maximize their joint payoff, compute the optimal cooperative strategies and optimal stock path of the fish biomass.

(iii) If these nations agree to share the excess of their gain equally along the optimal trajectory, obtain a time (optimal-trajectory-subgame) consistent solution.

Chapter 5
Dynamically Stable Cost-Saving Joint Venture

In this chapter, we consider a common economic activity involving cooperative optimization—joint venture. However, it is often observed that after a certain time of cooperation some firms in a joint venture may gain sufficient skills and technology that they would do better by breaking away from the joint operation. Analysis on time (optimal-trajectory subgame) consistent joint ventures are presented in the following sections.

As markets become increasingly globalized and firms become more multinational, corporate joint ventures are likely to yield opportunities to quickly create economies of scale and critical mass, incorporate new skills and technology, and facilitate rational resource sharing (see Bleeke and Ernst 1993). With joint ventures becoming a powerful force shaping global corporate strategy, partnerships between firms have significantly increased. Despite their purported benefits, however, joint ventures are highly unstable and have a consistently high rate of failure (Blodgett 1992; Parkhe 1993). In addition, other adverse effects, such as uncompensated transfers of technology, operational difficulties, disagreements, and anxiety over the loss of proprietary information, have been found (Hamel et al. 1989 and Gomes-Casseres 1987). D'Aspremont and Jacquemin (1988), Kamien et al. (1992), and Suzumura (1992) have studied cooperative R&D with spillovers in joint ventures under a static framework. Cellini and Lambertini (2002, 2004) considered cooperative solutions to investment in product differentiation in a dynamic approach.

A dynamic model of a corporate joint venture resulting in cost saving is presented in Sect. 5.1. Time (optimal-trajectory-subgame) consistent solutions are derived in Sect. 5.2 and an example is given in Sect. 5.3. The derivation of a Shapley value solution to the joint venture is given in Sect. 5.4 and the Shapley value profit sharing is analyzed in Sect. 5.5. An extension of the analysis to an infinite horizon is investigated in Sect. 5.6 and an example of an infinite-horizon joint venture is provided in Sect. 5.7.

D.W.K. Yeung, L.A. Petrosyan, *Subgame Consistent Economic Optimization*,
Static & Dynamic Game Theory: Foundations & Applications,
DOI 10.1007/978-0-8176-8262-0_5, © Springer Science+Business Media, LLC 2012

5.1 A Dynamic Model of Corporate Joint Venture

A fundamental premise is that joint ventures are formed primarily so that participating firms can readily gain core skills and technology that will be difficult for them to obtain on their own (Murray and Siehl 1989). Costs reduction often is an advantage gained by the firms in the joint venture. However, after a certain time of cooperation some firms may gain sufficient managerial and technological expertise that they would do better by breaking away from the joint venture. Thus a major source of instability is the lack of *dynamical stable* or *time consistent* cooperative solutions to the joint venture. Time (optimal-trajectory subgame) consistency is a fundamental element in dynamic cooperation, and it ensures that (i) the extension of the solution policy to a later starting time along the optimal trajectory will remain optimal, and (ii) all participating firms do not have an incentive to deviate from the initial plan. The absence of a formal mechanism to time consistent cooperative solutions has precluded the rigorous analysis of the problem of corporate joint ventures. Petrosyan and Zaccour (2003) provided a time consistent solution to a class of differential games involving pollution cost reduction. Yeung and Petrosyan (2006b) presented a dynamically stable joint venture involving cooperative R&D with spillovers. Yeung (2010) provided an analysis on time consistent cost-saving joint ventures.

In this section, we present a framework of a dynamic joint venture in which there are n firms. The venture horizon is $[t_0, T]$. The state dynamics of the ith firm is characterized by the set of vector-valued differential equations.

The state dynamics of the ith firm is characterized by the set of vector-valued differential equations

$$\dot{x}^i(s) = f^i[s, x^i(s), u_i(s)], \quad x^i(t_0) = x^{i(0)}, \quad \text{for } i \in N, \tag{5.1}$$

where $x^i(s) \in X^i \subset R^{m_i+}$ denotes the state variables of firm i, $u_i \in U_i \subset R^{\ell_i+}$ is firm i's investment in technology advancement. The state of firm i includes its capital stock, level of technology, special skills, and productive resources. The objective of firm i is

$$\int_{t_0}^{T} \left\{ g^i[s, x^i(s)] - c_i^{\{i\}}[u_i(s)] \right\} \exp\left[-\int_{t_0}^{s} r(y)\, dy \right] ds$$

$$+ \exp\left[-\int_{t_0}^{T} r(y)\, dy \right] q^i(x^i(T)), \tag{5.2}$$

for $i \in N$, where $\exp[-\int_{t_0}^{t} r(y)\, dy]$ is the discount factor, $g^i[s, x^i(s)]$ the instantaneous revenue, $c_i^{\{i\}}[u_i(s)]$ represents the costs of the firm's control $u_i(s)$ when it is operating on its own, and $q^i(x^i(T))$ is the terminal payment. In particular, the firm's revenue $g^i[s, x^i]$ is affected by the state variables, like capital stock, special skills, productive resources, and technologies.

Note that since the objectives and state dynamics of the firms in a noncooperative equilibrium are independent, the market outcome is represented by an n neoclassical theory of the firm problems. Let $V^{(t_0)i}(t, x^i)$ and $\phi_i^*(t, x^i)$ denote the payoff and investment strategies of firm i, for $i \in N$, by which a firm's equilibrium is characterized (see Theorem A.1 in the Technical Appendixes) as follows.

A set of investment strategies $\phi_i^*(t, x^i)$ for firm i constitutes an optimal solution to the neoclassical theory of the firm problem, which maximizes (5.2) subject to (5.1) if there exists a continuously differentiable function $V^{(t_0)i}(t, x^i)$ defined by $[t_0, T] \times R^{m_i} \to R$ and satisfying the following Bellman equation:

$$-V_t^{(t_0)i}(t, x^i) = \max_{u_i} \left\{ \left[g(t, x^i) - c_i^{\{i\}}(u_i) \right] \exp\left[-\int_{t_0}^t r(y)\, dy \right] \right.$$

$$\left. + V_x^{(t_0)i}(t, x) f^i\left[t, x^i, u_i \right] \right\} \quad \text{and}$$

$$V^{(t_0)i}(T, x^i) = q^i(x^i) \exp\left[-\int_{t_0}^T r(y)\, dy \right].$$

Let $V^{(\tau)i}(t, x^i)$ denote the payoff function of firm i in a game with the dynamics in (5.1) and the payoff of (5.2), which starts at time τ for $\tau \in [t_0, T)$. Note that the equilibrium feedback strategies are Markovian in the sense that they depend on the current time and the current state. Invoking Remark 2.1 of Chap. 2, one can obtain

$$\exp\left[\int_{t_0}^\tau r(y)\, dy \right] V^{(t_0)i}(t, x^i) = V^{(\tau)i}(t, x^i),$$

for $\tau \in [t_0, T]$ and $i \in N$.

Consider a joint venture consisting of all these n companies. The participating firms can gain core skills and technology that would be impossible for them to obtain on their own individually. Cost-saving opportunities are created under joint venture, for instance, savings in joint R&D, administration, marketing, customer services, purchasing, financing, and economy of scales and scope. The cost of control of firm j under the joint venture becomes $c_j^N[u_j(s)]$. With the absolute joint venture cost advantage we have

$$c_j^N(u_j) \le c_j^{\{j\}}(u_j), \quad \text{for } j \in N. \tag{5.3}$$

Moreover, marginal cost advantages lead to

$$\partial c_j^N(u_j)/\partial u_j \le \partial c_j^{\{j\}}(u_j)/\partial u_j, \quad \text{for } j \in N.$$

At time t_0, the joint venture would maximize the joint venture profit

$$\int_{t_0}^T \sum_{j=1}^n \{ g^j[s, x^j(s)] - c_j^N[u_j(s)] \} \exp\left[-\int_{t_0}^s r(y)\, dy \right] ds$$

$$+ \sum_{j=1}^n \exp\left[-\int_{t_0}^T r(y)\, dy \right] q^j(x^j(T)), \tag{5.4}$$

subject to (5.1).

The model adopted for analysis concentrates on the reduction of costs within the joint venture; the profit of an outside firm is not affected by the actions of the joint venture. Such an outcome would appear in scenarios in which firms are selling different products, in which firms are making vertical integrations, or in which there exists a sizable world market. For the sake of clarity in exposition, we consider the case where $m_i = 1$, for $i \in N$.

5.2 Time (Optimal-Trajectory-Subgame) Consistent Solution in Joint Venture

We begin with the characterization of the profit of the joint venture. Let x denote $\{x^1, x^2, \ldots, x^n\}$. Invoking Bellman's technique of dynamic programming in Sect. A.1 of the Technical Appendixes, the solution to the problem in (5.3) and (5.4) can be characterized as follows.

Corollary 5.1 *A set of controls $\{\psi_i^*(t, x), \text{for } i \in N \text{ and } t \in [t_0, T]\}$ provides an optimal solution to the control problem in (5.3) and (5.4) if there exists a continuously differentiable function $W^{(t_0)}(t, x) : [t_0, T] \times R^n \to R$ satisfying the following Bellman equation*:

$$
-W_t^{(t_0)}(t, x) = \max_{u_1, u_2, \ldots, u_n} \left\{ \sum_{j=1}^{n} [g^j(t, x^j) - c_j^N(u_j)] \exp\left[-\int_{t_0}^{t} r(y)\,dy\right] \right.
$$
$$
\left. + \sum_{j=1}^{n} W_{x^j}^{(t_0)}(t, x) f^j(t, x^j, u_j) \right\},
$$
(5.5)
$$
W^{(t_0)}(T, x) = \exp\left[-\int_{t_0}^{T} r(y)\,dy\right] \sum_{j=1}^{n} q^j(x^j).
$$

Hence the firms will adopt the cooperative control $\{\psi_i^*(t, x), \text{ for } i \in N \text{ and } t \in [t_0, T]\}$ to obtain the maximized level of joint profit. In a cooperative framework, the issue of the nonuniqueness of the optimal controls can be resolved by the agreement between the firms on a particular set of controls. Substituting this set of controls into (5.1) yields the dynamics of technology advancement under cooperation as

$$
\dot{x}^i(s) = f^i[s, x^i(s), \psi_i^*(s, x(s))], \quad x^i(t_0) = x_0^i, \text{ for } i \in N.
$$
(5.6)

Let $x^*(t) = \{x^{1*}(t), x^{2*}(t), \ldots, x^{n*}(t)\}$ denote the solution to (5.6). The optimal trajectory $\{x^*(t)\}_{t=t_0}^{T}$ can be expressed as

$$
x^{i*}(t) = x_0^i + \int_{t_0}^{t} f^i[s, x^{i*}(s), \psi_i^*(s, x^*(s))]\,ds, \quad \text{for } i \in N.
$$
(5.7)

For notational convenience, we use the terms $x^*(t)$ and x_t^* interchangeably.

The cooperative investment strategies for the cooperative game in (5.1) and (5.4) over the time interval $[t_0, T]$ can be expressed more precisely as

$$
\{\psi_i^*(t, x^*(t)), \text{ for } i \in N \text{ and } t \in [t_0, T]\}.
$$
(5.8)

Note that for group optimality to be achievable, the cooperative investment strategies $\{\psi_i^*(t, x^*(t)), \text{ for } i \in N \text{ and } t \in [t_0, T]\}$ must be exercised throughout the time interval $[t_0, T]$.

Along the cooperative investment path $\{x^*(t)\}_{t=t_0}^{T}$, the total venture profit over the interval $[t, T]$, for $t \in [t_0, T)$, can be expressed as

$$W^{(t_0)}(t, x_t^*) = \int_t^T \sum_{j=1}^n (g^j [s, x^{j*}(s)] - c_j^N [\psi_j^*(s, x^*(s))]) \exp\left[-\int_{t_0}^s r(y)\,dy\right] ds$$

$$+ \exp\left[-\int_{t_0}^T r(y)\,dy\right] \sum_{j=1}^n q^j (x^{j*}(T)). \tag{5.9}$$

Let $W^{(\tau)}(t, x_t^*)$ denote the total venture profit from the control problem with the dynamics in (5.1) and the payoff in (5.4), which begins at time $\tau \in [t_0, T]$ with initial state x_τ^*. Invoking Remark 3.1 of Chap. 3, one can readily obtain

$$\exp\left[\int_{t_0}^\tau r(y)\,dy\right] W^{(t_0)}(t, x_t^*) = W^{(\tau)}(t, x_t^*),$$

for $\tau \in [t_0, T]$ and $t \in [\tau, T)$.

Next, we consider an imputation scheme to share the total venture profit.

5.2.1 Imputation Scheme

The problem of profit sharing is inescapable in virtually every joint venture. Since the sizes and earning potentials of the firms in a corporate joint venture may vary significantly, we consider the case when the venture agrees to share the excess of the total cooperative payoff over the sum of individual noncooperative payoffs proportional to the firms' noncooperative payoffs.

The imputation scheme has to fulfill the following condition.

Condition 5.1 An imputation

$$\xi^{(t_0)i}(t_0, x_0) = V^{(t_0)i}(t_0, x_0^i) + \frac{V^{(t_0)i}(t_0, x_0^i)}{\sum_{j=1}^n V^{(t_0)j}(t_0, x_0^j)}$$

$$\times \left[W^{(t_0)}(t_0, x_0) - \sum_{j=1}^n V^{(t_0)j}(t_0, x_0^j) \right]$$

$$= \frac{V^{(t_0)i}(t_0, x_0^i)}{\sum_{j=1}^n V^{(t_0)j}(t_0, x_0^j)} W^{(t_0)}(t_0, x_0),$$

is assigned to firm i, for $i \in N$ at the outset and an imputation

$$\xi^{(\tau)i}\left(\tau, x_\tau^*\right) = \frac{V^{(\tau)i}(\tau, x_\tau^{i*})}{\sum_{j=1}^{n} V^{(\tau)j}(\tau, x_\tau^{j*})} W^{(\tau)}\left(\tau, x_\tau^*\right), \qquad (5.10)$$

is assigned to firm i, for $i \in N$ at time $\tau \in (t_0, T]$.

The imputation in (5.10) satisfies

(i) $\xi^{(\tau)i}(\tau, x_\tau^*) \geq V^{(\tau)i}(\tau, x_\tau^{i*})$, for $i \in N$ and $\tau \in [t_0, T]$; and

(ii) $\sum_{j=1}^{n} \xi^{(\tau)j}(\tau, x_\tau^*) = W^{(\tau)}(\tau, x_\tau^*)$ for $\tau \in [t_0, T]$.

Hence the imputation vector $\xi^{(\tau)}(\tau, x_\tau^*)$ in (5.10) satisfies individual rationality and group optimality throughout the game horizon $[t_0, T]$.

The solution to the optimality principle guiding the joint venture can then be expressed as

$$P\left(x_t^*, T - t\right) = \left\{\psi^*\left(s, x^*(s)\right) \text{ and } B(s) \text{ for } s \in [t, T], \xi^{(t)}\left(t, x_t^*\right)\right\},$$

for $t \in [t_0, T]$, where $\psi^*(s, x^*(s)) = \{\psi_1^*(s, x^*(s)), \psi_2^*(s, x^*(s)), \dots, \psi_n^*(s, x^*(s))\}$ is the vector of the cooperative investment strategies maximizing joint profit, $\xi^{(t)}(t, x_t^*)$ is the imputation scheme satisfying Condition 5.1, and $B(s)$ is a profit distribution mechanism that will lead to the realization of Condition 5.1.

All the participating firms in the joint venture will have no incentive to exit the venture if the agreed-upon optimality principle is maintained at every instant $t \in [t_0, T]$. A profit distribution mechanism that will lead to the realization of Condition 5.1 will be formulated in the next section.

5.2.2 Time (Optimal-Trajectory-Subgame) Consistent Venture Profit Distribution

To formulate a payoff distribution procedure over time so that the agreed imputations satisfy Condition 5.1 we first obtain the following.

Lemma 5.1 *A PDP with a terminal payment $q^i(x_T^*)$ at time T and an instantaneous payment at time $\tau \in [t_0, T]$*

$$B_i(\tau) = -\left[\xi_t^{(\tau)i}\left(t, x_t^*\right)|_{t=\tau}\right] - \sum_{h=1}^{n}\left[\xi_{x_t^{h*}}^{(\tau)i}\left(t, x_t^*\right)|_{t=\tau}\right] f^h\left[\tau, x_\tau^{h*}, \psi_h^*\left(\tau, x_\tau^*\right)\right]$$

$$= -\frac{\partial}{\partial t}\left[\frac{V^{(\tau)i}(t, x_t^{i*})}{\sum_{j=1}^{n} V^{(\tau)j}(t, x_t^{j*})} W^{(\tau)}\left(t, x_t^*\right)|_{t=\tau}\right]$$

$$-\sum_{h=1}^{n} \frac{\partial}{\partial x_\tau^{h*}} \left[\frac{V^{(\tau)i}(\tau, x_\tau^{i*})}{\sum_{j=1}^{n} V^{(\tau)j}(\tau, x_\tau^{j*})} W^{(\tau)}(\tau, x_\tau^*) \right]$$

$$\times f^h \left[\tau, x_\tau^{h*}, \psi_h^*(\tau, x_\tau^*) \right], \tag{5.11}$$

for $i \in N$, will lead to realization of the solution imputations $\xi^{(\tau)i}(\tau, x_\tau^*)$, for $i \in N$ and $\tau \in [t_0, T]$, satisfying Condition 5.1.

Proof Invoking Theorem 4.2 in Chap. 4, Lemma 5.1 follows.   □

A time (optimal-trajectory-subgame) consistent solution can be obtained as $P(x_0, T - t_0) = \{u(s) \text{ and } B(s) \text{ for } s \in [t_0, T] \text{ and } \xi^{(t_0)}(t_0, x_0)\}$ where

(i) $u(s)$ for $s \in [t_0, T]$ is the set of group optimal strategies $\psi^*(s, x_s^*)$ characterized in (5.6), and

(ii) $B(s) = \{B_1(s), B_2(s), \dots, B_n(s)\}$ for $s \in [t_0, T]$ is given as in (5.11).

With firms using the cooperative investment strategies $\{\psi_i^*(\tau, x_\tau^*)$, for $\tau \in [t_0, T]$ and $i \in N\}$, the instantaneous receipt of firm i at time instant τ is

$$\zeta_i(\tau) = g^i(\tau, x_\tau^{i*}) - c_i^N[\psi_i^*(\tau, x_\tau^*)], \tag{5.12}$$

for $\tau \in [t_0, T]$ and $i \in N$.

According to Lemma 5.1, the instantaneous payment that firm i should receive under the agreed-upon optimality principle is $B_i(\tau)$, for $\tau \in [t_0, T]$ and $i \in N$, as stated in (5.11). Hence an instantaneous transfer payment

$$\chi^i(\tau) = B_i(\tau) - \zeta_i(\tau) \tag{5.13}$$

has to be given or charged to firm i at time τ, for $i \in N$ and $\tau \in [t_0, T]$.

5.3 A Cost-Saving Joint Venture

Consider the case when there are three companies involved in a joint venture. The planning period is $[t_0, T]$. Company i's profit is

$$\int_{t_0}^{T} \left[P_i[x^i(s)]^{1/2} - c_i^{\{i\}} u_i(s) \right] \exp[-r(s - t_0)] \, ds + \exp[-r(T - t_0)] q_i [x^i(T)]^{1/2},$$

for $i \in \{1, 2, 3\}$, where $P_i, c_i^{\{i\}}$ and q_i are positive constants, r is the discount rate, $x_i(s) \subset R^+$ is the level of technology of company i at time s, and $u_i(s) \subset R^+$ is its physical investment in technological advancement. The term $P_i[x^i(s)]^{1/2}$ reflects the net operating revenue of company i at technology level $x_i(s)$, and $c_i u_i$ is the cost of the investment. The salvage value of company i's technology at time T is given by $q_i[x^i(T)]^{1/2}$. The evolution of the technology level of company i follows the dynamics

$$\dot{x}^i(s) = \left[\alpha_i [u^i(s) x_i(s)]^{1/2} - \delta x^i(s) \right], \quad \text{for } i \in \{1, 2, 3\}. \tag{5.14}$$

In the case when each of these three firms acts independently, and using Theorem A.1 in the Technical Appendixes, we obtain the Bellman equation as

$$-V_t^{(t_0)i}\left(t, x^i\right) = \max_{u_i}\left\{\left[P_i\left(x^i\right)^{1/2} - c_i^{\{i\}}u_i\right]\exp\left[-r(t - t_0)\right]\right.$$
$$\left. + V_{x^i}^{(\tau)i}\left(t, x^i\right)\left[\alpha_i\left(u^i x_i\right)^{1/2} - \delta x^i\right]\right\}, \qquad (5.15)$$
$$V^{(t_0)i}\left(T, x^i\right) = \exp\left[-r(T - t_0)\right]q_i\left(x^i\right)^{1/2}, \quad \text{for } i \in \{1, 2, 3\}.$$

Performing the indicated maximization in (5.15) yields

$$u_i = \frac{\alpha_i^2}{4(c_i^{\{i\}})^2}\left[V_{x^i}^{(t_0)i}\left(t, x^i\right)\exp\left[r(t - t_0)\right]\right]^2 x^i, \quad \text{for } i \in \{1, 2, 3\}.$$

Substituting u_i into the Bellman equation yields

$$-V_t^{(t_0)i}\left(t, x^i\right) = P_i\left(x^i\right)^{1/2}\exp\left[-r(t - t_0)\right]$$
$$- \frac{\alpha_i^2}{4c_i^{\{i\}}}\left[V_{x^i}^{(t_0)i}\left(t, x^i\right)\right]^2\exp\left[r(t - t_0)\right]x^i$$
$$+ \frac{\alpha_i^2}{2c_i}\left[V_{x^i}^{(t_0)i}\left(t, x^i\right)\right]^2\exp\left[r(t - \tau)\right]x^i - \delta V_{x^i}^{(t_0)i}\left(t, x^i\right)x_i,$$

for $i \in \{1, 2, 3\}$.

Solving the above system of partial differential equations yields

$$V^{(t_0)i}\left(t, x^i\right) = \left[A_i^{\{i\}}(t)\left(x^i\right)^{1/2} + C_i^{\{i\}}(t)\right]\exp\left[-r(\tau - t_0)\right],$$
$$\text{for } i \in \{1, 2, 3\}, \qquad (5.16)$$

where

$$\dot{A}_i^{\{i\}}(t) = \left(r + \frac{\delta}{2}\right)A_i^{\{i\}}(t) - P_i,$$
$$\dot{C}_i^{\{i\}}(t) = rC_i^{\{i\}}(t) - \frac{\alpha_i^2}{16c_i^{\{i\}}}\left[A_i^{\{i\}}(t)\right]^2, \qquad (5.17)$$
$$A_i^{\{i\}}(T) = q_i \quad \text{and} \quad C_i^{\{i\}}(T) = 0.$$

The first equation in the block-recursive system in (5.17) is a first-order linear differential equation in $A_i^{\{i\}}(t)$ that can be solved independently by standard techniques. Substituting the solution of $A_i^{\{i\}}(t)$ into the second equation of (5.17) yields a first-order linear differential equation in $C_i^{\{i\}}(t)$. The solution of $C_i^{\{i\}}(t)$ can be readily obtained by standard techniques.

Moreover, one can easily derive

$$V^{(\tau)i}\left(t, x^i\right) = \left[A_i^{\{i\}}(t)\left(x^i\right)^{1/2} + C_i^{\{i\}}(t)\right]\exp\left[-r(t - \tau)\right],$$

for $i \in \{1, 2, 3\}$ and $\tau \in [t_0, T]$.

After characterizing the outcome when each of these three firms acts independently, we investigate the outcome when these firms form a joint venture.

5.3.1 Joint Venture Profit and Cost Saving

Consider the case when all three firms agree to form a joint venture and share their joint profit proportionally to their noncooperative profits. Cost-saving opportunities are created under joint venture from joint R&D, administration, purchasing, financing, and economy of scales and scope. The cost of control of firm j under the joint venture becomes $c_j^{\{1,2,3\}} u_j(s)$. With joint venture cost advantage

$$c_j^{\{1,2,3\}} \le c_j^{\{j\}}, \quad \text{for } j \in N. \tag{5.18}$$

The profit of the joint venture is the sum of the participating firms' profits

$$\int_{t_0}^{T} \sum_{j=1}^{3} \left[P_j \left[x^j(s) \right]^{1/2} - c_j^{\{1,2,3\}} u_j(s) \right] \exp\left[-r(s - t_0) \right] ds$$

$$+ \sum_{j=1}^{3} \exp\left[-r(T - t_0) \right] q_j \left[x^j(T) \right]^{1/2}. \tag{5.19}$$

The firms in the joint venture then act cooperatively to maximize (5.19) subject to (5.14). In particular, (5.14) and (5.19) become an optimization problem under the three firms' cost-saving joint venture. Using Theorem A.1 in the Technical Appendixes, we obtain the Bellman equation as

$$- W_t^{(t_0)\{1,2,3\}} \left(t, x^1, x^2, x^3 \right)$$

$$= \max_{u_1, u_2, u_3} \left\{ \sum_{i=1}^{3} \left[P_i \left(x^i \right)^{1/2} - c_i u_i \right] \exp\left[-r(t - t_0) \right] \right.$$

$$\left. + \sum_{i=1}^{3} W_{x^i}^{(t_0)\{1,2,3\}} \left(t, x^1, x^2, x^3 \right) \left[\alpha_i \left(u_i x^i \right)^{1/2} - \delta x^i \right] \right\}, \tag{5.20}$$

$$W^{(t_0)\{1,2,3\}} \left(T, x^1, x^2, x^3 \right) = \sum_{j=1}^{3} \exp\left[-r(T - t_0) \right] q_j \left(x^j \right)^{1/2}.$$

Performing the indicated maximization yields

$$u_i = \frac{\alpha_i^2}{4(c_i^{\{1,2,3\}})^2} \left[W_{x^i}^{(t_0)\{1,2,3\}} \left(t, x^1, x^2, x^3 \right) \exp\left[r(t - t_0) \right] \right]^2 x^i, \tag{5.21}$$

for $i \in \{1, 2, 3\}$.

Substituting (5.21) into (5.20) yields

$$
-W_t^{(t_0)\{1,2,3\}}\left(t,x^1,x^2,x^3\right)
$$

$$
= \sum_{i=1}^{3}\left[P_i\left(x^i\right)^{1/2}\exp\left[-r(t-t_0)\right] - \frac{\alpha_i^2 x^i}{4c_i^{\{1,2,3\}}}\right.
$$

$$
\times\left[W_{x^i}^{(t_0)\{1,2,3\}}\left(t,x^1,x^2,x^3\right)\right]^2\exp\left[r(t-t_0)\right]\Big]
$$

$$
+ \sum_{i=1}^{3}W_{x_i}^{(t_0)\{1,2,3\}}(t,x_1,x_2,x_3)\left[\frac{\alpha_i^2}{2c_i}\left[W_{x_i}^{(t_0)i}(t,x_1,x_2,x_3)\right.\right.
$$

$$
\times\exp\left[r(t-t_0)\right]\big]x^i - \delta x^i\Big],
$$

and

$$
W^{(t_0)\{1,2,3\}}\left(T,x^1,x^2,x^3\right) = \sum_{j=1}^{3}\exp\left[-r(T-t_0)\right]q_j\left(x^j\right)^{1/2}. \qquad (5.22)
$$

Solving (5.22) yields

$$
W^{(t_0)\{1,2,3\}}\left(t,x^1,x^2,x^3\right)
$$

$$
= \left[A_1^{\{1,2,3\}}(t)\left(x^1\right)^{1/2} + A_2^{\{1,2,3\}}(t)\left(x^2\right)^{1/2} + A_3^{\{1,2,3\}}(t)\left(x^3\right)^{1/2}\right.
$$

$$
\left. + C^{\{1,2,3\}}(t)\right]\exp\left[-r(t-t_0)\right], \qquad (5.23)
$$

where $A_1^{\{1,2,3\}}(t)$, $A_2^{\{1,2,3\}}(t)$, $A_3^{\{1,2,3\}}(t)$, and $C^{\{1,2,3\}}(t)$ satisfy

$$
\dot{A}_i^{\{1,2,3\}}(t) = \left(r + \frac{\delta}{2}\right)A_i^{\{1,2,3\}}(t) - P_i
$$

for $i, j, h \in \{1, 2, 3\}$ and $i \neq j \neq h$,

$$
\dot{C}^{\{1,2,3\}}(t) = rC^{\{1,2,3\}}(t) - \sum_{i=1}^{3}\frac{\alpha_i^2}{16c_i^{\{1,2,3\}}}\left[A_i^{\{1,2,3\}}(t)\right]^2, \qquad (5.24)
$$

$$
A_i^{\{1,2,3\}}(T) = q_i \quad \text{for } i \in \{1, 2, 3\}, \text{ and } C^{\{1,2,3\}}(T) = 0.
$$

The first three equations in the block recursive system in (5.24) are a system of three linear differential equations that can be solved explicitly by standard techniques. Upon solving $A_i^{\{1,2,3\}}(t)$ for $i \in \{1, 2, 3\}$ and substituting them into the fourth equation of (5.24), one has a linear differential equation in $C^{\{1,2,3\}}(t)$.

The investment strategies of the grand coalition joint venture can be derived as

$$\psi_i^{\{1,2,3\}}(t,x) = \frac{\alpha_i^2}{16(c_i^{\{1,2,3\}})^2}\left[A_i^{\{1,2,3\}}(t)\right]^2, \quad \text{for } i \in \{1,2,3\}. \quad (5.25)$$

The dynamics of the technological progress of the joint venture over the time interval $s \in [t_0, T]$ can be expressed as

$$\dot{x}^i(s) = \frac{\alpha_i^2}{4c_i}A_i^{\{1,2,3\}}(t)\left[x^i(s)\right]^{1/2} - \delta x^i(s), \quad x^i(t_0) = x_0^i, \quad (5.26)$$

for $i \in \{1,2,3\}$.

Taking the transforming $y^i(s) = x^i(s)^{1/2}$, for $i \in \{1,2,3\}$, equation system in (5.26) can be expressed as

$$\dot{y}^i(s) = \frac{\alpha_i^2}{8c_i}A_i^{\{1,2,3\}}(t) - \frac{\delta}{2}y^i(s), \quad y^i(t_0) = (x_0^i)^{1/2}, \quad (5.27)$$

for $i \in \{1,2,3\}$.

Equation (5.27) is a system of linear differential equations that can be solved by standard techniques. Solving (5.27) yields the joint venture's state trajectory. Let $\{y^{1*}(t), y^{2*}(t), y^{3*}(t)\}$ denote the solution to (5.27). Transforming $x^i = (y^i)^2$, we obtain the state trajectories of the joint venture over the time interval $s \in [t_0, T]$ as

$$\left\{x^*(t)\right\}_{t=t_0}^T \equiv \left\{x^{1*}(t), x^{2*}(t), x^{3*}(t)\right\}_{t=t_0}^T$$

$$= \left\{\left[y^{1*}(t)\right]^2, \left[y^{2*}(t)\right]^2, \left[y^{3*}(t)\right]^2\right\}_{t=t_0}^T. \quad (5.28)$$

Once again, we use the terms $x^{i*}(t)$ and x_t^{i*} interchangeably.

Remark 5.1 One can readily verify that

$$W^{(t_0)\{1,2,3\}}(t, x^{1*}, x^{2*}, x^{3*}) = W^{(t)\{1,2,3\}}(t, x^{1*}, x^{2*}, x^{3*})\exp\left[-r(t-t_0)\right],$$

for $i \in \{1,2,3\}$.

5.3.2 Time (Optimal-Trajectory-Subgame) Consistent Venture Profit Sharing

Since the firms agree to share their joint profit proportionally to their noncooperative profits, the imputation scheme has to fulfill the following condition.

Condition 5.2 In the game $\Gamma_c(x_0, T - t_0)$, an imputation

$$\xi^{(t_0)i}(t_0, x_0) = \frac{V^{(t_0)i}(t_0, x_0^i)}{\sum_{j=1}^n V^{(t_0)j}(t_0, x_0^i)}W^{(t_0)\{1,2,3\}}(t_0, x_0^1, x_0^2, x_0^3),$$

is assigned to firm i, for $i \in \{1, 2, 3\}$, and in the subgame $\Gamma_c(x_\tau^*, T - \tau)$, for $\tau \in (t_0, T]$, an imputation

$$\xi^{(\tau)i}\left(\tau, x_\tau^*\right) = \frac{V^{(\tau)i}(\tau, x_\tau^{i*})}{\sum_{j=1}^{n} V^{(\tau)j}(\tau, x_\tau^{i*})} W^{(\tau)\{1,2,3\}}\left(\tau, x_\tau^{1*}, x_\tau^{2*}, x_\tau^{3*}\right), \qquad (5.29)$$

is assigned to firm i, for $i \in \{1, 2, 3\}$.

To formulate a payoff distribution procedure over time so that the agreed imputations in Condition 5.2 are satisfied we present the following proposition.

Proposition 5.1 *A PDP with a terminal payment $q^i(x_T^*))$ at time T and an instantaneous payment at time $\tau \in [t_0, T]$*

$$B_i(\tau) = -\frac{\partial}{\partial t}\left[\frac{V^{(\tau)i}(t, x_t^{i*})}{\sum_{j=1}^{3} V^{(\tau)j}(t, x_t^{j*})} W^{(\tau)\{1,2,3\}}\left(t, x_t^{1*}, x_t^{2*}, x_t^{3*}\right)\Big|_{t=\tau}\right]$$

$$- \sum_{\ell=1}^{n} \frac{\partial}{\partial x_\tau^{\ell*}}\left[\frac{V^{(\tau)i}(\tau, x_\tau^{i*})}{\sum_{j=1}^{3} V^{(\tau)j}(\tau, x_\tau^{j*})} W^{(\tau)\{1,2,3\}}\left(\tau, x_\tau^{1*}, x_\tau^{2*}, x_\tau^{3*}\right)\right]$$

$$\times \left[\frac{\alpha_\ell^2}{4c_\ell} A_\ell^{\{1,2,3\}}(\tau)\left(x_\tau^{\ell*}\right)^{1/2} - \delta x_\tau^{\ell*}\right], \quad \text{for } i \in \{1, 2, 3\}, \qquad (5.30)$$

would lead to the realization of the solution imputations $\xi^{(\tau)i}(\tau, x_\tau^*)$, for $i \in \{1, 2, 3\}$ and $\tau \in [t_0, T]$, satisfying Condition 5.2.

Proof Invoking Lemma 5.1, one obtains an equation similar to (5.11) with

$$i \in \{1, 2, 3\} \quad \text{and} \quad f^\ell\left[\tau, x_\tau^{\ell*}, \psi_\ell^*\left(\tau, x_\tau^*\right)\right] = \left[\frac{\alpha_\ell^2}{4c_\ell} A_\ell^{\{1,2,3\}}(\tau)\left(x_\tau^{\ell*}\right)^{1/2} - \delta x_\tau^{\ell*}\right].$$

Hence Proposition 5.1 follows.   □

In particular, from (5.17) and (5.24),

$$\frac{\partial}{\partial t}\left[\frac{V^{(\tau)i}(t, x_t^{i*})}{\sum_{j=1}^{3} V^{(\tau)j}(t, x_t^{j*})} W^{(\tau)\{1,2,3\}}\left(t, x_t^{1*}, x_t^{2*}, x_t^{3*}\right)\Big|_{t=\tau}\right]$$

$$= \left[A_1^{\{1,2,3\}}(\tau)\left(x_\tau^{1*}\right)^{1/2} + A_2^{\{1,2,3\}}(\tau)\left(x_\tau^{2*}\right)^{1/2} + A_3^{\{1,2,3\}}(\tau)\left(x_\tau^{3*}\right)^{1/2} + C^{\{1,2,3\}}(\tau)\right]$$

$$\times \left[\left(\sum_{j=1}^{3}[A_j^{\{j\}}(\tau)\left(x_\tau^{j*}\right)^{1/2} + C_j^{\{j\}}(\tau)]\right)\right]$$

$$\times \frac{(r[A_i^{\{i\}}(\tau)(x^{i*})^{1/2} + C_i^{\{i\}}(\tau)] + [\dot{A}_i^{\{i\}}(\tau)(x^{i*})^{1/2} + \dot{C}_i^{\{i\}}(\tau)])}{(\sum_{j=1}^{3}[A_j^{\{j\}}(\tau)(x_\tau^j)^{1/2} + C_j^{\{j\}}(\tau)])^2}$$

$$- \left[A_i^{\{i\}}(\tau)\big(x^{i*}\big)^{1/2} + C_i^{\{i\}}(\tau)\right]$$

$$\times \frac{\sum_{j=1}^{3}\left(r[A_j^{\{j\}}(\tau)(x^{j*})^{1/2} + C_j^{\{j\}}(\tau)] + [\dot{A}_j^{\{j\}}(\tau)(x^{j*})^{1/2} + \dot{C}_j^{\{j\}}(\tau)]\right)}{\left(\sum_{j=1}^{3}[A_j^{\{j\}}(\tau)(x_\tau^{j*})^{1/2} + C_j^{\{j\}}(\tau)]\right)^2}$$

$$+ \frac{[A_i^{\{i\}}(\tau)(x^{i*})^{1/2} + C_i^{\{i\}}(\tau)]}{\left(\sum_{j=1}^{3}[A_j^{\{j\}}(\tau)(x_\tau^{j*})^{1/2} + C_j^{\{j\}}(\tau)]\right)^2}$$

$$\times \big(r\big[A_1^{\{1,2,3\}}(\tau)\big(x_\tau^{1*}\big)^{1/2} + A_2^{\{1,2,3\}}(\tau)\big(x_\tau^{2*}\big)^{1/2}$$

$$+ A_3^{\{1,2,3\}}(\tau)\big(x_\tau^{3*}\big)^{1/2} + C^{\{1,2,3\}}(\tau)\big]$$

$$+ \big[\dot{A}_1^{\{1,2,3\}}(\tau)\big(x_\tau^{1*}\big)^{1/2} + \dot{A}_2^{\{1,2,3\}}(\tau)\big(x_\tau^{2*}\big)^{1/2}$$

$$+ \dot{A}_3^{\{1,2,3\}}(\tau)\big(x_\tau^{3*}\big)^{1/2} + \dot{C}^{\{1,2,3\}}(\tau)\big]\big);$$

and

$$\frac{\partial}{\partial x_\tau^{i*}}\left[\frac{V^{(\tau)i}(\tau, x_\tau^{i*})}{\sum_{j=1}^{3} V^{(\tau)j}(\tau, x_\tau^{j*})} W^{(\tau)\{1,2,3\}}\big(\tau, x_\tau^{1*}, x_\tau^{2*}, x_\tau^{3*}\big)\right]$$

$$= \left[A_1^{\{1,2,3\}}(\tau)\big(x_\tau^{1*}\big)^{1/2} + A_2^{\{1,2,3\}}(\tau)\big(x_\tau^{2*}\big)^{1/2} + A_3^{\{1,2,3\}}(\tau)\big(x_\tau^{3*}\big)^{1/2} + C^{\{1,2,3\}}(\tau)\right]$$

$$\times \left[-\frac{[A_i^{\{i\}}(\tau)(x^{i*})^{1/2} + C_i^{\{i\}}(\tau)]\frac{1}{2}A_h^{\{h\}}(\tau)(x^{h*})^{-1/2}}{\left(\sum_{j=1}^{3}[A_j^{\{j\}}(\tau)(x_\tau^{j*})^{1/2} + C_j^{\{j\}}(\tau)]\right)^2}\right]$$

$$+ \frac{[A_i^{\{i\}}(\tau)(x^{i*})^{1/2} + C_i^{\{i\}}(\tau)]}{\left(\sum_{j=1}^{3}[A_j^{\{j\}}(\tau)(x_\tau^{j*})^{1/2} + C_j^{\{j\}}(\tau)]\right)^2}\frac{1}{2}A_h^{\{1,2,3\}}(\tau)\big(x_\tau^{h*}\big)^{-1/2};$$

$$\frac{\partial}{\partial x_\tau^{h*}}\left[\frac{V^{(\tau)i}(\tau, x_\tau^{i*})}{\sum_{j=1}^{3} V^{(\tau)j}(\tau, x_\tau^{j*})} W^{(\tau)\{1,2,3\}}\big(\tau, x_\tau^{1*}, x_\tau^{2*}, x_\tau^{3*}\big)\right]$$

$$= \left[A_1^{\{1,2,3\}}(\tau)\big(x_\tau^{1*}\big)^{1/2} + A_2^{\{1,2,3\}}(\tau)\big(x_\tau^{2*}\big)^{1/2} + A_3^{\{1,2,3\}}(\tau)\big(x_\tau^{3*}\big)^{1/2} + C^{\{1,2,3\}}(\tau)\right]$$

$$\times \left[-\frac{[A_i^{\{i\}}(\tau)(x^{i*})^{1/2} + C_i^{\{i\}}(\tau)]\frac{1}{2}A_h^{\{h\}}(\tau)(x^{h*})^{-1/2}}{\left(\sum_{j=1}^{3}[A_j^{\{j\}}(\tau)(x_\tau^{j*})^{1/2} + C_j^{\{j\}}(\tau)]\right)^2}\right]$$

$$+ \frac{[A_i^{\{i\}}(\tau)(x^{i*})^{1/2} + C_i^{\{i\}}(\tau)]}{\left(\sum_{j=1}^{3}[A_j^{\{j\}}(\tau)(x_\tau^{j*})^{1/2} + C_j^{\{j\}}(\tau)]\right)^2}\frac{1}{2}A_h^{\{1,2,3\}}(\tau)\big(x_\tau^{h*}\big)^{-1/2},$$

for $h \neq i$.

A time (optimal-trajectory-subgame) consistent solution can be obtained as $P(x_0, T - t_0) = \{u(s) \text{ and } B(s) \text{ for } s \in [t_0, T] \text{ and } \xi^{(t_0)}(t_0, x_0)\}$ where

(i) $u(s)$ for $s \in [t_0, T]$ is the set of group optimal strategies $\psi^*(s, x_s^*)$ in (5.25), and
(ii) $B(s) = \{B_1(s), B_2(s), \ldots, B_n(s)\}$ for $s \in [t_0, T]$ is given as in (5.30).

Using the cooperative strategies the instantaneous receipt of firm i at time instant τ is

$$\zeta_i(\tau) = P_i\left(x_\tau^{i*}\right)^{1/2} - \frac{\alpha_i^2}{16(c_i^{\{1,2,3\}})}\left[A_i^{\{1,2,3\}}(\tau)\right]^2, \qquad (5.31)$$

for $\tau \in [t_0, T]$ and $i \in \{1, 2, 3\}$.

According to Proposition 5.1, the instantaneous payment that firm i should receive under the agreed-upon optimality principle is $B_i(\tau)$, for $\tau \in [t_0, T]$ and $i \in \{1, 2, 3\}$, as stated in (5.30). Hence an instantaneous transfer payment

$$\chi^i(\tau) = B_i(\tau) - \zeta_i(\tau), \qquad (5.32)$$

has to be given or charged to firm i at time τ, for $i \in \{1, 2, 3\}$ and $\tau \in [t_0, T]$.

5.4 A Shapley Value Solution to Joint Venture

Consider again the dynamic venture model in (5.1) and (5.2). If firms are allowed to form different coalitions consisting of a subset of companies $K \subseteq N$. There are k firms in the subset K. The participating firms in a coalition can gain core skills and technology from each other. In particular, they can obtain cost reduction and with absolute joint venture cost advantage

$$c_j^K\left[u_j(s)\right] \le c_j^L\left[u_j(s)\right], \quad \text{for } j \in L \subseteq K, \qquad (5.33)$$

where $c_j^K[u_j(s)]$ represents the costs of the controls of the firm j in the subset K and $c_j^L[u_j(s)]$ represents the costs of the controls of the firm j in the subset L.

Moreover, marginal cost advantages lead to

$$\partial c_j^K\left[u_j(s)\right]/\partial u_j(s) \le \partial c_j^L\left[u_j(s)\right]/\partial u_j(s), \quad \text{for } j \in L \subseteq K. \qquad (5.34)$$

At time t_0, the profit to the joint venture K becomes

$$\int_{t_0}^T \sum_{j \in K} \{g^j[s, x^j(s)] - c_j^K[u_j(s)]\} \exp\left[-\int_{t_0}^s r(y)\,dy\right] ds$$

$$+ \sum_{j \in K} \exp\left[-\int_{t_0}^T r(y)\,dy\right] q^j\left(x^j(T)\right), \quad \text{for } K \subseteq N. \qquad (5.35)$$

To compute the profit of the joint venture K we have to consider the optimal control problem $\varpi[K; t_0, x_0^K]$ which maximizes the joint venture profit in (5.35) subject to the technology accumulation dynamics in (5.1).

Invoking Bellman's technique of dynamic programming in Sect. A.1 of the Technical Appendixes, the solution to the optimal control problem $\varpi[K; t_0, x_0^K]$ can be characterized as follows.

Corollary 5.2 *A set of controls* $\{\psi_i^{K*}(t, x^K), \text{for } i \in K \text{ and } t \in [t_0, T]\}$, *provides an optimal solution to the control problem* $\varpi[K; t_0, x_0^K]$ *if there exists a continuously differentiable function* $W^{(t_0)K}(t, x^K) : [t_0, T] \times R^k \to R$ *satisfying the following Bellman equation*:

$$-W_t^{(t_0)K}\left(t, x^K\right) = \max_{u_K}\left\{\sum_{j \in K}[g^j\left(t, x^j\right) - c_j^K(u_j)]\exp\left[-\int_{t_0}^t r(y)\,dy\right]\right.$$

$$\left. + \sum_{j \in K} W_{x_j}^{(t_0)K}\left(t, x^K\right)f^j[s, x^j, u_j]\right\}, \qquad (5.36)$$

$$W^{(t_0)K}\left(T, x^K\right) = \exp\left[-\int_{t_0}^T r(y)\,dy\right]\sum_{j \in K}q^j\left(x^j\right).$$

Following Corollary 5.2, one can characterize the maximized payoff $W^{(\tau)K}(t, x^K)$ to the optimal control problem $\varpi[K; \tau, x_\tau^K]$ which maximizes

$$\int_\tau^T \sum_{j \in K}[g^j\left(s, x^j(s)\right) - c_j^K(u_j(s))]\exp\left[-\int_\tau^s r(y)\,dy\right]ds$$

$$+ \sum_{j \in K}\exp\left[-\int_\tau^T r(y)\,dy\right]q^j\left(x^j(T)\right), \qquad (5.37)$$

subject to

$$\dot{x}^j(s) = f^j[s, x^j(s), u_j(s)], \quad x^j(\tau) = x_\tau^j, \text{ for } j \in K. \qquad (5.38)$$

Invoking Remark 3.1 of Chap. 3, one can readily obtain $W^{(t_0)K}(t, x^K) = W^{(\tau)K}(t, x^K)\exp[-\int_{t_0}^\tau r(y)\,dy]$, for $\tau \in [t_0, T]$ and $t \in [\tau, T)$. Now consider the case of a grand coalition N in which all the n firms are in the coalition. In the grand coalition, firms will adopt the cooperative control $\{\psi_i^{N*}(t, x^N), \text{for } i \in N \text{ and } t \in [t_0, T]\}$, to obtain the maximized level of joint profit. The state dynamics of the grand coalition can be obtained as in (5.6) and the optimal trajectory $\{x^*(t)\}_{t=t_0}^T$ as in (5.7) $\{x^*(t)\}_{t=t_0}^T$. Note that for group optimality to be achievable the cooperative investment strategies $\{\psi_i^{N*}(t, x^*(t)), \text{for } i \in N \text{ and } t \in [t_0, T]\}$, must be exercised throughout time interval $[t_0, T]$.

Along the cooperative control path $\{x^*(t)\}_{t=t_0}^T$ the total venture profit over the interval $[t, T]$, for $t \in [t_0, T)$, can be expressed as

$$W^{(t_0)N}\left(t, x_t^*\right) = \int_t^T \sum_{j=1}^n g^j[s, x^{j*}(s), \psi_j^{N*}(s, x^*(s))]\exp\left[-\int_{t_0}^s r(y)\,dy\right]ds$$

$$+ \exp\left[-\int_{t_0}^T r(y)\,dy\right]\sum_{j=1}^n q^j\left(x^{j*}(T)\right). \qquad (5.39)$$

Moreover, the superadditivity of the coalition payoff can be demonstrated.

Proposition 5.2 *The coalition profits* $W^{(\tau)K}(t, x^K)$ *is superadditivity, that is,*

$$W^{(\tau)K}\left(\tau, x^K\right) \geq W^{(\tau)L}\left(\tau, x^L\right) + W^{(\tau)K\setminus L}\left(\tau, x^{K\setminus L}\right), \quad for\ L \subset K \subseteq N,$$

where $K \setminus L$ *is the relative complement of* L *in* K.

Proof See the Appendix of this chapter.   □

With joint profits under different venture coalitions characterized we proceed to consider the distribution of venture profits according to the Shapley Value (1953).

5.4.1 Dynamic Shapley Value Imputation

Consider a joint venture involving n firms. The member firms will maximize their joint profit and share their cooperative profits according to the Shapley Value (1953). The problem of profit sharing is inescapable in virtually every joint venture. The Shapley Value is one of the most commonly used sharing mechanisms in static co-operation games with transferable payoffs. Besides being individually rational and group rational, the Shapley Value is also unique. Specifically, the Shapley Value gives an imputation rule

$$\varphi^i(v) = \sum_{K \subseteq N} \frac{(k-1)!(n-k)!}{n!}\left[v(K) - v(K\setminus i)\right], \quad for\ i \in N, \qquad (5.40)$$

where $K\setminus i$ is the relative complement of i in K, $v(K)$ is the profit of coalition K, and $[v(K) - v(K\setminus i)]$ is the marginal contribution of firm i to the coalition K.

Though the Shapley Value is used as the profit allocation mechanism, there exist two features that do not conform with the standard Shapley Value analysis. The first is that the present analysis is dynamic so that, instead of a one-time allocation of the Shapley Value, we have to consider the maintenance of the Shapley Value imputation over the joint venture horizon. The second is that the profit $v(K)$ is the maximized profit to coalition K and is not a characteristic function (from the game in which coalition K is playing a zero-sum game against coalition $N\setminus K$). Applications of the Shapley Value in cost allocation usually do not follow the characteristic function approach. Moreover, since profit maximization by coalition K is not affected by firms outside the coalition, the analysis does not have to adopt arbitrary assumptions like those in Petrosyan and Zaccour (2003) in which the left-out players are assumed to stick with their feedback Nash strategies in computing a nonstandard characteristic function.

Consider the situation when the firms in the joint venture agree to adopt an optimality principle which (i) maximizes the joint venture profit, and (ii) shares the venture profit among participating firms according to the Shapley Value.

To maximize the joint venture's profits the firms will adopt the cooperative investment strategies $\{\psi_i^{N*}(t, x^*(t)), \text{ for } i \in N \text{ and } t \in [t_0, T]\}$, and the corresponding cooperative investment path $\{x^*(t)\}_{t=t_0}^T \equiv \{x^{N*}(t)\}_{t=t_0}^T$ in (5.38) would result.

To share the venture profit among participating firms according to the Shapley Value, the imputation has to satisfy the following condition.

Condition 5.3 In the game $\Gamma_c(x_0, T - t_0)$, an imputation

$$\xi^{(t_0)i}\left(t_0, x_0^N\right) = \sum_{K \subseteq N} \frac{(k-1)!(n-k)!}{n!}\left[W^{(t_0)K}\left(t_0, x_0^K\right) - W^{(t_0)K \setminus i}\left(t_0, x_0^{K \setminus i}\right)\right],$$

is assigned to firm i, for $i \in N$ and in the subgame $\Gamma_c(x_\tau^*, T - \tau)$, for $\tau \in (t_0, T]$, an imputation

$$\xi^{(\tau)i}\left(\tau, x_\tau^{N*}\right) = \sum_{K \subseteq N} \frac{(k-1)!(n-k)!}{n!}\left[W^{(\tau)K}\left(\tau, x_\tau^{K*}\right) - W^{(\tau)K \setminus i}\left(\tau, x_\tau^{K \setminus i*}\right)\right],$$

$$(5.41)$$

is assigned to firm i, for $i \in N$.

Note that $\xi^{(\tau)}(\tau, x_\tau^{N*}) = \xi^{(\tau)i}(\tau, x_\tau^{N*}), \xi^{(\tau)i}(\tau, x_\tau^{N*}), \dots, v^{(\tau)i}(\tau, x_\tau^{N*})]$, as specified in (5.41), satisfies the basic properties of an imputation vector as follows:

(i) $\quad \sum_{j=1}^{n} v^{(\tau)j}\left(\tau, x_\tau^{N*}\right) = W^{(\tau)N}\left(\tau, x_\tau^{N*}\right), \quad$ and

$$(5.42)$$

(ii) $\quad v^{(\tau)i}\left(\tau, x_\tau^{N*}\right) \geq W^{(\tau)i}\left(\tau, x_\tau^{N*}\right), \quad$ for $i \in N$ and $\tau \in [t_0, T]$.

Part (i) of (5.42) shows that $\xi^{(\tau)}(\tau, x_\tau^{N*})$ satisfies the property of Pareto optimality throughout the game interval. Part (ii) demonstrates that $\xi^{(\tau)}(\tau, x_\tau^{N*})$ guarantees individual rationality throughout the game interval. Crucial to the analysis is the formulation of a profit distribution mechanism that would lead to the realization of Condition 5.3. This will be done in the next section.

5.4.2 The PDP for Shapley Value

To formulate a payoff distribution procedure over time so that the agreed imputations satisfy the Shapley Value in Condition 5.3 we invoke Theorem 4.2 in Chap. 4 and obtain the following.

Lemma 5.2 *A PDP with a terminal payment $q^i(x_T^*)$ at time T and an instantaneous payment at time $\tau \in [t_0, T]$*

$$B_i(\tau) = -\sum_{K \subseteq N} \frac{(k-1)!(n-k)!}{n!} \left\{ \left[W_t^{(\tau)K}\left(t, x_t^{K*}\right)\big|_{t=\tau} \right] - \left[W_t^{(\tau)K \setminus i}\left(t, x_t^{K \setminus i*}\right)\big|_{t=\tau} \right] \right.$$

$$+ \sum_{h \in K} \left[\frac{\partial}{\partial x_\tau^{h*}} W^{(\tau)K}\left(\tau, x_\tau^{K*}\right) \right] f^h\left[\tau, x_\tau^{h*}, \psi_h^*\left(\tau, x_\tau^*\right)\right]$$

$$\left. - \sum_{h \in K \setminus i} \left[\frac{\partial}{\partial x_\tau^{h*}} W^{(\tau)K \setminus i}\left(\tau, x_\tau^{K \setminus i*}\right) \right] f^h\left[\tau, x_\tau^{h*}, \psi_h^*\left(\tau, x_\tau^*\right)\right] \right\}, \qquad (5.43)$$

for $i \in N$, will lead to the realization of the Shapley Value imputations $\xi^{(\tau)i}(\tau, x_\tau^{N*})$ in Condition 5.3.

Proof Invoking Theorem 4.2 in Chap. 4 one can obtain Lemma 5.2. $\square$

A time (optimal-trajectory-subgame) consistent solution can be obtained using the set of group optimal strategies $\psi^*(s, x_s^*)$ characterized in Corollary 5.2 and $B(s) = \{B_1(s), B_2(s), \ldots, B_n(s)\}$ in (5.43).

With firms using the cooperative investment strategies $\{\psi_i^*(\tau, x_\tau^*)$, for $\tau \in [t_0, T]$ and $i \in N\}$, the instantaneous receipt of firm i at time instant τ is

$$\zeta_i(\tau) = g^i\left(\tau, x_\tau^{i*}\right) - c_i^N\left[\psi_i^*\left(\tau, x_\tau^*\right)\right], \quad \text{for } \tau \in [t_0, T] \text{ and } i \in N. \quad (5.44)$$

According to Lemma 5.2, the instantaneous payment that firm i should receive under the agreed-upon optimality principle is $B_i(\tau)$, for $\tau \in [t_0, T]$ and $i \in N$, as stated in (5.43). Hence an instantaneous transfer payment

$$\chi^i(\tau) = B_i(\tau) - \zeta_i(\tau), \qquad (5.45)$$

would be given or charged to firm i at time τ, for $i \in N$ and $\tau \in [t_0, T]$.

5.5 A Joint Venture with Shapley Value Profit Sharing

Consider the joint venture with technology spillovers in Sect. 5.3. In particular, the participating firms would share their cooperative profits according to the Shapley Value (1953). In the case when each of these three firms acts independently, we follow the analysis in Sect. 5.3 and obtain

$$W^{(t_0)i}\left(t, x^i\right) = \left[A_i^{\{i\}}(t)\left(x^i\right)^{1/2} + C_i^{\{i\}}(t) \right] \exp[-r(\tau - t_0)], \quad \text{for } i \in \{1, 2, 3\},$$

$$(5.46)$$

where

$$\dot{A}_i^{\{i\}}(t) = \left(r + \frac{\delta}{2}\right) A_i^{\{i\}}(t) - P_i, \qquad \dot{C}_i^{\{i\}}(t) = r C_i^{\{i\}}(t) - \frac{\alpha_i^2}{16 c_i^{\{i\}}} \left[A_i^{\{i\}}(t)\right]^2,$$
(5.47)

$$A_i^{\{i\}}(T) = q_i, \quad \text{and} \quad C_i^{\{i\}}(T) = 0.$$

Moreover, one can easily derive for, $\tau \in [t_0, T]$,

$$W^{(\tau)i}(t, x^i) = \left[A_i^{\{i\}}(t)(x^i)^{1/2} + C_i^{\{i\}}(t)\right] \exp\left[-r(t - \tau)\right],$$
for $i \in \{1, 2, 3\}$ and $\tau \in [t_0, T]$.

5.5.1 Coalition Payoffs

Through knowledge diffusion, participating firms can gain core skills and technology that would be very difficult for them to obtain in a coalition. In particular, the cost savings in a joint venture are depicted as follows:

$$c_i^{\{i\}} \le c_i^{\{i,j\}}, \quad \text{for } i, j \in \{1, 2, 3\} \text{ and } i \ne j,$$

$$c_i^{\{i,j\}} \le c_i^{\{i,j,k\}}, \quad \text{for } i, j, k \in \{1, 2, 3\} \text{ and } i \ne j \ne k.$$
(5.48)

The firms in the joint venture maximize the sum of their profits

$$\int_{t_0}^{T} \sum_{j=1}^{3} \left[P_j[x^j(s)]^{1/2} - c_j^{\{1,2,3\}} u_j(s)\right] \exp\left[-r(s - t_0)\right] ds$$

$$+ \sum_{j=1}^{3} \exp\left[-r(T - t_0)\right] q_j [x^j(T)]^{1/2},$$
(5.49)

subject to (5.48).

Following the analysis in Sect. 5.3, one can obtain

$$W^{(t_0)\{1,2,3\}}(t, x^1, x^2, x^3)$$

$$= \left[A_1^{\{1,2,3\}}(t)(x^1)^{1/2} + A_2^{\{1,2,3\}}(t)(x^2)^{1/2} + A_3^{\{1,2,3\}}(t)(x^3)^{1/2} + C^{\{1,2,3\}}(t)\right]$$

$$\times \exp\left[-r(t - t_0)\right],$$
(5.50)

as in (5.24).

The investment strategies of the grand coalition joint venture can be derived as in (5.25) and the dynamics of technological progress of the joint venture over the time interval $s \in [t_0, T]$ can be expressed as in (5.28).

Once again, we denote the state trajectories of the joint venture over the time interval $s \in [t_0, T]$ as $\{x^{1*}(t), x^{2*}(t), x^{3*}(t)\}_{t=t_0}^{T} \equiv \{x^{*}(t)\}_{t=t_0}^{T}$, and use the terms $x^{i*}(t)$ and x_t^{i*} interchangeably.

For the computation of the dynamic in the Shapley Value, we consider cases when two of the firms form a coalition $\{i, j\} \subset \{1, 2, 3\}$ to maximize joint profit

$$
\int_{t_0}^{T} \left[P_i[x^i(s)]^{1/2} - c_i^{\{i,j\}} u_i(s) + P_j[x^j(s)]^{1/2} - c_j^{\{i,j\}} u_j(s) \right] \exp[-r(s - t_0)] \, ds
$$
$$
+ \exp[-r(T - t_0)]\{q_i[x^i(T)]^{1/2} + q_j[x^j(T)]^{1/2}\}, \tag{5.51}
$$

subject to

$$
\dot{x}^i(s) = \left[\alpha_i [u_i(s) x^i(s)]^{1/2} - \delta x_i(s) \right], \quad x^i(t_0) = x_0^i \in X^i, \tag{5.52}
$$

for $i, j \in \{1, 2, 3\}$ and $i \neq j$.

Following the above analysis, we obtain the following value functions:

$$
W^{(t_0)\{i,j\}}(t, x^i, x^j) = \left[A_i^{\{i,j\}}(t)(x^i)^{1/2} + A_j^{\{i,j\}}(t)(x^j)^{1/2} \right.
$$
$$
\left. + C^{\{i,j\}}(t) \right] \exp[-r(t - t_0)], \tag{5.53}
$$

for $i, j \in \{1, 2, 3\}$ and $i \neq j$, where $A_i^{\{i,j\}}(t)$, $A_j^{\{i,j\}}(t)$, and $C^{\{i,j\}}(t)$ satisfy

$$
\dot{A}_i^{\{i,j\}}(t) = \left(r + \frac{\delta}{2} \right) A_i^{\{i,j\}}(t) - P_i, \quad \text{and} \quad A_i^{\{i,j\}}(T) = q_i,
$$

for $i, j \in \{1, 2, 3\}$ and $i \neq j$;

$$
\dot{C}^{\{i,j\}}(t) = rC^{\{i,j\}}(t) - \sum_{h \in \{i,j\}} \frac{\alpha_h^2}{16 c_h^{\{i,j\}}} [A_h^{\{i,j\}}(t)]^2,
$$

$$
C^{\{i,j\}}(T) = 0.
$$

The block-recursive system in (5.54) can be solved readily by standard techniques. Moreover, one can easily derive for, $\tau \in [t_0, T]$,

$$
W^{(t_0)\{i,j\}}(t, x^i, x^j) = \exp[-r(\tau - t_0)] W^{(\tau)\{i,j\}}(t, x^i, x^j),
$$

for $i, j \in \{1, 2, 3\}$ and $i \neq j$.

5.5.2 PDP for Shapley Value

To formulate a payoff distribution procedure over time so that the agreed imputations satisfy the Shapley Value in Condition 5.3 we present the following.

Proposition 5.3 *A PDP with a terminal payment $q^i(x_T^*)$ at time T and an instantaneous payment at time $\tau \in [t_0, T]$*

$$B_i(\tau) = - \sum_{K \subseteq \{1,2,3\}} \frac{(k-1)!(3-k)!}{3!}$$

$$\times \left\{ \left[W_t^{(\tau)K}(t, x_t^{K*}) \big|_{t=\tau} \right] - \left[W_t^{(\tau)K\setminus i}(t, x_t^{K\setminus i*}) \big|_{t=\tau} \right] \right.$$

$$+ \sum_{h \in K} \left[\frac{\partial}{\partial x_\tau^{h*}} W^{(\tau)K}(\tau, x_\tau^{K*}) \right] \left[\frac{\alpha_h^2}{4c_h} A_h^{\{1,2,3\}}(\tau)(x_\tau^{i*})^{1/2} - \delta x_\tau^{h*} \right]$$

$$\left. - \sum_{h \in K\setminus i} \left[\frac{\partial}{\partial x_\tau^{h*}} W^{(\tau)K\setminus i}(\tau, x_\tau^{K\setminus i*}) \right] \left[\frac{\alpha_h^2}{4c_h} A_h^{\{1,2,3\}}(\tau)(x_\tau^{i*})^{1/2} - \delta x_\tau^{h*} \right] \right\},$$

for $i \in \{1, 2, 3\}$,   (5.54)

will lead to the realization of the Shapley Value imputations $\xi^{(\tau)i}(\tau, x_\tau^{N*})$ in Condition 5.3.

Proof Invoking Theorem 4.2 in Chap. 4 one can readily obtain Proposition 5.3.   □

Using (5.46), (5.50), and (5.53),

$$\left[W_t^{(\tau)i}(t, x_t^{i*}) \big|_{t=\tau} \right] = r \left[A_i^{\{i\}}(\tau)(x_\tau^{i*})^{1/2} + C_i^{\{i\}}(\tau) \right] + \left[\dot{A}_i^{\{i\}}(\tau)(x_\tau^{i*})^{1/2} + \dot{C}_i^{\{i\}}(\tau) \right],$$

for $i \in \{1, 2, 3\}$;

$$\left[W_t^{(\tau)\{i,j\}}(t, x_t^{i*}) \big|_{t=\tau} \right] = r \left[A_i^{\{i,j\}}(\tau)(x_\tau^{i*})^{1/2} + A_j^{\{i,j\}}(\tau)(x_\tau^{j*})^{1/2} + C^{\{i,j\}}(\tau) \right]$$

$$+ \left[\dot{A}_i^{\{i,j\}}(\tau)(x_\tau^{i*})^{1/2} + \dot{A}_j^{\{i,j\}}(\tau)(x_\tau^{j*})^{1/2} + \dot{C}^{\{i,j\}}(\tau) \right],$$

for $i, j \in \{1, 2, 3\}$ and $i \neq j$;

$$\left[W_t^{(\tau)\{1,2,3\}}(t, x_t^{i*}) \big|_{t=\tau} \right] = r \left[A_1^{\{1,2,3\}}(\tau)(x_\tau^{1*})^{1/2} + A_2^{\{1,2,3\}}(\tau)(x_\tau^{2*})^{1/2} \right.$$

$$+ A_3^{\{1,2,3\}}(\tau)(x_\tau^{3*})^{1/2} + C^{\{1,2,3\}}(\tau) \big]$$

$$+ \left[\dot{A}_1^{\{1,2,3\}}(\tau)(x_\tau^{1*})^{1/2} + \dot{A}_2^{\{1,2,3\}}(\tau)(x_\tau^{2*})^{1/2} \right.$$

$$+ \dot{A}_3^{\{1,2,3\}}(\tau)(x_\tau^{3*})^{1/2} + \dot{C}^{\{1,2,3\}}(\tau) \big]; \quad \text{and}$$

$$\left[\frac{\partial}{\partial x_\tau^h} W^{(\tau)K}(\tau, x_\tau^{K*}) \right] = \frac{1}{2} A_h^K(\tau)(x_\tau^{h*})^{-1/2}, \quad \text{for } h \in K \subseteq \{1, 2, 3\}.$$

A time (optimal-trajectory-subgame) consistent solution can be obtained using the set of group optimal strategies $\psi^*(s, x_s^*)$ and $B(s) = \{B_1(s), B_2(s), \ldots, B_n(s)\}$ in (5.54).

Finally, using the cooperative strategies, the instantaneous receipt of firm i at time instant τ is

$$\zeta_i(\tau) = P_i\left(x_\tau^{i*}\right)^{1/2} - \frac{\alpha_i^2}{16(c_i^{\{1,2,3\}})}\left[A_i^{\{1,2,3\}}(\tau)\right]^2 \qquad (5.55)$$

for $\tau \in [t_0, T]$ and $i \in \{1, 2, 3\}$.

According to Proposition 5.3, the instantaneous payment that firm i should receive under the agreed-upon optimality principle is $B_i(\tau)$, for $\tau \in [t_0, T]$ and $i \in \{1, 2, 3\}$, as stated in (5.54). Hence an instantaneous transfer payment

$$\chi^i(\tau) = B_i(\tau) - \zeta_i(\tau) \qquad (5.56)$$

has to be given or charged to firm i at time τ, for $i \in \{1, 2, 3\}$ and $\tau \in [t_0, T]$.

5.6 Infinite-Horizon Analysis

Consider the case when the horizon of the analysis approaches infinity. The state dynamics of the ith firm is characterized by the set of vector-valued differential equations

$$\dot{x}^i(s) = f^i\left[x^i(s), u_i(s)\right], \quad x^i(t_0) = x_0^i, \text{ for } i \in N. \qquad (5.57)$$

The objective of firm i to be maximized is

$$\int_{t_0}^{\infty} \left(g^i\left[x^i(s)\right] - c_i^{\{i\}}\left[u_i(s)\right]\right) \exp\left[-r(s - t_0)\right] ds, \qquad (5.58)$$

for $i \in N$.

Consider the alternative formulation of (5.57) and (5.58) as

$$\max_{u_i} \int_t^{\infty} \left(g^i\left[x^i(s)\right] - c_i^{\{i\}}\left[u_i(s)\right]\right) \exp\left[-r(s - t)\right] ds, \quad \text{for } i \in N, \qquad (5.59)$$

subject to

$$\dot{x}^i(s) = f_i\left[x^i(s), u_i(s)\right], \quad x^i(t) = x^i, \text{ for } i \in N. \qquad (5.60)$$

The infinite-horizon theory of the firm problem in (5.59) and (5.60) is independent of the choice of t and dependent only upon the state at the starting time.

Invoking Theorem 2.4 in Chap. 2, a noncooperative feedback Nash equilibrium solution can be characterized by a set of strategies $\{\phi_i^*(x^i), \text{ for } i \in N\}$, constituting a firm's equilibrium solution to the problem in (5.59) and (5.60), if there exist functionals $\hat{V}^i(x^i) : R^m \to R$ for $i \in N$, satisfying the following set of partial differential equations:

$$r\hat{V}^i\left(x^i\right) = \max_{u_i}\{g(x^i) - c_i^{\{i\}}(u_i) + \hat{V}_x^i(x)f\left[x^i, u_i\right]\}. \qquad (5.61)$$

5.6.1 Dynamic Joint Venture

Consider the case when all these n companies form a joint venture. The cost of control of firm j under the joint venture becomes $c_j^N(u_j)$. With the absolute joint venture cost advantage

$$c_j^N(u_j) \leq c_j^{\{j\}}(u_j), \quad \text{for } j \in N,$$

and

$$\partial c_j^N(u_j)/\partial u_j \leq \partial c_j^{\{j\}}(u_j)/\partial u_j, \quad \text{for } j \in N. \tag{5.62}$$

The joint venture would maximize the joint venture profit

$$\int_t^\infty \sum_{j=1}^n \left(g^j\left[x^j(s)\right] - c_j^N\left[u_j(s)\right]\right) \exp\left[-r(s-t)\right] ds, \tag{5.63}$$

subject to (5.60).

An optimal solution of the control problem in (5.60) and (5.63) can be characterized using Theorem A.2 in the Technical Appendixes as follows.

Corollary 5.3 *A set of control strategies $\{\psi_i^*(x)$ for $i \in N\}$ provides a solution to the control problem in (5.57) and (5.63), if there exist continuously differentiable functions $W(x): R^n \rightarrow R$, satisfying the following partial differential equation:*

$$r W(x) = \max_{u_1, u_2, \dots, u_n} \left\{ \sum_{j=1}^n [g^j(x^j) - c_j^N(u_j)] + \sum_{j=1}^n W_{x_j}(x) f^j(x^j, u_j) \right\}, \tag{5.64}$$

where $x = \{x^1, x^2, \dots, x^n\}$.

Hence the firms will adopt the cooperative control $\{\psi_i^*(x)$, for $i \in N\}$ to obtain the maximized level of joint profit. Substituting this set of control into (5.57) yields the dynamics of technology advancement under cooperation as

$$\dot{x}^i(s) = f^i\left[x^i(s), \psi_i^*(x(s))\right], \quad x^i(t_0) = x_0^i, \text{ for } i \in N. \tag{5.65}$$

Let $x^*(t) = \{x^{1*}(t), x^{2*}(t), \dots, x^{n*}(t)\}$ denote the solution to (5.65). The optimal trajectory $\{x^*(t)\}_{t=t_0}^\infty$ can be expressed as

$$x^{i*}(t) = x_0^i + \int_{t_0}^t f^i\left[x^{i*}(s), \psi_i^*(x^*(s))\right] ds, \quad \text{for } i \in N. \tag{5.66}$$

For notational convenience, we use the terms $x^*(t)$ and x_t^* interchangeably.

Substituting the optimal extraction strategies in $\{\psi_i^*(x), \text{ for } i \in N\}$ into (5.63) yields the venture profit as

$$W(x_t^*) = \int_t^{\infty} \sum_{j=1}^{n} (g^j[x^{j*}(s)] - c_j^N[\psi_j^*(x^*(s))]) \exp[-r(s-t)] \, ds. \quad (5.67)$$

5.6.2 Time (Optimal-Trajectory-Subgame) Consistent Venture Profit Sharing

Consider the case when the firms in the venture share the excess of the total cooperative payoff over the sum of individual noncooperative payoffs proportional to the firms' noncooperative payoffs.

The imputation scheme has to fulfill the following condition.

Condition 5.4 An imputation

$$\xi^{(\tau)i}(\tau, x_\tau^*) = \frac{\hat{V}^i(x_\tau^*)}{\sum_{i=1}^{n} \hat{V}^i(x_\tau^*)} W(x_\tau^*), \quad (5.68)$$

is assigned to firm i, for $i \in N$ at time $\tau \in [t_0, \infty)$.

To formulate a payoff distribution procedure over time so that the agreed imputations satisfy Condition 5.4 we obtain the following.

Proposition 5.4 *A PDP with an instantaneous payment at time* $\tau \in [t_0, \infty)$

$$B_i(\tau) = r \left[\frac{\hat{V}^i(x_\tau^{i*})}{\sum_{j=1}^{n} \hat{V}^j(x_\tau^{j*})} W(x_\tau^*) \right]$$

$$- \sum_{h=1}^{n} \frac{\partial}{\partial x_\tau^{h*}} \left[\frac{\hat{V}^i(x_\tau^{i*})}{\sum_{j=1}^{n} \hat{V}^j(x_\tau^{j*})} W(x_\tau^*) \right] f^h[x_\tau^{h*}, \psi_h^*(x_\tau^*)], \quad (5.69)$$

for $i \in N$, *will lead to the realization of the solution imputations in Condition 5.4.*

Proof Invoking Theorem 4.3 in Chap. 4 one can obtain Proposition 5.4. $\qquad \square$

A time (optimal-trajectory-subgame) consistent solution can be obtained as $P(\tau, x_\tau) = \{u(s) \text{ and } B(s), \text{ for } s \in [\tau, \infty) \text{ and } \xi^{(\tau)}(\tau, x_\tau)\}$, with

(i) $u(s)$ for $s \in [\tau, \infty)$ being the set of group optimal strategies $\psi^*(x_s^*)$ characterized in (5.64), and
(ii) $B(s) = \{B_1(s), B_2(s), \ldots, B_n(s)\}$ in (5.69).

With firms using the cooperative investment strategies $\{\psi_i^*(x_\tau^*), \text{for } i \in N\}$, the instantaneous receipt of firm i at time instant τ is

$$\zeta_i(\tau) = g^i(x_\tau^{i*}) - c_i^N[\psi_i^*(x_\tau^*)], \quad \text{for } \tau \in [t_0, \infty) \text{ and } i \in N.$$

According to Proposition 5.4, the instantaneous payment that firm i should receive under the agreed-upon optimality principle is $B_i(\tau)$, for $\tau \in [t_0, \infty)$ and $i \in N$, as stated in (5.69). Hence an instantaneous transfer payment

$$\chi^i(\tau) = B_i(\tau) - \zeta_i(\tau),$$

has to be given or charged to firm i at time τ, for $i \in N$.

5.6.3 Shapley Value Profit Sharing

Consider again the infinite-horizon dynamic venture model in (5.59) and (5.60). The member firms would maximize their joint profit and share their cooperative profits according to the Shapley Value. If firms are allowed to form different coalitions consisting of a subset of companies, $K \subseteq N$. There are k firms in the subset K. In particular, under the coalition they can obtain cost reduction and with absolute joint venture cost advantage

$$c_j^K[u_j(s)] \le c_j^L[u_j(s)], \quad \text{for } j \in L \subseteq K, \tag{5.70}$$

where $c_j^K[u_j(s)]$ represents the costs of the controls of the firm j in the subset K and $c_j^L[u_j(s)]$ represents the costs of the controls of the firm j in the subset L.

Moreover, marginal cost advantages lead to

$$\partial c_j^K[u_j(s)]/\partial u_j(s) \le \partial c_j^L[u_j(s)]/\partial u_j(s), \quad \text{for } j \in L \subseteq K.$$

The profit to the joint venture K becomes

$$\int_t^\infty \sum_{j \in K} \left(g^j[s, x^j(s)] - c_j^K[u_j(s)]\right) \exp[-r(s-t)]\, ds, \tag{5.71}$$

for $K \subseteq N$.

To compute the profit of the joint venture K we have to consider the optimal control problem in (5.70) and (5.71). Invoking Bellman's technique of dynamic programming as in Theorem A.2 of the Technical Appendixes, the solution to the optimal control problem can be characterized as follows.

Corollary 5.4 *A set of controls $\{\psi_i^{K*}(x^K), \text{for } i \in K \text{ and } t \in [t_0, \infty)\}$ provides an optimal solution to the control problem in (5.57) and (5.71) if there exists a continuously differentiable function $W^K(x^K) : R^k \to R$ satisfying the following Bellman* *equation:*

$$r W^K\left(x^K\right) = \max_{u_K}\left\{\sum_{j\in K}\left[g^j\left(x^j\right) - c_j^K\left(u_j\right)\right] + \sum_{j\in K} W_{x_j}^K\left(x^K\right) f^j\left[x^j, u_j\right]\right\}. \quad (5.72)$$

Now consider the case of a grand coalition N in which all the n firms are in the coalition. Using the result in Corollary 5.3, the cooperative state trajectory can be obtained as in (5.66).

To share the venture profit among participating firms according to the Shapley Value the imputation has to satisfy the following condition.

Condition 5.5 An imputation

$$\xi^{(\tau)i}\left(\tau, x_\tau^{N*}\right) = \sum_{K\subseteq N}\frac{(k-1)!(n-k)!}{n!}\left[W^K\left(x_\tau^{K*}\right) - W^{K\setminus i}\left(x_\tau^{K\setminus i*}\right)\right] \quad (5.73)$$

is assigned to firm i for $i \in N$ at time τ when the state is x_τ^*.

To formulate a payoff distribution procedure over time so that the agreed imputations satisfy the Shapley Value in Condition 5.5 we obtain the following.

Proposition 5.5 *A PDP with an instantaneous payment at time $\tau \in [t_0, \infty)$*

$$B_i(\tau) = -\sum_{K\subseteq N}\frac{(k-1)!(n-k)!}{n!}\left\{r W^{K\setminus i}\left(x_\tau^{K\setminus i*}\right) - r W^K\left(x_\tau^{K*}\right)\right.$$

$$+ \sum_{h\in K}\left[\frac{\partial}{\partial x_\tau^{h*}} W^K\left(x_\tau^{K*}\right)\right] f^h\left[x_\tau^{h*}, \psi_h^*\left(x_\tau^*\right)\right]$$

$$\left. - \sum_{h\in K\setminus i}\left[\frac{\partial}{\partial x_\tau^{h*}} W^{K\setminus i}\left(x_\tau^{K\setminus i*}\right)\right] f^h\left[x_\tau^{h*}, \psi_h^*\left(x_\tau^*\right)\right]\right\}, \quad \textit{for } i \in N, \quad (5.74)$$

will lead to the realization of the Shapley Value in Condition 5.5.

Proof Invoking Theorem 4.3 in Chap. 4 one can readily obtain Proposition 5.5. □

A time (optimal-trajectory-subgame) consistent solution can be obtained with the group optimal strategies $\psi^*(x_s^*)$ characterized in (5.72) and $B(s) = \{B_1(s), B_2(s), \dots, B_n(s)\}$ in (5.74).

5.7 An Infinite-Horizon Joint Venture

Consider the infinite-horizon version of the three-company joint venture in Sect. 5.3. The planning period is $[t_0, \infty)$. Company i's profit is

$$\int_{t_0}^{\infty} \left[P_i \left[x^i(s) \right]^{1/2} - c_i^{\{i\}} u_i(s) \right] \exp\left[-r(s - t_0) \right] ds, \qquad (5.75)$$

for $i \in N = \{1, 2, 3\}$.

The evolution of the technology level of company i follows the dynamics

$$\dot{x}^i(s) = \left[\alpha_i \left[u_i(s) x^i(s) \right]^{1/2} - \delta x^i(s) \right],$$
$$x^i(t_0) = x_0^i \in X^i, \quad \text{for } i \in \{1, 2, 3\}. \qquad (5.76)$$

In the case when each of these three firms acts independently, using Theorem A.2 in the Technical Appendixes, we obtain the Bellman equation as

$$r W^i \left(x^i \right) = \max_{u_i} \left\{ \left[P_i \left(x^i \right)^{1/2} - c_i^{\{i\}} u_i \right] + W_{x^i}^i \left(x^i \right) \left[\alpha_i \left(u^i x_i \right)^{1/2} - \delta x^i \right] \right\}, \qquad (5.77)$$

for $i \in \{1, 2, 3\}$.

Performing the indicated maximization yields

$$u_i = \frac{\alpha_i^2}{4(c_i^{\{i\}})^2} \left[V_{x^i}^i \left(x^i \right) \right]^2 x^i, \quad \text{for } i \in \{1, 2, 3\}.$$

Substituting u_i into the Bellman equation yields

$$r V^i \left(x^i \right) = P_i \left(x^i \right)^{1/2} - \frac{\alpha_i^2}{4 c_i^{\{i\}}} \left[V_{x^i}^i \left(x^i \right) \right]^2 x^i$$

$$+ \frac{\alpha_i^2}{2 c_i^{\{i\}}} \left[V_{x^i}^i \left(x^i \right) \right]^2 x^i - \delta V_{x^i}^i \left(x^i \right) x_i, \quad \text{for } i \in [1, 2, 3].$$

Solving the above system of partial differential equations yields

$$V^i \left(x^i \right) = \left[A_i^{\{i\}} \left(x^i \right)^{1/2} + C_i^{\{i\}} \right], \quad \text{for } i \in \{1, 2, 3\}, \qquad (5.78)$$

where

$$0 = \left(r + \frac{\delta}{2} \right) A_i^{\{i\}} - P_i, \qquad r C_i^{\{i\}} = \frac{\alpha_i^2}{16 c_i^{\{i\}}} \left(A_i^{\{i\}} \right)^2.$$

After obtaining the noncooperative outcome we move on to consider the formation of a joint venture with these three firms.

5.7.1 Joint Venture and Costs

Consider the case when all three firms agree to form a joint venture and share their joint profit proportional to their noncooperative profits. With joint venture cost advantage

$$c_j^{\{1,2,3\}} \leq c_j^{\{j\}}, \quad \text{for } j \in N, \tag{5.79}$$

the profit of the joint venture is the sum of the participating firms' profits

$$\int_{t_0}^{\infty} \sum_{j=1}^{3} \left[P_j \left[x^j(s) \right]^{1/2} - c_j^{\{1,2,3\}} u_j(s) \right] \exp\left[-r(s - t_0) \right] ds. \tag{5.80}$$

The firms in the joint venture then act cooperatively to maximize (5.80) subject to (5.76). Using Theorem A.2 in the Technical Appendixes, we obtain the Bellman equation as

$$rW^{\{1,2,3\}}\left(x^1, x^2, x^3\right) = \max_{u_1, u_2, u_3} \left\{ \sum_{i=1}^{3} \left[P_i \left(x^i\right)^{1/2} - c_i^{\{1,2,3\}} u_i \right] \right.$$

$$\left. + \sum_{i=1}^{3} W_{x^i}^{\{1,2,3\}}\left(x^1, x^2, x^3\right) \left[\alpha_i \left[u_i x^i \right]^{1/2} - \delta x^i \right] \right\}. \tag{5.81}$$

Performing the indicated maximization yields

$$u_i = \frac{\alpha_i^2}{4(c_i^{\{1,2,3\}})^2} \left[W_{x^i}^{\{1,2,3\}}\left(x^1, x^2, x^3\right) \right]^2 x^i, \quad \text{for } i \in \{1, 2, 3\}. \tag{5.82}$$

Substituting (5.82) into (5.81) yields

$$rW^{\{1,2,3\}}\left(x^1, x^2, x^3\right)$$

$$= \sum_{i=1}^{3} \left[P_i \left(x^i\right)^{1/2} - \frac{\alpha_i^2 x^i}{4c_i^{\{1,2,3\}}} \left[W_{x^i}^{\{1,2,3\}}\left(x^1, x^2, x^3\right) \right]^2 \right]$$

$$+ \sum_{i=1}^{3} W_{x^i}^{\{1,2,3\}}(x_1, x_2, x_3) \left[\frac{\alpha_i^2}{2(c_i^{\{1,2,3\}})^2} \left[W_{x^i}^{\{1,2,3\}}(x_1, x_2, x_3) \right] x^i - \delta x^i \right]. \tag{5.83}$$

Solving (5.83) yields

$$W^{\{1,2,3\}}\left(x^1, x^2, x^3\right) = \left[A_1^{\{1,2,3\}}\left(x^1\right)^{1/2} + A_2^{\{1,2,3\}}\left(x^2\right)^{1/2} \right.$$

$$\left. + A_3^{\{1,2,3\}}\left(x^3\right)^{1/2} + C^{\{1,2,3\}} \right], \tag{5.84}$$

where $A_1^{\{1,2,3\}}$, $A_2^{\{1,2,3\}}$, $A_3^{\{1,2,3\}}$, and $C^{\{1,2,3\}}$ satisfy

$$0 = \left(r + \frac{\delta}{2}\right)A_i^{\{1,2,3\}} - P_i$$

for $i, j, h \in \{1, 2, 3\}$ and $i \neq j \neq h$,

$$rC^{\{1,2,3\}} = \sum_{i=1}^{3} \frac{\alpha_i^2}{16 c_i^{\{1,2,3\}}} \left(A_i^{\{1,2,3\}}\right)^2. \tag{5.85}$$

The investment strategies of the grand coalition joint venture can be derived as

$$\psi_i^{\{1,2,3\}}(x) = \frac{\alpha_i^2}{16(c_i^{\{1,2,3\}})^2}\left(A_i^{\{1,2,3\}}\right)^2, \quad \text{for } i \in \{1, 2, 3\}. \tag{5.86}$$

The dynamics of the technological progress of the joint venture over the time interval $s \in [t_0, \infty)$ can be expressed as

$$\dot{x}^i(s) = \frac{\alpha_i^2}{4c_i}A_i^{\{1,2,3\}}\left[x^i(s)\right]^{1/2} - \delta x^i(s), \quad x^i(t_0) = x_0^i, \text{ for } i \in \{1, 2, 3\}. \tag{5.87}$$

Taking the transforming $y^i(s) = x^i(s)^{1/2}$, for $i \in \{1, 2, 3\}$, the equation system in (5.87) can be expressed as

$$\dot{y}^i(s) = \frac{\alpha_i^2}{8c_i}A_i^{\{1,2,3\}} - \frac{\delta}{2}y^i(s), \quad y^i(t_0) = \left(x_0^i\right)^{1/2}, \text{ for } i \in \{1, 2, 3\}. \tag{5.88}$$

Equation (5.88) is a system of linear differential equations that can be solved by standard techniques. Solving (5.88) yields the joint venture's state trajectory. Let $\{y^{1*}(t), y^{2*}(t), y^{3*}(t)\}$ denote the solution to (5.88). Transforming $x^i = (y^i)^2$, we obtain the state trajectories of the joint venture over the time interval $s \in [t_0, \infty)$ as

$$\begin{aligned}
\left\{x^*(t)\right\}_{t=t_0}^{\infty} &\equiv \left\{x^{1*}(t), x^{2*}(t), x^{3*}(t)\right\}_{t=t_0}^{T} \\
&= \left\{\left[y^{1*}(t)\right]^2, \left[y^{2*}(t)\right]^2, \left[y^{3*}(t)\right]^2\right\}_{t=t_0}^{T}.
\end{aligned} \tag{5.89}$$

Once again, we use the terms $x^{i*}(t)$ and x_t^{i*} interchangeably.

5.7.2 Time (Optimal-Trajectory-Subgame) Consistent Venture Profit Sharing

If the firms agree to share their joint profit proportional to their noncooperative profits the imputation scheme has to fulfill the following condition.

Condition 5.6 An imputation

$$\xi^{(\tau)i}\left(\tau, x_\tau^*\right) = \frac{\hat{V}^i(x_\tau^{i*})}{\sum_{j=1}^3 \hat{V}^j(x_\tau^{i*})} W^{\{1,2,3\}}\left(x_\tau^{1*}, x_\tau^{2*}, x_\tau^{3*}\right), \qquad (5.90)$$

is assigned to firm i, for $i \in \{1, 2, 3\}$ at time τ when the state is x_τ^*.

To formulate a payoff distribution procedure over time so that the agreed imputations in Condition 5.6 are satisfied we obtain the following.

Proposition 5.6 *A PDP with an instantaneous payment at time $\tau \in [t_0, \infty)$*

$$B_i(\tau) = r \frac{\hat{V}^i(x_\tau^{i*})}{\sum_{j=1}^3 \hat{V}^j(x_\tau^{i*})} W^{\{1,2,3\}}\left(x_\tau^{1*}, x_\tau^{2*}, x_\tau^{3*}\right)$$

$$- \sum_{h=1}^3 \frac{\partial}{\partial x_\tau^{h*}} \frac{\hat{V}^i(x_\tau^{i*})}{\sum_{j=1}^3 \hat{V}^j(x_\tau^{i*})} W^{\{1,2,3\}}\left(x_\tau^{1*}, x_\tau^{2*}, x_\tau^{3*}\right)$$

$$\times \left[\frac{\alpha_h^2}{4c_h} A_h^{\{1,2,3\}}\left(x_\tau^{h*}\right)^{1/2} - \delta x_\tau^{h*}(s)\right], \quad for \ i \in \{1, 2, 3\}, \qquad (5.91)$$

will lead to the realization of the imputation in Condition 5.6.

Proof Invoking Proposition 5.4 one can obtain Proposition 5.6.   □

In particular,

$$\frac{\hat{V}^i(x_\tau^{i*})}{\sum_{j=1}^3 \hat{V}^j(x_\tau^{i*})} W^{\{1,2,3\}}\left(x_\tau^{1*}, x_\tau^{2*}, x_\tau^{3*}\right)$$

$$= \frac{[A_i^{\{i\}}(x^{i*})^{1/2} + C_i^{\{i\}}]}{(\sum_{j=1}^3 [A_j^{\{j\}}(x_\tau^{j*})^{1/2} + C_j^{\{j\}}])}$$

$$\times \left[A_1^{\{1,2,3\}}\left(x_\tau^{1*}\right)^{1/2} + A_2^{\{1,2,3\}}\left(x_\tau^{2*}\right)^{1/2} + A_3^{\{1,2,3\}}\left(x_\tau^{3*}\right)^{1/2} + C^{\{1,2,3\}}\right];$$

$$\frac{\partial}{\partial x_\tau^{i*}} \frac{\hat{V}^i(x_\tau^{i*})}{\sum_{j=1}^3 \hat{V}^j(x_\tau^{i*})} W^{\{1,2,3\}}\left(x_\tau^{1*}, x_\tau^{2*}, x_\tau^{3*}\right)$$

$$= \left[A_1^{\{1,2,3\}}\left(x_\tau^{1*}\right)^{1/2} + A_2^{\{1,2,3\}}\left(x_\tau^{2*}\right)^{1/2} + A_3^{\{1,2,3\}}\left(x_\tau^{3*}\right)^{1/2} + C^{\{1,2,3\}}\right]$$

$$\times \left[\frac{(\sum_{j=1}^3 [A_j^{\{j\}}(x_\tau^{j*})^{1/2} + C_j^{\{j\}}])\frac{1}{2}A_i^{\{i\}}(x^{i*})^{-1/2}}{(\sum_{j=1}^3 [A_j^{\{j\}}(x_\tau^{j*})^{1/2} + C_j^{\{j\}}])^2}\right.$$

$$- \frac{[A_i^{\{i\}}(x^{i*})^{1/2} + C_i^{\{i\}}(t)]\frac{1}{2}A_i^{\{i\}}(x^{i*})^{-1/2}}{(\sum_{j=1}^{3}[A_j^{\{j\}}(x_\tau^{j*})^{1/2} + C_j^{\{j\}}])^2}\Bigg]$$

$$+ \frac{[A_i^{\{i\}}(x^{i*})^{1/2} + C_i^{\{i\}}]}{(\sum_{j=1}^{3}[A_j^{\{j\}}(x_\tau^{j*})^{1/2} + C_j^{\{j\}}])^2}\frac{1}{2}A_i^{\{1,2,3\}}(x_\tau^{i*})^{-1/2};$$

and

$$\frac{\partial}{\partial x_\tau^{h*}}\frac{\hat{V}^i(x_\tau^{i*})}{\sum_{j=1}^{3}\hat{V}^j(x_\tau^{i*})}W^{\{1,2,3\}}(x_\tau^{1*}, x_\tau^{2*}, x_\tau^{3*})$$

$$= [A_1^{\{1,2,3\}}(x_\tau^{1*})^{1/2} + A_2^{\{1,2,3\}}(x_\tau^{2*})^{1/2} + A_3^{\{1,2,3\}}(x_\tau^{3*})^{1/2} + C^{\{1,2,3\}}]$$

$$\times \left[-\frac{[A_i^{\{i\}}(x^{i*})^{1/2} + C_i^{\{i\}}]\frac{1}{2}A_h^{\{h\}}(x^{h*})^{-1/2}}{(\sum_{j=1}^{3}[A_j^{\{j\}}(x_\tau^{j*})^{1/2} + C_j^{\{j\}}])^2}\right]$$

$$+ \frac{[A_i^{\{i\}}(x^{i*})^{1/2} + C_i^{\{i\}}]}{(\sum_{j=1}^{3}[A_j^{\{j\}}(x_\tau^{j*})^{1/2} + C_j^{\{j\}}])^2}\frac{1}{2}A_h^{\{1,2,3\}}(x_\tau^{h*})^{-1/2},$$

for $h \neq i$.

A time (optimal-trajectory-subgame) consistent solution can be obtained with the group optimal strategies $\psi^*(x_s^*)$ characterized in (5.86) and $B(s) = \{B_1(s), B_2(s), \dots, B_n(s)\}$ in (5.91).

Using the cooperative strategies the instantaneous receipt of firm i at time instant τ is

$$\zeta_i(\tau) = P_i(x_\tau^{i*})^{1/2} - \frac{\alpha_i^2}{16(c_i^{\{1,2,3\}})}(A_i^{\{1,2,3\}})^2, \tag{5.92}$$

for $i \in \{1, 2, 3\}$ along the cooperative path $\{x^*(t)\}_{t=t_0}^{\infty}$.

According to Proposition 5.6, the instantaneous payment that firm i should receive under the agreed-upon optimality principle is $B_i(\tau)$, for $i \in \{1, 2, 3\}$, as stated in (5.91). Hence an instantaneous transfer payment

$$\chi^i(\tau) = B_i(\tau) - \zeta_i(\tau), \tag{5.93}$$

has to be given or charged to firm i at time τ, for $i \in \{1, 2, 3\}$ along the cooperative path $\{x^*(t)\}_{t=t_0}^{\infty}$.

5.7.3 Shapley Value Solution

Consider the case when the participating firms agree to share their cooperative profits according to the Shapley Value. For the computation of the dynamic in the Shapley Value we consider cases when two of the firms form a coalition $\{i, j\} \subset \{1, 2, 3\}$. In particular, they can obtain cost reduction, and with joint venture cost advantage

$$
\begin{aligned}
c_i^{\{i\}} &\leq c_i^{\{i,j\}}, \quad \text{for } i, j \in \{1, 2, 3\} \text{ and } i \neq j, \\
c_i^{\{i,j\}} &\leq c_i^{\{i,j,k\}}, \quad \text{for } i, j, k \in \{1, 2, 3\} \text{ and } i \neq j \neq k.
\end{aligned} \tag{5.94}
$$

To maximize the joint profit of coalition $\{i, j\}$, the firms consider the problem of maximizing

$$
\int_{t_0}^{\infty} \left[P_i[x^i(s)]^{1/2} - c_i^{\{i,j\}} u_i(s) + P_j[x^j(s)]^{1/2} - c_j^{\{i,j\}} u_j(s) \right] \exp\left[-r(s - t_0)\right] ds,
$$

$$\tag{5.95}$$

subject to (5.76).

Following the above analysis, we obtain the following value functions:

$$
W^{\{i,j\}}(x^i, x^j) = \left[A_i^{\{i,j\}}(x^i)^{1/2} + A_j^{\{i,j\}}(x^j)^{1/2} + C^{\{i,j\}} \right], \tag{5.96}
$$

for $i, j, \in \{1, 2, 3\}$ and $i \neq j$, where $A_i^{\{i,j\}}$, $A_j^{\{i,j\}}$, and $C^{\{i,j\}}$ satisfy

$$
0 = \left(r + \frac{\delta}{2} \right) A_i^{\{1,2\}} - P_i, \quad \text{for } i, j, \in \{1, 2, 3\} \text{ and } i \neq j, \quad \text{and}
$$

$$
rC^{\{i,j\}} = \sum_{h \in \{i,j\}} \frac{\alpha_h^2}{16c_h^{\{i,j\}}} \left(A_h^{\{i,j\}} \right)^2.
$$

To formulate a payoff distribution procedure over time so that the agreed imputations satisfy the Shapley Value in Condition 5.5 we obtain the following.

Proposition 5.7 *A PDP with an instantaneous payment at time $\tau \in [t_0, \infty)$*

$$
B_i(\tau) = - \sum_{K \subseteq \{1,2,3\}} \frac{(k-1)!(3-k)!}{3!} \left\{ r W^{K \setminus i}\left(x_\tau^{K \setminus i*}\right) - r W^K\left(x_\tau^{K*}\right) \right.
$$

$$
+ \sum_{h \in K} \left[\frac{\partial}{\partial x_\tau^{h*}} W^K\left(x_\tau^{K*}\right) \right] \left[\frac{\alpha_h^2}{4c_h} A_h^{\{1,2,3\}}\left(x_\tau^{i*}\right)^{1/2} - \delta x_\tau^{h*} \right]
$$

$$
- \sum_{h \in K \setminus i} \left[\frac{\partial}{\partial x_\tau^{h*}} W^{K \setminus i}\left(x_\tau^{K \setminus i*}\right) \right] \left[\frac{\alpha_h^2}{4c_h} A_h^{\{1,2,3\}}\left(x_\tau^{i*}\right)^{1/2} - \delta x_\tau^{h*} \right] \right\},
$$

for $i \in \{1, 2, 3\}$, $\tag{5.97}$

will lead to the realization of the Shapley Value in Condition 5.5.

Proof Invoking Proposition 5.5 the results in Proposition 5.7 follow.   □

In particular, $W^K(x_\tau^{K*})$ is given in (5.78), (5.84), and (5.96), and

$$\left[\frac{\partial}{\partial x_\tau^{h*}} W^K\left(x_\tau^{K*}\right)\right] = \frac{1}{2} A_h^K\left(x_\tau^{h*}\right)^{-1/2}, \quad \text{for } h \in K \subseteq \{1, 2, 3\}.$$

A time (optimal-trajectory-subgame) consistent solution can be obtained with the group optimal strategies $\psi^*(x_s^*)$ characterized in (5.86) and the PDP $B(s) = \{B_1(s), B_2(s), \ldots, B_n(s)\}$ in (5.97).

5.8 Exercises

5.1 Consider the case when there are three companies involved in a joint venture. The planning period is $[0, 2]$. We use $x^i(s)$ to denote the level of technology of company i at time $s \in [0, 2]$, and $u_i(s) \subset R^+$ is its physical investment in technological advancement. The discount rate is 0.05. The salvage values of the firms' technologies are $2[x^1(2)]^{1/2}$, $[x^2(2)]^{1/2}$, and $3[x^3(2)]^{1/2}$. If they act independently, the costs of the physical investment of these three companies are, respectively,

$$2u_1(s), \qquad 3u_2(s), \quad \text{and} \quad 2u_3(s).$$

The profits for companies 1, 2, and 3 are, respectively,

$$\int_0^2 \left[10\left[x^1(s)\right]^{1/2} - 2u_1(s)\right]\exp(-0.05s)\,ds + \exp\left[-0.05(4)\right]2\left[x^1(2)\right]^{1/2},$$

$$\int_0^2 \left[9\left[x^2(s)\right]^{1/2} - 3u_2(s)\right]\exp(-0.05s)\,ds + \exp\left[-0.05(4)\right]\left[x^2(2)\right]^{1/2}, \quad \text{and}$$

$$\int_0^2 \left[8\left[x^3(s)\right]^{1/2} - 2u_3(s)\right]\exp(-0.05s)\,ds + \exp\left[-0.05(4)\right]3\left[x^3(2)\right]^{1/2}.$$

The evolution of the technology level of company $i \in \{1, 2, 3\}$ follows the dynamics

$$\dot{x}^1(s) = \left[2\left[u_1(s)x^1(s)\right]^{1/2} - 0.01x^1(s)\right], \quad x^1(0) = 20,$$

$$\dot{x}^2(s) = \left[\left[u_2(s)x^2(s)\right]^{1/2} - 0.03x^2(s)\right], \quad x^2(0) = 10, \quad \text{and}$$

$$\dot{x}^3(s) = \left[1.5\left[u_3(s)x^3(s)\right]^{1/2} - 0.02x^3(s)\right], \quad x^3(0) = 15.$$

Compute a Nash equilibrium solution when these three firms act independently.

5.2 Consider the case when these three firms form a joint venture. The participating firms in a coalition can gain core skills and technology from each other. In particular, they can obtain cost reduction and, with absolute joint venture, cost advantage.

With joint venture cost advantage, the cost of investment of firm $j \in \{1, 2, 3\}$ under the joint venture becomes $c_j^{\{1,2,3\}} u_j(s)$, where $c_1^{\{1,2,3\}} = 1$, $c_2^{\{1,2,3\}} = 1.2$, and $c_3^{\{1,2,3\}} = 0.8$.

If the joint venture firms agree to maximize their joint profit and share the excess gain equally, characterize a time (optimal-trajectory-subgame) consistent solution.

5.3 Consider the joint venture in exercise 5.2. In particular, the firms would like to share the venture profit according to the Shapley Value. The costs under joint ventures in different coalitions $K \subseteq \{1, 2, 3\}$ are

$$c_1^{\{1,2,3\}} = 1, \qquad c_2^{\{1,2,3\}} = 1.2, \quad \text{and} \quad c_3^{\{1,2,3\}} = 0.8;$$

$$c_1^{\{1,2\}} = 1.5 \quad \text{and} \quad c_2^{\{1,2\}} = 2;$$

$$c_1^{\{1,3\}} = 1.3 \quad \text{and} \quad c_3^{\{1,3\}} = 1.2;$$

$$c_2^{\{2,3\}} = 1.8 \quad \text{and} \quad c_3^{\{2,3\}} = 1.1;$$

$$c_1^{\{1\}} = 2, \qquad c_2^{\{2\}} = 3, \quad \text{and} \quad c_3^{\{3\}} = 2.$$

Characterize a time (optimal-trajectory-subgame) consistent solution.

5.4 Prove that the coalition profits in Exercise 5.3 are superadditive.

Appendix: Proof of Proposition 5.2

To prove Proposition 5.2 we first use $\hat{x}^{j(L)}$, for $j \in L$, to denote the optimal trajectory of the optimal control problem $\varpi[L; \tau, x_\tau^L]$, which maximizes

$$\int_\tau^T \sum_{j \in L} \{g^j[s, x^j(s)] - c_j^L[u_j(s)]\} \exp\left[-\int_\tau^s r(y)\,dy\right] ds$$

$$+ \sum_{j \in L} \exp\left[-\int_{t_0}^T r(y)\,dy\right] q^j\left(x^j(T)\right)$$

subject to

$$\dot{x}^j(s) = f^j[s, x^j(s), u_j(s)], \quad x^j(\tau) = x_\tau^j, \ \text{for } j \in L.$$

Note that

$$W^{(\tau)L}\left(\tau, x_\tau^L\right)$$

$$= \int_\tau^T \sum_{j \in L} \{g^j[s, \hat{x}^{j(L)}(s)] - c_j^L[\psi_j^{(\tau)L*}(s, \hat{x}^{L(L)}(s))]\} \exp\left[-\int_\tau^s r(y)\,dy\right] ds$$

$$+ \sum_{j \in L} \exp\left[-\int_{\tau}^{T} r(y)\,dy\right] q^{j}\left(\hat{x}^{j(L)}(T)\right)$$

$$\leq \int_{\tau}^{T} \sum_{j \in L} \{g^{j}\left[s, \hat{x}^{j(L)}(s)\right] - c_{j}^{K}\left[\psi_{j}^{(\tau)L*}\left(s, \hat{x}^{L(L)}(s)\right)\right]\} \exp\left[-\int_{\tau}^{s} r(y)\,dy\right] ds$$

$$+ \sum_{j \in L} \exp\left[-\int_{\tau}^{T} r(y)\,dy\right] q^{j}\left(\hat{x}^{j(L)}(T)\right)$$

because $c_{j}^{K}\left[u_{j}(s)\right] \leq c_{j}^{L}\left[u_{j}(s)\right]$, for $j \in L \subseteq K$. \hfill (5.98)

Similarly, for the optimal control problem $\varpi[K \backslash L; \tau, x_{\tau}^{K \backslash L}]$, we have

$$W^{(\tau)K \backslash L}\left(\tau, x_{\tau}^{K \backslash L}\right)$$

$$= \int_{\tau}^{T} \sum_{j \in K \backslash L} \{g^{j}\left[s, \hat{x}^{j(K \backslash L)}(s)\right] - c_{j}^{K \backslash L}\left[\psi_{j}^{(\tau)K \backslash L*}\left(s, \hat{x}^{K \backslash L(K \backslash L)}(s)\right)\right]\}$$

$$\times \exp\left[-\int_{\tau}^{s} r(y)\,dy\right] ds$$

$$+ \sum_{j \in K \backslash L} \exp\left[-\int_{\tau}^{T} r(y)\,dy\right] q^{j}\left(\hat{x}^{j(K \backslash L)}(T)\right)$$

$$\leq \int_{\tau}^{T} \sum_{j \in K \backslash L} \{g^{j}\left[s, \hat{x}^{j(K \backslash L)}(s)\right] - c_{j}^{K}\left[\psi_{j}^{(\tau)K \backslash L*}\left(s, \hat{x}^{K \backslash L(K \backslash L)}(s)\right)\right]\}$$

$$\times \exp\left[-\int_{\tau}^{s} r(y)\,dy\right] ds$$

$$+ \sum_{j \in K \backslash L} \exp\left[-\int_{\tau}^{T} r(y)\,dy\right] q^{j}\left(\hat{x}^{j(K \backslash L)}(T)\right)$$

because $c_{j}^{K}\left[u_{j}(s)\right] \leq c_{j}^{K \backslash L}\left[u_{j}(s)\right]$, for $j \in K \backslash L \subseteq K$. \hfill (5.99)

Now consider the optimal control problem $\varpi[K; \tau, x_{\tau}^{K}]$ that maximizes

$$\int_{\tau}^{T} \sum_{j \in K} \{g^{j}\left[s, x^{j}(s)\right] - c_{j}^{K}\left[u_{j}(s)\right]\} \exp\left[-\int_{\tau}^{s} r(y)\,dy\right] ds$$

$$+ \sum_{j \in K} \exp\left[-\int_{t_{0}}^{T} r(y)\,dy\right] q^{j}\left(x^{j}(T)\right),$$

subject to

$$\dot{x}^{j}(s) = f^{j}\left[s, x^{j}(s), u_{j}(s)\right], \quad x^{j}(\tau) = x_{\tau}^{j}, \text{ for } j \in K.$$

Since $\psi_j^{(\tau)K*}(s, \hat{x}^{K(K)}(s))$ and $\hat{x}^{K(K)}(s)$ are, respectively, the optimal control and optimal state trajectory of the control problem $\varpi[K; \tau, x_\tau^K]$,

$$W^{(\tau)K}\left(\tau, x_\tau^K\right)$$

$$= \int_\tau^T \sum_{j \in K} \{g^j[s, \hat{x}^{j(K)}(s)] - c_j^K[\psi_j^{(\tau)K*}(s, \hat{x}^{K(K)}(s))]\} \exp\left[-\int_\tau^s r(y)\,dy\right] ds$$

$$+ \sum_{j \in K} \exp\left[-\int_\tau^T r(y)\,dy\right] q^j\left(\hat{x}^{j(K)}(T)\right)$$

$$\geq \int_\tau^T \sum_{j \in L} \{g^j[s, \hat{x}^{j(L)}(s)] - c_j^K[\psi_j^{(\tau)L*}(s, \hat{x}^{L(L)}(s))]\} \exp\left[-\int_\tau^s r(y)\,dy\right] ds$$

$$+ \sum_{j \in L} \exp\left[-\int_\tau^T r(y)\,dy\right] q^j\left(\hat{x}^{j(L)}(T)\right)$$

$$+ \int_\tau^T \sum_{j \in K \backslash L} \{g^j[s, \hat{x}^{j(K \backslash L)}(s)] - c_j^K[\psi_j^{(\tau)K \backslash L*}(s, \hat{x}^{K \backslash L(K \backslash L)}(s))]\}$$

$$\times \exp\left[-\int_\tau^s r(y)\,dy\right] ds$$

$$+ \sum_{j \in K \backslash L} \exp\left[-\int_\tau^T r(y)\,dy\right] q^j\left(\hat{x}^{j(K \backslash L)}(T)\right). \tag{5.100}$$

Invoking (5.98), (5.99), and (5.100), we have

$$W^{(\tau)K}\left(\tau, x_\tau^K\right) \geq W^{(\tau)L}\left(\tau, x_\tau^L\right) + W^{(\tau)K \backslash L}\left(\tau, x_\tau^{K \backslash L}\right).$$

Hence Proposition 5.2 follows.

Chapter 6
Collaborative Environmental Management

After decades of rapid technological advancement and economic growth, alarming levels of pollution and environmental degradation are emerging globally. Due to the geographical diffusion of pollutants, the unilateral response of one nation or region is often ineffective. Reports portray the situation as an industrial civilization on the verge of suicide, destroying its environmental conditions of existence, with people being held as prisoners on a runaway catastrophe-bound train. Though global cooperation in environmental control holds out the best promise of effective action, limited success has been observed. This is the result of many hurdles, ranging from commitment, monitoring, and sharing of costs to disparities in future development under the cooperative plans. One finds it hard to be convinced that multinational joint initiatives, like the Kyoto Protocol, can offer a long-term solution because there is no guarantee that participants will always be better off within the entire extent of the agreement. More than anything else, it is due to the lack of these kinds of incentives that current cooperative schemes fail to provide an effective means to avert disaster. This is a "classic" game-theoretic problem.

To construct a theoretical framework capturing the essence of a transboundary industrial pollution paradigm a differential game approach is adopted. Differential games provide an effective tool to study pollution control problems and to analyze the interactions between the participants' strategic behaviors and the dynamic evolution of pollution. Applications of noncooperative differential games in environmental studies can be found in Yeung (1992), Dockner and Long (1993), Tahvonen (1994), Stimming (1999), Feenstra et al. (2001), and Dockner and Leitmann (2001). Cooperative differential games in environmental control are presented by Dockner and Long (1993), Jørgensen and Zaccour (2001), Fredj et al. (2004), Breton et al. (2005, 2006), Petrosyan and Zaccour (2003), Yeung (2007), and Yeung and Petrosyan (2008).

To formulate the foundation for an effective policy to tackle one of the gravest problems facing the global market economy this chapter presents a cooperative initiative involving a set of environmental policy instruments including taxes, subsidies, and pollution abatement activities. The implementation of such a scheme will inevitably bring about different implications in cost and benefit to each of the participating nations. To construct a cooperative solution that every party will commit

D.W.K. Yeung, L.A. Petrosyan, *Subgame Consistent Economic Optimization*,   147
Static & Dynamic Game Theory: Foundations & Applications,
DOI 10.1007/978-0-8176-8262-0_6, © Springer Science+Business Media, LLC 2012

to from beginning to end, the arrangements must guarantee that every participant
will be better off and the originally agreed-upon arrangement remains effective at
any time within the cooperative period along the cooperative trajectory.

An analytical framework for studying transboundary industrial pollution management is established in Sect. 6.1. Noncooperative outcomes appear in Sect. 6.2.
Cooperative arrangements, cooperative state trajectory, and time consistent imputations are derived in Sect. 6.3. Benefit distributions leading to a time (optimal-
trajectory-subgame) consistent collaborative environmental management scheme
are obtained in Sect. 6.4. Policy implications are examined in the following section.
An explicitly solvable model of transboundary industrial pollution management is
given in Sect. 6.6 and a time (optimal-trajectory-subgame) consistent solution is
presented in Sect. 6.7.

6.1 An Analytical Framework

In this section we present an analytical framework to study transboundary industrial
pollution management.

6.1.1 The Industrial Sector

Consider an international economy with n nations. At time instant s the demand
system of the outputs of the nations is

$$P_i(s) = f^i[q_1(s), q_2(s), \ldots, q_n(s), s], \quad i \in N \equiv \{1, 2, \ldots, n\}, \qquad (6.1)$$

where $q_j(s)$ is the output of nation j and $P_i(s)$ is the price of the output of nation i.
The demand system of (6.1) shows that the multination economy is a form of a
generalized differentiated products oligopoly.

Industrial profits of nation i at time s can be expressed as

$$f^i[q_1(s), q_2(s), \ldots, q_n(s), s]q_i(s) - c^i[q_i(s), v_i(s)], \quad \text{for } i \in N, \qquad (6.2)$$

where $v_i(s)$ is the set of environmental policy instruments of government i. Policy
instruments may include tools like taxes, subsidies, technology choices, and pollution legislations. The cost of producing $q_i(s)$ under policy $v_i(s)$ is $c^i[q_i(s), v_i(s)]$.

Profit maximization by the industrial sectors yields

$$f^i[q_1(s), q_2(s), \ldots, q_n(s), s] + f^i_{q_i}[q_1(s), q_2(s), \ldots, q_n(s), s]q_i(s)$$

$$- c^i_{q_i}[q_i(s), v_i(s)] = 0, \quad \text{for } i \in N. \qquad (6.3)$$

Equation (6.3) is a system of implicit functions in $q(s) = [q_1(s), q_2(s), \ldots, q_n(s)]$
with government policies $v(s) = [v_1(s), v_2(s), \ldots, v_n(s)]$ being regarded as parameters. The existence of a market equilibrium reflects the satisfaction of the Implicit

Function Theorem in (6.3) and nation i's instantaneous market equilibrium output can be expressed as

$$q_i^*(s) = \hat{q}^i\big[v_1(s), v_2(s), \ldots, v_n(s), s\big] \equiv \hat{q}^i\big[v(s), s\big], \quad \text{for } i \in N. \qquad (6.4)$$

One can readily observe from (6.4) that each nation's output decision depends on the government's environmental policies.

6.1.2 Impacts and Accumulation Dynamics of Pollutants

Industrial production emits pollutants into the environment and the amount of pollution created by different nations' outputs may be different. For an output of $q_i(s)$ produced by nation i, there will be an instantaneous damaging environmental impact of $\varepsilon_i^i[q_i(s)]$ on nation i itself and a damaging impact of $\varepsilon_j^i[q_i(s)]$ on its adjacent nation j for $j \in \underline{K}^i$. On the other hand, nation i will receive instantaneous damaging environmental impacts from its adjacent nations measured as $\varepsilon_i^j[q_j(s)]$ for $j \in \bar{K}^i$. This first type of externality is typical in static analysis and the second in dynamic analysis.

Moreover, the pollutant will then add to the stock of existing pollution. Each government adopts its own pollution abatement policy to reduce the pollution stock. Let $x(s) \subset R^m$ denote the level of pollution at time s, the dynamics of pollution stock is governed by the differential equation

$$\dot{x}(s) = \sum_{j=1}^{n} a_j\big[q_j(s), v_j(s)\big] - \sum_{j=1}^{n} b_j\big[u_j(s), x(s)\big] - \delta\big[x(s)\big]x(s),$$

$$x(t_0) = x_{t_0}, \qquad (6.5)$$

where $a_j[q_j(s), v_j(s)]$ is the amount of pollution created by $q_j(s)$ (the amount of output produced under policy $v_i(s)$, $u_j(s)$ is the pollution abatement effort of nation j), $b_j[u_j(s), x(s)]$ is the amount of pollution removed by $u_j(s)$ (the unit of abatement effort of nation j), and $\delta[x(s)]$ is the natural rate of decay of the pollutants. Moreover, $\delta(x)$ is negatively related to x, reflecting the phenomenon that the natural rate of decay declines as the level of pollution stock rises.

6.1.3 The Governments' Objectives

The governments have to promote business interests and at the same time handle the financing of the costs brought about by pollution. In particular, each government maximizes the net gains in the industrial sector, plus tax revenue, minus expenditures on pollution abatement and damages from pollution. A lump-sum income tax is levied on the industrial sector to balance the government budget. The last item turns out to be a net transfer between the government and the public (with no effect on industrial output). The instantaneous objective of government i at time s can be expressed as

$$f^i[q_1(s), q_2(s), \ldots, q_n(s), s]q_i(s) - c^i[q_i(s), v_i(s)] - c^P_i[v_i(s)] - c^a_i[u_i(s)]$$

$$- \varepsilon^i_i[q_i(s)] - \sum_{j \in \bar{K}^i} \varepsilon^j_i[q_j(s)] - h_i[x(s)], \quad i \in N, \tag{6.6}$$

where $c^P_i[v_i(s)]$ is the cost of implementing the vector policy instrument $v_i(s)$, $c^a_i[u_i(s)]$ is the cost of employing the u_i amount of pollution abatement effort, and $h_i[x(s)]$ is the value of damage to country i from an $x(s)$ amount of pollution.

The governments' planning horizon is $[t_0, T]$. It is possible that T may be very large. The discount rate is r. At time T, the terminal appraisal of pollution damage is $g^i[x(T)]$, where $\partial g^i / \partial x < 0$. Each one of the n governments seeks to maximize the integral of its instantaneous objective found in (6.6) over the planning horizon subject to the pollution dynamics in (6.5) with controls on the level of abatement effort and output tax.

Substitute $q_i(s)$, for $i \in N$, from (6.4) into (6.5) and (6.6) one obtains a differential game in which government $i \in N$ seeks to

$$\max_{v_i(s), u_i(s)} \left\{ \int_{t_0}^T \left[f^i\{\hat{q}^1[v(s), s], \hat{q}^2[v(s), s], \ldots, \hat{q}^n[v(s), s], s\}\hat{q}^i[v(s), s] \right. \right.$$

$$- c^i\{\hat{q}^i[v(s), s], v_i(s)\} - c^P_i[v_i(s)] - c^a_i[u_i(s)] - \varepsilon^i_i\{\hat{q}^i[v(s), s]\}$$

$$\left. \left. - \sum_{j \in \bar{K}^i} \varepsilon^j_i\{\hat{q}^j[v(s), s]\} - h_i[x(s)] \right] e^{-r(s-t_0)} ds + g^i[x(T)]e^{-r(T-t_0)} \right\}, \tag{6.7}$$

subject to

$$\dot{x}(s) = \sum_{j=1}^n a_j\{\hat{q}^j[v(s), s], v_j(s)\} - \sum_{j=1}^n b_j[u_j(s), x(s)] - \delta[x(s)]x(s),$$

$$x(t_0) = x_{t_0}. \tag{6.8}$$

Thus the economic interactions among nations in industrial production, pollution emission, and abatement are characterized as a differential game with the payoffs in (6.7) and pollution dynamics in (6.8).

6.2 Noncooperative Outcomes

In this section we discuss the solution to the noncooperative game in (6.7) and (6.8). Since the payoffs of nations are measured in monetary terms, the game is a transferable payoff game. Invoking Theorem 2.3 in Chap. 2, a feedback Nash equilibrium solution can be characterized as follows.

Corollary 6.1 *A set of feedback strategies* $\{u_i^*(t) = \mu_i(t, x), v_i^*(t) = \phi_i(t, x),$ *for* $i \in N\}$, *provides a feedback Nash equilibrium solution to the game in* (6.7) *and* (6.8) *if there exist suitably smooth functions* $V^{(t_0)i}(t, x) : [t_0, T] \times R^m \to R, i \in N,$ *satisfying the following partial differential equations:*

$$-V_t^{(t_0)i}(t, x) = \max_{v_i, u_i} \left\{ \left[f^i \{ \hat{q}^1 [v_i, \phi_{\neq i}(t, x), t], \hat{q}^2 [v_i, \phi_{\neq i}(t, x), t], \dots, \right. \right.$$

$$\hat{q}^n [v_i, \phi_{\neq i}(t, x), t] \}$$

$$\times \hat{q}^i [v_i, \phi_{\neq i}(t, x), t] - c \{ \hat{q}^i [v_i, \phi_{\neq i}(t, x), t], v_i \}$$

$$- c_i^P [v_i] - c_i^a [u_i] - \varepsilon_i^i \{ \varphi^i [v_i, \phi_{\neq i}(t, x), t] \}$$

$$\left. - \sum_{j \in \bar{K}^i} \varepsilon_i^j \{ \hat{q}^j [v_i, \phi_{\neq i}(t, x), t] \} - h_i(x) \right] e^{-r(t-t_0)}$$

$$+ V_x^{(t_0)i} \left[\sum_{j=1}^n a_j \{ \hat{q}^j [v_i, \phi_{\neq i}(t, x), t], v_j \} - b_i(u_i, x) \right.$$

$$\left. \left. - \sum_{\substack{j=1 \\ j \neq i}}^n b_j [\mu_j(t, x), x] - \delta(x)x \right] \right\}, \tag{6.9}$$

$$V^{(t_0)i}(T, x) = g^i [x] e^{-r(T-t_0)}, \tag{6.10}$$

where

$$\phi_{\neq i}(t, x) = [\phi^1(t, x), \phi^2(t, x), \dots, \phi^{i-1}(t, x), \phi^{i+1}(t, x), \dots, \phi^n(t, x)].$$

In a prevailing Nash equilibrium the function $V^{(t_0)i}(t, x)$ is then the integral

$$\int_t^T \left[f^i \{ \hat{q}^1 [\phi(s, x(s)), s], \hat{q}^2 [\phi(s, x(s)), s], \dots, \hat{q}^n [\phi(s, x(s)), s], s \} \right.$$

$$\times \hat{q}^i [\phi(s, x(s)), s] - c^i \{ \hat{q}^i [\phi(s, x(s)), s], \phi_i(s, x(s)) \}$$

$$- c_i^P [\phi_i(s, x(s))] - c_i^a [\mu_i(s, x(s))] - \varepsilon_i^i \{ \hat{q}^i [\phi(s, x(s)), s] \}$$

$$- \sum_{j \in \bar{K}^i} \varepsilon_i^j \{ \hat{q}^j [\phi(s, x(s)), s] \}$$

$$\left. - h_i [x(s)] \right] e^{-r(s-t_0)} \, ds + g^i [x(T)] e^{-r(T-t_0)} \Big|_{x(t)=x}, \quad \text{for } i \in N. \tag{6.11}$$

The game equilibrium dynamics then becomes

$$\dot{x}(s) = \sum_{j=1}^{n} a_j \{ \hat{q}^j [\phi(s, x(s)), s], \phi_j(s, x(s)) \}$$

$$- \sum_{j=1}^{n} b_j [\mu_j(s, x(s)), x(s)] - \delta[x(s)]x(s),$$

$$x(t_0) = x_{t_0}. \tag{6.12}$$

Remark 6.1 One can readily verify that $V^{(\tau)i}(t, x_t) = V^i(t, x_t)e^{r(\tau - t_0)}$, for $\tau \in [t_0, T]$, is the value function to nation i at time $t \in [\tau, T]$ when the state $x(t) = x_t$ in the game of (6.7) and (6.8), which starts at time τ.

With negative externalities in pollution a noncooperative outcome is suboptimal. International collaboration in industrial pollution management is an effective way out of this situation.

6.3 Cooperative Arrangement

Now consider the case when all the nations want to cooperate and agree to act so that an international optimum can be achieved. For the cooperative scheme to be upheld throughout the game horizon both group rationality and individual rationality are required to be satisfied at any time. Group optimality ensures that all potential gains from cooperation are captured. The failure to fulfill group optimality leads to the condition where the participants prefer to deviate from the agreed-upon solution plan to extract the unexploited gains. Individual rationality is required to hold so that the payoff allocated to a nation under cooperation will be no less than its noncooperative payoff. The failure to guarantee individual rationality leads to a condition where the concerned participants will reject the agreed-upon solution plan and play noncooperatively. In the absence of a punishment scheme, the cooperative plan will dissolve if any of the nations deviate from the agreed-upon plan.

In addition, as mentioned above, an agreeable optimality principle must uphold group optimality and individual rationality throughout the period of cooperation.

6.3.1 Group Optimality and Cooperative State Trajectory

Consider the collaborative environmental scheme with the participating nations' payoff structure in (6.7) and the pollution dynamics in (6.8). To secure group optimality the participating nations seek to maximize their joint payoff by solving the following control problem:

$$
\begin{aligned}
\max_{v_1, v_2, \ldots, v_n; u_1, u_2, \ldots, u_n} & \left\{ \int_{t_0}^{T} \left[\sum_{i=1}^{n} f^i \left\{ \left[\hat{q}^1[v(s), s], \hat{q}^2[v(s), s], \ldots, \hat{q}^n[v(s), s], s \right] \right\} \right. \right. \\
& \times \hat{q}^i[v(s), s] - c^i \left\{ \hat{q}^i[v(s), s], v_i(s) \right\} \\
& - c_i^P[v_i(s)] - c_i^a[u_i(s)] - \varepsilon_i^i \left\{ \hat{q}^i[v(s), s] \right\} - \sum_{j \in \bar{K}^i} \varepsilon_i^j \left\{ \hat{q}^j[v(s), s] \right\} \\
& \left. \left. - h_i[x(s)] \right] e^{-r(t-t_0)} \, ds + \sum_{i=1}^{n} g^i[x(T)] e^{-r(T-t_0)} \right\},
\end{aligned}
\tag{6.13}
$$

subject to (6.8).

Invoking Theorem A.1 in the Technical Appendixes, a set of controls $\{[v_i^{**}(t),$ $u_i^{**}(t)] = [\psi_i(t, x), \varpi_i(t, x)],$ for $i \in N\}$, constitutes an optimal solution to the control problem in (6.13) and (6.8) if there exists a continuously differentiable function $W^{(t_0)}(t, x) : [t_0, T] \times R^m \to R, i \in N$, satisfying the following partial differential equations:

$$
\begin{aligned}
- W_t^{(t_0)}(t, x) = & \max_{v_1, v_2, \ldots, v_n; u_1, u_2, \ldots, u_n} \left\{ \sum_{i=1}^{n} f^i \left[\hat{q}^1(v, t), \hat{q}^2(v, t), \ldots, \hat{q}^n(v, t), t \right] \right. \\
& \times \hat{q}^i(v, t) - c^i \left[\hat{q}^i(v, t), v_i \right] - c_i^P(v_i) - c_i^a(u_i) - \varepsilon_i^i \left[\hat{q}^i(v, t) \right] \\
& - \sum_{j \in \bar{K}^i} \varepsilon_i^j \left[\hat{q}^j(v, t) \right] - h_i(x) \right] e^{-r(t-t_0)} + W_x^{(t_0)}(t, x) \\
& \times \left[\sum_{j=1}^{n} a_j \left[\hat{q}^j(v, t), v_j \right] - \sum_{j=1}^{n} b_j(u_j, x) - \delta(x)x \right] \right\}, \quad \text{and}
\end{aligned}
\tag{6.14}
$$

$$
W^{(t_0)}(T, x) = \sum_{i=1}^{n} g^i(x) e^{-r(T-t_0)}.
$$

Hence the nations will adopt the cooperative control $\{[\psi_i(t, x), \varpi_i(t, x)],$ for $i \in N$ and $t \in [t_0, T]\}$. The optimal trajectory under cooperation becomes

$$
\dot{x}(s) = \sum_{j=1}^{n} a_j \left\{ \hat{q}^j[\psi(s, x(s)), s], \psi_j(s, x(s)) \right\}
$$

$$
- \sum_{j=1}^{n} b_j \left[\varpi_j(s, x(s)), x(s) \right] - \delta[x(s)] x(s),
$$

$$
x(t_0) = x_{t_0}.
\tag{6.15}
$$

The solution to (6.15) can be expressed as

$$x^*(t) = x_0 + \int_{t_0}^{t} \left\{ \sum_{j=1}^{n} a_j \{\hat{q}^j [\psi(s, x^*(s)), s], \psi_j(s, x^*(s))\} \right.$$

$$\left. - \sum_{j=1}^{n} b_j [\varpi_j(s, x^*(s)), x^*(s)] - \delta[x^*(s)]x^*(s) \right\} ds. \qquad (6.16)$$

We use $\{x^*(t)\}_{t=t_0}^{T}$ to denote the solution path generated by (6.16). The terms x_t^* and $x^*(t)$ are used interchangeably.

The cooperative control for the game $\Gamma_c(x_0, T - t_0)$ over the time interval $[t_0, T]$ can be expressed more precisely as

$$\psi_i(t, x^*(t)), \qquad \varpi_i(t, x^*(t)), \quad \text{for } t \in [t_0, T] \text{ and } i \in N. \qquad (6.17)$$

Note that, for group optimality to be achievable, the cooperative controls in (6.17) must be exercised throughout time interval $[t_0, T]$.

The value function $W^{(t_0)i}(t, x)$ is then the integral

$$\int_{t}^{T} \left[\sum_{i=1}^{n} f^i \{\hat{q}^1 [\psi(s, x^*(s)), s], \hat{q}^2 [\psi(s, x^*(s)), s], \ldots, \hat{q}^n [\psi(s, x^*(s)), s], s\} \right.$$

$$\times \hat{q}^i [\psi(s, x^*(s)), s] - c^i \{\hat{q}^i [\psi(s, x^*(s)), s], \psi_i(s, x^*(s))\}$$

$$- c_i^P [\psi_i(s, x^*(s))] - c_i^a [\varpi_i(s, x^*(s))] - \varepsilon_i^i \{\hat{q}^i [\psi(s, x^*(s)), s]\}$$

$$\left. - \sum_{j \in \bar{K}^i} \varepsilon_i^j \{\hat{q}^j [\psi(s, x^*(s)), s]\} - h_i [x^*(s)] \right] e^{-r(s-t_0)} ds$$

$$+ \sum_{i=1}^{n} g^i [x^*(T)] e^{-r(T-t_0)}, \quad \text{for } i \in N, \qquad (6.18)$$

where $\psi^*(s, x^*(s)) = \{\psi_1^*(s, x^*(s)), \psi_2^*(s, x^*(s)), \ldots, \psi_n^*(s, x^*(s))\}$.

Remark 6.2 One can readily verify that $W^{(\tau)}(t, x_t^*) = W^{(t_0)}(t, x_t^*) e^{r(\tau - t_0)}$, for $\tau \in [t_0, T]$, is the value function at time $t \in [\tau, T]$ of the control problem in (6.8) and (6.13) which starts at time τ with $x(t) = x_t^*$.

6.3.2 Individually Rational and Time (Optimal-Trajectory-Subgame) Consistent Imputation

An agreed-upon optimality principle must be sought to allocate the cooperative payoff. In a dynamic framework individual rationality has to be maintained at every instant of time within the cooperative duration $[t_0, T]$ along the cooperative path $\{x_\tau^*\}_{\tau=t_0}^T$. For $\tau \in [t_0, T]$, let $\xi^{(\tau)i}(\tau, x_\tau^*)$ denote the imputation (payoff according to the agreed-upon under optimality principle) over the period $[\tau, T]$ to nation $i \in N$ along the cooperative path. Individual rationality along the cooperative trajectory requires

$$\xi^{(\tau)i}(\tau, x_\tau^*) \geq V^{(\tau)i}(\tau, x_\tau^*), \quad \text{for } i \in N, x_\tau^* \in X_\tau^* \text{ and } \tau \in [t_0, T]. \quad (6.19)$$

Since nations are asymmetric and the number of nations may be large, a reasonable optimality principle for gain distribution is to share the gain from cooperation proportional to the nations' relative sizes of noncooperative payoffs. As mentioned before, time (optimal-trajectory-subgame) consistency is required for a credible cooperative solution under a dynamic framework. In particular, the agreed-upon optimality principle must be maintained in any subgame that starts at a later time along the cooperative state trajectory so that no nation has the incentive to deviate from the previously adopted optimal behavior throughout the game.

Hence the solution imputation scheme $\{\xi^{(\tau)i}(\tau, x_\tau^*); \text{ for } i \in N\}$ has to satisfy the following condition:

Condition 6.1

$$\xi^{(\tau)i}(\tau, x_\tau^*) = \frac{V^{(\tau)i}(\tau, x_\tau^*)}{\sum_{j=1}^n V^{(\tau)j}(\tau, x_\tau^*)} W^{(\tau)}(\tau, x_\tau^*), \quad (6.20)$$

for $i \in N, x_\tau^* \in X_\tau^*$ and $\tau \in [t_0, T]$.

The imputation scheme in Condition 6.1 satisfies individual rationality. Crucial to the analysis is the formulation of a payment distribution mechanism that will lead to the realization of Condition 6.1. This will be done in the next section.

6.4 Benefit Distribution in Collaborative Environmental Management

The solution to the optimality principle guiding the multinational collaboration in environmental management can then be expressed as

$$P(x_t^*, T - t) = \{[\psi(s, x^*(s))]_{s=t}^T, [\varpi_i(s, x^*(s))]_{s=t}^T, [B(s)]_{s=t}^T \xi^{(t)}(t, x_t^*)\}$$
for $t \in [t_0, T]$,

where $\psi(s, x^*(s)) = \{\psi_1(s, x^*(s)), \psi_2(s, x^*(s)), \ldots, \psi_n(s, x^*(s))\}$ is a set of joint payoffs maximizing the environmental policy instruments, and

$$\varpi(s, x^*(s)) = \{\varpi_1(s, x^*(s)), \varpi_2(s, x^*(s)), \ldots, \varpi_n(s, x^*(s))\}$$

is the set of joint payoffs maximizing pollution abatement efforts in the collaborative environmental scheme; $\xi^{(t)}(t, x_t^*)$ is an imputation scheme satisfying Condition 6.1 and $B(s)$ is the payoff distribution procedure leading to Condition 6.1.

All the participating nations in a collaborative scheme will have no incentive to exit the venture if the agreed-upon optimality principle is maintained at every instant $t \in [t_0, T]$.

To formulate a payoff distribution procedure over time so that the agreed imputations satisfy Condition 6.1 we obtain the following.

Proposition 6.1 *A distribution scheme with a terminal payment* $-g^i[x_T^* - \bar{x}^i]$ *at time T and an instantaneous payment at time* $\tau \in [t_0, T]$

$$
\begin{aligned}
B_i(\tau) = &-\frac{\partial}{\partial t}\left[\frac{V^{(\tau)i}(t, x_t^*)}{\sum_{j=1}^n V^{(\tau)j}(t, x_t^*)} W^{(\tau)}(t, x_t^*)|_{t=\tau}\right] \\
&-\frac{\partial}{\partial x_\tau^*}\left[\frac{V^{(\tau)i}(\tau, x_\tau^{i*})}{\sum_{j=1}^n V^{(\tau)j}(\tau, x_\tau^{j*})} W^{(\tau)}(\tau, x_\tau^*)\right] \\
&\times \left[\sum_{j=1}^n a_j\{\hat{q}^j[\psi(\tau, x_\tau^*), \tau], \psi_j(\tau, x_\tau^*)\}\right. \\
&\left.-\sum_{j=1}^n b_j[\varpi_j(\tau, x_\tau^*), x_\tau^*] - \delta(x_\tau^*)x_\tau^*\right],
\end{aligned}
$$

for $i \in N$, (6.21)

will lead to a realization of the imputations $\xi^{(\tau)i}(\tau, x_\tau^*)$, *for* $i \in N$ *and* $\tau \in [t_0, T]$, *satisfying Condition 6.1.*

Proof Invoking Theorem 4.3 in Chap. 4 the results in Proposition 6.1 follow. □

A time (optimal-trajectory-subgame) consistent solution can be obtained using the set of group optimal strategies $\psi^*(s, x_s^*)$ characterized in (6.14) and $B(s) = \{B_1(s), B_2(s), \ldots, B_n(s)\}$ in (6.21).

With the nations using the cooperative environmental policy instruments $\psi(s, x^*(s))$ and pollution abatement efforts $\varpi(s, x^*(s))$, the instantaneous receipt of nation i at time instant τ is

$$
\begin{aligned}
\zeta_i(\tau) = &f^i\{\hat{q}^1[\psi(\tau, x_\tau^*), \tau], \hat{q}^2[\psi(\tau, x_\tau^*), \tau], \ldots, \hat{q}^n[\psi(\tau, x_\tau^*), \tau], s\} \\
&\times \hat{q}^i[\psi(\tau, x_\tau^*), \tau] - c^i\{\hat{q}^i[\psi(\tau, x_\tau^*), \tau], \psi_i(\tau, x_\tau^*)\} \\
&- c_i^P[\psi_i(\tau, x_\tau^*)] - c_i^a[\varpi_i(s, x^*(s))] - \varepsilon_i^i\{\hat{q}^i[\psi(s, x^*(s)), s]\} \\
&- \sum_{j \in \bar{K}^i} \varepsilon_i^j\{\hat{q}^j[\psi(\tau, x_\tau^*), \tau]\} - h_i(x_\tau^*),
\end{aligned}
$$
(6.22)

for $\tau \in [t_0, T]$ and $i \in N$.

According to Proposition 6.1, the instantaneous payment that firm i should receive under the agreed-upon optimality principle is $B_i(\tau)$, for $\tau \in [t_0, T]$ and $i \in N$, as stated in (6.21). Hence an instantaneous transfer payment

$$\chi^i(\tau) = B_i(\tau) - \zeta_i(\tau) \tag{6.23}$$

would be given or charged to firm i at time τ, for $i \in N$ and $\tau \in [t_0, T]$.

6.5 Policy Implications

Facing an increasing demand for a sustainable solution, the international community has responded to the deteriorating problem of global pollution. Over a decade ago most countries joined an international treaty—the United Nations Framework Convention on Climate Change (UNFCCC)—to consider solutions to reduce global warming and to cope with whatever temperature increases are inevitable. Recently, a number of nations approved an addition to the treaty: the Kyoto Protocol, which has more powerful and legally binding measures. In brief, the Kyoto Protocol is an international agreement, which builds on the United Nations Framework Convention on Climate Change, and sets legally binding targets and timetables for cutting the greenhouse-gas emissions of industrialized countries. Conditions for entry are that some UNFCCC parties cut greenhouse-gas emissions of at least 5% from 1990 levels in the commitment period 2008–2012. As of December 2006, 169 countries and other governmental entities ratified the agreement. Notable exceptions include the United States. Other countries, like India and China, which have ratified the protocol, are not required to reduce carbon emissions under the present agreement despite their relatively large industrial production activities.

As mentioned before, placing a constraint just on certain types of pollution emissions cannot offer a long-term solution because the plans are limited to a confined set of controls, like gas emissions and permits, which is unlikely to be able to offer an effective means to reverse the accelerating trend of environmental deterioration. In addition, there is no guarantee that participants will always be better off and hence be committed within the entire duration of the agreement. Guided by the analysis shown previously, a grand coalition of all nations should be formed to pursue a comprehensive cooperative scheme of industrial pollution abatement. In particular, the entire set of policy instruments available—including environmental taxes and charges, a subsidy for the replacement of polluting techniques, and the restoration and preservation of the natural ecosystem—will be used to achieve an optimal cooperative outcome. A payment distribution mechanism has to be formulated so that cooperative gains will be shared according to the proportions of the nations' relative sizes of noncooperative payoffs throughout the planning horizon. In sum, the appropriate policy coordination will lead to the enhancement of economic performance and the realization of a cleaner environment.

This analysis opens up a novel policy forum for the international community. A particularly relevant instance would be the formation of a United Nations Agency to coordinate international cooperative actions on pollution and climate change. The proposed agency is to be comprised of three divisions. An executive branch will be established to coordinate the adoption and development of clean technology, pollution abatement activities, use of materials, waste disposal, mode of resource extraction, and cooperation in environmental R&D. A financial branch (or FUND) would be set up to handle pollution charges, clean technology subsidies, and allocate payoff distributions so that the agreed-upon optimality principle will be realized throughout the cooperative period. Lastly, a legislative body would be in place to enact regulations on the activities damaging the environment and in violation of the cooperative agreement. Finally, a large-scale scheme is in order for research in the mechanism design theory initiated by Hurwicz (1973) and refined and applied by Myerson (1989) and Maskin (1999). In particular, the mechanism designs for conventional markets in the face of the impact from a comprehensive set of environmental policy instruments, including taxes, subsidies, technology choices, pollution abatement activities, pollution legislations, and green technology R&D, have to be considered. In addition, the mechanism designs for intergovernment transfers, institution formation, like-market, and beyond-conventional market arrangements have also to be investigated.

6.6 A Model of Transboundary Industrial Pollution Management

As an illustration of a collaborative scheme for transboundary industrial pollution management we consider the deterministic version of the Yeung and Petrosyan (2008) game.

6.6.1 A Multinational Economy with Industrial Pollution

We first present a multinational economy with n asymmetric nations or regions. Industrial pollution is generated via the production process.

6.6.1.1 The Industrial Economy

Consider a multinational economy that is comprised of n nations. To allow different degrees of substitutability among the nations' outputs a differentiated products oligopoly model has to be adopted. The differentiated oligopoly model used by Dixit (1979) and Singh and Vives (1984) in industrial organizations is adopted to characterize the interactions in this international market. In particular, the nations' outputs may range from a homogeneous product to n unrelated products. Specifically, the inverse demand function of the output of nation $i \in N$ at time instant s is

$$P_i(s) = \alpha^i - \sum_{j=1}^{n} \beta_j^i q_j(s), \qquad (6.24)$$

where $P_i(s)$ is the price of the output of nation i, $q_j(s)$ is the output of nation j, and α^i and β_j^i for $i \in N$ and $j \in N$ are positive constants. The output choice $q_j(s) \in [0, \bar{q}_j]$ is nonnegative and bounded by a maximum output constraint $\bar{q}_j$. The output price is equal to zero if the right-hand side of (6.24) becomes negative. The demand system in (6.24) shows that the economy is a form of differentiated products oligopoly with substitute goods. In the case when $\alpha^i = \alpha^j$ and $\beta_j^i = \beta_i^j$ for all $i \in N$ and $j \in N$, the industrial outputs resemble a homogeneous good. In the case when $\beta_j^i = 0$ for $i \neq j$, the n nations produce n unrelated products. Industrial profits of nation i at time s can be expressed as

$$\pi_i(s) = \left[\alpha^i - \sum_{j=1}^{n} \beta_j^i q_j(s) \right] q_i(s) - c_i q_i(s) - v_i(s) q_i(s), \quad \text{for } i \in N, \qquad (6.25)$$

where $v_i(s) \geq 0$ is the tax rate imposed by government i on its industrial output at time s and c_i is the unit cost of production. At each time instant s, the industrial sector of nation $i \in N$ seeks to maximize (6.25). Note that each industrial sector would consider the information on the demand structure, each other's cost structures, and tax policies. The first-order condition for a Nash equilibrium for the n nations economy yields

$$\sum_{j=1}^{n} \beta_j^i q_j(s) + \beta_i^i q_i(s) = \alpha^i - c_i - v_i(s), \quad \text{for } i \in N. \qquad (6.26)$$

With output tax rates $v(s) = \{v_1(s), v_2(s), \ldots, v_n(s)\}$ being regarded as parameters, (6.26) becomes a system of equations linear in $q(s) = \{q_1(s), q_2(s), \ldots, q_n(s)\}$. Solving (6.26) yields an industry equilibrium

$$q_i(s) = \phi_i(v(s)) = \bar{\alpha}^i + \sum_{j \in N} \bar{\beta}_j^i v_j(s), \qquad (6.27)$$

where $\bar{\alpha}^i$ and $\bar{\beta}_j^i$, for $i \in N$ and $j \in N$, are constants involving the model parameters $\{\beta_1^1, \beta_2^1, \ldots, \beta_n^1; \beta_1^2, \beta_2^2, \ldots, \beta_n^2; \ldots; \beta_1^n, \beta_2^n, \ldots, \beta_n^n\}$, $\{\alpha^1, \alpha^2, \ldots, \alpha^n\}$, and $\{c_1, c_2, \ldots, c_n\}$.

The industry equilibrium generated by this oligopoly model is computable and fully tractable. One can readily observe from (6.26) that an increase in the tax rate has the same effect as of an increase in cost. *Ceteris paribus*, an increase in nation i's tax rate would depress the output of industrial sector i and vice versa. Given that outputs are substitutable products and the linear demand functions of (6.24), industrial sector i's output and nation j's tax rate, where $j \neq i$, are positively related.

6.6.1.2 Local and Global Environmental Impacts

Industrial production emits pollutants into the environment. The emitted pollutants cause short-term local impacts on neighboring areas of the origin of production in forms like passing-by waste in waterways, wind-driven suspended particles in the air, unpleasant odor, noise, dust, and heat. For an output of $q_i(s)$ produced by nation i there will be a short-term local environmental impact (cost) of $\varepsilon_i^i q_i(s)$ on nation i itself and a local impact of $\varepsilon_j^i q_i(s)$ on its neighbor nation j. In particular, ε_j^i is a positive constant. Nation i will receive short-term local environmental impacts from its adjacent nations measured as $\varepsilon_i^j q_j(s)$ for $j \in \bar{K}^i$. Thus $\bar{K}^i$ is the subset of nations whose outputs produce local environmental impacts to nation i. Moreover, industrial production will also create long-term global environmental impacts by building up existing pollution stocks, like greenhouse gases, chlorofluorocarbons (CFC), and atmospheric particulates. Each government adopts its own pollution abatement policy to reduce the pollution stock. Let $x(s) \subset R^+$ denote the level of pollution at time s, the dynamics of the pollution stock is governed by the differential equation

$$\dot{x}(s) = \sum_{j=1}^{n} a_j q_j(s) - \sum_{j=1}^{n} b_j u_j(s)[x(s)]^{1/2} - \delta x(s), \quad x(t_0) = x_{t_0}, \quad (6.28)$$

where $a_j > 0, b_j > 0, \delta > 0$ are positive constants, $a_j q_j$ is the amount added to the pollution stock by a unit of nation j's output, $u_j(s)$ is the pollution abatement effort of nation j, $b_j u_j(s)[x(s)]^{1/2}$ is the amount of pollution removed by the $u_j(s)$ unit of abatement effort of nation j, and δ is the natural rate of decay of the pollutants.

6.6.1.3 The Governments' Objectives

The governments have to promote business interests and at the same time handle the financing of the costs brought about by pollution. In particular, each government maximizes the net gains in the industrial sector minus the sum of expenditures on pollution abatement and damages from pollution. The instantaneous objective of government i at time s can be expressed as

$$\left[\alpha^i - \sum_{j=1}^{n} \beta_j^i q_j(s) \right] q_i(s) - c_i q_i(s) - c_i^a [u_i(s)]^2 - \sum_{j \in \bar{K}^i} \varepsilon_i^j [q_j(s)] - h_i x(s),$$

for $i \in N$,   (6.29)

where $c_i^a > 0$ and $h_i > 0$ are constants, $c_i^a [u_i(s)]^2$ is the cost of employing a u_i amount of the pollution abatement effort, and $h_i x(s)$ is the value of damage to country i from an $x(s)$ amount of pollution.

The governments' planning horizon is $[t_0, T]$. It is possible that T may be very large. At time T, the terminal appraisal associated with the state of pollution is $g^i[\bar{x}^i - x(T)]$, where $g^i \geq 0$ and $\bar{x}^i \geq 0$. The discount rate is r. Each one of the n

governments seeks to maximize the integral of its instantaneous objective in (6.29) over the planning horizon subject to the pollution dynamics in (6.28) with controls on the level of the abatement effort and output tax.

By substituting $q_i(s)$, for $i \in N$, from (6.27) into (6.28) and (6.29) one obtains a differential game in which government $i \in N$ seeks to

$$
\max_{v_i(s), u_i(s)} \left\{ \int_{t_0}^{T} \left[\left(\alpha^i - \sum_{j=1}^{n} \beta_j^i \left[\bar{\alpha}^j + \sum_{h \in N} \bar{\beta}_h^j v_h(s) \right] \right) \left[\bar{\alpha}^i + \sum_{h \in N} \bar{\beta}_h^i v_h(s) \right] \right.\right.
$$

$$
- c_i \left[\bar{\alpha}^i + \sum_{j \in N} \bar{\beta}_j^i v_j(s) \right] - c_i^a [u_i(s)]^2 - \sum_{j \in \bar{K}^i} \varepsilon_i^j \left[\bar{\alpha}^j + \sum_{\ell \in N} \bar{\beta}_\ell^j v_\ell(s) \right]
$$

$$
\left.\left. - h_i x(s) \right] e^{-r(s-t_0)} \, ds - g^i \left[x(T) - \bar{x}^i \right] e^{-r(T-t_0)} \right\}, \tag{6.30}
$$

subject to

$$
\dot{x}(s) = \sum_{j=1}^{n} a_j \left[\bar{\alpha}^j + \sum_{h \in N} \bar{\beta}_h^j v_h(s) \right] - \sum_{j=1}^{n} b_j u_j(s) [x(s)]^{1/2} - \delta x(s),
$$

$$
x(t_0) = x_{t_0}. \tag{6.31}
$$

In the game in (6.30) and (6.31) one can readily observe that government i's tax policy $v_i(s)$ is not only explicitly reflected in its own output but also on the outputs of other nations. This modeling formulation allows some intriguing scenarios to arise. For instance, an increase of $v_i(s)$ may just cause a minor drop in nation i's industrial profit, but may cause significant increases in its neighbors' outputs, which produce large, local, negative environmental impacts to nation i. This results in the nations' reluctance to increase or impose taxes on industrial outputs.

6.6.2 Noncooperative Outcomes

In this section we discuss the solution to the noncooperative game in (6.30) and (6.31). Invoking Theorem 2.3 of Chap. 2 a feedback Nash equilibrium solution can be characterized as follows.

Corollary 6.2 *A set of strategies* $\{u_i^*(t) = \mu_i(t, x), v_i^*(t) = \phi_i(t, x), \text{for } i \in N\}$ *provides a feedback Nash equilibrium solution to the game in* (6.30) *and* (6.31) *if there exist suitably smooth functions* $V^{(t_0)i}(t, x) : [t_0, T] \times R \rightarrow R, i \in N$, *satisfying the* *following partial differential equations:*

$$
-V_t^{(t_0)i}(t, x) = \max_{v_i, u_i} \left\{ \left[\left(\alpha^i - \sum_{j=1}^n \beta_j^i \left[\bar{\alpha}^j + \sum_{\substack{h \in N \\ h \neq i}} \bar{\beta}_h^j \phi_h(t, x) + \bar{\beta}_i^j v_i \right] \right) \right. \right.
$$

$$
\times \left[\bar{\alpha}^i + \sum_{\substack{h \in N \\ h \neq i}} \bar{\beta}_h^i \phi_h(t, x) + \bar{\beta}_i^i v_i \right]
$$

$$
- c_i \left[\bar{\alpha}^i + \sum_{\substack{j \in N \\ j \neq i}} \bar{\beta}_j^i \phi_j(t, x) + \bar{\beta}_i^i v_i \right] - c_i^a [u_i]^2
$$

$$
- \sum_{j \in \bar{K}^i} \varepsilon_i^j \left[\bar{\alpha}^j + \sum_{\substack{\ell \in N \\ \ell \neq i}} \bar{\beta}_\ell^j v_\ell(s) + \bar{\beta}_i^j v_i \right] - h_i x \right] e^{-r(t - t_0)}
$$

$$
+ V_x^{(t_0)i}(t, x) \left[\sum_{j=1}^n a_j \left[\bar{\alpha}^j + \sum_{\substack{h \in N \\ h \neq i}} \bar{\beta}_h^j \phi_h(t, x) + \bar{\beta}_i^j v_i \right] \right.
$$

$$
\left. \left. - \sum_{\substack{j=1 \\ j \neq i}}^n b_j \mu_j(t, x) x^{1/2} - b_i u_i x^{1/2} - \delta x \right] \right\}, \tag{6.32}
$$

$$
V^{(t_0)i}(T, x) = -g^i \left[x - \bar{x}^i \right] e^{-r(T - t_0)}. \tag{6.33}
$$

Performing the indicated maximization in (6.32) yields

$$
\mu_i(t, x) = -\frac{b_i}{2c_i^a} V_x^{(t_0)i}(t, x) e^{r(t - t_0)} x^{1/2}, \tag{6.34}
$$

$$
\left(\alpha^i - \sum_{j=1}^n \beta_j^i \left[\bar{\alpha}^j + \sum_{h \in N} \bar{\beta}_h^j \phi_h(t, x) \right] \right) \bar{\beta}_i^i - \left[\sum_{j=1}^n \beta_j^i \bar{\beta}_i^j \right] \left[\bar{\alpha}^i + \sum_{h \in N} \bar{\beta}_h^i \phi_h(t, x) \right]
$$

$$
- c_i \bar{\beta}_i^i - \sum_{j \in \bar{K}^i} \varepsilon_i^j \bar{\beta}_i^j + V_x^{(t_0)i}(t, x) \sum_{j=1}^n a_j \bar{\beta}_i^j e^{r(t - t_0)} = 0, \tag{6.35}
$$

for $t \in [t_0 < T]$ and $i \in N$.

     *The system in (6.35) forms a set of equations linear in $\{\phi_1(t, x), \phi_2(t, x), \ldots,$
$\phi_n(t, x)\}$ with $\{V_x^{(t_0)1}(t, x) e^{r(t - t_0)}, V_x^{(t_0)2}(t, x) e^{r(t - t_0)}, \ldots, V_x^{(t_0)n}(t, x) e^{r(t - t_0)}\}$ being taken as a set of parameters. Solving (6.35) yields*

$$
\phi_i(t, x) = \hat{\alpha}^i + \sum_{j \in N} \hat{\beta}_j^i V_x^{(t_0)j}(t, x) e^{r(t - t_0)}, \quad i \in N, \tag{6.36}
$$

where $\hat{\alpha}^i$ and $\hat{\beta}^i_j$, for $i \in N$ and $j \in N$, are constants involving the constant coefficients in (6.35). Substituting the results in (6.34) and (6.36) into (6.32) and (6.33) we obtain the following.

Proposition 6.2 *The system in* (6.32) *and* (6.33) *admits a solution*

$$V^{(t_0)i}(t,x) = \left[A_i(t)x + C_i(t) \right] e^{-r(t-t_0)}, \quad for \ i \in N, \tag{6.37}$$

where $\{ A_1(t), A_2(t), \dots, A_n(t) \}$ *satisfies the following set of constant coefficient quadratic ordinary differential equations*:

$$\dot{A}_i(t) = (r+\delta)A_i(t) - \frac{b_i^2}{4c_i^a}\left[A_i(t) \right]^2 - A_i(t) \sum_{\substack{j=1 \\ j \neq i}}^{n} \frac{b_j^2}{2c_j^a} A_j(t) + h_i, \tag{6.38}$$

$$A_i(T) = -g^i: \quad for \ i \in N,$$

and $\{ C_i(t); i \in N \}$ *is given by*

$$C_i(t) = e^{r(t-t_0)}\left[\int_{t_0}^{t} F_i(y)e^{-r(y-t_0)}\,dy + C_i^0 \right], \tag{6.39}$$

where

$$C_i^0 = g^i \bar{x}^i e^{-r(T-t_0)} - \int_{t_0}^{T} F_i(y)e^{-r(y-t_0)}\,dy,$$

$$F_i(t) = -\left(\alpha^i - \sum_{j=1}^{n} \beta^i_j \left\{ \bar{\alpha}^j + \sum_{h \in Ni}^{n} \bar{\beta}^j_h \left[\hat{\alpha}^h + \sum_{k \in N} \hat{\beta}^h_k A_k(t) \right] \right\} \right)$$

$$\times \left(\bar{\alpha}^i + \sum_{h \in N} \bar{\beta}^i_h \left[\hat{\alpha}^h + \sum_{k \in N} \hat{\beta}^h_k A_k(t) \right] \right)$$

$$+ c_i \left\{ \bar{\alpha}^i - \sum_{j \in N} \bar{\beta}^i_j \left[\hat{\alpha}^j + \sum_{k \in N} \hat{\beta}^j_k A_k(t) \right] \right\}$$

$$+ \sum_{j \in \bar{K}^i} \varepsilon^j_i \left\{ \bar{\alpha}^j + \sum_{\ell \in N} \bar{\beta}^j_\ell \left[\hat{\alpha}^\ell + \sum_{k \in N} \hat{\beta}^\ell_k A_k(t) \right] \right\}$$

$$- A_i(t) \left[\sum_{j=1}^{n} a_j \left\{ \bar{\alpha}^j + \sum_{h \in N} \bar{\beta}^j_h \left[\hat{\alpha}^h + \sum_{k \in N} \hat{\beta}^h_k A_k(t) \right] \right\} \right].$$

Proof See Appendix 1 in this chapter's appendixes.   □

The corresponding feedback Nash equilibrium strategies of the game in (6.30) and (6.31) can be obtained as

$$
\mu_i(t, x) = -\frac{b_i}{2c_i^a} A_i(t) x^{1/2} \quad \text{and} \quad \phi_i(t, x) = \hat{\alpha}^i + \sum_{j \in N} \hat{\beta}_j^i A_j(t), \qquad (6.40)
$$

for $i \in N$ and $t \in [t_0, T]$.

A remark that will be utilized in the subsequent analysis is given below.

Remark 6.3 Let $V^{(\tau)i}(t, x_t)$ denote the value function of nation i in a game with the payoffs in (6.30) and dynamics in (6.31) that starts at time τ. One can readily verify that $V^{(\tau)i}(t, x_t) = V^{(t_0)i}(t, x_t) e^{r(\tau - t_0)}$, for $\tau \in [t_0, T]$.

6.7 Collaborative Scheme in Transboundary Industrial Pollution Management

Now consider the case when all the nations want to cooperate and agree to act so that an international optimum can be achieved. Since nations are asymmetric and the number of nations may be large, a reasonable optimality principle for gain distribution is to share the gain from cooperation proportional to the relative sizes of the nations' noncooperative payoffs. As mentioned before, to ensure that the cooperative solution is time consistent, the above optimality principle must be maintained throughout the game.

6.7.1 Cooperative Optimization and State Trajectory

To secure group optimality the participating nations seek to maximize their joint payoff by solving the following optimal control problem:

$$
\max_{v_1, v_2, \ldots, v_n; u_1, u_2, \ldots, u_n} \left\{ \int_{t_0}^T \sum_{\ell=1}^n \left[\left(\alpha^\ell - \sum_{j=1}^n \beta_j^\ell \left[\bar{\alpha}^j + \sum_{h \in N} \bar{\beta}_h^j v_h(s) \right] \right) \right. \right.
$$
$$
\times \left[\bar{\alpha}^\ell + \sum_{h \in N} \bar{\beta}_h^\ell v_h(s) \right] - c_\ell \left[\bar{\alpha}^\ell + \sum_{j \in N} \bar{\beta}_j^\ell v_j(s) \right] - c_\ell^a [u_\ell(s)]^2
$$
$$
\left. - \sum_{j \in \bar{K}^\ell} \varepsilon_\ell^j \left[\bar{\alpha}^j + \sum_{k \in N} \bar{\beta}_k^j v_k(s) \right] - h_\ell x(s) \right] e^{-r(s-t_0)} \, ds
$$
$$
\left. - \sum_{\ell=1}^n g^\ell [x(T) - \bar{x}^\ell] e^{-r(T-t_0)} \right\}, \qquad (6.41)
$$

subject to (6.31).

Invoking Theorem A.1 in the Technical Appendixes, a set of controls $\{[v_i^{**}(t),$ $u_i^{**}(t)] = [\psi_i(t, x), \varpi_i(t, x)],$ for $i \in N\}$ constitutes an optimal solution to the control problem in (6.41) and (6.31) if there exists a continuously differentiable function $W^{(t_0)}(t, x) : [t_0, T] \times R \to R, i \in N,$ satisfying the following partial differential equations:

$$-W_t^{(t_0)}(t, x)$$

$$= \max_{v_1, v_2, \ldots, v_n; u_1, u_2, \ldots, u_n} \left\{ \sum_{\ell=1}^n \left[\left(\alpha^\ell - \sum_{j=1}^n \beta_j^\ell \left[\bar{\alpha}^j + \sum_{h \in N} \bar{\beta}_h^j v_h \right] \right) \left[\bar{\alpha}^\ell + \sum_{h \in N} \bar{\beta}_h^\ell v_h \right] \right. $$

$$\left. - c_\ell \left[\bar{\alpha}^\ell + \sum_{j \in N} \bar{\beta}_j^\ell v_j \right] - c_\ell^a [u_\ell]^2 - \sum_{j \in \bar{K}^\ell} \varepsilon_\ell^j \left[\bar{\alpha}^j + \sum_{k \in N} \bar{\beta}_k^j v_k \right] - h_\ell x \right] e^{-r(s-t_0)}$$

$$+ W_x^{(t_0)}(t, x) \left[\sum_{j=1}^n a_j \left[\bar{\alpha}^j + \sum_{h \in N} \bar{\beta}_h^j v_h \right] - \sum_{j=1}^n b_j u_j x^{1/2} - \delta x \right] \right\}, \tag{6.42}$$

$$W^{(t_0)}(T, x) = -\sum_{i=1}^n g^i \left[x(T) - \bar{x}^i \right] e^{-r(T-t_0)}. \tag{6.43}$$

Performing the indicated maximization in (6.42) yields the optimal controls under cooperation as

$$\varpi_i(t, x) = -\frac{b_i}{2c_i^a} W_x^{(t_0)}(t, x) e^{r(t-t_0)} x^{1/2}, \quad \text{for } i \in N; \tag{6.44}$$

$$\sum_{\ell=1}^n \left[\left(\alpha^\ell - \sum_{j=1}^n \beta_j^\ell \left[\bar{\alpha}^j + \sum_{h \in N} \bar{\beta}_h^j \psi_h(t, x) \right] \right) \bar{\beta}_i^\ell \right.$$

$$\left. - \left[\sum_{j=1}^n \beta_j^\ell \bar{\beta}_i^j \right] \left[\bar{\alpha}^\ell + \sum_{h \in N} \bar{\beta}_h^\ell \psi_h(t, x) \right] \right]$$

$$- \sum_{\ell=1}^n \left[c_\ell \bar{\beta}_i^\ell + \sum_{j \in \bar{K}^i} \varepsilon_\ell^j \bar{\beta}_i^j \right] + W_x^{(t_0)} \sum_{j=1}^n a_j \bar{\beta}_i^j e^{r(t-t_0)} = 0, \quad \text{for } i \in N. \tag{6.45}$$

The system in (6.45) can be viewed as a set of equations linear in $\{\psi_1(t, x),$ $\psi_2(t, x), \ldots, \psi_n(t, x)\}$ with $W_x(t, x) e^{r(t-t_0)}$ being taken as a parameter. Solving (6.45) yields

$$\psi_i(t, x) = \hat{\alpha}^i + \hat{\beta}^i W_x^{(t_0)}(t, x) e^{r(t-t_0)}, \tag{6.46}$$

where $\hat{\alpha}^i$ and $\hat{\beta}^i$, for $i \in N$, are constants involving the model parameters.

Proposition 6.3 *The system in (6.42) and (6.43) admits a solution*

$$W^{(t_0)}(t, x) = \left[A^*(t)x + C^*(t)\right]e^{-r(t-t_0)}, \tag{6.47}$$

with

$$A^*(t) = A_*^P + \Phi^*(t)\left[\bar{C}^* - \int_{t_0}^t \sum_{j=1}^n \frac{b_j^2}{2c_j^a}\Phi^*(y)\,dy\right]^{-1}, \quad and$$

$$C^*(t) = e^{r(t-t_0)}\left[\int_{t_0}^t F^*(y)e^{-r(y-t_0)}\,dy + C_*^0\right],$$

where

$$\Phi^*(t) = \exp\left\{\int_{t_0}^t \left[\sum_{j=1}^n \frac{b_j^2}{2c_j^a}A_*^P + (r+\delta)\right]dy\right\},$$

$$\bar{C}^* = \frac{-\Phi^*(T)}{(A_*^P + \sum_{j=1}^n g^j)} + \int_{t_0}^T \sum_{j=1}^n \frac{b_j^2}{2c_j^a}\Phi^*(y)\,dy,$$

$$A_*^P(t) = \left\{(r+\delta) - \left[(r+\delta)^2 + 4\sum_{j=1}^n \frac{b_j^2}{2c_j^a}\sum_{j=1}^n h_j\right]^{1/2}\right\}\bigg/\sum_{j=1}^n \frac{b_j^2}{c_j^a},$$

$$\begin{aligned}
F^*(t) = &-\sum_{\ell=1}^n \Bigg[\left(\alpha^\ell - \sum_{j=1}^n \beta_j^\ell\left\{\bar{\alpha}^j + \sum_{h\in N}\bar{\beta}_h^j[\hat{\alpha}^h + \hat{\beta}^h A^*(t)]\right\}\right)\left\{\bar{\alpha}^\ell\right. \\
&\left. + \sum_{h\in N}\bar{\beta}_h^\ell[\hat{\alpha}^h + \hat{\beta}^h A^*(t)]\right\} - c_\ell\left\{\bar{\alpha}^\ell + \sum_{j\in N}\bar{\beta}_j^\ell[\hat{\alpha}^j + \hat{\beta}^j A^*(t)]\right\} \\
&- \sum_{j\in\bar{K}^\ell}\varepsilon_\ell^j\left\{\bar{\alpha}^j + \sum_{k\in N}\bar{\beta}_k^j[\hat{\alpha}^k + \hat{\beta}^{kj}A^*(t)]\right\}\Bigg] \\
&- A_x^*(t)\left[\sum_{j=1}^n a_j\left\{\bar{\alpha}^j + \sum_{h\in N}\bar{\beta}_h^j[\hat{\alpha}^h + \hat{\beta}^h A^*(t)]\right\}\right], \quad and
\end{aligned}$$

$$C_*^0 = \sum_{j=1}^n g^j \bar{x}^j e^{-r(T-t_0)} - \int_{t_0}^T F^*(y)e^{-r(y-t_0)}\,dy.$$

Proof See Appendix 2 in this chapter's appendixes.   □

Using (6.44), (6.46), and (6.47), the control strategy under cooperation can be obtained as

$$\psi_i(t,x) = \hat{\alpha}^i + \hat{\beta}^i A^*(t) \quad \text{and} \quad \varpi_i(t,x) = -\frac{b_i}{2c_i^a} A^*(t) x^{1/2}, \qquad (6.48)$$

for $t \in [t_0 < T]$ and $i = 1, 2, \ldots, n$.

Substituting the optimal control strategy from (6.48) into (6.26) yields the dynamics of pollution accumulation under cooperation. Solving the cooperative pollution dynamics yields the cooperative state trajectory

$$
\begin{aligned}
x^*(t) = e^{\left[\int_{t_0}^{t}\left[\sum_{j=1}^{n}\frac{b_j^2}{2c_j^a}A^*(s)-\delta\right]ds\right]} \\
\times \left[x_{t_0} + \int_{t_0}^{t} \sum_{j=1}^{n} a_j \left\{ \bar{\alpha}^j + \sum_{h\in N} \bar{\beta}_h^j [\hat{\alpha}^h + \hat{\beta}^h A^*(s)] \right\} \right. \\
\left. \times e^{\left[\int_{t_0}^{s}[\delta-\sum_{j=1}^{n}\frac{b_j^2}{2c_j^a}A^*(\tau)]d\tau\right]} \, ds \right],
\end{aligned} \qquad (6.49)
$$

for $t \in [t_0, T]$.

We use X_t^* to denote the set of realizable values of $x^*(t)$ at time t generated by (6.49). The term x_t^* is used to denote an element in the set X_t^*.

A remark that will be utilized in the subsequent analysis is given below.

Remark 6.4 Let $W^{(\tau)}(t, x_t)$ denote the value function of the optimal control problem with the objective in (6.41) and dynamics in (6.31), which starts at time τ. One can readily verify that

$$W^{(\tau)}(t, x_t^*) = W^{(t_0)}(t, x_t^*) e^{r(\tau - t_0)}, \quad \text{for } \tau \in [t_0, T].$$

A group optimal scenario will be realized with the nations adopting the cooperative strategies in (6.48). To construct a cooperative solution that every party will commit to throughout cooperation period, a time consistent solution has to be sought. This will be investigated in the following subsection.

6.7.2 Consistent Imputation and Benefit Distribution

To satisfy the property of time (optimal-trajectory-subgame) consistency, the optimality principle has to remain in effect throughout the cooperation period along the cooperative trajectory $\{x^*(\tau)\}_{\tau=t_0}^{T}$. Hence the solution imputation scheme $\{\xi^{(\tau)i}(\tau, x_\tau^*); \text{ for } i \in N\}$ has to satisfy the following:

Condition 6.2

$$\xi^{(\tau)i}\left(\tau, x_\tau^*\right) = \frac{V^{(\tau)i}(\tau, x_\tau^*)}{\sum_{j=1}^{n} V^{(\tau)j}(\tau, x_\tau^*)} W^{(\tau)}\left(\tau, x_\tau^*\right), \tag{6.50}$$

for $i \in N, x_\tau^* \in X_\tau^*$ and $\tau \in [t_0, T]$.

To formulate a payoff distribution procedure over time so that the agreed imputations satisfy Condition 6.2 we obtain the following.

Proposition 6.4 *A distribution scheme with a terminal payment* $-g^i[x_T^* - \bar{x}^i]$ *at time T and an instantaneous payment at time* $\tau \in [t_0, T]$

$$B_i(\tau) = -\frac{\partial}{\partial t}\left[\frac{V^{(\tau)i}(t, x_t^*)}{\sum_{j=1}^{n} V^{(\tau)j}(t, x_t^*)} W^{(\tau)}\left(t, x_t^*\right)|_{t=\tau}\right]$$

$$- \frac{\partial}{\partial x_\tau^*}\left[\frac{V^{(\tau)i}(\tau, x_\tau^{i*})}{\sum_{j=1}^{n} V^{(\tau)j}(\tau, x_\tau^{j*})} W^{(\tau)}\left(\tau, x_\tau^*\right)\right]$$

$$\times \left[\sum_{j=1}^{n} a_j\left[\bar{\alpha}^j + \sum_{h \in N} \bar{\beta}_h^j \psi_h\left(\tau, x_\tau^*\right)\right] - \sum_{j=1}^{n} b_j \varpi_j\left(\tau, x_\tau^*\right)\left(x_\tau^*\right)^{1/2} - \delta x_\tau^*\right], \tag{6.51}$$

for $i \in N$, *will lead to the realization of the solution imputations* $\xi^{(\tau)i}(\tau, x_\tau^*)$, *for* $i \in N$ *and* $\tau \in [t_0, T]$, *satisfying Condition 6.2.*

Proof Invoking Theorem 4.2 in Chap. 4, Proposition 6.4 follows.   □

Using Propositions 6.2, 6.3, and (6.48), one can express $B_i(\tau)$ in Proposition 6.4 as

$$B_i(\tau) = \frac{-[A_i(\tau)x_\tau^* + C_i(\tau)]}{(\sum_{j=1}^{2}[A_j(\tau)x_\tau^* + C_j(\tau)])}\{[\dot{A}(\tau)x_\tau^* + \dot{C}(\tau)] - r[A(\tau)x_\tau^* + C(\tau)]\}$$

$$- \frac{[A(\tau)x_\tau^* + C(\tau)]}{(\sum_{j=1}^{2}[A_j(\tau)x_\tau^* + C_j(\tau)])}$$

$$\times \{[\dot{A}_i(\tau)x_\tau^* + \dot{C}_i(\tau)] - r[A_i(\tau)x_\tau^* + C_i(\tau)]\}$$

$$+ \frac{[A_i(\tau)x_\tau^* + C_i(\tau)][A(\tau)x_\tau^* + C(\tau)]}{(\sum_{j=1}^{2}[A_j(\tau)x_\tau^* + C_j(\tau)])^2}$$

$$\times \sum_{j=1}^{2}\{[\dot{A}_j(\tau)x_\tau^* + \dot{C}_j(\tau)] - r[A_j(\tau)x_\tau^* + C_j(\tau)]\}$$

$$+ \left[\frac{[A_i(\tau)x^*_\tau + C_i(\tau)][A(\tau)x^*_\tau + C(\tau)]}{(\sum_{j=1}^2 [A_j(\tau)x^*_\tau + C_j(\tau)])^2} \left(\sum_{j=1}^2 A_j(\tau) \right) \right]$$

$$\times \left[\sum_{j=1}^n a_j \left[\bar{\alpha}^j + \sum_{h \in N} \bar{\beta}^j_h (\hat{\alpha}^h + \hat{\beta}^h A^*(\tau)) \right] + \sum_{j=1}^n \frac{b_j^2}{2c^a_j} A^*(\tau)x^*_\tau - \delta x^*_\tau \right],$$

$$(6.52)$$

for $i \in N$.

A time (optimal-trajectory-subgame) consistent solution to the multinational collaboration in environmental management can then be expressed as

$$P(x^*_t, T - t) = \{ [\psi(s, x^*(s))]^T_{s=t}, [\varpi(s, x^*(s))]^T_{s=t}, [B(s)]^T_{s=t}, \xi^{(t)}(t, x^*_t) \},$$

for $t \in [t_0, T]$,

where

$$\psi_i(s, x^*(s)) = \hat{\alpha}^i + \hat{\beta}^i A^*(s), \qquad \varpi_i(s, x^*(s)) = -\frac{b_i}{2c^a_i} A^*(s) [x^*(s)]^{1/2},$$

$B_i(\tau)$ is given as in (6.52), and $\xi^{(t)}(t, x^*_t)$ is an imputation scheme satisfying Condition 6.2.

When all nations are adopting the cooperative strategies the rate of instantaneous payment that nation $\ell \in N$ will realize at time t with the state being x^*_t can be expressed as

$$\Re_\ell(t) = \left(\alpha^\ell - \sum_{j=1}^n \beta^\ell_j \left\{ \bar{\alpha}^j + \sum_{h \in N} \bar{\beta}^j_h [\hat{\alpha}^h + \hat{\beta}^h A^*(t)] \right\} \right)$$

$$\times \left\{ \bar{\alpha}^\ell + \sum_{h \in N} \bar{\beta}^\ell_h [\hat{\alpha}^h + \hat{\beta}^h A^*(t)] \right\}$$

$$- c_\ell \left\{ \bar{\alpha}^\ell + \sum_{j \in N} \bar{\beta}^\ell_j [\hat{\alpha}^j + \hat{\beta}^j A^*(t)] \right\} - c^a_\ell \left[\frac{b_\ell}{2c^a_\ell} A^*(t) \right]^2 x^*_t$$

$$- \sum_{j \in \bar{K}^\ell} \varepsilon^j_\ell \left\{ \bar{\alpha}^j + \sum_{k \in N} \bar{\beta}^j_k [\hat{\alpha}^k + \hat{\beta}^{kj} A^*(t)] \right\} - h_\ell x^*_t. \qquad (6.53)$$

Since, according to Proposition 6.4 under the cooperative scheme, an instantaneous payment to nation ℓ equaling $B_\ell(t)$ at time t, and a side payment of the value $B_\ell(t) - \Re_\ell(t)$ will be offered to nation ℓ.

6.8 Exercises

6.1 Consider a two-nation international economy in which transboundary pollution is generated by the production process. The planning horizon is $[0, 2]$. The inverse demand function of the outputs of nations 1 and 2 at time instant $s \in [0, 2]$ are, respectively,

$$P_1(s) = 100 - 2q_1(s) - 0.5q_2(s) \quad \text{and} \quad P_2(s) = 120 - q_1(s) - 0.5q_2(s),$$

where $P_i(s)$ is the price of the output and $q_i(s)$ is the output of the nation. The unit costs of production in these nations are $c_1 = 1$ and $c_2 = 1.5$. The instantaneous industrial profits of nations 1 and 2 at time s can be expressed as

$$\pi_1(s) = \left[100 - 2q_1(s) - 0.5q_2(s)\right]q_1(s) - q_1(s) - v_1(s)q_1(s) \quad \text{and}$$

$$\pi_2(s) = \left[120 - 0.5q_1(s) - q_2(s)\right]q_2(s) - 1.5q_2(s) - v_2(s)q_2(s),$$

where $v_i(s) \geq 0$ is the tax rate imposed by government i on its industrial output at time s. At each time instant s, the industrial sector of nation $i \in \{1, 2\}$ seeks to maximize its instantaneous profit.

Derive the market equilibrium at time instant s.

6.2 Industrial production emits pollutants into the environment. The emitted pollutants cause short-term local impacts on neighboring areas of the origin of production in forms like passing-by waste in waterways, wind-driven suspended particles in the air, unpleasant odor, noise, dust, and heat. For an output of $q_1(s)$ produced by nation 1 there will be a short-term local environmental impact (cost) of $0.2q_1(s)$ on nation 1 itself and a local impact of $0.1q_1(s)$ on its neighbor nation 2. On the other hand, for an output of $q_2(s)$ produced by nation 2, there will be a short-term local environmental impact (cost) of $0.3q_2(s)$ on nation 2 itself and a local impact of $0.25q_2(s)$ on its neighbor nation 1.

Moreover, industrial production will also create long-term global environmental impacts by building up existing pollution stocks, like greenhouse gases, CFC, and atmospheric particulates. Each government adopts its own pollution abatement policy to reduce the pollution stock. Let $x(s)$ denote the level of pollution and $u_j(s)$ the pollution abatement effort of nation j at time s, the dynamics of the pollution stock is governed by the differential equation

$$\dot{x}(s) = 4q_1(s) + 3q_2(s) - 0.05u_1(s)\left[x(s)\right]^{1/2} - 0.02u_2(s)\left[x(s)\right]^{1/2} - 0.04x(s),$$

$$x(0) = 100.$$

The governments have to promote business interests and at the same time handle the financing of the costs brought about by pollution. The damages to countries 1 and 2 from an $x(s)$ amount of pollution are $0.15x(s)$ and $0.1x(s)$. In particular, each government maximizes the net gains in the industrial sector minus the sum of the expenditures on pollution abatement and damages from pollution. The instantaneous objective of governments 1 and 2 at time s are, respectively,

$$[100 - 2q_1(s) - 0.5q_2(s)]q_1(s) - q_1(s) - 0.2q_1(s) - 0.25q_2(s) - 0.15x(s), \quad \text{and}$$
$$[120 - q_1(s) - q_2(s)]q_2(s) - 1.5q_2(s) - 0.1q_1(s) - 0.3q_2(s) - 0.1x(s).$$

The governments' planning horizon is $[0, 2]$. At terminal time 2, the terminal appraisal associated with the state of pollution is $5[60 - x(2)]$ for nation 1 and $4[40 - x(2)]$ for nation 2. The discount rate is 0.05. Each government seeks to maximize the integral of its instantaneous objective over the planning horizon subject to pollution stock dynamics.

Construct a differential game of noncooperative pollution management by these two nations. Obtain a feedback Nash equilibrium solution for the game.

6.3 Consider the case when both nations want to cooperate and agree to act so that an international optimum can be achieved.

Obtain the optimal cooperative levels of outputs and abatement efforts.

6.4 These cooperating nations adopt an optimality principle that distributes the gain from cooperation proportional to the relative sizes of the nations' noncooperative payoffs. Characterize a time (optimal-trajectory-subgame) consistent solution.

Appendix 1: Proof of Proposition 6.2

Using (6.34), (6.36), and (6.37), the system in (6.32) and (6.33) can be expressed as

$$r\left[A_i(t)x + C_i(t)\right] - \left[\dot{A}_i(t)x + \dot{C}_i(t)\right]$$

$$= \left[\left(\alpha^i - \sum_{j=1}^{n}\beta_j^i\left\{\bar{\alpha}^j + \sum_{h \in Ni}\bar{\beta}_h^j\left[\hat{\alpha}^h + \sum_{k \in N}\hat{\beta}_k^h A_k(t)\right]\right\}\right.$$

$$\times \left(\bar{\alpha}^i + \sum_{h \in N}\bar{\beta}_h^i\left[\hat{\alpha}^h + \sum_{k \in N}\hat{\beta}_k^h A_k(t)\right]\right)$$

$$- c_i\left\{\bar{\alpha}^i + \sum_{j \in N}\bar{\beta}_j^i\left[\hat{\alpha}^j + \sum_{k \in N}\hat{\beta}_k^j A_k(t)\right]\right\}$$

$$- c_i^a\left[\frac{b_i}{2c_i^a}A_i(t)\right]^2 x - \sum_{j \in \bar{K}i}\varepsilon_i^j\left\{\bar{\alpha}^j + \sum_{\ell \in N}\bar{\beta}_\ell^j\left[\hat{\alpha}^\ell + \sum_{k \in N}\hat{\beta}_k^\ell A_k(t)\right]\right\} - h_i x\right]$$

$$+ A_i(t)\left[\sum_{j=1}^{n}a_j\left\{\bar{\alpha}^j + \sum_{h \in N}\bar{\beta}_h^j\left[\hat{\alpha}^h + \sum_{k \in N}\hat{\beta}_k^h A_k(t)\right]\right\}\right.$$

$$+ \sum_{j=1}^{n}b_j\frac{b_j}{2c_j^a}A_j(t)x - \delta x\right], \qquad (6.54)$$

$$\left[A_i(T)x + C_i(T)\right] = -g^i\left(x - \bar{x}^i\right), \quad \text{for } i \in N. \tag{6.55}$$

For (6.54) and (6.55) to hold, it is required that

$$\dot{A}_i(t) = (r+\delta)A_i(t) - A_i(t)\sum_{\substack{j=1 \\ j \neq i}}^{n}\frac{b_j^2}{2c_j^a}A_j(t) - \frac{b_i^2}{4c_i^a}\left[A_i(t)\right]^2 + h_i, \tag{6.56}$$

$$A_i(T) = -g^i, \tag{6.57}$$

$$\dot{C}_i(t) = rC_i(t) - \left(\alpha^i - \sum_{j=1}^{n}\beta_j^i\left\{\bar{\alpha}^j + \sum_{h \in Ni}\bar{\beta}_h^j\left[\hat{\alpha}^h + \sum_{k \in N}\hat{\beta}_k^h A_k(t)\right]\right\}\right)$$

$$\times \left(\bar{\alpha}^i + \sum_{h \in N}\bar{\beta}_h^i\left[\hat{\alpha}^h + \sum_{k \in N}\hat{\beta}_k^h A_k(t)\right]\right)$$

$$+ c_i\left\{\bar{\alpha}^i - \sum_{j \in N}\bar{\beta}_j^i\left[\hat{\alpha}^j + \sum_{k \in N}\hat{\beta}_k^j A_k(t)\right]\right\}$$

$$+ \sum_{j \in \bar{K}^i}\varepsilon_i^j\left\{\bar{\alpha}^j + \sum_{\ell \in N}\bar{\beta}_\ell^j\left[\hat{\alpha}^\ell + \sum_{k \in N}\hat{\beta}_k^\ell A_k(t)\right]\right\}$$

$$- A_i(t)\left[\sum_{j=1}^{n}a_j\left\{\bar{\alpha}^j + \sum_{h \in N}\bar{\beta}_h^j\left[\hat{\alpha}^h + \sum_{k \in N}\hat{\beta}_k^h A_k(t)\right]\right\}\right]$$

$$= rC_i(t) + F_i(t), \tag{6.58}$$

$$C_i(T) = g^i\bar{x}^i. \tag{6.59}$$

Equations (6.56)–(6.59) form a block recursive system of differential equations, with (6.56) and (6.57) being independent of (6.58) and (6.59).

Solving $\{A_1(t), A_2(t), \ldots, A_n(t)\}$ in (6.56) and (6.57) and substituting them into (6.58) and (6.59) yields a system of linear first-order differential equations

$$\dot{C}_i(t) = rC_i(t) + F_i(t), \tag{6.60}$$

$$C_i(T) = g^i\bar{x}^i, \quad \text{for } i \in N. \tag{6.61}$$

Since $C_i(t)$ is independent of $C_j(t)$ for $i \neq j$, $C_i(t)$ can be solved as

$$C_i(t) = e^{r(t-t_0)}\left[\int_{t_0}^{t}F_i(y)e^{-r(y-t_0)}\,dy + C_i^0\right], \tag{6.62}$$

where

$$C_i^0 = g^i\bar{x}^i e^{-r(T-t_0)} - \int_{t_0}^{T}F_i(y)e^{-r(y-t_0)}\,dy. \tag{6.63}$$

Hence Proposition 6.2 follows.

Appendix 2: Proof of Proposition 6.3

Substituting (6.44) and (6.46) into (6.42) and using (6.47) one obtains

$$
r\left[A^*(t)x + C^*(t)\right] - \left[\dot{A}^*(t)x + \dot{C}^*(t)\right]
$$

$$
= \sum_{\ell=1}^{n}\left[\left(\alpha^\ell - \sum_{j=1}^{n}\beta_j^\ell\left\{\bar{\alpha}^j + \sum_{h\in N}\bar{\beta}_h^j[\hat{\alpha}^h + \hat{\beta}^h A^*(t)]\right\}\right)\right.
$$

$$
\times\left\{\bar{\alpha}^\ell + \sum_{h\in N}\bar{\beta}_h^\ell[\hat{\alpha}^h + \hat{\beta}^h A^*(t)]\right\}
$$

$$
- c_\ell\left\{\bar{\alpha}^\ell + \sum_{j\in N}\bar{\beta}_j^\ell[\hat{\alpha}^j + \hat{\beta}^j A^*(t)]\right\} - c_\ell^a\left[\frac{b_\ell}{2c_\ell^a}A^*(t)\right]^2 x
$$

$$
- \sum_{j\in\bar{K}^\ell}\varepsilon_\ell^j\left\{\bar{\alpha}^j + \sum_{k\in N}\bar{\beta}_k^j[\hat{\alpha}^k + \hat{\beta}^{kj} A^*(t)]\right\} - h_\ell x\right]
$$

$$
+ A_x^*(t)\left[\sum_{j=1}^{n}a_j\left\{\bar{\alpha}^j + \sum_{h\in N}\bar{\beta}_h^j[\hat{\alpha}^h + \hat{\beta}^h A^*(t)]\right\}\right.
$$

$$
+ \sum_{j=1}^{n}\frac{b_j^2}{2c_j^a}A^*(t)x - \delta x\right], \tag{6.64}
$$

$$
\left[A^*(T)x + C^*(T)\right] = -\sum_{i=1}^{n}g^i\left[x(T) - \bar{x}^i\right]. \tag{6.65}
$$

For (6.64) and (6.65) to hold, it is required that

$$
\dot{A}^*(t) = (r + \delta)A^*(t) - \sum_{j=1}^{n}\frac{b_j^2}{2c_j^a}[A^*(t)]^2 + \sum_{j=1}^{n}h_j, \tag{6.66}
$$

$$
A^*(T) = -\sum_{j=1}^{n}g^j; \tag{6.67}
$$

$$
\dot{C}^*(t) = rC^*(t)
$$

$$
- \sum_{\ell=1}^{n}\left[\left(\alpha^\ell - \sum_{j=1}^{n}\beta_j^\ell\left\{\bar{\alpha}^j + \sum_{h\in N}\bar{\beta}_h^j[\hat{\alpha}^h + \hat{\beta}^h A^*(t)]\right\}\right)\right.
$$

$$
\times\left\{\bar{\alpha}^\ell + \sum_{h\in N}\bar{\beta}_h^\ell[\hat{\alpha}^h + \hat{\beta}^h A^*(t)]\right\}
$$

$$- c_\ell \left\{ \bar{\alpha}^\ell + \sum_{j \in N} \bar{\beta}_j^\ell [\hat{\alpha}^j + \hat{\beta}^j A^*(t)] \right\}$$

$$- \sum_{j \in \bar{K}^\ell} \varepsilon_\ell^j \left\{ \bar{\alpha}^j + \sum_{k \in N} \bar{\beta}_k^j [\hat{\alpha}^k + \hat{\beta}^{kj} A^*(t)] \right\} \Bigg]$$

$$- A_x^*(t) \left[\sum_{j=1}^n a_j \left\{ \bar{\alpha}^j + \sum_{h \in N} \bar{\beta}_h^j [\hat{\alpha}^h + \hat{\beta}^h A^*(t)] \right\} \right] \Bigg]$$

$$= r C^*(t) + F^*(t), \tag{6.68}$$

$$C^*(T) = \sum_{j=1}^n g^j \bar{x}^j. \tag{6.69}$$

Equations (6.66)–(6.69) form a block recursive system of differential equations, with (6.66) and (6.67) being independent of (6.68) and (6.69). Moreover, (6.68) and (6.69) are a Riccati equation with constant coefficients, the solution to which can be obtained by standard methods as

$$A^*(t) = A_*^P + \Phi^*(t) \left[\bar{C}^* - \int_{t0}^t \sum_{j=1}^n \frac{b_j^2}{2c_j^a} \Phi^*(y) \, dy \right]^{-1}, \tag{6.70}$$

where

$$\Phi^*(t) = \exp \left\{ \int_{t0}^t \left[\sum_{j=1}^n \frac{b_j^2}{2c_j^a} A_*^P + (r + \delta) \right] dy \right\},$$

$$\bar{C}^* = \frac{-\Phi^*(T)}{(A_*^P + \sum_{j=1}^n g^j)} + \int_{t0}^T \sum_{j=1}^n \frac{b_j^2}{2c_j^a} \Phi^*(y) \, dy, \quad \text{and}$$

$$A_*^P = \left\{ (r + \delta) - \left[(r + \delta)^2 + 4 \sum_{j=1}^n \frac{b_j^2}{2c_j^a} \sum_{j=1}^n h_j \right]^{1/2} \right\} \Bigg/ \sum_{j=1}^n \frac{b_j^2}{c_j^a}$$

is a particular solution of (6.66).

Substituting $A^*(t)$ above into (6.68), the system in (6.68) and (6.69) becomes a system of linear first-order differential equations

$$\dot{C}^*(t) = r C^*(t) + F^*(t), \tag{6.71}$$

$$C^*(T) = \sum_{j=1}^n g^j \bar{x}^j. \tag{6.72}$$

Solving (6.71) and (6.72) yields

$$C^*(t) = e^{r(t-t_0)} \left[\int_{t_0}^{t} F^*(y) e^{-r(y-t_0)} \, dy + C_*^0 \right],$$

(6.73)

where

$$C_*^0 = \sum_{j=1}^{n} g^j \bar{x}^j e^{-r(T-t_0)} - \int_{t_0}^{T} F^*(y) e^{-r(y-t_0)} \, dy.$$

Hence Proposition 6.3 follows.

Chapter 7
Dynamically Stable Dormant Firm Cartel

In this chapter, the optimization by cartels that restricts outputs to enhance their joint profit is examined. In particular, we consider oligopolies in which firms agree to form a cartel to restrain output and enhance their profits. Some firms have cost disadvantages that force them to become dormant partners. In Sect. 7.1 a dynamic oligopoly in which there are cost differentials among firms is presented. Pareto optimal output path, imputation schemes, profit sharing arrangements, and time (optimal-trajectory-subgame) consistent solution are derived for a dormant firm cartel in Sect. 7.2. An illustration is shown in the following section. The case when the planning horizon becomes infinite is analyzed in Sect. 7.4, including an illustration with an explicit solution following in the subsequent section.

7.1 A Dynamic Oligopoly

For analytical purposes we develop a dynamic model of oligopoly in which cost differentials among firms are present.

7.1.1 Basic Settings

Consider an oligopoly in which n firms are allowed to extract a renewable resource within the duration $[t_0, T]$. Among the n firms, n_1 of them have cost advantages over the other $n_2 = n - n_1$ firms. For notational convenience, the firms with cost advantages are numbered from 1 to n_1 and the firms with cost disadvantages are numbered from $n_1 + 1$ to n. The subset of firms with cost advantages is denoted by N_1 and that of firms with cost disadvantages is denoted by N_2. The firms with cost advantages are identical and so are the firms with cost disadvantages.

D.W.K. Yeung, L.A. Petrosyan, *Subgame Consistent Economic Optimization*, 177
Static & Dynamic Game Theory: Foundations & Applications,
DOI 10.1007/978-0-8176-8262-0_7, © Springer Science+Business Media, LLC 2012

The dynamics of the resource is characterized by the differential equations

$$\dot{x}(s) = f\left[s, x(s), \sum_{j=1}^{n} u_j(s)\right] = f\left[s, x(s), \sum_{j^1 \in N_1} u_{j^1}(s) + \sum_{j^2 \in N_2} u_{j^2}(s)\right],$$

$$x(t_0) = x_0 \in X, \tag{7.1}$$

where $u_j \in U_j$ is the (nonnegative) amount of resource extracted by firm i, for $i \in N$, and $x(s)$ is the resource stock.

The extraction cost depends on the quantity of resource extracted $u^i(s)$ and the resource stock size $x(s)$. In particular, the extraction cost for the n_1 firms with cost advantages is

$$c^{j^1}\left[u_{j^1}(s), x(s)\right], \quad \text{for } j^1 \in N_1,$$

and the extraction cost for the n_1 firms with cost advantages is

$$c^{j^2}\left[u_{j^2}(s), x(s)\right], \quad \text{for } j^2 \in N_2.$$

This formulation of unit cost follows from two assumptions: (i) the cost of extraction is positively related to the extraction effort, and (ii) the amount of the resource extracted, seen as the output of a production function of two inputs (effort and stock level), is increasing in both inputs (see Clark 1976). In particular, firm $j^1 \in N_1$ has cost advantage so that

$$\partial c^{j^1}\left[u_{j^1}(s), x(s)\right]/\partial u_{j^1}(s) < \partial c^{j^2}\left[u_{j^2}(s), x(s)\right]/\partial u_{j^2}(s),$$

for all levels of $u_{j^1} \in U_{j^1}$ and $u_{j^2} \in U_{j^2}$ at any $x \in X$.

The market price of the resource depends on the total amount extracted and supplied to the market. The price-output relationship at time s is given by the following downward-sloping inverse demand curve $P(s) = g[Q(s)]$, where $Q(s) = \sum_{j^1 \in N_1} u_{j^1}(s) + \sum_{j^2 \in N_2} u_{j^2}(s)$ is the total amount of the resource extracted and marketed at time s. At time T, firm $j^1 \in N_1$ will receive a termination bonus $q^{j^1}[x(T)]$ and firm $j^2 \in N_2$ will receive a termination bonus $q^{j^2}[x(T)]$. There exists a discount rate r, and the profits received at time t have to be discounted by the factor $\exp[-r(t - t_0)]$.

At time t_0, firm $j^1 \in N_1$, which has cost advantages, seeks to maximize its profit

$$\int_{t_0}^{T} \left(g\left[\sum_{h \in N_i} u_h(s) + \sum_{\ell \in N_2} u_\ell(s)\right] u_{j^1}(s) - c^{j^1}\left[u_{j^1}(s), x(s)\right]\right) \exp[-r(s - t_0)] ds$$

$$+ \exp[-r(T - t_0)]q^{j^1}[x(T)], \tag{7.2}$$

subject to (7.1).

At time t_0, firm $j^2 \in N_2$, which has cost disadvantages, seeks to maximize profit

$$\int_{t_0}^{T} \left(g\left[\sum_{h \in N_i} u_h(s) + \sum_{\ell \in N_2} u_\ell(s) \right] u_{j^2}(s) - c^{j^2}\left[u_{j^2}(s), x(s) \right] \right) \exp[-r(s - t_0)] \, ds$$

$$+ \exp[-r(T - t_0)] q^{j^2}[x(T)], \tag{7.3}$$

subject to (7.1).

The noncooperative market equilibrium of the oligopoly game in (7.1)–(7.3) will be characterized next.

7.1.2 Market Outcome

We use $\Gamma(x_0, T - t_0)$ to denote the game in (7.1)–(7.3) and $\Gamma(x_\tau, T - \tau)$ to denote an alternative game with state dynamics in (7.1) and the payoff structures in (7.2) and (7.3), which starts at time $\tau \in [t_0, T]$ with initial state $x_\tau \in X$. Invoking Theorem 2.3 in Chap. 2, a noncooperative Nash equilibrium solution of the game $\Gamma(x_\tau, T - \tau)$ can be characterized as follows.

Corollary 7.1 *A set of feedback strategies $\{\phi_{j^1}^*(t, x)$ for $j^1 \in N_1$ and $\phi_{j^2}^*(t, x)$ for $j^2 \in N_2\}$ provides a Nash equilibrium solution to the game $\Gamma(x_\tau, T - \tau)$ if there exist continuously differentiable functions $V^{(\tau)j^1}(t, x) : [\tau, T] \times R^m \to R$ for $j^1 \in N_1$ and $V^{(\tau)j^2}(t, x) : [\tau, T] \times R^m \to R$ for $j^2 \in N_2$, satisfying the following partial differential equations:*

$$-V_t^{(\tau)j^1}(t, x)$$

$$= \max_{u_{j^1}} \left\{ \left(g\left[\sum_{\substack{h \in N_i \\ h \neq j^1}} \phi_h^*(t, x) + u_{j^1} + \sum_{\ell \in N_2} \phi_\ell^*(t, x) \right] u_{j^1}, -c^{j^1}(u_{j^1}, x) \right) \right.$$

$$\times \exp[-r(t - \tau)] + V_x^{(\tau)j^1}(t, x)$$

$$\left. \times f\left[t, x, \sum_{\substack{h \in N_i \\ h \neq j^1}} \phi_h^*(t, x) + u_{j^1} + \sum_{\ell \in N_2} \phi_\ell^*(t, x) \right] \right\}, \quad and$$

$$V^{(\tau)j^1}(T, x) = \exp[-r(T - t_0)] q^{j^1}(x), \quad for \ j^1 \in N_1; \tag{7.4}$$

$$-V_t^{(\tau)j^2}(t,x)$$

$$= \max_{u_{j^2}} \left\{ \left(g\left[\sum_{h\in N_i} \phi_h^*(t,x) + \sum_{\substack{\ell\in N_2 \\ \ell\neq j^2}} \phi_\ell^*(t,x) + u_{j^2} \right] u_{j^2} - c^{j^2}(u_{j^2},x) \right) \right.$$

$$\times \exp\left[-r(t-\tau)\right] + V_x^{(\tau)j^2}(t,x)$$

$$\left. \times f\left[t,x, \sum_{h\in N_i} \phi_h^*(t,x) + \sum_{\substack{\ell\in N_2 \\ \ell\neq j^2}} \phi_\ell^*(t,x) + u_{j^2} \right] \right\}, \quad and$$

$$V^{(\tau)j^2}(T,x) = \exp\left[-r(T-t_0)\right]q^{j^2}(x), \quad for\ j^2\in N_2.$$

Conditions satisfying the indicated maximization in (7.5) yield

$$\left\{ g\left(\sum_{\substack{h\in N_i \\ h\neq j^1}} \phi_h^*(t,x) + u_{j^1} + \sum_{\ell\in N_2} \phi_\ell^*(t,x) \right) \right.$$

$$+ g'\left(\sum_{\substack{h\in N_i \\ h\neq j^1}} \phi_h^*(t,x) + u_{j^1} + \sum_{\ell\in N_2} \phi_\ell^*(t,x) \right) u_{j^1}$$

$$\left. - \frac{\partial}{\partial u_{j^1}} c^{j^1}(u_{j^1},x) \right\} \exp\left[-r(t-\tau)\right]$$

$$+ V_x^{(\tau)j^1}(t,x)\frac{\partial}{\partial u_{j^1}} f\left[t,x, \sum_{\substack{h\in N_i \\ h\neq j^1}} \phi_h^*(t,x) + u_{j^1} + \sum_{\ell\in N_2} \phi_\ell^*(t,x) \right] = 0,$$

for $j^1\in N_1$;   (7.5)

$$\left\{ g\left(\sum_{h\in N_i} \phi_h^*(t,x) + \sum_{\substack{\ell\in N_2 \\ \ell\neq j^2}} \phi_\ell^*(t,x) + u_{j^2} \right) \right.$$

$$+ g'\left(\sum_{h\in N_i} \phi_h^*(t,x) + \sum_{\substack{\ell\in N_2 \\ \ell\neq j^2}} \phi_\ell^*(t,x) + u_{j^2} \right) u_{j^2}$$

$$\left. - \frac{\partial}{\partial u_{j^2}} c^{j^2}(u_{j^2},x) \right\} \exp\left[-r(t-\tau)\right]$$

$$+ V_x^{(\tau)j^2}(t, x) \frac{\partial}{\partial u_{j^2}} f\left[t, x, \sum_{h \in N_i} \phi_h^*(t, x) + \sum_{\substack{\ell \in N_2 \\ \ell \neq j^2}} \phi_\ell^*(t, x) + u_{j^2}\right] = 0,$$

for $j^2 \in N_2$.

The profits of firm $j^1 \in N_1$, which has cost advantages, can be expressed as

$$V^{(\tau)j^1}(t, x_\tau) = \int_\tau^T \left(g\left[\sum_{h \in N_i} \phi_h^*[s, x(s)] + \sum_{\ell \in N_2} \phi_\ell^*[s, x(s)]\right] \phi_{j^1}^*[s, x(s)] \right.$$

$$\left. - c^{j^1}\left[\phi_{j^1}^*(s, x(s)), x(s)\right]\right) \exp[-r(s - \tau)] \, ds$$

$$+ \exp[-r(T - \tau)] q^{j^1}[x(T)],$$

for $j^1 \in N_1$. The profits of firm $j^2 \in N_2$, which has cost disadvantages, can be expressed as

$$V^{(\tau)j^2}(t, x_\tau) = \int_\tau^T \left(g\left[\sum_{h \in N_i} \phi_h^*[s, x(s)] + \sum_{\ell \in N_2} \phi_\ell^*[s, x(s)]\right] \phi_{j^2}^*[s, x(s)] \right.$$

$$\left. - c^{j^2}\left[\phi_{j^2}^*(s, x(s)), x(s)\right]\right) \exp[-r(s - \tau)] \, ds$$

$$+ \exp[-r(T - \tau)] q^{j^2}[x(T)],$$

for $j^2 \in N_2$, where

$$\dot{x}(s) = f\left[s, x(s), \sum_{j^i \in N_1} u_{j^1}(s) + \sum_{j^2 \in N_2} u_{j^2}(s)\right], \quad x(\tau) = x_\tau \in X.$$

The dynamic oligopoly model presented above is an extension of the dormant firm duopoly model in Yeung (2005).

7.2 Time (Optimal-Trajectory-Subgame) Consistent Cartel

Assume that the firms in the oligopoly agree to form a cartel to restrain output and enhance their profits.

7.2.1 Pareto Optimal Output Path

To achieve a group optimum these firms are required to solve the following joint profit maximization problem:

$$
\begin{aligned}
\max_{u_1,u_2,\dots,u_n} & \left\{ \int_{t_0}^{T} \left(g\left[\sum_{h\in N_i} u_h(s) + \sum_{\ell\in N_2} u_\ell(s) \right]\left[\sum_{h\in N_i} u_h(s) + \sum_{\ell\in N_2} u_\ell(s) \right] \right.\right. \\
& \left. - \left[\sum_{h\in N_i} c^h\big[u_h(s), x(s)\big] + \sum_{\ell\in N_2} c^\ell\big[u_\ell(s), x(s)\big] \right] \right) \exp[-r(s-t_0)]\, ds \\
& + \exp[-r(T-t_0)]\left[\sum_{h\in N_i} q^h\big[x(T)\big] + \sum_{\ell\in N_2} q^\ell\big[x(T)\big] \right],
\end{aligned}
\tag{7.6}
$$

subject to (7.1).

An optimal solution of the problem in (7.1) and (7.6) can be characterized using Theorem A.1 in the Technical Appendixes as follows.

Corollary 7.2 *A set of control strategies $\{\psi_{j^1}^*(t,x)\ \text{for}\ j^1 \in N_1\ \text{and}\ \psi_{j^2}^*(t,x)$ for $j^2 \in N_2\}$ provides a solution to the control problem in (7.1) and (7.6), if there exist continuously differentiable functions $W^{(t_0)}(t,x) : [\tau, T] \times R^m \to R$, satisfying the following partial differential equation:*

$$
\begin{aligned}
-W_t^{(t_0)}(t,x) = \max_{u_1,u_2,\dots,u_n} & \left\{ \left(g\left[\sum_{h\in N_i} u_h + \sum_{\ell\in N_2} u_\ell \right]\left[\sum_{h\in N_i} u_h + \sum_{\ell\in N_2} u_\ell \right] \right.\right. \\
& \left. - \left[\sum_{h\in N_i} c^h[u_h, x] + \sum_{\ell\in N_2} c^\ell[u_\ell, x] \right] \right) \exp[-r(t-t_0)] \\
& + W_x^{(t_0)}(t,x)\, f\left[t, x, \sum_{h\in N_1} u_h + \sum_{\ell\in N_2} u_\ell \right] \right\}, \quad and
\end{aligned}
\tag{7.7}
$$

$$
W^{(t_0)}(T,x) = \exp[-r(T-t_0)]\left[\sum_{h\in N_i} q^h x + \sum_{\ell\in N_2} q^\ell x \right].
$$

Conditions satisfying the indicated maximization in (7.7) include

$$
\begin{aligned}
& \left\{ g\left[\sum_{h\in N_i} u_h + \sum_{\ell\in N_2} u_\ell \right] + g'\left[\sum_{h\in N_i} u_h + \sum_{\ell\in N_2} u_\ell \right]\left[\sum_{h\in N_i} u_h + \sum_{\ell\in N_2} u_\ell \right] \right. \\
& \left. - \frac{\partial}{\partial u_{j^1}} c^{j^1}(u_{j^1}, x) \right\} \exp[-r(t-t_0)] + W_x^{(t_0)}(t,x), \\
& \frac{\partial}{\partial u_{j^1}} f\left[t, x, \sum_{h\in N_1} u_h + \sum_{\ell\in N_2} u_\ell \right] \le 0, \quad u_{j^1} \ge 0,
\end{aligned}
$$

and if $u_{j^1} > 0$, the equality sign must hold, for $j^1 \in N_1$;

$$\left\{ g\left[\sum_{h \in N_i} u_h + \sum_{\ell \in N_2} u_\ell \right] + g'\left[\sum_{h \in N_i} u_h + \sum_{\ell \in N_2} u_\ell \right]\left[\sum_{h \in N_i} u_h + \sum_{\ell \in N_2} u_\ell \right] \right.$$

$$\left. - \frac{\partial}{\partial u_{j2}} c^{j^2}(u_{j2}, x) \right\} \exp\left[-r(t - t_0) \right] + W_x^{(t_0)}(t, x), \tag{7.8}$$

$$\frac{\partial}{\partial u_{j2}} f\left[t, x, \sum_{h \in N_1} u_h + \sum_{\ell \in N_2} u_\ell \right] \leq 0, \quad u_{j2} \geq 0,$$

and if $u_{j2} > 0$, the equality sign must hold, for $j^2 \in N_2$.

Since $\frac{\partial}{\partial u_{j^1}} c^{j^1}(u_{j^1}, x) < \frac{\partial}{\partial u_{j^2}} c^{j^2}(u_{j^2}, x)$, all the firms that have cost disadvantages will refrain from extraction. The optimal extraction strategies under cooperation become

$$u_{j^1}^*(t) = \psi_{j^1}^*(t, x) \quad \text{for } j^1 \in N_1 \quad \text{and} \quad u_{j^2}^*(t) = 0 \quad \text{for } j^2 \in N_2. \tag{7.9}$$

The optimal cooperative state dynamics follows

$$\dot{x}(s) = f\left[s, x(s), \sum_{j^i \in N_1} \psi_{j^1}^*(s, x(s)) \right], \quad x(t_0) = x_0. \tag{7.10}$$

The solution to (7.10) yields a group optimal trajectory, which can be expressed as

$$x^*(t) = x_0 + \int_{t_0}^t f\left[s, x^*(s), \sum_{j^i \in N_1} \psi_{j^1}^*(s, x^*(s)) \right] ds. \tag{7.11}$$

Substituting the optimal extraction strategies in (7.9) into (7.6) yields the cartel profit as

$$W^{(t_0)}(t_0, x_0) = \int_{t_0}^T \left(g\left[\sum_{h \in N_i} \psi_h^*[s, x^*(s)] \right]\left[\sum_{h \in N_i} \psi_h^*[s, x^*(s)] \right] \right.$$

$$\left. - \left[\sum_{h \in N_i} c^h\left[\psi_h^*(s, x^*(s)), x^*(s) \right] \right] \right) \exp\left[-r(s - t_0) \right]$$

$$+ \exp\left[-r(T - t_0) \right]\left[\sum_{h \in N_i} q^h[x^*(T)] + \sum_{\ell \in N_2} q^\ell[x^*(T)] \right]. \tag{7.12}$$

Let $W^{(\tau)}(t, x_t^*)$ denote the total venture profit from the control problem with dynamics in (7.1) and payoff in (7.6), which begins at time $\tau \in [t_0, T]$ with initial state x_τ^*. Invoking Remark 3.1 of Chap. 3, one can readily obtain

$$\exp\left[\int_{t_0}^{\tau} r(y)\,dy\right] W^{(t_0)}\left(t, x_t^*\right) = W^{(\tau)}\left(t, x_t^*\right),$$

for $\tau \in [t_0, T]$ and $t \in [\tau, T)$.

The design of an imputation scheme to share cartel profits over time will be examined next.

7.2.2 Imputation Scheme and Cartel Profit Sharing

In a dormant firm, cartel firms having cost disadvantages will refrain from extraction to enhance the cartel's profit to a group optimum. Compensation must be made to the dormant firms for stopping their production activities. Since there are cost differentials among firms in the cartel, the sizes and earning potentials of the firms cannot be identical. Consider the case when the firms in the cartel agree to share the excess of the total cooperative payoff over the sum of individual noncooperative payoffs proportional to the firms' noncooperative payoffs.

The imputation scheme has to fulfill the following condition.

Condition 7.1 An imputation

$$\xi^{(t_0)j^1}(t_0, x_0) = \frac{V^{(t_0)j^1}(t_0, x_0)}{\sum_{h \in N_1} V^{(t_0)h}(t_0, x_0) + \sum_{\ell \in N_2} V^{(t_0)\ell}(t_0, x_0)} W^{(t_0)}(t_0, x_0),$$

is assigned to firm j^1, for $j^1 \in N_1$ at the outset. An imputation

$$\xi^{(t_0)j^2}(t_0, x_0) = \frac{V^{(t_0)j^2}(t_0, x_0)}{\sum_{h \in N_1} V^{(t_0)h}(t_0, x_0) + \sum_{\ell \in N_2} V^{(t_0)\ell}(t_0, x_0)} W^{(t_0)}(t_0, x_0),$$

is assigned to firm j^2, for $j^2 \in N_2$ at the outset. An imputation

$$\xi^{(\tau)j^1}\left(\tau, x_\tau^*\right) = \frac{V^{(\tau)j^1}(\tau, x_\tau^*)}{\sum_{h \in N_1} V^{(\tau)h}(\tau, x_\tau^*) + \sum_{\ell \in N_2} V^{(\tau)\ell}(\tau, x_\tau^*)} W^{(\tau)}\left(\tau, x_\tau^*\right)$$

is assigned to firm j^1, for $j^1 \in N_1$ at time $\tau \in (t_0, T]$. An imputation

$$\xi^{(\tau)j^2}\left(\tau, x_\tau^*\right) = \frac{V^{(\tau)j^2}(\tau, x_\tau^*)}{\sum_{h \in N_1} V^{(\tau)h}(\tau, x_\tau^*) + \sum_{\ell \in N_2} V^{(\tau)\ell}(\tau, x_\tau^*)} W^{(\tau)}\left(\tau, x_\tau^*\right) \quad (7.13)$$

is assigned to firm j^2, for $j^2 \in N_2$ at time $\tau \in (t_0, T]$.

The optimality principle guiding the cartel can then be expressed as

$$P\left(x_t^*, T - t\right) = \left\{\left[\psi^*\left(s, x^*(s)\right)\right]_{s=t}^T, \xi^{(t)}\left(t, x_t^*\right)\right\} \quad \text{for } t \in [t_0, T],$$

where $\psi^*(s, x^*(s)) = \{\psi_{j^1}^*(t, x) \text{ for } j^1 \in N_1\}$ and $\xi^{(t)}(t, x_t^*)$ is an imputation scheme satisfying Condition 7.1.

To formulate a payoff distribution procedure over time so that the agreed imputations satisfy Condition 7.1 we invoke Theorem 4.2 in Chap. 4 to obtain the following.

Corollary 7.3 *A PDP with a terminal payment $q^i(x_T^*)$ at time T and an instantaneous payment at time $\tau \in [t_0, T]$*

$$B_{j^1}(\tau) = -\frac{\partial}{\partial t}\left[\frac{V^{(\tau)j^1}(t, x_t^*)}{\sum_{h \in N_1} V^{(\tau)h}(t, x_t^*) + \sum_{\ell \in N_2} V^{(\tau)\ell}(t, x_t^*)} W^{(\tau)}(t, x_t^*)|_{t=\tau}\right]$$

$$-\frac{\partial}{\partial x_\tau^*}\left[\frac{V^{(\tau)j^1}(\tau, x_\tau^*)}{\sum_{h \in N_1} V^{(\tau)h}(\tau, x_\tau^*) + \sum_{\ell \in N_2} V^{(\tau)\ell}(\tau, x_\tau^*)} W^{(\tau)}(\tau, x_\tau^*)\right]$$

$$\times f\left[s, x^*(s), \sum_{j^i \in N_1} \psi_{j^1}^*\left(s, x^*(s)\right)\right], \quad \text{for } j^1 \in N_1;$$

$$\tag{7.14}$$

$$B_{j^2}(\tau) = -\frac{\partial}{\partial t}\left[\frac{V^{(\tau)j^2}(t, x_t^*)}{\sum_{h \in N_1} V^{(\tau)h}(t, x_t^*) + \sum_{\ell \in N_2} V^{(\tau)\ell}(t, x_t^*)} W^{(\tau)}(t, x_t^*)|_{t=\tau}\right]$$

$$-\frac{\partial}{\partial x_\tau^*}\left[\frac{V^{(\tau)j^2}(\tau, x_\tau^*)}{\sum_{h \in N_1} V^{(\tau)h}(\tau, x_\tau^*) + \sum_{\ell \in N_2} V^{(\tau)\ell}(\tau, x_\tau^*)} W^{(\tau)}(\tau, x_\tau^*)\right]$$

$$\times f\left[s, x^*(s), \sum_{j^i \in N_1} \psi_{j^1}^*\left(s, x^*(s)\right)\right], \quad \text{for } j^2 \in N_1,$$

leads to the realization of the solution imputations $\xi^{(\tau)i}(\tau, x_\tau^*)$, for $i \in N$ and $\tau \in [t_0, T]$, satisfying Condition 7.1.

A time (optimal-trajectory-subgame) consistent solution can be obtained using the set of group optimal strategies $\psi^*(s, x_s^*)$ characterized in (7.9) and $B(s) = \{B_1(s), B_2(s), \ldots, B_n(s)\}$ in (7.14).

With firms having cost advantages producing an output $\psi_{j^1}^*(t, x)$ for $j^1 \in N_1$ and firms having cost disadvantages refraining from production, the instantaneous receipt of firm i at time instant τ is

$$\zeta_{j^1}(\tau) = g\left[\sum_{h \in N_i} \psi_h^*\left(\tau, x_\tau^*\right)\right]\psi_{j^1}^*\left(\tau, x_\tau^*\right) - c^{j^1}\left[\psi_{j^1}^*\left(\tau, x_\tau^*\right), x_\tau^*(s)\right],$$

for $\tau \in [t_0, T]$ and $j^1 \in N_1$, and $\zeta_{j^2}(\tau) = 0$, for $\tau \in [t_0, T]$ and $j^2 \in N_2$.

According to Corollary 7.3, the instantaneous payment that firm i should receive under the agreed-upon optimality principle is $B_{j^1}(\tau)$ for $j^1 \in N_1$ and $B_{j^2}(\tau)$ for $j^2 \in N_2$ as stated in (7.14). Hence an instantaneous transfer payment

$$\chi^{j^1}(\tau) = \zeta_{j^1}(\tau) - B_{j^1}(\tau), \quad \text{for firm } j^1 \in N_1 \text{ and } \tau \in [t_0, T],$$

$$\chi^{j^2}(\tau) = B_{j^1}(\tau), \quad \text{for firm } j^2 \in N_2 \text{ and } \tau \in [t_0, T], \tag{7.15}$$

would have to be arranged.

7.3 A Dormant-Firm Cartel

Consider the deterministic version of the dormant-firm duopoly game example in Yeung (2005) in which two firms are allowed to extract a renewable resource within the duration $[t_0, T]$. The dynamics of the resource is characterized by the differential equations

$$\dot{x}(s) = ax(s)^{1/2} - bx(s) - u_1(s) - u_2(s), \quad x(t_0) = x_0 \in X, \tag{7.16}$$

where $u_i \in U_i$ is the (nonnegative) amount of the resource extracted by firm i, for $i \in \{1, 2\}$, a and b are positive constants.

The extraction cost for firm $i \in \{1, 2\}$ depends on the quantity of the resource extracted $u_i(s)$, the resource stock size $x(s)$, and a parameter c_i. In particular, firm i's extraction cost can be specified as $c_i u_i(s) x(s)^{-1/2}$. This formulation of unit cost follows from two assumptions: (i) the cost of extraction is proportional to the extraction effort, and (ii) the amount of resource extracted, seen as the output of a production function of two inputs (effort and stock level), is increasing in both inputs (see Clark 1976). In particular, firm 1 has absolute and marginal cost advantages with $c_1 < c_2$.

The market price of the resource depends on the total amount extracted and supplied to the market. The price-output relationship at time s is given by the following downward-sloping inverse demand curve $P(s) = Q(s)^{-1/2}$, where $Q(s) = u_1(s) + u_2(s)$ is the total amount of the resource extracted and marketed at time s. At time T, firm i will receive a termination bonus $q_i x(T)^{1/2}$, where q_i is nonnegative. There exists a discount rate r, and the profits received at time t have to be discounted by the factor $\exp[-r(t - t_0)]$.

At time t_0 the profit of firm $i \in \{1, 2\}$ is

$$\int_{t_0}^{T} \left[\frac{u_i(s)}{[u_1(s) + u_2(s)]^{1/2}} - \frac{c_i}{x(s)^{1/2}} u_i(s) \right] \exp[-r(s - t_0)] \, ds$$

$$+ \exp[-r(T - t_0)] q_i x(T)^{\frac{1}{2}}. \tag{7.17}$$

A set of strategies $\{u_i^*(t) = \phi_i^*(t, x), \text{ for } i \in \{1, 2\}\}$, provides a feedback Nash equilibrium solution to the game in (7.16) and (7.17) if there exist continuously differentiable functions $V^{(t_0)i}(t,x) : [t_0, T] \times R \to R, i \in \{1,2\}$, satisfying the following partial differential equations:

$$-V_t^{(t_0)i}(t,x) = \max_{u_i}\left\{\left[\frac{u_i}{(u_i + \phi_j^*(t,x))^{1/2}} - \frac{c_i}{x^{1/2}}u_i\right]\exp[-r(t - t_0)]\right.$$

$$\left. + V_x^{(t_0)i}(t,x)\left[ax^{1/2} - bx - u_i - \phi_j^*(t,x)\right]\right\}, \quad \text{and} \qquad (7.18)$$

$$V^{(t_0)i}(T,x) = q_i x^{1/2}\exp[-r(T - t_0)], \quad \text{for } i \in \{1,2\}, j \in \{1,2\} \text{ and } j \neq i.$$

Performing the indicated maximization yields

$$\phi_1^*(t,x) = \frac{x}{4[c_1 + V_x^{(t_0)1}\exp[r(t - t_0)]x^{1/2}]^2}, \quad \text{and}$$

$$\phi_2^*(t,x) = \frac{x}{4[c_2 + V_x^{(t_0)2}\exp[r(t - t_0)]x^{1/2}]^2}. \qquad (7.19)$$

Proposition 7.1 *The value function of firm i in the game in (7.16) and (7.17) is*

$$V^{(t_0)i}(t,x) = \exp[-r(t - t_0)]\left[A_i(t)x^{1/2} + C_i(t)\right], \qquad (7.20)$$

for $i \in \{1,2\}$ and $t \in [t_0, T]$, where $A_i(t), C_i(t), A_j(t)$, and $C_j(t)$, for $i \in \{1,2\}$ and $j \in \{1,2\}$ and $i \neq j$, satisfy

$$\dot{A}_i(t) = \left[r + \frac{b}{2}\right]A_i(t) - \left(\frac{3}{2}\right)\frac{[2c_j - c_i + A_j(t) - A_i(t)/2]}{[c_1 + c_2 + A_1(t)/2 + A_2(t)/2]^2}$$

$$+ \left(\frac{3}{2}\right)^2\frac{c_i[2c_j - c_i + A_j(t) - A_i(t)/2]}{[c_1 + c_2 + A_1(t)/2 + A_2(t)/2]^3}$$

$$+ \left(\frac{9}{8}\right)\frac{A_i(t)}{[c_1 + c_2 + A_1(t)/2 + A_2(t)/2]^2},$$

$$A_i(T) = q_i,$$

$$\dot{C}_i(t) = rC_i(t) - \frac{a}{2}A_i(t), \quad \text{and} \quad C_i(T) = 0.$$

Proof First, substitute the results in (7.19) and $V^{(t_0)1}(t,x), V_x^{(t_0)1}(t,x), V^{(t_0)2}(t,x),$ and $V_x^{(t_0)2}(t,x)$ obtained via (7.20) into the set of partial differential equations in (7.19). One can readily show that for this set of equations to be satisfied Proposition 7.1 has to hold.   □

One can readily verify that

$$V^{(\tau)i}(t,x) = \exp[-r(t - \tau)]\left[A_i(t)x^{1/2} + C_i(t)\right], \quad \text{for } i \in \{1,2\} \text{ and } t \in [\tau, T].$$

Assume that the firms agree to form a cartel and seek to solve the following joint profit maximization problem to achieve a group optimum

$$\max_{u_1,u_2} \int_{t_0}^{T} \left[[u_1(s) + u_2(s)]^{1/2} - \frac{c_1 u_1(s) + c_2 u_2(s)}{x(s)^{1/2}} \right] \exp[-r(s - t_0)] \, ds$$

$$+ \exp[-r(T - t_0)][q_1 + q_2] x(T)^{1/2}, \tag{7.21}$$

subject to the dynamics in (7.16).

A set of strategies $[\psi_1^*(s, x), \psi_2^*(s, x)]$, for $s \in [t_0, T]$, provides an optimal control solution to the problem in (7.16) and (7.21) if there exist a continuously differentiable function $W^{(t_0)}(t, x) : [t_0, T] \times R \to R$ satisfying the following partial differential equations:

$$-W_t^{(t_0)}(t, x) = \max_{u_1,u_2} \{ [(u_1 + u_2)^{1/2} - (c_1 u_1 + c_2 u_2) x^{-1/2}] \exp[-r(t - t_0)]$$

$$+ W_x^{(t_0)}(t, x)[ax^{1/2} - bx - u_1 - u_2] \}, \quad \text{and} \tag{7.22}$$

$$W^{(t_0)}(T, x) = (q_1 + q_2) x^{1/2} \exp[-r(T - t_0)].$$

Performing the indicated maximization operation in (7.23) yields

$$\psi_1^*(t, x) = \frac{x}{4[c_1 + W_x \exp[r(t - t_0)]x^{1/2}]^2} \quad \text{and} \quad \psi_2^*(t, x) = 0. \tag{7.23}$$

Along the optimal trajectory, firm 2 has to refrain from extraction. The more efficient firm (firm 1) would buy the less efficient firm (firm 2) out from the resource extraction process. Firm 2 becomes a dormant firm under cooperation.

Proposition 7.2 *The value function of the control problem in (7.16) and (7.21) can be obtained as*

$$W^{(t_0)}(t, x) = \exp[-r(t - t_0)][A(t)x^{1/2} + C(t)], \tag{7.24}$$

where $A(t)$ and $B(t)$ satisfy

$$\dot{A}(t) = \left[r + \frac{b}{2} \right] A(t) - \frac{1}{4[c_1 + A(t)/2]},$$

$$A(T) = q_1 + q_2,$$

$$\dot{C}(t) = rC(t) - \frac{a}{2} A(t), \quad \text{and} \quad B(T) = 0.$$

Proof First substitute the results in (7.23), and $W^{(t_0)}(t, x)$ and $W_x^{(t_0)}(t, x)$ obtained via (7.24) into the set of partial differential equations in (7.23). One can readily show that for this set of equations to be satisfied Proposition 7.2 has to hold. □

Again, one can readily verify that

$$W^{(\tau)}(t, x) = \exp[-r(t - \tau)][A(t)x^{1/2} + B(t)].$$

Substituting $\psi_1^*(t, x)$ and $\psi_2^*(t, x)$ into (7.16) yields the optimal cooperative state dynamics as

$$\dot{x}(s) = \left[ax(s)^{1/2} - bx(s) - \frac{x(s)}{4[c_1 + A(s)/2]^2}\right], \quad x(t_0) = x_0 \in X. \quad (7.25)$$

The solution to (7.25) yields a Pareto optimal trajectory, which can be expressed as

$$x^*(t) = \left\{ \Phi(t, t_0)\left[x_0^{1/2} + \int_{t_0}^{t} \Phi^{-1}(s, t_0)\frac{a}{2}\, ds\right] \right\}^2, \quad (7.26)$$

where

$$\Phi(t, t_0) = \exp\left[\int_{t_0}^{t}\left(\frac{-b}{2} - \frac{1}{8[c_1 + A(s)/2]^2}\right) ds\right].$$

Consider the case when the firms in the cartel agree to share the excess of the total cooperative payoff over the sum of individual noncooperative payoffs proportional to the firms' noncooperative payoffs.

The imputation scheme has to fulfill the following condition:

Condition 7.2 $\xi^{(\tau)i}(\tau, x_\tau^*) = \dfrac{V^{(\tau)i}(\tau, x_\tau^*)}{\sum_{j=1}^{2} V^{(\tau)j}(\tau, x_\tau^*)} W^{(\tau)}(\tau, x_\tau^*)$, for $i \in \{1, 2\}$ and $\tau \in [t_0, T]$.

To formulate a payoff distribution procedure over time so that the agreed imputations satisfy Condition 7.2 we invoke Theorem 4.2 in Chap. 4 to obtain the following.

Corollary 7.4 *A PDP with a terminal payment $q^i(x_T^*)$ at time T and an instantaneous payment at time $\tau \in [t_0, T]$*

$$B_i(\tau) = -\frac{\partial}{\partial t}\left[\frac{V^{(\tau)i}(t, x_t^*)}{\sum_{j=1}^{2} V^{(\tau)j}(t, x_t^*)} W^{(\tau)}(t, x_t^*)|_{t=\tau}\right]$$

$$- \frac{\partial}{\partial x_\tau^*}\left[\frac{V^{(\tau)i}(\tau, x_\tau^*)}{\sum_{j=1}^{2} V^{(\tau)j}(\tau, x_\tau^*)} W^{(\tau)}(\tau, x_\tau^*)\right]$$

$$\times \left[a(x_\tau^*)^{1/2} - bx_\tau^* - \frac{x_\tau^*}{4[c_1 + A(\tau)/2]^2}\right], \quad (7.27)$$

for $i \in \{1, 2\}$, leads to the realization of the solution imputations $\xi^{(\tau)i}(\tau, x_\tau^*)$, for $i \in [1, 2]$ and $\tau \in [t_0, T]$, satisfying Condition 7.2.

In particular, from (7.20) and (7.24), we have

$$
\frac{\partial}{\partial t}\left[\frac{V^{(\tau)i}(t,x_t^*)}{\sum_{j=1}^2 V^{(\tau)j}(t,x_t^*)}W^{(\tau)}(t,x_t^*)\Big|_{t=\tau}\right]
$$

$$
=\left[A(\tau)\big(x_\tau^*\big)^{1/2}+C(\tau)\right]
$$

$$
\times\left[\left(\sum_{j=1}^2[A_j(\tau)\big(x_\tau^*\big)^{1/2}+C_j(\tau)]\right)\right.
$$

$$
\times\frac{(r[A_i(\tau)(x_\tau^*)^{1/2}+C_i(\tau)]+[\dot A_i(\tau)(x_\tau^*)^{1/2}+\dot C_i(\tau)])}{(\sum_{j=1}^2[A_j(\tau)(x_\tau^*)^{1/2}+C_j(\tau)])^2}
$$

$$
-\left[A_i(\tau)\big(x_\tau^*\big)^{1/2}+C_i(\tau)\right]
$$

$$
\left.\times\frac{\sum_{j=1}^2(r[A_j(\tau)(x_\tau^*)^{1/2}+C_j(\tau)]+[\dot A_j(\tau)(x_\tau^*)^{1/2}+\dot C_j(\tau)])}{(\sum_{j=1}^2[A_j(\tau)(x_\tau^*)^{1/2}+C_j(\tau)])^2}\right]
$$

$$
+\frac{[A_i(\tau)(x_\tau^*)^{1/2}+C_i(\tau)]}{(\sum_{j=1}^2[A_j(\tau)(x_\tau^*)^{1/2}+C_j(\tau)])}
$$

$$
\times\left(r\left[A(\tau)\big(x_\tau^*\big)^{1/2}+C(\tau)\right]+\left[\dot A(\tau)\big(x_\tau^*\big)^{1/2}+\dot C(\tau)\right]\right);
$$

and

$$
\frac{\partial}{\partial x_\tau^*}\left[\frac{V^{(\tau)i}(\tau,x_\tau^*)}{\sum_{j=1}^2 V^{(\tau)j}(\tau,x_\tau^*)}W^{(\tau)}(\tau,x_\tau^*)\right]
$$

$$
=\left[A(\tau)\big(x_\tau^*\big)^{1/2}+C(\tau)\right]
$$

$$
\times\left[\frac{(\sum_{j=1}^2[A_j(\tau)(x_\tau^*)^{1/2}+C_j(\tau)])\frac12 A_i(\tau)(x_\tau^*)^{-1/2}}{(\sum_{j=1}^2[A_j(\tau)(x_\tau^*)^{1/2}+C_j(\tau)])^2}\right.
$$

$$
\left.-\frac{[A_i(\tau)(x_\tau^*)^{1/2}+C_i(\tau)]\sum_{j=1}^2\frac12[A_j(\tau)(x_\tau^*)^{-1/2}]}{(\sum_{j=1}^2[A_j(\tau)(x_\tau^*)^{1/2}+C_j(\tau)])^2}\right]
$$

$$
+\frac{[A_i(\tau)(x_\tau^*)^{1/2}+C_i(\tau)]}{(\sum_{j=1}^2[A_j(\tau)(x_\tau^*)^{1/2}+C_j(\tau)])}\frac12 A(\tau)\big(x_\tau^*\big)^{-1/2}.
$$

A time (optimal-trajectory-subgame) consistent solution can be obtained using the set of group optimal strategies $\psi^*(s,x_s^*)$ in (7.23) and $B(s)=\{B_1(s),B_2(s),\dots,B_n(s)\}$ in (7.27).

Under cooperation, firm 1 would derive a payoff

$$
W^{(t_0)1}(t_0,x_0)=\int_{t_0}^T\left[\left[\psi_1^*\big(s,x^*(s)\big)\right]^{1/2}-\frac{c_1}{x^*(s)^{1/2}}\psi_1^*\big(s,x^*(s)\big)\right]
$$

$$
\times\exp[-r(s-t_0)]\,ds+\exp[-r(T-t_0)]q_1 x^*(T)^{\frac12},
$$

where $\psi_1^*(s, x^*(s)) = \frac{x^*(s)}{4[c_1 + A(s)/2]^2}$, and firm 2 would derive a payoff

$$W^{(t_0)2}(t_0, x_0) = 0 \quad \text{for being dormant.} \tag{7.28}$$

The instantaneous receipt of firm 1 at time instant τ is

$$\zeta_1(\tau) = \frac{(x_\tau^*)^{1/2}}{2[c_1 + A(\tau)/2]} - \frac{c_1(x_\tau^*)^{1/2}}{4[c_1 + A(\tau)/2]^2},$$

for $\tau \in [t_0, T]$.

The instantaneous receipt of firm 2 at time instant τ is

$$\zeta_2(\tau) = 0, \quad \text{for } \tau \in [t_0, T].$$

According to Corollary 7.4, the instantaneous payment that firm i should receive under the agreed-upon optimality principle is $B_1(\tau)$ and $B_2(\tau)$ as stated in (7.27). Hence an instantaneous transfer payment

$$
\begin{aligned}
\chi^1(\tau) &= \zeta_1(\tau) - B_1(\tau), \quad \text{for firm 1 at time } \tau \in [t_0, T], \\
\chi^2(\tau) &= B_2(\tau), \quad \text{for firm 2 at time } \tau \in [t_0, T],
\end{aligned}
\tag{7.29}
$$

would be arranged.

7.4 Infinite-Horizon Cartel

In this section we consider the Dormant-Firm Cartel in Sect. 7.1 with an infinite horizon. Consider an oligopoly in which n firms are given extraction rights of a renewable resource.

The dynamics of the resource is characterized by the differential equations

$$\dot{x}(s) = f\left[x(s), \sum_{j=1}^n u_j(s)\right] = f\left[x(s), \sum_{j^i \in N_1} u_{j^1}(s) + \sum_{j^2 \in N_2} u_{j^2}(s)\right],$$

$$x(t_0) = x_0 \in X, \tag{7.30}$$

where $u_j \in U_j$ is the (nonnegative) amount of resource extracted by firm i, for $i \in N$, and $x(s)$ is the resource stock.

The extraction cost for the n_1 firms with cost advantages is

$$c^{j^1}\left[u_{j^1}(s), x(s)\right], \quad \text{for } j^1 \in N_1,$$

and the extraction cost for the n_1 firms with cost advantages is

$$c^{j^2}\left[u_{j^2}(s), x(s)\right], \quad \text{for } j^2 \in N_2.$$

In particular, firm $j^1 \in N_1$ has absolute and marginal cost advantage so that

$$c^{j^1}\left[u_{j^1}(s), x(s)\right] < c^{j^2}\left[u_{j^2}(s), x(s)\right] \quad \text{and}$$

$$\partial c^{j^1}\left[u_{j^1}(s), x(s)\right]/\partial u_{j^1}(s) < \partial c^{j^2}\left[u_{j^2}(s), x(s)\right]/\partial u_{j^2}(s),$$

for any levels of u_{j^1}, u_{j^2}, and x.

The market price of the resource is governed by the following downward-sloping inverse demand curve

$$P(s) = g\left[Q(s)\right],$$

where $Q(s) = \sum_{j^i \in N_1} u_{j^1}(s) + \sum_{j^2 \in N_2} u_{j^2}(s)$ is the total amount of the resource extracted and marketed at time s. There exists a discount rate r, and the profits received at time t have to be discounted by the factor $\exp[-r(t - t_0)]$.

At time t_0, firm $j^1 \in N_1$, which has cost advantages, seeks to maximize its profit

$$\int_{t_0}^{\infty}\left(g\left[\sum_{h \in N_i} u_h(s) + \sum_{\ell \in N_2} u_\ell(s)\right]u_{j^1}(s) - c^{j^1}\left[u_{j^1}(s), x(s)\right]\right)\exp\left[-r(s - t_0)\right]ds,$$

(7.31)

subject to (7.30).

At time t_0, firm $j^2 \in N_2$, which has cost disadvantages, seeks to maximize profit

$$\int_{t_0}^{\infty}\left(g\left[\sum_{h \in N_i} u_h(s) + \sum_{\ell \in N_2} u_\ell(s)\right]u_{j^2}(s) - c^{j^2}\left[u_{j^2}(s), x(s)\right]\right)\exp\left[-r(s - t_0)\right]ds,$$

(7.32)

subject to (7.30).

Consider the alternative formulation of (7.30)–(7.32) as

$$\max_{u_{j^1}}\left\{\int_t^{\infty}\left(g\left[\sum_{h \in N_i} u_h(s) + \sum_{\ell \in N_2} u_\ell(s)\right]u_{j^1}(s) - c^{j^1}\left[u_{j^1}(s), x(s)\right]\right)\right.$$

$$\left. \times \exp\left[-r(s - t)\right]ds\right\},$$

(7.33)

for $j^1 \in N_1$, and

$$\max_{u_{j^2}}\left\{\int_t^{\infty}\left(g\left[\sum_{h \in N_i} u_h(s) + \sum_{\ell \in N_2} u_\ell(s)\right]u_{j^2}(s) - c^{j^2}\left[u_{j^2}(s), x(s)\right]\right)\right.$$

$$\left. \times \exp\left[-r(s - t)\right]ds\right\}$$

(7.34)

subject to the state dynamics

$$\dot{x}(s) = f\left[x(s), \sum_{j^i \in N_1} u_{j^1}(s) + \sum_{j^2 \in N_2} u_{j^2}(s)\right], \quad x(t) = x. \tag{7.35}$$

The infinite-horizon autonomous game in (7.33)–(7.35) is independent of the choice of t and dependent only upon the state at the starting time, that is, x.

Invoking Theorem 2.4 in Chap. 2, a noncooperative feedback Nash equilibrium solution can be characterized by a set of strategies $\{\phi^*_{j^1}(x)$ for $j^1 \in N_1$ and $\phi^*_{j^2}(x)$ for $j^2 \in N_2\}$ constitutes a feedback Nash equilibrium solution to the game in (7.33)–(7.35) if there exist functionals $\hat{V}^{j^1}(x) : R^m \to R$ for $j^1 \in N_1$ and $\hat{V}^{j^2}(x) : R^m \to R$ for $j^2 \in N_2$, satisfying the following set of partial differential equations:

$$r\hat{V}^{j^1}(x) = \max_{u_{j^1}} \left\{ \left(g\left[\sum_{\substack{h \in N_i \\ h \neq j^1}} \phi^*_h(x) + u_{j^1} + \sum_{\ell \in N_2} \phi^*_\ell(x)\right] u_{j^1} - c^{j^1}(u_{j^1}, x)\right) \right.$$

$$\left. + \hat{V}^{j^1}_x(x) f\left[x, \sum_{\substack{h \in N_i \\ h \neq j^1}} \phi^*_h(x) + u_{j^1} + \sum_{\ell \in N_2} \phi^*_\ell(x)\right]\right\},$$

for $j^1 \in N_1$;   and   (7.36)

$$r\hat{V}^{j^2}(x) = \max_{u_{j^2}} \left\{ \left(g\left[\sum_{h \in N_i} \phi^*_h(x) + \sum_{\substack{\ell \in N_2 \\ \ell \neq j^2}} \phi^*_\ell(x) + u_{j^2}\right] u_{j^2} - c^{j^2}(u_{j^2}, x)\right) \right.$$

$$\left. + \hat{V}^{j^2}_x(x) f\left[x, \sum_{h \in N_i} \phi^*_h(x) + \sum_{\substack{\ell \in N_2 \\ \ell \neq j^2}} \phi^*_\ell(x) + u_{j^2}\right]\right\}, \quad \text{for } j^2 \in N_2.$$

In particular, the profits of firm $j^1 \in N_1$, which has cost advantages, can be expressed as

$$\hat{V}^{j^1}(x)$$

$$= \int_t^\infty \left(g\left[\sum_{h \in N_i} \phi^*_h[x(s)] + \sum_{\ell \in N_2} \phi^*_\ell[x(s)]\right] \phi^*_{j^1}[x(s)] - c^{j^1}[\phi^*_{j^1}(x(s)), x(s)]\right)$$

$$\times \exp[-r(s - \tau)]\, ds,$$

for $j^1 \in N_1$; the profits of firm $j^2 \in N_2$, which has cost disadvantages, can be expressed as

$$\hat{V}^{j^2}(x_\tau)$$

$$= \int_\tau^\infty \left(g\left[\sum_{h \in N_i} \phi_h^*[x(s)] + \sum_{\ell \in N_2} \phi_\ell^*[x(s)] \right] \phi_{j^2}^*[x(s)] - c^{j^2}\left[\phi_{j^2}^*(x(s)), x(s) \right] \right)$$

$$\times \exp[-r(s-\tau)] \, ds,$$

for $j^2 \in N_2$; where $\dot{x}(s) = f[x(s), \sum_{h \in N_1} \phi_h^*(x(s)) + \sum_{\ell \in N_2} \phi_\ell^*(x(s))]$, $x(t) = x \in X$.

7.4.1 Pareto Optimal Trajectory

Assume that the firms agree to form a cartel to restrain output and enhance their profits. To achieve a group optimum, these firms are required to solve the following joint profit maximization problem:

$$\max_{u_1, u_2, \dots, u_n} \left\{ \int_t^\infty \left(g\left[\sum_{h \in N_i} u_h(s) + \sum_{\ell \in N_2} u_\ell(s) \right] \left[\sum_{h \in N_i} u_h(s) + \sum_{\ell \in N_2} u_\ell(s) \right] \right. \right.$$

$$\left. \left. - \left[\sum_{h \in N_i} c^h[u_h(s), x(s)] + \sum_{\ell \in N_2} c^\ell[u_\ell(s), x(s)] \right] \right) \exp[-r(s-t)] \, ds \right\} \quad (7.37)$$

subject to (7.35).

An optimal solution of the problem in (7.35) and (7.37) can be characterized using Theorem A.2 in the Technical Appendixes as follows.

Corollary 7.5 *A set of control strategies $\{\psi_{j^1}^*(x) \text{ for } j^1 \in N_1 \text{ and } \psi_{j^2}^*(x) \text{ for } j^2 \in N_2\}$ provides a solution to the control problem in (7.35) and (7.37) if there exist continuously differentiable functions $W(x) : R^m \to R$, satisfying the following partial differential equation:*

$$rW(x) = \max_{u_1, u_2, \dots, u_n} \left\{ \left(g\left[\sum_{h \in N_i} u_h + \sum_{\ell \in N_2} u_\ell \right] \left[\sum_{h \in N_i} u_h + \sum_{\ell \in N_2} u_\ell \right] \right. \right.$$

$$\left. - \left[\sum_{h \in N_i} c^h[u_h, x] + \sum_{\ell \in N_2} c^\ell[u_\ell, x] \right] \right)$$

$$\left. + W_x(x) f\left[t, x, \sum_{h \in N_1} u_h + \sum_{\ell \in N_2} u_\ell \right] \right\}. \quad (7.38)$$

Conditions satisfying the indicated maximization in (7.38) include

$$
\left\{ g\left[\sum_{h\in N_i} u_h + \sum_{\ell\in N_2} u_\ell \right] + g'\left[\sum_{h\in N_i} u_h + \sum_{\ell\in N_2} u_\ell \right]\left[\sum_{h\in N_i} u_h + \sum_{\ell\in N_2} u_\ell \right]\right.
$$

$$
\left. -\frac{\partial}{\partial u_{j1}} c^{j^1}(u_{j1},x)\right\} + W_x(x)\frac{\partial}{\partial u_{j1}} f\left[x, \sum_{h\in N_1} u_h + \sum_{\ell\in N_2} u_\ell \right] \le 0,
$$

$$
u_{j1} \ge 0,
$$

and if $u_{j1} > 0$, the equality sign must hold, for $j^1 \in N_1$;

$$
\left\{ g\left[\sum_{h\in N_i} u_h + \sum_{\ell\in N_2} u_\ell \right] + g'\left[\sum_{h\in N_i} u_h + \sum_{\ell\in N_2} u_\ell \right]\left[\sum_{h\in N_i} u_h + \sum_{\ell\in N_2} u_\ell \right]\right.
$$

$$
\left. -\frac{\partial}{\partial u_{j2}} c^{j^2}(u_{j1},x)\right\} + W_x(x)\frac{\partial}{\partial u_{j2}} f\left[x, \sum_{h\in N_1} u_h + \sum_{\ell\in N_2} u_\ell \right] \le 0,
$$

$$
u_{j2} \ge 0, \tag{7.39}
$$

and if $u_{j2} > 0$, the equality sign must hold, for $j^2 \in N_2$.

Since $\frac{\partial}{\partial u_{j1}} c^{j^1}(u_{j1},x) < \frac{\partial}{\partial u_{j2}} c^{j^2}(u_{j1},x)$, all the firms that have cost disadvantages will refrain from extraction. The optimal extraction strategies under cooperation become

$$
u^*_{j1}(t) = \psi^*_{j1}(x) \quad \text{for } j^1 \in N_1 \quad \text{and} \quad u^*_{j2}(t) = 0 \quad \text{for } j^2 \in N_2. \tag{7.40}
$$

The optimal cooperative state dynamics follows

$$
\dot{x}(s) = f\left[x(s), \sum_{j^i\in N_1} \psi^*_{j1}(x(s))\right], \quad x(t_0) = x_0. \tag{7.41}
$$

The solution to (7.41) yields a group optimal trajectory, which can be expressed as

$$
x^*(t) = x_0 + \int_{t_0}^{t} f\left[x^*(s), \sum_{j^i\in N_1} \psi^*_{j1}(x^*(s))\right] ds. \tag{7.42}
$$

Substituting the optimal extraction strategies in (7.40) into (7.35) yields the cartel profit as

$$
W(x) = \int_{t}^{\infty} \left(g\left[\sum_{h\in N_i} \psi^*_h[x^*(s)]\right]\left[\sum_{h\in N_i} \psi^*_h[x^*(s)]\right]\right.
$$

$$
\left. -\left[\sum_{h\in N_i} c^h[\psi^*_h(x^*(s)),x^*(s)]\right]\right) \exp[-r(s-t)]. \tag{7.43}
$$

7.4.2 Imputation Scheme and Cartel Profit Sharing

In a dormant-firm cartel, firms having cost disadvantages will refrain from extraction to enhance the cartel's profit to a group optimum. Consider the case when the firms in the cartel agree to share the excess of the total cooperative payoff over the sum of individual noncooperative payoffs proportional to the firms' noncooperative payoffs.

The imputation scheme has to fulfill the following condition.

Condition 7.3 An imputation

$$\xi^{(\tau)j^1}\left(\tau, x_\tau^*\right) = \frac{\hat{V}^{j^1}(x_\tau^*)}{\sum_{h \in N_1} \hat{V}^h(x_\tau^*) + \sum_{\ell \in N_2} \hat{V}^\ell(x_\tau^*)} W\left(x_\tau^*\right),$$

is assigned to firm j^1, for $j^1 \in N_1$ at time $\tau \in [t_0, \infty)$ and an imputation

$$\xi^{(\tau)j^2}\left(\tau, x_\tau^*\right) = \frac{\hat{V}^{j^2}(x_\tau^*)}{\sum_{h \in N_1} \hat{V}^h(x_\tau^*) + \sum_{\ell \in N_2} \hat{V}^\ell(x_\tau^*)} W\left(x_\tau^*\right), \qquad (7.44)$$

is assigned to firm j^2, for $j^2 \in N_2$ at time $\tau \in [t_0, \infty)$.

To formulate a payoff distribution procedure over time so that the agreed imputations satisfy Condition 7.3 we invoke Theorem 4.3 in Chap. 4 to obtain the following.

Corollary 7.6 *A PDP with instantaneous payments at time* $\tau \in [t_0, \infty)$ *equaling*

$$B_{j^1}(\tau) = r\xi^{(\tau)j^1}\left(\tau, x_\tau^*\right) - \xi_{x_\tau^*}^{(\tau)j^1}\left(\tau, x_\tau^*\right) f\left[x_\tau^*, \sum_{j^i \in N_1} \psi_{j^1}^*(x_\tau^*)\right]$$

$$= \frac{r\hat{V}^{j^1}(x_\tau^*)}{\sum_{h \in N_1} \hat{V}^h(x_\tau^*) + \sum_{\ell \in N_2} \hat{V}^\ell(x_\tau^*)} W\left(x_\tau^*\right)$$

$$- \frac{\partial}{\partial x_\tau^*}\left[\frac{\hat{V}^{j^1}(x_\tau^*)}{\sum_{h \in N_1} \hat{V}^h(x_\tau^*) + \sum_{\ell \in N_2} \hat{V}^\ell(x_\tau^*)} W\left(x_\tau^*\right)\right]$$

$$\times f\left[x_\tau^*, \sum_{j^i \in N_1} \psi_{j^1}^*(x_\tau^*)\right], \quad \text{for } j^1 \in N_1, \qquad (7.45)$$

$$B_{j^2}(\tau) = r\xi^{(\tau)j^2}\left(\tau, x_\tau^*\right) - \xi_{x_\tau^*}^{(\tau)j^2}\left(\tau, x_\tau^*\right) f\left[x_\tau^*, \sum_{j^i \in N_1} \psi_{j^1}^*(x_\tau^*)\right]$$

$$= \frac{r\hat{V}^{j^2}(x_\tau^*)}{\sum_{h \in N_1} \hat{V}^h(x_\tau^*) + \sum_{\ell \in N_2} \hat{V}^\ell(x_\tau^*)} W\left(x_\tau^*\right)$$

$$-\frac{\partial}{\partial x_\tau^*}\left[\frac{\hat{V}^{j^2}(x_\tau^*)}{\sum_{h\in N_1}\hat{V}^h(x_\tau^*)+\sum_{\ell\in N_2}\hat{V}^\ell(x_\tau^*)}W(x_\tau^*)\right]$$

$$\times f\left[x_\tau^*,\sum_{j^i\in N_1}\psi_{j^1}^*(x_\tau^*)\right],\quad\text{for }j^2\in N_2$$

yields a consistent solution to the cooperative game $\Gamma_c(x_0)$ with the imputation as specified in Condition 7.3.

A time (optimal-trajectory-subgame) consistent solution can be obtained using the set of group optimal strategies $\psi^*(s,x_s^*)$ characterized in (7.39) and $B(s)=\{B_1(s),B_2(s),\ldots,B_n(s)\}$ in (7.45).

With the firms having cost advantages producing an output $\psi_{j^1}^*(x)$ for $j^1\in N_1$ and the firms having cost disadvantages refraining from production, the instantaneous receipt of firm i at time instant τ is

$$\zeta_{j^1}(\tau)=g\left[\sum_{h\in N_i}\psi_h^*(x_\tau^*)\right]\psi_{j^1}^*(x_\tau^*)-c^{j^1}\left[\psi_{j^1}^*(x_\tau^*),x_\tau^*(s)\right],$$

for $\tau\in[t_0,\infty)$ and $j^1\in N_1$, and $\zeta_{j^2}(\tau)=0$, for $\tau\in[t_0,\infty)$ and $j^2\in N_2$.

According to Corollary 7.6, the instantaneous payment that firm i should receive under the agreed-upon optimality principle is $B_{j^1}(\tau)$ for $j^1\in N_1$ and $B_{j^2}(\tau)$ for $j^2\in N_2$ as stated in (7.46). Hence an instantaneous transfer payment

$$\chi^{j^1}(\tau)=\zeta_{j^1}(\tau)-B_{j^1}(\tau),\quad\text{for firm }j^1\in N_1\text{ and }\tau\in[t_0,T],$$
$$\chi^{j^2}(\tau)=B_{j^2}(\tau),\quad\text{for firm }j^2\in N_2\text{ and }\tau\in[t_0,T],$$

(7.46)

would have to be arranged.

7.5 An Infinite-Horizon Dormant-Firm Cartel

Consider an infinite-horizon version of the game in Sect. 7.3. The dynamics of the resource is characterized by the differential equations

$$\dot{x}(s)=ax(s)^{1/2}-bx(s)-u_1(s)-u_2(s),\quad x(t_0)=x_0\in X,\tag{7.47}$$

where $u_i\in U_i$ is the (nonnegative) amount of the resource extracted by firm i, for $i\in\{1,2\}$, where a and b are positive constants.

At time t_0 the profit of firm $i\in\{1,2\}$ is

$$\int_{t_0}^\infty\left[\frac{u_i(s)}{[u_1(s)+u_2(s)]^{1/2}}-\frac{c_i}{x(s)^{1/2}}u_i(s)\right]\exp[-r(s-t_0)]\,ds,$$

$$\text{where }c_1<c_2.\tag{7.48}$$

Consider the alternative game problem

$$\max_{u_i} \int_t^\infty \left[\frac{u_i(s)}{[u_1(s) + u_2(s)]^{1/2}} - \frac{c_i}{x(s)^{1/2}} u_i(s) \right] \exp[-r(s-t)] \, ds, \quad (7.49)$$

subject to

$$\dot{x}(s) = ax(s)^{1/2} - bx(s) - u_1(s) - u_2(s), \quad x(t) = x \in X. \quad (7.50)$$

A set of strategies $\{\phi_i^*(x), \text{ for } i \in \{1, 2\}\}$ provides a feedback Nash equilibrium solution to the game in (7.49) and (7.50) if there exist continuously differentiable functions $\hat{V}^i(x) : R \to R, i \in \{1, 2\}$, satisfying the following partial differential equations:

$$r\hat{V}^i(x) = \max_{u_i} \left\{ \left[\frac{u_i}{(u_i + \phi_j^*(x))^{1/2}} - \frac{c_i}{x^{1/2}} u_i \right] \right.$$

$$\left. + \hat{V}_x^i(t, x)\left[ax^{1/2} - bx - u_i - \phi_j^*(x) \right] \right\},$$

for $i \in \{1, 2\}, j \in \{1, 2\}$ and $j \neq i$. $\quad (7.51)$

Performing the indicated maximization yields

$$\phi_1^*(x) = \frac{x}{4[c_1 + \hat{V}_x^1 x^{1/2}]^2}, \quad \text{and}$$

$$\phi_2^*(t, x) = \frac{x}{4[c_2 + \hat{V}_x^2 x^{1/2}]^2}. \quad (7.52)$$

Proposition 7.3 *The value function of firm i in the game in (7.49) and (7.50) is*

$$\hat{V}^i(x) = \left[A_i x^{1/2} + C_i \right], \quad \text{for } i \in \{1, 2\}, \quad (7.53)$$

where $A_i, C_i, A_j,$ and $C_j,$ for $i \in \{1, 2\}, j \in \{1, 2\},$ and $i \neq j$ satisfy

$$0 = \left[r + \frac{b}{2} \right] A_i - \left(\frac{3}{2} \right) \frac{[2c_j - c_i + A_j - A_i/2]}{[c_1 + c_2 + A_1/2 + A_2/2]^2}$$

$$+ \left(\frac{3}{2} \right)^2 \frac{c_i[2c_j - c_i + A_j - A_i/2]}{[c_1 + c_2 + A_1/2 + A_2/2]^3} + \left(\frac{9}{8} \right) \frac{A_i}{[c_1 + c_2 + A_1/2 + A_2/2]^2},$$

$$rC_i = \frac{a}{2} A_i.$$

Proof First, substitute the results in (7.52), and $\hat{V}^1(x), \hat{V}_x^1(x), \hat{V}^2(x),$ and $\hat{V}_x^2(x)$, obtained via (7.53), into the set of partial differential equations in (7.51). One can readily show that for this set of equations to be satisfied Proposition 7.3 has to hold. □

A noncooperative market equilibrium to the infinite-horizon resource extraction duopoly in (7.47) and (7.48) is obtained with the game equilibrium strategies in (7.52) and the value functions in Proposition 7.3.

7.5.1 Cartel Output and Optimal Resource Path

Assume that the firms agree to form a cartel and seek to solve the following joint profit maximization problem to achieve a group optimum

$$\max_{u_1,u_2} \int_{t_0}^{\infty} \left[[u_1(s) + u_2(s)]^{1/2} - \frac{c_1 u_1(s) + c_2 u_2(s)}{x(s)^{1/2}} \right] \exp[-r(s - t_0)] \, ds \quad (7.54)$$

subject to the dynamics in (7.47).

Consider the alternative optimal control problem

$$\max_{u_1,u_2} \int_{t}^{\infty} \left[[u_1(s) + u_2(s)]^{1/2} - \frac{c_1 u_1(s) + c_2 u_2(s)}{x(s)^{1/2}} \right] \exp[-r(s - t)] \, ds \quad (7.55)$$

subject to (7.50).

A set of strategies $[\psi_1^*(x), \psi_2^*(x)]$ provides an optimal control solution to the problem in (7.55) and (7.50) if there exist a continuously differentiable function $W(x) : R \to R$ satisfying the following partial differential equations:

$$rW(x) = \max_{u_1,u_2} \left\{ \left[(u_1 + u_2)^{1/2} - (c_1 u_1 + c_2 u_2) x^{-1/2} \right] \right.$$
$$\left. + W_x(x) \left[ax^{1/2} - bx - u_1 - u_2 \right] \right\}. \quad (7.56)$$

Performing the indicated maximization operation in (7.56) yields

$$\psi_1^*(x) = \frac{x}{4[c_1 + W_x x^{1/2}]^2} \quad \text{and} \quad \psi_2^*(x) = 0. \quad (7.57)$$

Along the optimal trajectory, firm 2 has to refrain from extraction. The more efficient firm (firm 1) would buy the less efficient firm (firm 2) out from the resource extraction process. Firm 2 becomes a dormant firm under cooperation.

Proposition 7.4 *The value function of the control problem in (7.55) and (7.50) can be obtained as*

$$W(x) = \left[Ax^{1/2} + C \right], \quad (7.58)$$

where A and B satisfy

$$0 = \left[r + \frac{b}{2} \right] A - \frac{1}{4[c_1 + A/2]} \quad \text{and} \quad rC = \frac{a}{2} A.$$

Proof First substitute the results in (7.57) and $W(x)$ and $W_x(x)$, obtained via (7.58), into the partial differential equation in (7.56). One can readily show that for this equation to be satisfied Proposition 7.4 has to hold.   □

Substituting $\psi_1^*(x)$ and $\psi_2^*(x)$ into (7.46) yields the optimal cooperative state dynamics as

$$\dot{x}(s) = \left[ax(s)^{1/2} - bx(s) - \frac{x(s)}{4[c_1 + A/2]^2} \right], \qquad x(t_0) = x_0 \in X. \qquad (7.59)$$

The solution to (7.59) yields a Pareto optimal trajectory, which can be expressed as

$$x^*(t) = \left\{ \Phi(t, t_0) \left[x_0^{1/2} + \int_{t_0}^t \Phi^{-1}(s, t_0) \frac{a}{2} \, ds \right] \right\}^2, \qquad (7.60)$$

where

$$\Phi(t, t_0) = \exp\left[\int_{t_0}^t \left(\frac{-b}{2} - \frac{1}{8[c_1 + A/2]^2} \right) ds \right].$$

7.5.2 Sharing of Cartel Profits

Consider the case when the firms in the cartel agree to share the excess of the total cooperative payoff over the sum of individual noncooperative payoffs proportional to the firms' noncooperative payoffs.

The imputation scheme has to fulfill Condition 7.3, that is,

$$\xi^{(\tau)i}\left(\tau, x_\tau^*\right) = \frac{\hat{V}^i(x_\tau^*)}{\sum_{j=1}^2 \hat{V}^j(x_\tau^*)} W\left(x_\tau^*\right),$$

for $i \in \{1, 2\}$ along the cooperative path $\{x_\tau^*\}_{\tau=t_0}^\infty$.

To formulate a payoff distribution procedure over time so that the agreed imputations satisfy Condition 7.3 we invoke Theorem 4.3 in Chap. 4 and obtain the following.

Corollary 7.7 *A PDP with an instantaneous payment along the cooperative path* $\{x_\tau^*\}_{\tau=t_0}^\infty$ *equaling*

$$B_i(\tau) = r \frac{\hat{V}^i(x_\tau^*)}{\sum_{j=1}^2 \hat{V}^j(x_\tau^*)} W\left(x_\tau^*\right) - \frac{\partial}{\partial x_\tau^*} \left[\frac{\hat{V}^i(x_\tau^*)}{\sum_{j=1}^2 \hat{V}^j(x_\tau^*)} W\left(x_\tau^*\right) \right]$$

$$\times \left[a\left(x_\tau^*\right)^{1/2} - bx_\tau^* - \frac{x_\tau^*}{4[c_1 + A/2]^2} \right], \qquad (7.61)$$

for $i \in \{1, 2\}$, *leads to the realization of the solution imputations satisfying Condition 7.3.*

In particular,

$$\frac{\hat{V}^i(x_\tau^*)}{\sum_{j=1}^2 \hat{V}^j(x_\tau^*)} W(x_\tau^*) = \frac{A_i(x_\tau^*)^{1/2} + C_i}{\sum_{j=1}^2 [A_j(x_\tau^*)^{1/2} + C_j]}[A(x_\tau^*)^{1/2} + C],$$

$$\frac{\partial}{\partial x_\tau^*}\left[\frac{\hat{V}^i(x_\tau^*)}{\sum_{j=1}^2 \hat{V}^j(x_\tau^*)} W(x_\tau^*)\right] = [A(x_\tau^*)^{1/2} + C]$$

$$\times \left[\frac{(\sum_{j=1}^2 [A_j(x_\tau^*)^{1/2} + C_j])\frac{1}{2}A_i(x_\tau^*)^{-1/2}}{(\sum_{j=1}^2 [A_j(x_\tau^*)^{1/2} + C_j])^2}\right.$$

$$\left. - \frac{[A_i(x_\tau^*)^{1/2} + C_i]\sum_{j=1}^2 \frac{1}{2}[A_j(x_\tau^*)^{-1/2}]}{(\sum_{j=1}^2 [A_j(x_\tau^*)^{1/2} + C_j])^2}\right]$$

$$+ \frac{[A_i(x_\tau^*)^{1/2} + C_i]}{(\sum_{j=1}^2 [A_j(x_\tau^*)^{1/2} + C_j])}\frac{1}{2}A(x_\tau^*)^{-1/2}.$$

A time (optimal-trajectory-subgame) consistent solution can be obtained using the set of group optimal strategies $\psi^*(s, x_s^*)$ in (7.57) and $B(s) = \{B_1(s), B_2(s), \dots, B_n(s)\}$ in (7.61).

The instantaneous receipt of firm 1 at time instant τ is

$$\zeta_1(\tau) = \frac{(x_\tau^*)^{1/2}}{2[c_1 + A/2]} - \frac{c_1(x_\tau^*)^{1/2}}{4[c_1 + A/2]^2}, \quad \text{for } \tau \in [t_0, \infty).$$

The instantaneous receipt of firm 2 at time instant τ is

$$\zeta_2(\tau) = 0, \quad \text{for } \tau \in [t_0, \infty).$$

According to Corollary 7.7, the instantaneous payment that firm i should receive under the agreed-upon optimality principle is $B_1(\tau)$ and $B_2(\tau)$ as stated in (7.61). Hence an instantaneous transfer payment

$$\chi^1(\tau) = \zeta_1(\tau) - B_1(\tau) \quad \text{for firm 1} \quad \text{and}$$

$$\chi^2(\tau) = B_2(\tau) \quad \text{for firm 2 along the cooperative path } \{x_\tau^*\}_{\tau=t_0}^\infty$$

would be arranged.

7.6   Exercises

7.1 Consider a duopoly in which two firms are allowed to extract a renewable resource within the duration $[0, 3]$. The dynamics of the resource is characterized by

$$\dot{x}(s) = 10x(s)^{1/2} - x(s) - u_1(s) - u_2(s), \quad x(0) = 100,$$

where $x(s)$ is the resource biomass and $u_i(s)$ is the amount of resource extracted by firm i at time $s \in [0, 3]$, for $i \in \{1, 2\}$.

The extraction cost for firms 1 and 2 are, respectively, $u_1(s)x(s)^{-1/2}$ and $2u_1(s)x(s)^{-1/2}$. The market price of the resource depends on the total amount extracted and supplied to the market. The price-output relationship at time s is given by the following downward-sloping inverse demand curve $P(s) = Q(s)^{-1/2}$, where $Q(s) = u_1(s) + u_2(s)$ is the total amount of the resource extracted and marketed at time s. At time 3, firm 1 will receive a termination bonus $2x(3)^{1/2}$ and firm 2 a bonus $x(3)^{1/2}$. The discount factor is 0.05.

At time 0, the profits of firms 1 and 2 then are

$$\int_0^3 \left[\frac{u_1(s)}{[u_1(s) + u_2(s)]^{1/2}} - \frac{u_1(s)}{x(s)^{1/2}} \right] \exp(-0.05s)\, ds$$

$$+ \exp\left[-0.05(3)\right]2x(3)^{\frac{1}{2}}, \quad \text{and}$$

$$\int_0^3 \left[\frac{u_2(s)}{[u_1(s) + u_2(s)]^{1/2}} - \frac{2u_2(s)}{x(s)^{1/2}} \right] \exp(-0.05s)\, ds + \exp\left[-0.05(3)\right]x(3)^{\frac{1}{2}}.$$

Obtain a feedback Nash equilibrium solution when these firms act independently.

7.2 If these two firms form a cartel show that firm 2 has to be dormant. Derive the optimal output strategies of the cartel and the optimal output trajectory.

7.3 Consider the case when the firms in the cartel agree to share the excess of the total cooperative profit over the sum of the individual noncooperative profit proportional to the firms' noncooperative profits. Characterize a time (optimal-trajectory-subgame) consistent solution.

7.4 Consider the case when the firms in the cartel agree to share the excess of the total cooperative profit over the sum of individual noncooperative profits equally. Characterize a time (optimal-trajectory-subgame) consistent solution.

Chapter 8
Subgame Consistent Economic Optimization Under Uncertainty

In many economic problems, uncertainty prevails. An essential characteristic of time—and hence decision making over time—is that though the individual may, through the expenditure of resources, gather past and present information, the future is inherently unknown and therefore (in the mathematical sense) uncertain. There is no escape from this fact, regardless of what resources the individual should choose to devote to obtaining data, information, and to forecasting. An empirically meaningful theory must therefore incorporate time-uncertainty in an appropriate manner. This development establishes a framework or paradigm for modeling game-theoretic situations with stochastic dynamics and uncertain environments over time. Again, the noncooperative stochastic differential games discussed in Chap. 2 fail to reflect all the facets of optimal behavior in n-person market games. Therefore cooperative optimization will generally lead to improved outcomes. Moreover, similar to cooperative differential game solutions, dynamically stable solutions of cooperative stochastic differential games have to be consistent over time. In the presence of stochastic elements, a very stringent condition—that of subgame consistency—is required for a credible cooperative solution. In particular, the optimality principle agreed upon at the outset must remain effective in any subgame starting at a later time with a realizable state brought about by prior optimal behavior.

In Sect. 8.1 of this chapter, a general framework of economic optimization under uncertainty is formulated. The principle of subgame consistency is discussed and examined in Sect. 8.2. Section 8.3 presents payoff distribution procedures leading to subgame consistent solutions. An illustration in cooperative fishery under uncertainty is given in Sect. 8.4. Infinite-horizon subgame consistent economic optimization is investigated in Sect. 8.5 and an illustration appears in the following section. Though the analysis in this chapter focuses on the standard stochastic differential game framework, it is worth noting that the development of applications of the recently emerging robust control techniques (see Hansen and Sargent 2008) in stochastic differential games should prove to be fruitful.

D.W.K. Yeung, L.A. Petrosyan, *Subgame Consistent Economic Optimization*, 203
Static & Dynamic Game Theory: Foundations & Applications,
DOI 10.1007/978-0-8176-8262-0_8, © Springer Science+Business Media, LLC 2012

8.1 Dynamic Economic Optimization Under Uncertainty

Consider again the situation when the economic agents agree to optimize cooperatively in a dynamic context. Let $\Gamma_c(x_0, T - t_0)$ denote a cooperative stochastic differential game in which economic agent i's expected payoff is

$$
E_{t_0} \left\{ \int_{t_0}^{T} g^i \big[s, x(s), u_1(s), u_2(s), \ldots, u_n(s) \big] \exp\left[-\int_{t_0}^{s} r(y)\, dy \right] ds \right.
$$
$$
\left. + \exp\left[-\int_{t_0}^{T} r(y)\, dy \right] q^i \big(x(T) \big) \right\}, \quad \text{for } i \in N, \tag{8.1}
$$

with $E_{t_0}\{\cdot\}$ denoting the expectation operation taken at time t_0, and the dynamics of the state is

$$
dx(s) = f\big[s, x(s), u_1(s), u_2(s), \ldots, u_n(s) \big] ds + \sigma\big[s, x(s) \big] dz(s), \quad x(t_0) = x_0, \tag{8.2}
$$

where $\sigma[s, x(s)]$ is a $m \times \Theta$ matrix and $z(s)$ is a Θ-dimensional Wiener process and the initial state x_0 is given. Let $\Omega[s, x(s)] = \sigma[s, x(s)]\sigma[s, x(s)]'$ denote the covariance matrix with its element in row h and column ζ denoted by $\Omega^{h\zeta}[s, x(s)]$. Moreover, $E[dz_\varpi] = 0$, $E[dz_\varpi\, dt] = 0$, and $E[(dz_\varpi)^2] = dt$, for $\varpi \in [1, 2, \ldots, \Theta]$, and $E[dz_\varpi\, dz_\omega] = 0$, for $\varpi \in [1, 2, \ldots, \Theta]$, $\varpi \in [1, 2, \ldots, \Theta]$, and $\varpi \neq \omega$.

The participating agents agree to act according to an agreed-upon optimality principle. The solution generated by the agreed-upon optimality principle includes agreements on how to act cooperatively and allocate cooperative payoff.

Let there be an optimality principle agreed upon by all agents in the cooperative game $\Gamma_c(x_0, T - t_0)$. Based on this optimality principle, the solution $P(x_0, T - t_0)$ of the game $\Gamma_c(x_0, T - t_0)$ at time t_0 includes the following.

(i) A set of cooperative strategies

$$
u^{(t_0)*}(s, x_s) = \big[u_1^{(t_0)*}(s, x_s), u_2^{(t_0)*}(s, x_s), \ldots, u_n^{(t_0)*}(s, x_s) \big], \quad \text{for } s \in [t_0, T]
$$

given that the state is x_s at time s.
(ii) An imputation vector $\xi^{(t_0)}(t_0, x_0) = [\xi^{(t_0)1}(t_0, x_0), \xi^{(t_0)2}(t_0, x_0), \ldots, \xi^{(t_0)n}(t_0, x_0)]$ to allocate the cooperative payoff to the agents.
(iii) A payoff distribution procedure $B^{t_0}(s, x_s) = [B_1^{t_0}(s, x_s), B_2^{t_0}(s, x_s), \ldots, B_n^{t_0}(s, x_s)]$ for $s \in [t_0, T]$, where $B_i^{t_0}(s, x_s)$ is the instantaneous payments for agent i at time s given that the state is x_s. In particular,

$$
\xi^{(t_0)i}(t_0, x_0) = E_{t_0}\left\{ \int_{t_0}^{T} B_i^{t_0}(s, x_s) \exp\left[-\int_{t_0}^{s} r(y)\, dy \right] ds \right.
$$
$$
\left. + q^i(x_T) \exp\left[-\int_{t_0}^{T} r(y)\, dy \right] \right\}, \quad \text{for } i \in N. \tag{8.3}
$$

This means that the agents agree at the outset on a set of cooperative strategies $u^{(t_0)*}(s, x_s)$, an imputation $\xi^{(t_0)i}(t_0, x_0)$ of the gains to the ith agent covering the time interval $[t_0, T]$, and a payoff distribution procedure $\{B^{t_0}(s, x_s)\}_{s=t_0}^T$ to allocate payments to the agents over the game interval.

Recall that group optimality is an essential element in dynamic cooperation. Any optimality principle has to require the agents to maximize their expected joint payoff

$$
E_{t_0} \left\{ \sum_{j=1}^n \int_{t_0}^T g^j \left[s, x(s), u_1(s), u_2(s), \ldots, u_n(s)\right] \exp\left[-\int_{t_0}^s r(y)\,dy \right] ds \right.
$$
$$
\left. + \sum_{j=1}^n \exp\left[-\int_{t_0}^T r(y)\,dy \right] q^i\left(x(T)\right) \right\},
\tag{8.4}
$$

subject to (8.1).

Invoking Theorem A.5 in the Technical Appendixes yields the characterization of the solution of the problem in (8.2) and (8.4) as follows.

Corollary 8.1 *A set of controls $\{u_i^*(t) = \psi_i^*(t, x), \text{ for } i \in N\}$ constitutes an optimal solution to the stochastic control problem in (8.2) and (8.4) if there exist continuously twice differentiable functions $W^{(t_0)}(t, x) : [t, T_0] \times R^m \to R$, satisfying the following partial differential equation:*

$$
-W_t^{(t_0)}(t, x) - \frac{1}{2} \sum_{h, \zeta=1}^m \Omega^{h\zeta}(t, x) W_{x^h x^\zeta}^{(t_0)}(t, x)
$$
$$
= \max_{u_1, u_2, \ldots, u_n} \left\{ \sum_{j=1}^n g^j[t, x, u_1, u_2, \ldots, u_n] \exp\left[-\int_{t_0}^t r(y)\,dy \right] \right.
$$
$$
\left. + W_x^{(t_0)}(t, x) f[t, x, u_1, u_2, \ldots, u_n] \right\}
\tag{8.5}
$$
$$
= \sum_{j=1}^n g^j\left[t, x, \psi_1^*(t, x), \psi_2^*(t, x), \ldots, \psi_n^*(t, x)\right]
$$
$$
+ W_x^{(t_0)}(t, x) f\left[t, x, \psi_1^*(t, x), \psi_2^*(t, x), \ldots, \psi_n^*(t, x)\right], \quad and
$$
$$
W^{(t_0)}(T, x) = \sum_{j=1}^n q^j(x) \exp\left[-\int_{t_0}^T r(y)\,dy \right].
$$

Hence the agents will adopt the cooperative control $\{\psi_i^*(t, x), \text{ for } i \in N \text{ and } t \in [t_0, T]\}$, to obtain the maximized level of expected joint profit. Substituting this set of controls into (8.1) yields the dynamics of the optimal (cooperative) trajectory as

$$
dx(s) = f\left[s, x(s), \psi_1^*\left(s, x(s)\right), \psi_2^*\left(s, x(s)\right), \ldots, \psi_n^*\left(s, x(s)\right)\right] ds + \sigma\left[s, x(s)\right] dz(s)
$$
$$
x(t_0) = x_0.
\tag{8.6}
$$

The solution to (8.6) can be expressed as

$$x^*(t) = x_0 + \int_{t_0}^{t} f\left[s, x^*(s), \psi_1^*\left(s, x^*(s)\right), \psi_2^*\left(s, x^*(s)\right), \ldots, \psi_n^*\left(s, x^*(s)\right)\right] ds$$

$$+ \int_{t_0}^{t} \sigma\left[s, x^*(s)\right] dz(s). \tag{8.7}$$

We use X_t^* to denote the set of realizable values of $x^*(t)$ at time t generated by (8.7). The term $x_t^* \in X_t^*$ is used to denote an element in X_t^*. We use the terms $x^*(t)$ and x_t^* interchangeably in cases where there is no ambiguity.

The cooperative control for the game in (8.2) and (8.4) over the time interval $[t_0, T]$ can be expressed more precisely as

$$\left\{\psi_i^*\left(t, x_t^*\right), \text{ for } i \in N \text{ and } t \in [t_0, T] \text{ when } x_t^* \in X_t^* \text{ is realized}\right\}. \tag{8.8}$$

Note that, for group optimality to be achievable, the cooperative controls $\{\psi_i^*(t, x_t^*), \text{ for } i \in N \text{ and } t \in [t_0, T]\}$ must be exercised throughout the time interval $[t_0, T]$.

The expected cooperative payoff over the interval $[t, T]$, for $t \in [t_0, T)$, can be expressed as

$$W^{(t_0)}\left(t, x_t^*\right)$$

$$= E_{t_0}\Bigg\{\int_t^T \sum_{j=1}^n g^j\left[s, x^*(s), \psi_1^*\left(s, x^*(s)\right), \psi_2^*\left(s, x^*(s)\right), \ldots, \psi_n^*\left(s, x^*(s)\right)\right]$$

$$\times \exp\left[-\int_{t_0}^s r(y)\,dy\right] ds + \exp\left[-\int_{t_0}^T r(y)\,dy\right]$$

$$\times \sum_{j=1}^n q^j\left(x^*(T)\right) \Bigg| x^*(t) = x_t^* \in X_t^* \Bigg\}. \tag{8.9}$$

To verify whether the agent would find it optimal to adopt the cooperative controls of (8.8) throughout the cooperative duration, we consider a stochastic control problem with the dynamics in (8.2) and payoff in (8.4), which begins at time $\tau \in [t_0, T]$ with initial state $x_\tau^* \in X_\tau^*$. At time τ, the optimality principle ensuring group rationality requires the agents to solve the problem

$$\max_{u_1, u_2, \ldots, u_n} E_\tau\Bigg\{\int_\tau^T \sum_{j=1}^n g^j\left[s, x(s), u_1(s), u_2(s), \ldots, u_n(s)\right] \exp\left[-\int_\tau^s r(y)\,dy\right] ds$$

$$+ \exp\left[-\int_\tau^T r(y)\,dy\right] \sum_{j=1}^n q^j\left(x(T)\right)\Bigg\}, \tag{8.10}$$

subject to

$$dx(s) = f\big[s, x(s), u_1(s), u_2(s), \ldots, u_n(s)\big]\,ds + \sigma\big[s, x(s)\big]\,dz(s),$$
$$x(\tau) = x_\tau^* \in X_t^*. \tag{8.11}$$

Note that

$$\max_{u_1, u_2, \ldots, u_n} E_{t_0}\left\{\left[\int_\tau^T \sum_{j=1}^n g^j\big[s, x(s), u_1(s), u_2(s), \ldots, u_n(s)\big]\exp\left[-\int_{t_0}^s r(y)\,dy\right]ds\right.\right.$$
$$\left.\left. + \exp\left[-\int_{t_0}^T r(y)\,dy\right]\sum_{j=1}^n q^j\big(x(T)\big)\right]\,\bigg|\,x(\tau) = x_\tau^* \in X_\tau^*\right\}$$

$$= \max_{u_1, u_2, \ldots, u_n} E_{t_0}\left\{\exp\left[-\int_{t_0}^\tau r(y)\,dy\right]\right.$$
$$\times \left(\int_\tau^T \sum_{j=1}^n g^j\big[s, x(s), u_1(s), u_2(s), \ldots, u_n(s)\big]\exp\left[-\int_\tau^s r(y)\,dy\right]ds\right.$$
$$\left.\left. + \exp\left[-\int_\tau^T r(y)\,dy\right]\sum_{j=1}^n q^j\big(x(T)\big)\right)\,\bigg|\,x(\tau) = x_\tau^* \in X_\tau^*\right\}$$

$$= \exp\left[-\int_{t_0}^\tau r(y)\,dy\right]$$
$$\times \max_{u_1, u_2, \ldots, u_n} E_\tau\left\{\left(\left[\int_\tau^T \sum_{j=1}^n g^j\big[s, x(s), u_1(s), u_2(s), \ldots, u_n(s)\big]\right.\right.\right.$$
$$\times \exp\left[-\int_\tau^s r(y)\,dy\right]ds$$
$$\left.\left.\left. + \exp\left[-\int_\tau^T r(y)\,dy\right]\sum_{j=1}^n q^j\big(x(T)\big)\right)\,\bigg|\,x(\tau) = x_\tau^*\right\}. \tag{8.12}$$

Hence the stochastic optimal controls strategies for the problem in (8.10) and (8.11) are analogous to the controls strategies for the problem in (8.2) and (8.4) in the time interval $[t, T]$.

A remark that will be utilized in the subsequent analysis is given below.

Remark 8.1 Let $W^{(\tau)}(t, x_t^*)$ denote the expected cooperative payoff function of the control problem in (8.10) and (8.11). One can readily verify that

$$\exp\left[\int_{t_0}^\tau r(y)\,dy\right]W^{(t_0)}(t, x_t^*) = \exp\left[\int_{t_0}^\tau r(y)\,dy\right]W^{(\tau)}(t, x_t^*),$$

for $\tau \in [t_0, T]$ and $t \in [\tau, T)$ and $x_\tau^* \in X_t^*$.

For notational sake we use $\Gamma_c(x_t^*, T - t)$ to denote the cooperative game with agent payoffs in (8.1) and dynamics in (8.2), which starts at time $t \in [t_0, T)$ given the state $x(\tau) = x_\tau^* \in X_\tau^*$. Let there exist the solution $P(x_t^*, T - t) \neq \emptyset$ under the agreed-upon optimality principle $t_0 \leq t \leq T$ along the optimal trajectory $\{x^*(t)\}_{t=t_0}^{T}$. If this condition is not satisfied it is impossible for the agents to adhere to the chosen principle of optimality since, at the very first instant t, when $P(x_t^*, T - t) = \emptyset$, the agents cannot follow this optimality principle.

Moreover, for $\xi^{(t)}(t, x_t^*), t \in [t_0, T]$ to be valid imputations it is required that both group optimality and individual rationality have to be satisfied. Hence a valid optimality principle would yield a solution $P(x_t^*, T - t)$ that contains

(i) $\sum_{j=1}^{n} \xi^{(t)j}(t, x_t^*) = W^{(t)}(t, x_t^*)$, for $t \in [t_0, T]$, and

(ii) $\xi^{(t)i}(t, x_t^*) \geq V^{(t)i}(t, x_t^*)$, for $i \in N$ and $t \in [t_0, T]$.

Part (i) yields group optimality, which guarantees that the sum of the expected imputation equals the maximum expected joint payoffs. Part (ii) yields individual rationality so that the expected payoff allocated to an economic agent under cooperation will be no less than its expected noncooperative payoff.

8.2 Principle of Subgame Consistency

Under a stochastic environment, it is known that stochastic dynamic cooperation represents, perhaps, decision making in its most complex form. Interactions between strategic behavior, dynamic evolution, and stochastic elements have to be considered simultaneously in the process, thereby leading to enormous difficulties in analysis. To ensure stability in dynamic cooperation over time in a stochastic framework, a stringent condition—subgame consistency—is required. In particular, the optimality principle agreed upon at the outset must remain effective in a subgame with a later starting time and any realizable state brought about by prior optimal behavior. Therefore, a cooperative solution is subgame consistent if an extension of the solution policy to a situation with a later starting time and any realizable state brought about by prior optimal behavior will remain optimal under the agreed-upon optimal principle.

Assume that at the start of the game the agents execute the solution $P(x_0, T - t_0)$ generated by an agreed-upon optimality principle (which includes a set of cooperative strategies, an imputation to distribute the cooperative payoff, and a payoff distribution procedure). When the game proceeds to time t and the state becomes $x_t^* \in X_t^*$, the continuation of the scheme in $P(x_0, T - t_0)$ has to be consistent with the solution $P(x_t^*, T - t)$ to the game $\Gamma_c(x_t^*, T - t)$ under the same optimality principle. If this consistency condition is violated, some of the agents will have an incentive to deviate from the initial agreement and instability arises. In particular, the dynamic stability of a solution of a cooperative stochastic differential game is the property that, when the game proceeds with any realizable state at each instant of time the agents are guided by the same optimality principle, and therefore do not have any incentive to deviate from the previously adopted optimal behavior.

A recent advancement in the study of cooperative stochastic differential games can be found in Yeung and Petrosyan (2004). In particular, a generalized theorem for the derivation of an analytically tractable "payoff distribution procedure" that will lead to subgame consistent solutions had been developed.

8.2.1 Subgame Consistent Solution

Let there be an optimality principle agreed upon by all agents in the cooperative stochastic differential game $\Gamma_c(x_0, T - t_0)$. At time t_0, the solution generated by this optimality principle is $P(x_0, T - t_0)$. At time $t \in (t_0, T]$ when the state is $x_t^* \in X_t^*$, we have the game $\Gamma_c(x_t^*, T - t)$. According to the agreed-upon optimality principle the solution of the game $\Gamma_c(x_t^*, T - t)$ is $P(x_t^c, T - t)$.

A cooperative game $\Gamma_c(x_0, T - t_0)$ has a subgame consistent solution $P(x_0, T - t_0)$ if the continuation of the scheme from the solution $P(x_0, T - t_0) = \{\psi^*(s, x_s^*)$ and $B^{t_0}(s, x_s^*)$ for $s \in [t_0, T]$; $\xi^{(t_0)}(t_0, x_0)\}$ over the time period $[t, T]$ coincides with the solution $P(x_t^*, T - t) = \{\psi^*(s, x_s^*)$ and $B^t(s, x_s^*)$ for $s \in [t, T]$; $\xi^{(t)}(t, x_t^*)\}$ generated by the same agreed-upon optimality principle at any time instant $t \in [t_0, T]$ along the optimal trajectory $\{x_s^*\}_{s=t_0}^T$.

If this coincidence does not appear, there is no guarantee that the agents will not abandon the solution $P(x_0, T - t_0)$ and switch to $P(x_t^*, T - t)$. Dynamical instability will arise as participants find that their agreed-upon optimality principle cannot be maintained after cooperation has gone on for some time.

To verify whether the solution $P(x_0, T - t_0)$ is indeed time consistent, one has to verify whether the agreed-upon cooperative strategies, payoff distribution procedures, and imputations are all time consistent. Since in Sect. 8.1 it was shown that joint payoff maximizing strategies are subgame consistent, we have to examine the subgame consistent imputation and payoff distribution procedure.

8.2.2 Subgame Consistency in Imputation and Payoff Distribution Procedure

Now we consider subgame consistency in imputation and payoff distribution procedure. At time t_0 when the initial state is x_0, according to the solution $P(x_0, T - t_0)$ generated by the agreed-upon optimality principle, the economic agents will use the payoff distribution procedure $\{B^{t_0}(s, x_s^*)\}_{s=t_0}^T$ to bring about an imputation to agent i as

$$\xi^{(t_0)i}(t_0, x_0) = E_{t_0}\left\{\int_{t_0}^T B_i^{t_0}(s, x_s^*) \exp\left[-\int_{t_0}^s r(y)\,dy\right] ds\right.$$

$$\left. + q^i\left(x^*(T)\right) \exp\left[-\int_{t_0}^T r(y)\,dy\right]\right\}, \quad \text{for } i \in N. \quad (8.13)$$

When the game proceeds to time $t \in (t_0, T]$, the current state is $x_t^* \in X_t^*$. According to the solution $P(x_0, T - t_0)$, agent i will receive an imputation (in the present value viewed at time t_0) equaling

$$\xi^{(t_0)i}(t, x_t^*) = E_{t_0}\left\{\left[\int_t^T B_i^{t_0}(s, x_s^*)\exp\left[-\int_{t_0}^s r(y)\,dy\right]ds\right.\right.$$

$$\left.\left. + q^i(x^*(T))\exp\left[-\int_{t_0}^T r(y)\,dy\right]\right]\middle| x(t) = x_t^*\right\}, \qquad (8.14)$$

over the time interval $[t, T]$.

At time $t \in (t_0, T]$ when the current state is $x_t^* \in X_t^*$, we have a cooperative game $\Gamma_c(x_t^*, T - t)$. According to the solution $P(x_t^*, T - t)$ generated by the agreed-upon optimality principle, the economic agents will use the payoff distribution procedure $\{B^t(s, x_s^*)\}_{s=t}^T$ to bring about an imputation to agent i as

$$\xi^{(t)i}(t, x_t^*) = E_t\left\{\left[\int_t^T B_i^t(s, x_s^*)\exp\left[-\int_t^s r(y)\,dy\right]ds\right.\right.$$

$$\left.\left. + q^i(x^*(T))\exp\left[-\int_t^T r(y)\,dy\right]\right]\right\}, \qquad \text{for } i \in N. \qquad (8.15)$$

For the imputation and payoff distribution procedure from $P(x_0, T - t_0)$ to be consistent with those from $P(x_t^c, T - t)$, it is essential that

$$\exp\left[\int_{t_0}^t r(y)\,dy\right]\xi^{(t_0)}(t, x_t^*) = \xi^{(t)}(t, x_t^*) \in P(x_t^*, T - t), \qquad \text{for } t \in [t_0, T].$$

In addition, at time t_0 when the initial state is x_0, according to the solution $P(x_0, T - t_0)$ generated by the agreed-upon optimality principle, the payoff distribution procedure is

$$B^{t_0}(s, x_s^*) = \left[B_1^{t_0}(s, x_s^*), B_2^{t_0}(s, x_s^*), \ldots, B_n^{t_0}(s, x_s^*)\right], \qquad \text{for } s \in [t_0, T].$$

Consider the case when the game has proceeded to time t and the state variable became $x_t^* \in X_t^*$. Then one has a cooperative game $\Gamma_c(x_t^*, T - t)$ which starts at time t with initial state x_t^*. According to the solution $P(x_t^*, T - t)$ generated by the agreed-upon optimality principle, the payoff distribution procedure

$$B^t(s, x_s^*) = \left[B_1^t(s, x_s^*), B_2^t(s, x_s^*), \ldots, B_n^t(s, x_s^*)\right], \qquad \text{for } s \in [t, T],$$

will be adopted.

For the continuation of the payoff distribution procedure $B^{t_0}(s, x_s^*)$ under $P(x_0, T - t_0)$ to be consistent with $B^t(s, x_s^*) \in P(x_t^*, T - t)$, it is required that

$$B^{t_0}(s, x_s^*) = B^t(s, x_s^*), \qquad \text{for } s \in [t, T] \text{ and } t \in [t_0, T].$$

Therefore a formal definition on subgame consistent imputation and payoff distribution procedures can be presented as below.

Definition 8.1 The imputation and payoff distribution procedure $\{\xi^{(t_0)}(t_0, x_0)$ and $B^{t_0}(s, x_s^*)$ for $s \in [t_0, T]\} \in P(x_0, T - t_0)$ are subgame consistent if

(i)
$$\exp\left[\int_{t_0}^{t} r(y)\,dy\right] \xi^{(t_0)i}(t, x_t^*)$$

$$\equiv \exp\left[\int_{t_0}^{t} r(y)\,dy\right] E\left\{\left[\int_{t}^{T} B_i^{t_0}(s, x_s^*)\exp\left[-\int_{t}^{s} r(y)\,dy\right]ds\right.\right.$$

$$\left.\left. + q^i(x^*(T))\exp\left[-\int_{t_0}^{T} r(y)\,dy\right]\right]\Big| x(t) = x_t^*\right\} = \xi^{(t)i}(t, x_t^*)$$

$$\equiv E_t\left\{\left[\int_{t}^{T} B_i^{t}(s, x_s^*)\exp\left[-\int_{t}^{s} r(y)\,dy\right]ds + q^i(x^*(T))\right.\right.$$

$$\left.\left. \times \exp\left[-\int_{t}^{T} r(y)\,dy\right]\right]\right\}$$

$$\in P(x_t^*, T - t), \quad \text{for } i \in N \text{ and } t \in [t_0, T].$$

(ii) The payoff distribution procedure $B^{t_0}(s, x_s^*) = [B_1^{t_0}(s, x_s^*), B_2^{t_0}(s, x_s^*), \ldots,$ $B_n^{t_0}(s, x_s^*)]$ for $s \in [t, T]$ is identical to $B^t(s, x_s^*) = [B_1^t(s, x_s^*), B_2^t(s, x_s^*), \ldots,$ $B_n^t(s, x_s^*)] \in P(x_t^*, T - t)$.

Definition 8.1 will be used to guide the derivation of subgame consistent solutions in the following section.

8.3 Payoff Distribution Procedure and Subgame Consistent Solutions

Crucial to obtaining a subgame consistent solution is the derivation of a payoff distribution procedure satisfying Definition 8.1 in Sect. 8.2.

8.3.1 Payoff Distribution Procedures Leading to Subgame Consistent Solutions

Invoking part (ii) of Definition 8.1, we have $B^{t_0}(s, x_s^*) = B^t(s, x_s^*)$ for $t \in [t_0, T]$ and $s \in [t, T]$. We use $B(s, x_s^*) = \{B_1(s, x_s^*), B_2(s, x_s^*), \ldots, B_n(s, x_s^*)\}$ to denote $B^t(s, x_s^*)$ for all $t \in [t_0, T]$. Along the conditional optimal trajectory $\{x^*(s)\}_{s=t_0}^{T}$ we then have

$$\xi^{(\tau)i}(\tau, x_\tau^*) = E_\tau\left\{\left[\int_{\tau}^{T} B_i^{\tau}(s, x^*(s))\exp\left[-\int_{\tau}^{s} r(y)\,dy\right]ds\right.\right.$$

$$\left.\left. + q^i(x_T^*)\exp\left[-\int_{\tau}^{T} r(y)\,dy\right]\right]\Big| x^*(\tau) = x_\tau^* \in X_\tau^*\right\}, \quad (8.16)$$

for $i \in N$ and $\tau \in [t_0, T]$.

Moreover, for $t \in [\tau, T]$, we use the term

$$\xi^{(\tau)i}\left(t, x_t^*\right) = E_\tau \left\{ \left[\int_t^T B_i^\tau\left(s, x^*(s)\right) \exp\left[-\int_\tau^s r(y)\,dy \right] ds \right. \right.$$

$$\left. \left. + q^i\left(x_T^*\right) \exp\left[-\int_\tau^T r(y)\,dy \right] \right| \, x^*(t) = x_t^* \in X_t^* \right\} \qquad (8.17)$$

to denote the expected present value (with the initial time being τ) of agent i's expected payoff under cooperation over the time interval $[t, T]$ according to the solution $P(x_\tau^*, T - \tau)$ along the cooperative state trajectory.

Invoking (8.16) and (8.17) we have

$$\xi^{(\tau)i}\left(t, x_t^c\right) = \exp\left[-\int_\tau^t r(y)\,dy \right] \xi^{(t)i}\left(t, x_t^*\right), \qquad (8.18)$$

for $i \in N$ and $\tau \in [t_0, T]$ and $t \in [\tau, T]$.

One can readily verify that a payoff distribution procedure $\{B(s, x_s^*)\}_{s=t_0}^T$ that satisfies (8.18) will give rise to time consistent imputations satisfying part (i) of Definition 8.1. The next task is the derivation of a payoff distribution procedure $\{B(s, x_s^*)\}_{s=t_0}^T$ that leads to the realization of (8.16)–(8.18).

We first consider the following condition concerning the imputation $\xi^{(\tau)}(t, x_t^*)$, for $\tau \in [t_0, T]$ and $t \in [\tau, T]$.

Condition 8.1 For $i \in N, t \in [\tau, T]$, and $\tau \in [t_0, T]$, the imputation $\xi^{(\tau)i}(t, x_t^*)$, for $i \in N$, is a function that is twice continuously differentiable in t and $x_t^* \in X_t^*$.

A theorem characterizing a formula for $B_i(s, x_s^*)$, for $s \in [t_0, T], x_s^* \in X_s^*$ and $i \in N$, which yields (8.16)–(8.18), can be provided as follows.

Theorem 8.1 *If Condition 8.1 is satisfied, a PDP with a terminal payment $q^i(x_T^*)$ at time T and an instantaneous payment at time $s \in [\tau, T]$*

$$B_i\left(s, x_s^*\right) = -\left[\xi_t^{(s)i}\left(t, x_t^*\right)\big|_{t=s} \right]$$

$$\quad - \left[\xi_{x_t^*}^{(s)i}\left(t, x_t^*\right)\big|_{t=s} \right] f\left[s, x_s^*, \psi_1^*(s, x_s^*), \psi_2^*(s, x_s^*) \ldots, \psi_n^*(s, x_s^*) \right]$$

$$\quad - \frac{1}{2} \sum_{h,\zeta=1}^m \Omega^{h\zeta}\left(s, x_s^*\right)\left[\xi_{x_t^h x_t^\zeta}^{(s)i}\left(t, x_t^*\right)\big|_{t=s} \right],$$

for $i \in N$ and $x_s^* \in X_s^*$, $\qquad (8.19)$

yields the imputation vector $\xi^{(\tau)}(\tau, x_\tau^*)$, for $\tau \in [t_0, T]$, which satisfy (8.16)–(8.18).

Proof Invoking (8.16)–(8.18), one can obtain

$$
\xi^{(v)i}\left(v, x_v^*\right)
$$

$$
= E_v \left\{ \int_v^{v+\Delta t} B_i\left(s, x_s^*\right) \exp\left[-\int_v^s r(y)\,dy\right] ds \right.
$$

$$
\left. + \exp\left[-\int_v^{v+\Delta t} r(y)\,dy\right] \xi^{(v+\Delta t)i}\left(v+\Delta t, x_v^* + \Delta x_v^*\right) \,\middle|\, x(v) = x_v^* \in X_v^* \right\},
$$

for $v \in [\tau, T]$ and $i \in N$, (8.20)

where

$$
\Delta x_v^* = f\left[v, x_v^c, \psi_1^*\left(v, x_v^*\right), \psi_2^*\left(v, x_v^*\right), \ldots, \psi_n^*\left(v, x_v^*\right)\right]\Delta t
$$

$$
+ \sigma\left[v, x_v^*\right]\Delta z_v + o(\Delta t),
$$

$$
\Delta z_v = Z(v + \Delta t) - z(v), \quad \text{and} \quad E_v\left[o(\Delta t)\right]/\Delta t \to 0 \quad \text{as } \Delta t \to 0.
$$

From (8.17) and (8.20), one obtains

$$
E_v \left\{ \int_v^{v+\Delta t} B_i\left(s, x_s^*\right) \exp\left[-\int_v^s r(y)\,dy\right] ds \,\middle|\, x(v) = x_v^* \right\}
$$

$$
= E_v \left\{ \xi^{(v)i}\left(v, x_v^*\right) - \exp\left[-\int_v^{v+\Delta t} r(y)\,dy\right] \xi^{(v+\Delta t)i}\left(v+\Delta t, x_v^* + \Delta x_v^*\right) \right\}
$$

$$
= E_v \left\{ \xi^{(v)i}\left(v, x_v^*\right) - \xi^{(v)i}\left(v+\Delta t, x_v^* + \Delta x_v^*\right) \right\}, \quad (8.21)
$$

for all $v \in [t_0, T]$ and $i \in N$.

If the imputations $\xi^{(v)}(t, x_t^*)$, for $v \in [t_0, T]$, satisfy Condition 8.1, as $\Delta t \to 0$, one can express (8.21) as

$$
E_v \left\{ B_i\left(v, x_v^*\right)\Delta t + o(\Delta t) \right\}
$$

$$
= E_v \left\{ -\left[\xi_t^{(v)i}\left(t, x_t^c\right)\big|_{t=v}\right]\Delta t \right.
$$

$$
- \left[\xi_{x_v^c}^{(v)i}\left(v, x_v^c\right)\right] f\left[v, x_v^c, \psi_1^*\left(v, x_v^c\right), \psi_2^*\left(v, x_v^c\right), \ldots, \psi_n^*\left(v, x_v^c\right)\right]\Delta t
$$

$$
- \frac{1}{2} \sum_{h,\zeta=1}^m \Omega^{h\zeta}\left(v, x_v^*\right)\left[\xi_{x_t^h x_t^\zeta}^{(v)i}\left(t, x_t^*\right)\big|_{t=v}\right]
$$

$$
\left. - \left[\xi_{x_v^c}^{(v)i}\left(v, x_v^c\right)\right]\sigma\left[v, x_v^*\right]\Delta z_v - o(\Delta t). \right. \quad (8.22)
$$

Dividing (8.22) throughout by Δt, with $\Delta t \to 0$, and taking the expectation yields (8.19). Thus the payoff distribution procedure in $B_i(s, x_s^*)$ in (8.19) will lead to the realization of $\xi^{(\tau)i}(\tau, x_\tau^c)$, for $\tau \in [t_0, T]$, which satisfies (8.16)–(8.18). □

Assigning the instantaneous payments according to the payoff distribution procedure in (8.19) leads to the realization of the imputation $\xi^{(\tau)}(\tau, x_\tau^*) \in P(x_\tau^*, T - \tau)$ for $\tau \in [t_0, T]$ and $x_\tau^* \in X_\tau^*$.

8.3.2 Subgame Consistent Solution

Consider the following optimality principle.

Principle PI It is an optimality principle that entails (i) group optimality and individual rationality, and (ii) the distribution of the total cooperative payoff according to an imputation that equals $\xi^{(\tau)}(\tau, x_\tau^*)$ for $\tau \in [t_0, T]$ over the game duration. Moreover, the function $\xi^{(\tau)}(\tau, x_\tau^*)$ is continuously differentiable in τ and x_τ^*.

A theorem characterizing a subgame consistent solution for the cooperative game $\Gamma_c(x_0, T - t_0)$ under optimality Principle PI is presented below.

Theorem 8.2 *For the cooperative game $\Gamma_c(x_0, T - t_0)$ with optimality Principle* PI, *the solution*

$$P(x_0, T - t_0) = \{u(s, x_s^*) \text{ and } B(s, x_s^*) \text{ for } s \in [t_0, T] \text{ and } \xi^{(t_0)}(t_0, x_0)\}$$

in which

(i) $u(s, x_s^*)$ *for $s \in [t_0, T]$ is the set of group optimal strategies $\psi^*(s, x_s^*)$ for the game $\Gamma_c(x_0, T - t_0)$, and*

(ii) *the imputation distribution procedure*

$$B(s, x_s^*) = \{B_1(s, x_s^*), B_2(s, x_s^*), \dots, B_n(s, x_s^*)\} \quad \text{for } s \in [t_0, T]$$

where

$$
\begin{aligned}
B_i(s, x_s^*) = &-\left[\xi_t^{(s)i}(t, x_t^*)\big|_{t=s}\right] \\
&- \left[\xi_{x_t^*}^{(s)i}(t, x_t^*)\big|_{t=s}\right] f\left[s, x_s^*, \psi_1^*(s, x_s^*), \psi_2^*(s, x_s^*), \dots, \psi_n^*(s, x_s^*)\right] \\
&- \frac{1}{2} \sum_{h,\zeta=1}^{m} \Omega^{h\zeta}(s, x_s^*)\left[\xi_{x_t^h x_t^\zeta}^{(s)i}(t, x_t^*)\big|_{t=s}\right],
\end{aligned}
$$

for $i \in N$ and $x_s^* \in X_s^*$, $\qquad\qquad$ (8.23)

and $\xi^{(s)}(s, x_s^*) = [\xi^{(s)1}(s, x_s^*), \xi^{(s)2}(s, x_s^*), \dots, \xi^{(s)n}(s, x_s^*)] \in P(x_s^*, T - s)$ is the imputation at time $s \in [t_0, T]$ with the state being $x_s^* \in X_s^*$ according to optimality Principle PI, *is subgame consistent.*

Proof Following the algorithm that specifies $P(x_0, T - t_0)$ as the solution to the game $\Gamma_c(x_0, T - t_0)$ under Principle PI one can readily obtain the solution of the cooperative game $\Gamma_c(x_\upsilon^*, T - \upsilon)$, for $\upsilon \in [t_0, T]$, as

$$P\big(x_\upsilon^*, T - \upsilon\big) = \big\{u\big(s, x_s^*\big) \text{ and } B\big(s, x_s^*\big) \text{ for } s \in [t_0, T] \text{ and } \xi^{(\upsilon)}\big(\upsilon, x_\upsilon^*\big)\big\}$$

in which

(i) $u(s, x_s^*)$ for $s \in [\upsilon, T]$ is the set of group optimal strategies $\psi^*(s, x_s^*)$ for the game $\Gamma_c(x_\upsilon^*, T - \upsilon)$, and

(ii)

$$B\big(s, x_s^*\big) = \big\{B_1\big(s, x_s^*\big), B_2\big(s, x_s^*\big), \ldots, B_n\big(s, x_s^*\big)\big\} \quad \text{for } s \in [\upsilon, T]$$

where

$$
\begin{aligned}
B_i\big(s, x_s^*\big) = &-\big[\xi_t^{(s)i}\big(t, x_t^*\big)|_{t=s}\big] \\
&- \big[\xi_{x_t^*}^{(s)i}\big(t, x_t^*\big)|_{t=s}\big] f\big[s, x_s^*, \psi_1^*\big(s, x_s^*\big), \psi_2^*\big(s, x_s^*\big), \ldots, \psi_n^*\big(s, x_s^*\big)\big] \\
&- \frac{1}{2} \sum_{h,\zeta=1}^{m} \Omega^{h\zeta}\big(s, x_s^*\big)\big[\xi_{x_t^h x_t^\zeta}^{(s)i}\big(t, x_t^*\big)|_{t=s}\big],
\end{aligned}
$$

for $i \in N$ and $x_s^* \in X_s^*$,   (8.24)

and $\xi^{(s)}(s, x_s^*) = [\xi^{(s)1}(s, x_s^*), \xi^{(s)2}(s, x_s^*), \ldots, \xi^{(s)n}(s, x_s^*)] \in P(x_s^*, T - s)$ is the imputation at time $s \in [\upsilon, T]$ with the state being $x_s^* \in X_s^*$ according to optimality Principle PI.

As shown before, the group optimal strategies $\psi^*(s, x_s^*)$ for the cooperative game $\Gamma_c(x_0, T - t_0)$ over the time interval $[\upsilon, T]$ are identical to the group optimal strategies for the cooperative game $\Gamma_c(x_\upsilon^*, T - \upsilon)$ over the same time interval.

Comparing (8.23) and (8.24) one can show that the payoff distribution procedure $B(s, x_s^*)$ for the cooperative game $\Gamma_c(x_0, T - t_0)$ over the time interval $[\upsilon, T]$ is identical to the payoff distribution procedure $B(s, x_s^*)$ for the cooperative game $\Gamma_c(x_\upsilon^*, T - \upsilon)$ over the same time interval.

Invoking Theorem 8.1 one can show that the payoff distribution procedure $B(s, x_s^*)$ in (8.23) will yield

$$
\begin{aligned}
\xi^{(\upsilon)i}\big(\upsilon, x_\upsilon^*\big) = E_\upsilon\bigg\{ &\int_\upsilon^T B_i\big(s, x_s^*\big) \exp\bigg[-\int_\upsilon^s r(y)\,\mathrm{d}y\bigg]\,\mathrm{d}s \\
&+ q^i\big(x_T^*\big)\exp\bigg[-\int_\upsilon^T r(y)\,\mathrm{d}y\bigg] \bigg| x(\upsilon) = x_\upsilon^* \bigg\} \in P\big(x_\upsilon^*, T - \upsilon\big),
\end{aligned}
$$

for $i \in N$, and $\upsilon \in [\tau, T]$.

Hence

$$\exp\bigg[\int_{t_0}^\upsilon r(y)\,\mathrm{d}y\bigg]\xi^{(t_0)i}\big(\upsilon, x_\upsilon^*\big) = \xi^{(\upsilon)i}\big(\upsilon, x_\upsilon^*\big) \in P\big(x_\upsilon^*, T - \upsilon\big).$$

In sum, the continuation of the solution $P(x_0, T - t_0)$ over the time interval $[v, T]$ coincides with the solution $P(x_v^*, T - v)$ of the game $\Gamma_c(x_v^*, T - v)$ under optimality Principle PI. Thus the solution $P(x_0, T - t_0)$ in Theorem 8.2 is indeed subgame consistent.   □

8.3.3 Instantaneous Transfer Payments

With agents using the cooperative strategies $\{\psi_i^*(\tau, x_\tau^*), \text{ for } \tau \in [t_0, T] \text{ and } i \in N\}$, the instantaneous payment received by agent i at time instant τ is

$$\zeta_i(\tau, x_\tau^*) = g^i[\tau, x_\tau^*, \psi_1^*(\tau, x_\tau^*), \psi_2^*(\tau, x_\tau^*), \dots, \psi_n^*(\tau, x_\tau^*)],$$

$$\text{for } \tau \in [t_0, T], x_\tau^* \in X_\tau^* \text{ and } i \in N. \tag{8.25}$$

According to Theorem 8.2, the instantaneous payment that agent i should receive under the agreed-upon optimality principle is $B_i(\tau, x_\tau^*)$, as stated in (8.23). Hence an instantaneous transfer payment

$$\chi^i(\tau, x_\tau^*) = B_i(\tau, x_\tau^*) - \zeta_i(\tau, x_\tau^*) \tag{8.26}$$

has to be given to agent i at time τ, for $i \in N$ and $\tau \in [t_0, T]$, when the state is $x_\tau^* \in X_\tau^*$.

Under a subgame consistent solution, an extension of the solution policy to a situation with a later starting time and any realizable state brought about by prior optimal behavior would remain optimal. Examples of cooperative stochastic differential games with solutions satisfying subgame consistency can be found in Yeung (2005, 2007) and Yeung and Petrosyan (2004, 2006a, 2006b, 2007a, 2007b). Theorem 8.2 can be applied to obtain a subgame consistent cooperative solution for existing differential games in economic analysis.

8.4 An Illustration in Cooperative Fishery Under Uncertainty

Consider the stochastic resource extraction game in Sect. 2.4.2 of Chap. 2 with two asymmetric extractors.

The resource stock $x(s) \in X \subset R$ follows the stochastic dynamics

$$dx(s) = [ax(s)^{1/2} - bx(s) - u_1(s) - u_2(s)]ds + \sigma x(s)dz(s),$$

$$x(t_0) = x_0 \in X, \tag{8.27}$$

where $u_i(s)$ is the harvest rate of extractor $i \in \{1, 2\}$. The instantaneous payoffs at time $s \in [t_0, T]$ for extractors 1 and 2 are, respectively, $[u_1(s)^{1/2} - \frac{c_1}{x(s)^{1/2}}u_1(s)]$ and $[u_2(s)^{1/2} - \frac{c_2}{x(s)^{1/2}}u_2(s)]$, where c_1 and c_2 are constants and $c_1 \neq c_2$. At time T, each extractor will receive a termination bonus $qx(T)^{1/2}$. Payoffs are transferable between extractors and over time. Given the constant discount rate r, the values received at time t are discounted by the factor $\exp[-r(t-t_0)]$.

At time t_0, the expected payoff of extractor i is

$$E_{t_0}\left\{\int_{t_0}^{T}\left[u_i(s)^{1/2}-\frac{c_i}{x(s)^{1/2}}u_i(s)\right]\exp\left[-r(t-t_0)\right]ds\right.$$

$$\left.+\exp\left[-r(T-t_0)\right]qx(T)^{\frac{1}{2}}\right\},\quad\text{for }i\in\{1,2\}.\tag{8.28}$$

Let $[\phi_1^*(t,x),\phi_2^*(t,x)]$ for $t\in[t_0,T]$ denote a set of strategies that provide a feedback Nash equilibrium solution to the game in (8.27) and (8.28), and $V^{(t_0)i}(t,x):[t_0,T]\times R^n\to R$ denote the value function of extractor $i\in\{1,2\}$ that satisfies the equations

$$-V_t^{(t_0)i}(t,x)-\frac{1}{2}\sigma^2x^2V_{xx}^{(t_0)i}(t,x)$$

$$=\max_{u_i}\left\{\left[(u_i)^{1/2}-\frac{c_i}{x^{1/2}}u_i\right]\exp\left[-r(t-t_0)\right]\right.$$

$$\left.+V_x^{(t_0)i}(t,x)\left[ax^{1/2}-bx-u_i-\phi_j^{(t_0)*}(t,x)\right]\right\},\quad\text{and}\tag{8.29}$$

$$V^{(t_0)i}(T,x)=\exp\left[-r(T-t_0)\right]qx(T)^{\frac{1}{2}},\quad\text{for }i\in\{1,2\},\ j\in\{1,2\},\text{ and }j\neq i.$$

Performing the indicated maximization in (8.29) yields the game equilibrium strategies

$$\phi_i^*(t,x)=\frac{x}{4[c_i+V_x^{(t_0)i}\exp[r(t-t_0)]x^{1/2}]^2},\quad\text{for }i\in\{1,2\}.$$

Proposition 8.1 *The value functions of extractor $i\in\{1,2\}$ in the game in (8.27) and (8.28) is*

$$V^{(t_0)i}(t,x)=\exp\left[-r(t-t_0)\right]\left[A_i(t)x^{1/2}+C_i(t)\right],\tag{8.30}$$

where for $i,j\in\{1,2\}$ and $i\neq j$, $A_i(t),B_i(t),A_j(t)$, and $B_j(t)$ satisfy

$$\dot{A}_i(t)=\left[r+\frac{1}{8}\sigma^2+\frac{b}{2}\right]A_i(t)-\frac{1}{2[c_i+A_i(t)/2]}+\frac{c_i}{4[c_i+A_i(t)/2]^2}$$

$$+\frac{A_i(t)}{8[c_i+A_i(t)/2]^2}+\frac{A_i(t)}{8[c_j+A_j(t)/2]^2},$$

$$\dot{C}_i(t)=rC_i(t)-\frac{a}{2}A_i(t),$$

$$A_i(T)=q,\quad\text{and}\quad C_i(T)=0.$$

Proof First substitute $\phi_1^*(t,x)$ and $\phi_2^*(t,x),V^{(t_0)i}(t,x)$ from (8.30) and the corresponding derivatives $V_t^{(t_0)i}(t,x),V_x^{(t_0)i}(t,x)$, and $V_{xx}^{(t_0)i}(t,x)$ into (8.29). Upon solving (8.29) one obtains Proposition 8.1.   □

Invoking Remark 2.2 in Chap. 2, we can obtain the value function of agent i in a game with the dynamics in (8.27) and expected payoffs in (8.28), which starts at time τ for $\tau \in [t_0, T)$, as

$$V^{(\tau)i}(t, x) = \exp\left[-r(t - \tau)\right]\left[A_i(t)x^{1/2} + B_i(t)\right], \quad \text{for } i \in \{1, 2\}.$$

Substituting the relevant derivatives of the value functions into the game equilibrium strategies yields a feedback Nash equilibrium solution to the game in (8.27) and (8.28).

8.4.1 Cooperative Extraction Under Uncertainty

Now consider the case when the resource extractors agree to act cooperatively and follow the optimality principle under which they would (i) maximize their joint expected payoffs and (ii) share the excess of the total expected cooperative payoff over the sum of expected individual noncooperative payoffs proportional to the agents' expected noncooperative payoffs.

Hence the extractors maximize the sum of their expected profits

$$E_{t_0}\left\{ \int_{t_0}^{T} \left(\left[u_1(s)^{1/2} - \frac{c_1}{x(s)^{1/2}}u_1(s)\right] + \left[u_2(s)^{1/2} - \frac{c_2}{x(s)^{1/2}}u_2(s)\right] \right) \right.$$
$$\left. \times \exp\left[-r(t - t_0)\right] ds + 2\exp\left[-r(T - t_0)\right]qx(T)^{\frac{1}{2}} \right\}, \tag{8.31}$$

subject to the stochastic dynamics in (8.27).

Invoking Theorem A.5 in the Technical Appendixes yields the characterization of the solution of the problem in (8.27) and (8.31) as follows.

Corollary 8.2 *A set of controls $\{u_i^*(t) = \psi_i^*(t, x), \text{ for } i \in \{1, 2\}\}$ constitutes an optimal solution to the stochastic control problem in (8.27) and (8.31) if there exist continuously twice differentiable functions $W^{(t_0)}(t, x) : [t_0, T] \times R^m \to R$, satisfying the following partial differential equation:*

$$-W_t^{(t_0)}(t, x) - \frac{1}{2}\sigma^2 x^2 W_{xx}^{(t_0)}(t, x)$$
$$= \max_{u_1, u_2}\left\{ \left(\left[u_1^{1/2} - \frac{c_1}{x^{1/2}}u_1\right] + \left[u_2^{1/2} - \frac{c_2}{x^{1/2}}u_2\right] \right) \exp\left[-r(t - t_0)\right] \right.$$
$$\left. + W_x^{(t_0)}(t, x)\left[ax^{1/2} - bx - u_1 - u_2\right] \right\}, \quad and \tag{8.32}$$

$$W^{(t_0)}(T, x) = 2\exp\left[-r(T - t_0)\right]qx^{\frac{1}{2}}.$$

Performing the indicated maximization we obtain

$$\psi_1^{(t_0)*}(t, x) = \frac{x}{4[c_1 + W_x^{(t_0)} \exp[r(t - t_0)]x^{1/2}]^2}, \quad \text{and}$$

$$\psi_2^{(t_0)*}(t, x) = \frac{x}{4[c_2 + W_x^{(t_0)} \exp[r(t - t_0)]x^{1/2}]^2}. \tag{8.33}$$

The maximized expected joint profit function can be derived as follows.

Proposition 8.2 *The maximized expected joint profit function is*

$$W^{(t_0)}(t, x) = \exp[-r(t - t_0)][A(t)x^{1/2} + C(t)],$$

$$\text{where } \dot{A}(t) = \left[r + \frac{\sigma^2}{8} + \frac{b}{2}\right]A(t) - \frac{1}{2[c_1 + A(t)/2]} - \frac{1}{2[c_2 + A(t)/2]}$$

$$+ \frac{c_1}{4[c_1 + A(t)/2]^2} + \frac{c_2}{4[c_2 + A(t)/2]^2} \tag{8.34}$$

$$+ \frac{A(t)}{8[c_1 + A(t)/2]^2} + \frac{A(t)}{8[c_2 + A(t)/2]^2},$$

$$\dot{C}(t) = rC(t) - \frac{a}{2}A(t), \qquad A(T) = 2q, \quad \text{and} \quad C(T) = 0.$$

Proof Upon substituting the optimal strategies in (8.33), $W^{(t_0)}(t, x)$ in (8.34), and the relevant derivatives $W_t^{(t_0)}(t, x)$, $W_x^{(t_0)}(t, x)$, and $W_{xx}^{(t_0)}(t, x)$ into (8.32) yields the results in Proposition 8.2.   □

The optimal cooperative controls can then be obtained as

$$\psi_1^*(t, x) = \frac{x}{4[c_1 + A(t)/2]^2}, \quad \text{and} \quad \psi_2^*(t, x) = \frac{x}{4[c_2 + A(t)/2]^2}. \tag{8.35}$$

Substituting these control strategies into (8.27) yields the dynamics of the state trajectory under cooperation

$$dx(s) = \left[ax(s)^{1/2} - bx(s) - \frac{x(s)}{4[c_1 + A(s)/2]^2} - \frac{x(s)}{4[c_2 + A(s)/2]^2}\right] ds$$

$$+ \sigma x(s) dz(s), \quad x(t_0) = x_0. \tag{8.36}$$

Solving (8.36) yields the optimal cooperative state trajectory as

$$x^*(s) = \varpi(t_0, s)^2 \left[x_0^{1/2} + \int_{t_0}^s \varpi^{-1}(t_0, t)H_1 dt\right]^2, \quad \text{for } s \in [t_0, T], \tag{8.37}$$

where

$$\varpi(t_0, s) = \exp\left[\int_{t_0}^{s}\left[H_2(\tau) - \frac{\sigma^2}{8}\right]dv + \int_{t_0}^{s}\frac{\sigma}{2}\,dz(v)\right],$$

$$H_1 = \frac{1}{2}a, \quad \text{and} \quad H_2(s) = -\left[\frac{1}{2}b + \frac{1}{8[c_1 + A(s)/2]^2} + \frac{1}{8[c_2 + A(s)/2]^2} + \frac{\sigma^2}{8}\right].$$

The cooperative control for the game $\Gamma_c(x_0, T - t_0)$ over the time interval $[t_0, T]$ along the optimal trajectory can be expressed as

$$\psi_1^*(t, x_t^*) = \frac{x_t^*}{4[c_1 + A(t)/2]^2}, \quad \text{and} \quad \psi_2^*(t, x_t^*) = \frac{x_t^*}{4[c_2 + A(t)/2]^2},$$

$$\text{for } t \in [t_0, T] \text{ and } x_t^* \in X_t^*. \tag{8.38}$$

8.4.2 Subgame Consistent Cooperative Extraction

The agreed-upon optimality principle requires the extractors to share the excess of the total expected cooperative payoff over the sum of individual noncooperative payoffs proportional to the agents' expected noncooperative payoffs. Therefore, the following imputation has to be satisfied.

Condition 8.2 An imputation

$$\xi^{(\tau)i}(\tau, x_\tau^*) = \frac{V^{(\tau)i}(\tau, x_\tau^*)}{\sum_{j=1}^{2} V^{(\tau)j}(\tau, x_\tau^*)}\,W^{(\tau)}(\tau, x_\tau^*)$$

$$= \frac{[A_i(\tau)(x_\tau^*)^{1/2} + C_i(\tau)]}{\sum_{j=1}^{2}[A_j(\tau)(x_\tau^*)^{1/2} + C_j(\tau)]}\,[A(\tau)(x_\tau^*)^{1/2} + C(\tau)] \tag{8.39}$$

is assigned to extractor i, for $i \in \{1, 2\}$ if $x_\tau^* \in X_\tau^*$ occurs at time $\tau \in [t_0, T]$.

Applying Theorem 8.2, a subgame consistent solution under the above optimal principle for the cooperative game $\Gamma_c(x_0, T - t_0)$ can be obtained as $P(x_0, T - t_0) = \{u(s, x_s^*) \text{ and } B(s, x_s^*) \text{ for } s \in [t_0, T] \text{ and } \xi^{(t_0)}(t_0, x_0)\}$ in which

(i) $u(s, x_s^*)$ for $s \in [t_0, T]$ is the set of group optimal strategies

$$\psi_1^*(s, x_s^*) = \frac{x_s^*}{4[c_1 + A(s)/2]^2}, \quad \text{and} \quad \psi_2^*(s, x_s^*) = \frac{x_s^*}{4[c_2 + A(s)/2]^2},$$

and

(ii) $B(s, x_s^*) = \{B_1(s, x_s^*), B_2(s, x_2^*)\}$ for $s \in [t_0, T]$ where

$$B_i\left(s, x_s^*\right) = -\left[\xi_t^{(s)i}\left(t, x_t^*\right)\big|_{t=s}\right] - \left[\xi_{x_s^*}^{(s)i}\left(s, x_s^*\right)\right]\left[a\left(x_s^*\right)^{1/2} - bx_s^*\right.$$

$$-\frac{x_s^*}{4[c_1 + A(s)/2]^2} - \frac{x_s^*}{4[c_2 + A(s)/2]^2}\bigg]$$

$$-\frac{1}{2}\sigma^2\left(x_s^*\right)^2\left[\xi_{x_s^* x_s^*}^{(s)i}\left(s, x_s^*\right)\right], \quad \text{for } i \in \{1, 2\}, \qquad (8.40)$$

where

$$\left[\xi_t^{(s)i}\left(t, x_t^*\right)\big|_{t=s}\right]$$

$$= \frac{[A_i(s)(x_s^*)^{1/2} + C_i(s)]}{\left(\sum_{j=1}^2 [A_j(s)(x_s^*)^{1/2} + C_j(s)]\right)}$$

$$\times \left\{\left[\dot{A}(s)\left(x_s^*\right)^{1/2} + \dot{C}(s)\right] - r\left[A(s)\left(x_s^*\right)^{1/2} + C(s)\right]\right\}$$

$$+ \frac{[A(s)(x_s^*)^{1/2} + B(s)]}{\left(\sum_{j=1}^2 [A_j(s)(x_s^*)^{1/2} + B_j(s)]\right)}$$

$$\times \left\{\left[\dot{A}_i(s)\left(x_s^*\right)^{1/2} + \dot{B}_i(s)\right] - r\left[A_i(s)\left(x_s^*\right)^{1/2} + B_i(s)\right]\right\}$$

$$- \frac{[A_i(s)(x_s^*)^{1/2} + B_i(s)][A(s)(x_s^*)^{1/2} + B(s)]}{\left(\sum_{j=1}^2 [A_j(s)(x_s^*)^{1/2} + B_j(s)]\right)^2}$$

$$\times \sum_{j=1}^2 \left\{\left[\dot{A}_j(s)\left(x_s^*\right)^{1/2} + \dot{C}_j(s)\right] - r\left[A_j(s)\left(x_s^*\right)^{1/2} + C_j(s)\right]\right\};$$

$$\left[\xi_{x_s^*}^{(s)i}\left(s, x_s^*\right)\right]$$

$$= \frac{[A_i(s)(x_s^*)^{1/2} + C_i(s)]A(s)(x_s^*)^{-1/2} + [A(s)(x_s^*)^{1/2} + C(s)]A_i(s)(x_s^*)^{-1/2}}{2\sum_{j=1}^2 [A_j(s)(x_s^*)^{1/2} + C_j(s)]}$$

$$- \frac{[A_i(s)(x_s^*)^{1/2} + C_i(s)][A(s)(x_s^*)^{1/2} + C(s)]}{\left(\sum_{j=1}^2 [A_j(s)(x_s^*)^{1/2} + C_j(s)]\right)^2}\left(\frac{1}{2}\sum_{j=1}^2 A_j(s)\left(x_s^*\right)^{-1/2}\right);$$

and

$$\left[\xi_{x_s^* x_s^*}^{(s)i}\left(s, x_s^*\right)\right]$$

$$= -\frac{C_i(s)A(s)(x_s^*)^{-3/2} + C(s)A_i(s)(x_s^*)^{-3/2}}{4\sum_{j=1}^2 [A_j(s)(x_s^*)^{1/2} + C_j(s)]}$$

$$- \frac{[A_i(s)(x_s^*)^{1/2} + C_i(s)]A(s)(x_s^*)^{-1/2} + [A(s)(x_s^*)^{1/2} + C(s)]A_i(s)(x_s^*)^{-1/2}}{(2\sum_{j=1}^{2}[A_j(s)(x_s^*)^{1/2} + C_j(s)])^2}$$

$$\times \sum_{j=1}^{2}[A_j(s)(x_s^*)^{-1/2}] + \frac{[A_i(s)(x_s^*)^{1/2} + C_i(s)][A(s)(x_s^*)^{1/2} + C(s)]}{(\sum_{j=1}^{2}[A_j(s)(x_s^*)^{1/2} + C_j(s)])^2}$$

$$\times \left(\frac{1}{4}\sum_{j=1}^{2}A_j(s)(x_s^*)^{-3/2}\right) - \left(\frac{1}{2}\sum_{j=1}^{2}A_j(s)(x_s^*)^{-1/2}\right)$$

$$\times \left[\frac{A_i(s)A(s) + \frac{1}{2}[A_i(s)C(s) + A(s)C_i(\tau)](x_s^*)^{-1/2}}{(\sum_{j=1}^{2}[A_j(s)(x_s^*)^{1/2} + C_j(s)])^2}\right.$$

$$\left. - \frac{[A_i(s)(x_s^*)^{1/2} + C_i(s)][A(s)(x_s^*)^{1/2} + C(s)]}{(\sum_{j=1}^{2}[A_j(s)(x_s^*)^{1/2} + C_j(s)])^3}\sum_{j=1}^{2}A_j(s)(x_s^*)^{-1/2}\right].$$

With the extractors using the cooperative strategies in (8.38), the instantaneous receipt of agent i at time instant τ is

$$\zeta_i(\tau, x_\tau^*) = \frac{(x_\tau^*)^{1/2}}{2[c_i + A(\tau)/2]} - \frac{c_i(x_\tau^*)^{1/2}}{4[c_i + A(\tau)/2]^2},$$

$$\text{for } \tau \in [t_0, T], x_\tau^* \in X_\tau^*, \text{ and } i \in \{1, 2\}. \tag{8.41}$$

Under cooperation the instantaneous payment that agent $i \in \{1, 2\}$ should receive $B_i(\tau, x_\tau^*)$ in (8.40). Hence an instantaneous transfer payment

$$\chi^i(\tau, x_\tau^*) = B_i(\tau, x_\tau^*) - \zeta_i(\tau, x_\tau^*) \tag{8.42}$$

has to be given to agent i at time τ, for $i \in \{1, 2\}$ and $\tau \in [t_0, T]$ when the state is $x_\tau^* \in X_\tau^*$.

8.5 Infinite-Horizon Consistent Economic Optimization Under Uncertainty

As discussed in Chap. 2, in many economic situations, the terminal time T is either very far in the future or unknown to the agents. A way to resolve the problem, as suggested by Dockner et al. (2000), is to set $T = \infty$. In this section, we examine infinite-horizon economic optimization and the corresponding subgame consistent solutions.

Consider again the infinite-horizon problem in which economic agent i seeks to

$$\max_{u_i} E_\tau \left\{ \int_\tau^\infty g^i[x(s), u_1(s), u_2(s), \ldots, u_n(s)] \exp[-r(s - \tau)]\, ds \right\},$$

$$\text{for } i \in N, \tag{8.43}$$

subject to the stochastic dynamics

$$dx(s) = f\big[x(s), u_1(s), u_2(s), \ldots, u_n(s)\big]ds + \sigma\big[x(s)\big]dz(s), \quad x(\tau) = x_\tau.$$

$$(8.44)$$

Consider the alternative game that starts at time $t \in [t_0, \infty)$ with initial state $x(t) = x$

$$\max_{u_i} E_t\left\{\int_t^\infty g^i\big[x(s), u_1(s), u_2(s), \ldots, u_n(s)\big]\exp\big[-r(s-t)\big]ds\right\},$$
$$\text{for } i \in N,$$

$$(8.45)$$

subject to the stochastic dynamics

$$dx(s) = f\big[x(s), u_1(s), u_2(s), \ldots, u_n(s)\big]ds + \sigma\big[x(s)\big]dz(s), \quad x(t) = x_t.$$

$$(8.46)$$

Let $\Omega[x(s)] = \sigma[x(s)]\sigma[x(s)]^{\mathsf{T}}$ denote the covariance matrix with its element in row h and column ζ denoted by $\Omega^{h\zeta}[x(s)]$.

The infinite-horizon autonomous game in (8.45) and (8.46) is independent of the choice of t and dependent only upon the state at the starting time, that is, x.

A Nash equilibrium solution for the infinite-horizon stochastic differential game in (8.45) and (8.45) is characterized in Theorem 2.6 of Chap. 2.

The game equilibrium strategies $\{\phi_i^*(x) \in U^i; i \in N\}$ and the agents value functions $\hat{V}^i(x)$ for $i \in N$ can be obtained if an equilibrium exists.

Now consider the case when the economic agents agree to act cooperatively. Let $\Gamma_c(\tau, x_\tau)$ denote a cooperative game in which agent i's payoff is (8.43) and the state dynamics is (8.44). The agents agree to act according to an agreed-upon optimality principle. As noted before, group optimality is an essential factor in cooperation and we let the agreed-upon optimality principle be as follows.

Principle PII It is an optimality principle that entails (i) group optimality, and (ii) the distribution of the total cooperative payoff according to an imputation that equals $\xi^{(\upsilon)}(\upsilon, x_\upsilon^*)$ for $\upsilon \in [\tau, \infty)$ over the game duration. Moreover, the function $\xi^{(\upsilon)i}(\upsilon, x_\upsilon^*)$, for $i \in N$, is continuously differentiable in υ and x_υ^*.

The solution $P(\tau, x_\tau)$ of the cooperative game $\Gamma_c(\tau, x_\tau)$ under optimality Principle PII includes the following.

(i) A set of cooperative strategies

$$u^{(\tau)*}\big(s, x_s^*\big) = \big[u_1^{(\tau)*}\big(s, x_s^*\big), u_2^{(\tau)*}\big(s, x_s^*\big), \ldots, u_n^{(\tau)*}\big(s, x_s^*\big)\big], \quad \text{for } s \in [\tau, \infty).$$

(ii) An imputation vector

$$\xi^{(\tau)}(\tau, x_\tau) = \big[\xi^{(\tau)1}(\tau, x_\tau), \xi^{(\tau)2}(\tau, x_\tau), \ldots, \xi^{(\tau)n}(\tau, x_\tau)\big]$$

to allocate the cooperative payoff to the agents.

(iii) A payoff distribution procedure $B^\tau(s, x_s^*) = [B_1^\tau(s, x_s^*), B_2^\tau(s, x_s^*), \ldots,$
$B_n^\tau(s, x_s^*)]$ for $s \in [\tau, \infty)$, where $B_i^\tau(s, x_s^*)$ is the instantaneous payments for
agent i at time s when the state is $x_s^* \in X_s^*$. In particular,

$$\xi^{(\tau)i}(\tau, x_\tau) = E_\tau \left\{ \int_\tau^\infty B_i^\tau(s, x_s^*) \exp[-r(s-\tau)] ds \right\}, \quad \text{for } i \in N. \quad (8.47)$$

The derivation of a subgame consistent solution satisfying optimality Principle PII
will be performed in Sects. 8.5.1 to 8.5.4.

8.5.1 Group Optimal Cooperative Strategies

To ensure group rationality the agents maximize the sum of their expected payoffs,
the agents solve the problem

$$\max_{u_1, u_2, \ldots, u_n} E_\tau \left\{ \int_\tau^\infty \sum_{j=1}^n g^j[x(s), u_1(s), u_2(s), \ldots, u_n(s)] \exp[-r(s-\tau)] ds \right\},$$
$$(8.48)$$

subject to (8.44).

Invoking Theorem A.6 in the Technical Appendixes, a set of controls $\{\psi_i^*(x) \in U^i; i \in N\}$ constitutes an optimal solution to the infinite-horizon stochastic control
problem in (8.44) and (8.48) if there exists continuously twice differentiable func-
tion $W(x)$ defined on $R^m \to R$, which satisfies the following equation:

$$rW(x) - \frac{1}{2} \sum_{h,\zeta=1}^m \Omega^{h\zeta}(x) W_{x^h x^\zeta}(x)$$

$$= \max_{u_1, u_2, \ldots, u_n} \left\{ \sum_{j=1}^n g^j[x, u_1, u_2, \ldots, u_n] + W_x(x) f[x, u_1, u_2, \ldots, u_n] \right\}. \quad (8.49)$$

Hence the agents will adopt the cooperative control $\{\psi_i^*(x), \text{ for } i \in N\}$ to obtain
the maximized level of the expected joint profit. Substituting this set of control into
(8.71) yields the dynamics of the optimal (cooperative) trajectory as

$$dx(s) = f[x(s), \psi_1^*(x(s)), \psi_2^*(x(s)), \ldots, \psi_n^*(x(s))] ds + \sigma[x(s)] dz(s),$$
$$x(\tau) = x_\tau. \quad (8.50)$$

The solution to (8.50) can be expressed as

$$x^*(s) = x_\tau + \int_\tau^s f[x^*(\upsilon), \psi_1^*(x^*(\upsilon)), \psi_2^*(x^*(\upsilon)), \ldots, \psi_n^*(x^*(\upsilon))] d\upsilon$$

$$+ \int_\tau^s \sigma[x^*(\upsilon)] dz(\upsilon). \quad (8.51)$$

We use X_s^* to denote the set of realizable values of $x^*(s)$ at time s generated by (8.50). The term $x_s^* \in X_s^*$ is used to denote an element in X_s^*. We use the terms $x^*(s)$ and x_s^* interchangeably in cases where there is no ambiguity.

The expected cooperative payoff can be expressed as

$$W(x_\tau^*) = E_\tau \left\{ \int_\tau^\infty \sum_{j=1}^n g^j \left[x^*(s), \psi_1^*(x^*(s)), \psi_2^*(x^*(s)), \dots, \psi_n^*(x^*(s)) \right] \right.$$

$$\left. \times \exp[-r(s-\tau)] \, ds \, \middle| \, x^*(\tau) = x_\tau^* \right\}.$$

Moreover, one can easily verify that the joint payoff maximizing controls for the cooperative game $\Gamma_c(\tau, x_\tau)$ over the time interval $[t, \infty)$ is identical to the joint payoff maximizing controls for the cooperative game $\Gamma_c(t, x_t^*)$ over the same time interval.

8.5.2 Subgame Consistent Imputation and Payoff Distribution Procedure

Let $P(\tau, x_\tau)$ denote the solution to the cooperative game $\Gamma_c(\tau, x_\tau)$ under the agreed-upon optimality Principle PII. According to $P(\tau, x_\tau)$, the economic agents would use the Payoff Distribution Procedure $\{B^\tau(s, x_s^*)\}_{s=\tau}^\infty$ to bring about an imputation to agent i as

$$\xi^{(\tau)i}(\tau, x_\tau) = E_\tau \left\{ \int_\tau^\infty B_i^\tau(s, x_s^*) \exp[-r(s-\tau)] \, ds \right\}, \quad \text{for } i \in N.$$

We define

$$\xi^{(\tau)i}(t, x_t^*) = E_\tau \left\{ \int_t^\infty B_i^\tau(s, x_s^*) \exp[-r(s-\tau)] \, ds \, \middle| \, x(t) = x_t^* \in X_t^* \right\},$$

$$\text{for } i \in N, \tag{8.52}$$

where $t > \tau$ and $x_t^* \in \{x^*(s)\}_{s=\tau}^\infty$.

According to $P(\tau, x_\tau)$, agent i is supposed to receive a payoff $\xi^{(\tau)i}(t, x_t^*)$ over the remaining time interval $[t, \infty)$ if the state is $x_t^* \in X_t^*$.

Consider the case when the game has proceeded to time t and the state variable became $x_t^* \in X_t^*$. Then one has a cooperative game $\Gamma_c(t, x_t^*)$, which starts at time t with initial state x_t^*. According to the solution $P(t, x_t^*)$, an imputation

$$\xi^{(t)i}(t, x_t^*) = E_t \left\{ \int_t^\infty B_i^t(s, x_s^*) \exp[-r(s-t)] \, ds \, \middle| \, x(t) = x_t^* \in X_t^* \right\},$$

will be allotted to agent i, for $i \in N$.

However, according to the solution $P(\tau, x_\tau)$, the imputation (in the present value viewed at time τ) to agent i over the period $[t, \infty)$ is (8.52).

For the imputation from $P(\tau, x_\tau)$ to be consistent with those from $P(t, x_t^*)$, it is essential that

$$\exp[r(t - \tau)]\xi^{(\tau)i}(t, x_t^*) = \xi^{(t)i}(t, x_t^*) \in P(t, x_t^*), \quad \text{for } t \in (\tau, \infty). \quad (8.53)$$

In addition, at time τ when the initial state is x_τ, according to the solution $P(\tau, x_\tau)$ generated by optimality Principle PII, the payoff distribution procedure is

$$B^\tau(s, x_s^*) = [B_1^\tau(s, x_s^*), B_2^\tau(s, x_s^*), \ldots, B_n^\tau(s, x_s^*)], \quad \text{for } s \in [\tau, \infty) \text{ and } x_s^* \in X_s^*.$$

When the game proceeds to time t the state variable becomes $x_t^* \in X_t^*$. According to the solution $P(t, x_t^*)$ generated by optimality Principle PII, the payoff distribution procedure

$$B^t(s, x_s^*) = [B_1^t(s, x_s^*), B_2^t(s, x_s^*), \ldots, B_n^t(s, x_s^*)], \quad \text{for } s \in [t, \infty) \text{ and } x_s^* \in X_s^*,$$

will be adopted.

For the continuation of the payoff distribution procedure $B^\tau(s, x_s^*) \in P(\tau, x_\tau)$ to be consistent with $B^t(s, x_t^*) \in P(t, x_t^*)$, it is required that

$$B^{t0}(s, x_s^*) = B^t(s, x_s^*), \quad \text{for } s \in [t, \infty) \text{ and } t \in [\tau, \infty) \text{ and } x_s^* \in X_s^*.$$

Therefore, we have the following definition.

Definition 8.2 The imputation and payoff distribution procedure $\{\xi^{(\tau)}(\tau, x_\tau)$ and $B^\tau(s, x_s^*)$ for $s \in [\tau, \infty)\} \in P(\tau, x_\tau)$ are subgame consistent if

(i)

$$\exp[r(t - \tau)]\xi^{(\tau)i}(t, x_t^*)$$

$$\equiv \exp[r(t - \tau)]E_\tau \left\{ \int_t^\infty B_i^\tau(s, x_s^*) \exp[-r(s - \tau)] ds \mid x(t) = x_t^* \in X_t^* \right\}$$

$$= \xi^{(t)i}(t, x_t^*) \in P(t, x_t^*), \quad \text{for } t \in (\tau, \infty) \text{ and } i \in N; \quad (8.54)$$

and

(ii) the payoff distribution procedure $B^\tau(s, x_s^*)$ for $s \in [t, \infty)$ is identical to $B^t(s, x_s^*) \in P(t, x_t^*)$.

Thus a payoff distribution procedure leading to a subgame consistent imputation has to satisfy Definition 8.2.

8.5.3 Payoff Distribution Procedure Leading to Subgame Consistency

To derive a payoff distribution procedure leading to a subgame consistent imputation we invoke Definition 8.2 and obtain $B_i^\tau(s, x_s^*) = B_i^t(s, x_s^*) = B_i(s, x_s^*)$, for $s \in [\tau, \infty), x_s^* \in X_s^*$ and $t \in [\tau, \infty)$ and $i \in N$.

Therefore, along the cooperative trajectory,

$$\xi^{(\tau)i}(\tau, x_\tau) = E_\tau \left\{ \int_\tau^\infty B_i(s, x_s^*) \exp[-r(s - \tau)] \, ds \right\}, \quad \text{for } i \in N, \quad \text{and}$$

$$\xi^{(\upsilon)i}(\upsilon, x_\upsilon^*) = E_\upsilon \left\{ \int_\upsilon^\infty B_i(s, x_s^*) \exp[-r(s - \upsilon)] \, ds \, \bigg| \, x(\upsilon) = x_\upsilon^* \in X_\upsilon^* \right\},$$

for $i \in N$, and

$$\xi^{(t)i}(t, x_t^*) = E_t \left\{ \int_t^\infty B_i(s, x_s^*) \exp[-r(s - t)] \, ds \, \bigg| \, x(t) = x_t^* \in X_t^* \right\}, \quad (8.55)$$

for $i \in N$ and $t \geq \upsilon \geq \tau$.

Moreover, for $i \in N$ and $t \in [\tau, \infty)$, we define the term

$$\xi^{(\upsilon)i}(t, x_t^*) = E_\upsilon \left\{ \left(\int_t^\infty B_i(s, x_s^*) \exp[-r(s - \upsilon)] \, ds \right) \bigg| \, x(t) = x_t^* \right\}, \quad (8.56)$$

to denote the present value of agent i's cooperative payoff over the time interval $[t, \infty)$, given that the state is x_t^* at time $t \in [\upsilon, \infty]$, under the solution $P(\upsilon, x_\upsilon^*)$.

Invoking (8.55) and (8.56) one can readily verify that

$$\exp[r(t - \tau)] \xi^{(\tau)i}(t, x_t^*) = \xi^{(t)i}(t, x_t^*), \quad \text{for } i \in N \text{ and } \tau \in [t_0, T] \text{ and } t \in [\tau_0, T].$$

The next task is to derive $B_i(s, x_s^*)$, for $s \in [\tau, \infty)$ and $t \in [\tau, \infty)$ so that (8.55) can be realized. Consider again the following condition.

Condition 8.3 For $i \in N$ and $t \geq \upsilon$ and $\upsilon \in [\tau, T]$, the term $\xi^{(\upsilon)i}(t, x_t^*)$ is a function that is continuously differentiable in t and x_t^*.

Lemma 8.1 *If Condition 8.3 is satisfied, a PDP with instantaneous payments at time s with the state being $x_s^* \in X_s^*$ equaling*

$$B_i(s, x_s^*) = -[\xi_t^{(s)i}(t, x_t^*)|_{t=s}]$$
$$- [\xi_{x_t^*}^{(s)i}(t, x_t^*)|_{t=s}] f[x_s^*, \psi_1^*(s, x_s^*), \psi_2^*(s, x_s^*), \dots, \psi_n^*(s, x_s^*)]$$
$$- \frac{1}{2} \sum_{h,\zeta=1}^m \Omega^{h\zeta}(x_s^*) [\xi_{x_t^h x_t^\zeta}^{(s)i}(t, x_t^*)|_{t=s}],$$

for $i \in N$ and $s \in [\upsilon, \infty)$, \hfill (8.57)

yields imputation $\xi^{(\upsilon)i}(\upsilon, x_\upsilon^*)$ for $\upsilon \in [\tau, \infty)$ and $x_\upsilon^* \in X_\upsilon^*$, which satisfies (8.55).

Proof Note that along the cooperative trajectory

$$\xi^{(v)i}(t, x_t^*) = E_v\left\{\int_t^\infty B_i(s, x_s^*)\exp[-r(s-v)]\,ds \,\middle|\, x(t) = x_t^* \in X_t^*\right\}$$

$$= \exp[-r(t-v)]\xi^{(t)i}(t, x_t^*), \quad \text{for } i \in N \text{ and } t \in [v, \infty). \quad (8.58)$$

For $\Delta t \to 0$, (8.55) can be expressed as

$$\xi^{(v)i}(v, x_v^*) = E_v\left\{\int_v^\infty B_i(s, x_s^*)\exp[-r(s-v)]\,ds\right\}$$

$$= E_v\left\{\int_v^{v+\Delta t} B_i(s, x_s^*)\exp[-r(s-v)]\,ds\right.$$

$$\left. + \xi^{(v)i}(v + \Delta t, x_v^* + \Delta x_v^*)\right\}, \quad (8.59)$$

where

$$\Delta x_v^* = f[x_v^*, \psi_1^*(x_v^*), \psi_2^*(x_v^*), \ldots, \psi_n^*(x_v^*)]\Delta t + \sigma(x_v^*)\Delta z_v + o(\Delta t),$$

$$\Delta z_v = Z(v + \Delta t) - z(v), \quad \text{and} \quad E_v[o(\Delta t)]/\Delta t \to 0 \quad \text{as } \Delta t \to 0.$$

Replacing the term $x_v^* + \Delta x_v^*$ with $x_{v+\Delta t}^*$ and rearranging (8.59) yields

$$E_v\left\{\int_v^{v+\Delta t} B_i(s)\exp[-r(s-v)]\,ds\right\}$$

$$= E_v\left\{\xi^{(v)i}(v, x_v^*) - \xi^{(v)i}(v + \Delta t, x_{v+\Delta t}^*)\right\},$$

$$\text{for all } v \in [\tau, \infty) \text{ and } i \in N. \quad (8.60)$$

With Condition 8.3 holding and $\Delta t \to 0$, (8.60) can be expressed as

$$E_v\left\{B_i(s, x_s^*)\Delta t + o(\Delta t)\right\}$$

$$= E_v\left\{-[\xi_t^{(s)i}(t, x_t^*)|_{t=s}]\Delta t\right.$$

$$- [\xi_{x_t^*}^{(s)i}(t, x_t^*)|_{t=s}]f[x_s^*, \psi_1^*(s, x_s^*), \psi_2^*(s, x_s^*), \ldots, \psi_n^*(s, x_s^*)]\Delta t$$

$$- \frac{1}{2}\sum_{h,\zeta=1}^m \Omega^{h\zeta}(x_s^*)[\xi_{x_t^h x_t^\zeta}^{(s)i}(t, x_t^*)|_{t=s}]\Delta t$$

$$\left. - [\xi_{x_t^*}^{(s)i}(t, x_t^*)|_{t=s}]\sigma(x_v^*)\Delta z_v - o(\Delta t)\right\}. \quad (8.61)$$

Dividing (8.61) throughout by Δt, with $\Delta t \to 0$ and taking the expectation, yields (8.57). Thus the payoff distribution procedure in $B_i(v, x_v^*)$ in (8.57) would lead to the realization of the imputations that satisfy (8.55). □

Since the payoff distribution procedure in $B_i(\tau)$ in (8.57) leads to the realization of (8.55), it would yield subgame consistent imputations satisfying Definition 8.2.

A more succinct form of Lemma 8.1 can be derived as follows.

If Condition 8.3 is satisfied, a PDP with instantaneous payments at time s equaling

$$B_i\left(s, x_s^*\right) = r\xi^{(s)i}\left(s, x_s^*\right) - \xi_{x_s^*}^{(s)i}\left(s, x_s^*\right) f\left[x_s^*, \psi_1^*\left(x_s^*\right), \psi_2^*\left(x_s^*\right), \ldots, \psi_n^*\left(x_s^*\right)\right]$$

$$-\frac{1}{2}\sum_{h,\zeta=1}^{m} \Omega^{h\zeta}\left(x_s^*\right)\left[\xi_{x_t^h x_t^\zeta}^{(s)i}\left(t, x_t^*\right)\big|_{t=s}\right],$$

for $i \in N$, $x_s^* \in X_s^*$ and $s \in [v, \infty)$, $\quad\quad\quad\quad\quad\quad\quad\quad\quad$ (8.62)

yields the imputation $\xi^{(v)i}\left(v, x_v^c\right)$, for $v \in [\tau, \infty)$ which satisfies (8.55).

To demonstrate that (8.62) is an alternative form for (8.57) in Lemma 8.1, we first define

$$\hat{\xi}^i\left(x_v^*\right) = E_v\left\{\int_v^\infty B_i(s)\exp\left[-r(s-v)\right]ds \,\bigg|\, x(v) = x_v^*\right\} = \xi^{(v)i}\left(\tau, x_v^*\right),\quad \text{and}$$

$$\hat{\xi}^i\left(x_t^*\right) = E_t\left\{\int_t^\infty B_i(s)\exp\left[-r(s-t)\right]ds \,\bigg|\, x(t) = x_t^*\right\} = \xi^{(t)i}\left(t, x_t^*\right),$$

for $i \in N$, and $v \in [\tau, \infty)$ and $t \in [v, \infty)$ along the optimal cooperative trajectory $\{x_s^*\}_{s=\tau}^\infty$.

We then have

$$\xi^{(v)i}\left(t, x_t^*\right) = \exp\left[-r(t-v)\right]\hat{\xi}^i\left(x_t^*\right).$$

Differentiating the above condition with respect to t yields

$$\left[\xi_t^{(v)i}\left(t, x_t^*\right)\big|_{t=v}\right] = -r\exp\left[-r(t-v)\right]\hat{\xi}^i\left(x_t^*\right) = -r\xi^{(v)i}\left(t, x_t^*\right).$$

At $t = v$, $\xi^{(v)i}(t, x_t^*) = \xi^{(v)i}(v, x_v^*)$, therefore,

$$\left[\xi_t^{(v)i}\left(t, x_t^*\right)\big|_{t=v}\right] = r\xi^{(v)i}\left(t, x_t^*\right) = r\xi^{(v)i}\left(v, x_v^*\right). \quad\quad\quad (8.63)$$

Substituting (8.63) into (8.57) yields (8.62).

Using (8.62), a subgame consistent solution in an infinite-horizon framework is characterized in the section below.

8.5.4 Subgame Consistent Solution

A theorem characterizing a subgame consistent solution $P(\tau, x_\tau)$ for the cooperative game $\Gamma_c(\tau, x_\tau)$ under optimality Principle PII is presented below.

Theorem 8.3 *For the cooperative game* $\Gamma_c(\tau, x_\tau)$ *with optimality Principle* PII *the solution* $P(\tau, x_\tau) = \{u(s, x_s^*) \text{ and } B(s, x_s^*) \text{ for } s \in [\tau, \infty) \text{ and } \xi^{(\tau)}(\tau, x_\tau)\}$—*in which*

(i) $u(s, x_s^*)$ *for* $s \in [\tau, \infty)$ *is the set of group optimal strategies* $\psi^*(x_s^*)$ *for the game* $\Gamma_c(\tau, x_\tau)$, *and*

(ii) *the imputation distribution procedure*

$$B(s, x_s^*) = \{B_1(s, x_s^*), B_2(s, x_s^*), \ldots, B_n(s, x_s^*)\} \quad \text{for } s \in [\tau, \infty)$$

where

$$B_i(s, x_s^*) = r\xi^{(s)i}(s, x_s^*) - \xi_{x_s^*}^{(s)i}(s, x_s^*) f[x_s^*, \psi_1^*(x_s^*), \psi_2^*(x_s^*), \ldots, \psi_n^*(x_s^*)]$$

$$- \frac{1}{2} \sum_{h,\zeta=1}^{m} \Omega^{h\zeta}(x_s^*) [\xi_{x_t^h x_t^\zeta}^{(s)i}(t, x_t^*)|_{t=s}], \quad \text{for } i \in N, \tag{8.64}$$

and $\xi^{(s)}(s, x_s^*) = [\xi^{(s)1}(s, x_s^*), \xi^{(s)2}(s, x_s^*), \ldots, \xi^{(s)n}(s, x_s^*)] \in P(s, x_s^*)$ *is the imputation at time* $s \in [\tau, \infty)$ *with the state being* $x_s^* \in \{x^*(t)\}_{t \geq \tau}$ *under optimality Principle* PII—*is subgame consistent.*

Proof Following the algorithm that specifies $P(\tau, x_\tau)$ as the solution to the game $\Gamma_c(\tau, x_\tau)$ one can readily obtain the solution of the cooperative game $\Gamma_c(\upsilon, x_\upsilon^*)$, for $\upsilon > \tau$, as $P(\upsilon, x_\upsilon^*) = \{u(s, x_s^*) \text{ and } B(s, x_s^*) \text{ for } s \in [\upsilon, \infty) \text{ and } \xi^{(\upsilon)}(\upsilon, x_\upsilon^*)\}$ in which

(i) $u(s, x_s^*)$ *for* $s \in [\upsilon, \infty)$ *is the set of group optimal strategies* $\psi^*(x_s^*)$ *for the game* $\Gamma_c(\upsilon, x_\upsilon^*)$, *and*

(ii)

$$B(s, x_s^*) = \{B_1(s, x_s^*), B_2(s, x_s^*), \ldots, B_n(s, x_s^*)\} \quad \text{for } s \in [\upsilon, \infty)$$

where

$$B_i(s, x_s^*) = r\xi^{(s)i}(s, x_s^*) - \xi_{x_s^*}^{(s)i}(s, x_s^*) f[x_s^*, \psi_1^*(x_s^*), \psi_2^*(x_s^*), \ldots, \psi_n^*(x_s^*)]$$

$$- \frac{1}{2} \sum_{h,\zeta=1}^{m} \Omega^{h\zeta}(x_s^*) [\xi_{x_t^h x_t^\zeta}^{(s)i}(t, x_t^*)|_{t=s}], \quad \text{for } i \in N, \tag{8.65}$$

and $\xi^{(s)}(s, x_s^*) = [\xi^{(s)1}(s, x_s^*), \xi^{(s)2}(s, x_s^*), \ldots, \xi^{(s)n}(s, x_s^*)] \in P(s, x_s^*)$ *is the imputation at time* $s \in [\upsilon, \infty)$ *with the state being* $x_s^* \in \{x^*(t)\}_{t \geq \upsilon}$.

Using the characterization of optimal control strategies in (8.49), one can show that the group optimal joint expected payoff maximizing strategies $\psi^*(x_s^*)$ for the cooperative game $\Gamma_c(\tau, x_\tau)$ over the time interval $[\upsilon, \infty)$ is identical to the joint payoff maximizing strategies controls for the cooperative game $\Gamma_c(\upsilon, x_\upsilon^*)$ over the same time interval.

Comparing (8.64) and (8.65) one can show that the payoff distribution procedure $B(s, x_s^*)$ for the cooperative game $\Gamma_c(\tau, x_\tau)$ over the time interval $[\upsilon, \infty)$ is identical to the payoff distribution procedure $B(s, x_s^*)$ for the cooperative game $\Gamma_c(\upsilon, x_\upsilon^*)$ over the same time interval.

Invoking Lemma 8.1 and (8.62), one can show that the payoff distribution procedure $B(s, x_s^*) = \{B_1(s, x_s^*), B_2(s), \ldots, B_n(s, x_s^*)\}$ in (8.64) will yield

$$\xi^{(\upsilon)i}\left(\upsilon, x_\upsilon^*\right) = E_\upsilon\left\{\int_\upsilon^\infty B_i\left(s, x_s^*\right)\exp\left[-r(s-\upsilon)\right]ds\right\} \in P\left(\upsilon, x_\upsilon^*\right),$$

for $i \in N$, and $\upsilon \in [\tau, \infty)$.

Hence

$$\exp\left[r(\upsilon - \tau)\right]\xi^{(\tau)i}\left(\upsilon, x_\upsilon^*\right)$$

$$\equiv \exp\left[r(\upsilon - \tau)\right]E_\tau\left\{\int_\upsilon^\infty B_i(s)\exp\left[-r(s-\tau)\right]ds \,\middle|\, x(\upsilon) = x_\upsilon^*\right\}$$

$$= \xi^{(\upsilon)i}\left(\upsilon, x_\upsilon^*\right)P\left(\upsilon, x_\upsilon^*\right), \quad \text{for } i \in N \text{ and } \upsilon \in [\tau, \infty).$$

In sum, the continuation of the solution $P(\tau, x_\tau)$ over the time interval $[\upsilon, \infty)$ is consistent with the solution $P(\upsilon, x_\upsilon^*)$ of the game $\Gamma_c(\upsilon, x_\upsilon^*)$ under optimality Principle PII. Thus the solution $P(\tau, x_\tau)$ in Theorem 8.3 is indeed subgame consistent. $\square$

With agents using the cooperative strategies $\{\psi_i^*(x_\upsilon^*), \text{ for } i \in N \text{ and } \upsilon \in [\tau, \infty)\}$, the instantaneous receipt of agent i at time instant υ is

$$\zeta_i\left(\upsilon, x_\upsilon^*\right) = g^i\left[x_\upsilon^*, \psi_1^*\left(x_\upsilon^*\right), \psi_2^*\left(x_\upsilon^*\right), \ldots, \psi_n^*\left(x_\upsilon^*\right)\right], \quad \text{for } i \in N, \qquad (8.66)$$

when the state is $x_\upsilon^* \in X_\upsilon^*$.

According to Theorem 8.3, the instantaneous payment that agent i should receive under the agreed-upon optimality principle is $B_i(\upsilon, x_\upsilon^*)$ as stated in (8.64). Hence an instantaneous transfer payment

$$\chi^i\left(\upsilon, x_\upsilon^*\right) = B_i\left(\upsilon, x_\upsilon^*\right) - \zeta_i\left(\upsilon, x_\upsilon^*\right), \quad \text{for } i \in N, \qquad (8.67)$$

has to be given to agent i at time υ when the state is $x_\upsilon^* \in X_\upsilon^*$.

8.6 Infinite-Horizon Cooperative Fishery Under Uncertainty

Consider an infinite-horizon version of the cooperative fishery in Sect. 8.5. At time τ, the expected payoff function of extractors 1 and 2 are, respectively,

$$E_\tau\left\{\int_\tau^\infty\left[u_1(s)^{1/2} - \frac{c_1}{x(s)^{1/2}}u_1(s)\right]\exp[-r(t-\tau)]\,ds\right\},$$

and

$$E_\tau\left\{\int_\tau^\infty\left[u_2(s)^{1/2} - \frac{c_2}{x(s)^{1/2}}u_2(s)\right]\exp[-r(t-\tau)]\,ds\right\}. \tag{8.68}$$

The fish resource stock $x(s) \in X \subset R$ follows the stochastic dynamics

$$dx(s) = \left[ax(s)^{1/2} - bx(s) - u_1(s) - u_2(s)\right]ds + \sigma x(s)\,dz(s), \quad x(\tau) = x_\tau. \tag{8.69}$$

Invoking Theorem 2.6 in Chap. 2, we let $[\phi_1^*(x), \phi_2^*(x)]$ for $t \in [t_0, T]$ denote a set of strategies that provides a feedback Nash equilibrium solution to the game in (8.68) and (8.69) can be characterized by

$$r\hat{V}^i(x) - \frac{1}{2}\sigma^2 x^2 \hat{V}_{xx}^i(x)$$

$$= \max_{u_i}\left\{u_i^{1/2} - \frac{c_i}{x^{1/2}}u_i + \hat{V}_x^i(x)\left[ax^{1/2} - bx - u_i - \phi_j^*(x)\right]\right\},$$

for $i, j \in \{1, 2\}$ and $i \neq j$. \qquad (8.70)

Performing the indicated maximization in (8.70) yields

$$\phi_i^*(x) = \frac{x}{4[c_i + \hat{V}_x^i(x)x^{1/2}]^2}, \quad \text{for } i \in \{1, 2\}.$$

Substituting $\phi_1^*(x)$ and $\phi_2^*(x)$ above into (8.70) and upon solving (8.70) one obtains the value function of agent $i \in \{1, 2\}$ as

$$\hat{V}^i(t, x) = \left[A_i x^{1/2} + C_i\right], \tag{8.71}$$

where for $i, j \in \{1, 2\}$ and $i \neq j$, A_i, C_i, A_j, and C_j satisfy

$$\left[r + \frac{\sigma^2}{8} + \frac{b}{2}\right]A_i - \frac{1}{2[c_i + A_i/2]} + \frac{c_i}{4[c_i + A_i/2]^2}$$

$$+ \frac{A_i}{8[c_i + A_i/2]^2} + \frac{A_i}{8[c_j + A_j/2]^2} = 0, \quad \text{and}$$

$$C_i = \frac{a}{2}A_i.$$

The game equilibrium strategies can be obtained as

$$\phi_1^*(x) = \frac{x}{4[c_1 + A_1/2]^2}, \quad \text{and} \quad \phi_2^*(x) = \frac{x}{4[c_2 + A_2/2]^2}.$$

8.6.1 Cooperative Extraction

Consider the case when these two nations agree to act according to an agreed-upon optimality principle which entails (i) group optimality, and (ii) the distribution of the excess of the total expected cooperative payoff over the sum of expected individual noncooperative payoffs proportional to the agents' expected noncooperative payoffs.

To maximize their joint expected payoff for group optimality, the nations have to solve the stochastic control problem of maximizing

$$E_t \left\{ \int_t^\infty \left(\left[u_1(s)^{1/2} - \frac{c_1}{x(s)^{1/2}} u_1(s) \right] + \left[u_2(s)^{1/2} - \frac{c_2}{x(s)^{1/2}} u_2(s) \right] \right) \right.$$
$$\left. \times \exp\left[-r(t-t) \right] ds \right\}, \tag{8.72}$$

subject to (8.69).

Invoking Theorem A.6 in the Technical Appendixes yields the characterization of the solution of the problem in (8.69) and (8.72) as follows.

Corollary 8.3 *A set of controls $\{\psi_i^*(x), \text{ for } i \in \{1, 2\}\}$ constitutes an optimal solution to the stochastic control problem in (8.69) and (8.72), if there exist continuously twice differentiable functions $W(x) : R^m \to R$, satisfying the following partial differential equation*

$$rW(x) - \frac{1}{2}\sigma^2 x^2 W_{xx}(x) = \max_{u_1, u_2} \left\{ \left(\left[u_1^{1/2} - \frac{c_1}{x^{1/2}} u_1 \right] + \left[u_2^{1/2} - \frac{c_2}{x^{1/2}} u_2 \right] \right) \right.$$
$$\left. + W_x(x) \left[ax^{1/2} - bx - u_1 - u_2 \right] \right\}. \tag{8.73}$$

Performing the indicated maximization in (8.73) we obtain

$$\psi_1^*(x) = \frac{x}{4[c_1 + W_x(x)x^{1/2}]^2}, \quad \text{and}$$
$$\psi_2^*(x) = \frac{x}{4[c_2 + W_x(x)x^{1/2}]^2}. \tag{8.74}$$

The maximized expected joint profit function can be derived as follows.

Proposition 8.3 *The maximized expected joint profit function is*

$$W(x) = [Ax^{1/2} + C],$$ (8.75)

where

$$\left[r + \frac{\sigma^2}{8} + \frac{b}{2}\right]A - \frac{1}{2[c_1 + A/2]} - \frac{1}{2[c_2 + A/2]}$$

$$+ \frac{c_1}{4[c_1 + A/2]^2} + \frac{c_2}{4[c_2 + A/2]^2} + \frac{A}{8[c_1 + A/2]^2} + \frac{A}{8[c_2 + A/2]^2} = 0, \quad and$$

$$C = \frac{a}{2r}A.$$

Proof Substituting the optimal strategies in (8.74), $W(x)$ in (8.75), and the relevant derivatives $W_x(x)$ and $W_{xx}(x)$ into (8.73) yields the results in Proposition 8.3. □

The optimal cooperative controls can then be obtained as

$$\psi_1^*(x) = \frac{x}{4[c_1 + A/2]^2}, \quad and \quad \psi_2^*(x) = \frac{x}{4[c_2 + A/2]^2}.$$ (8.76)

Substituting these control strategies into (8.69) yields the dynamics of the state trajectory under cooperation

$$dx(s) = \left[ax(s)^{1/2} - bx(s) - \frac{x(s)}{4[c_1 + A/2]^2} - \frac{x(s)}{4[c_2 + A/2]^2}\right]ds + \sigma x(s)\,dz(s),$$

$$x(t_0) = x_0.$$ (8.77)

Solving (8.77) yields the optimal cooperative state trajectory as

$$x^*(s) = \varpi(t_0, s)^2 \left[x_0^{1/2} + \int_{t_0}^s \varpi^{-1}(t_0, t)H_1\,dt\right]^2, \quad \text{for } s \in [t_0, T], \quad (8.78)$$

where

$$\varpi(t_0, s) = \exp\left[\int_{t_0}^s \left[H_2(\tau) - \frac{\sigma^2}{8}\right]dv + \int_{t_0}^s \frac{\sigma}{2}\,dz(v)\right],$$

$$H_1 = \frac{1}{2}a, \quad and \quad H_2(s) = -\left[\frac{1}{2}b + \frac{1}{8[c_1 + A(s)/2]^2} + \frac{1}{8[c_2 + A(s)/2]^2} + \frac{\sigma^2}{8}\right].$$

The cooperative control for the game can be expressed as

$$\psi_1^*(x_t^*) = \frac{x_t^*}{4[c_1 + A/2]^2}, \quad and \quad \psi_2^*(x_t^*) = \frac{x_t^*}{4[c_2 + A/2]^2},$$

$$\text{for } x_t^* \in X_t^*.$$ (8.79)

8.6.2 Subgame Consistent Payoff Distribution

With the extractors using the cooperative strategies in (8.79) along the stochastic cooperative path, they agree to share the excess of the total expected cooperative payoff over the sum of individual noncooperative payoffs proportional to the agents' expected noncooperative payoffs. Therefore, the following imputation has to be satisfied.

Condition 8.4 An imputation

$$\xi^{(\upsilon)i}\left(\upsilon, x_{\upsilon}^{*}\right) = \frac{\hat{V}^{i}(x_{\upsilon}^{*})}{\sum_{j=1}^{2}\hat{V}^{j}(x_{\upsilon}^{*})} W\left(x_{\upsilon}^{*}\right) = \frac{[A_{i}(x_{\upsilon}^{*})^{1/2}+C_{i}]}{\sum_{j=1}^{2}[A_{j}(x_{\upsilon}^{*})^{1/2}+C_{j}]}\left[A\left(x_{\upsilon}^{*}\right)^{1/2}+C\right]$$

(8.80)

is assigned to extractor i, for $i \in \{1, 2\}$ if $x_{\upsilon}^{*} \in X_{\upsilon}^{*}$ occurs at time $\upsilon \in [\tau, \infty)$.

Applying Theorem 8.3, a subgame consistent solution for the cooperative game $\Gamma_{c}(\tau, x_{\tau})$ can be obtained as

$$P(\tau, x_{\tau}) = \left\{u\left(s, x_{s}^{*}\right) \text{ and } B\left(s, x_{s}^{*}\right) \text{ for } s \in [\tau, \infty) \text{ and } \xi^{(\tau)}(\tau, x_{\tau})\right\}$$

in which

(i) $u(s, x_{s}^{*})$ for $s \in [\tau, \infty)$ is the set of group optimal strategies

$$\psi_{1}^{*}\left(x_{s}^{*}\right) = \frac{x_{s}^{*}}{4[c_{1}+A/2]^{2}} \quad \text{and} \quad \psi_{2}^{*}\left(x_{s}^{*}\right) = \frac{x_{s}^{*}}{4[c_{2}+A/2]^{2}};$$

and

(ii) $B(s, x_{s}^{*}) = \{B_{1}(s, x_{s}^{*}), B_{2}(s, x_{s}^{*}), \ldots, B_{n}(s, x_{s}^{*})\}$ for $s \in [\tau, \infty)$ with

$$B_{i}\left(s, x_{s}^{*}\right) = r\xi^{(s)i}\left(s, x_{s}^{*}\right)$$

$$- \xi_{x_{s}^{*}}^{(s)i}\left(s, x_{s}^{*}\right)\left[a\left(x_{s}^{*}\right)^{1/2} - bx_{s}^{*} - \frac{x_{s}^{*}}{4[c_{1}+A/2]^{2}} - \frac{x_{s}^{*}}{4[c_{2}+A/2]^{2}}\right]$$

$$- \frac{1}{2}\sigma^{2}\left(x_{s}^{*}\right)^{2}\xi_{x_{s}^{h}x_{s}^{\zeta}}^{(\tau)i}\left(s, x_{s}^{*}\right), \quad \text{for } i \in \{1, 2\},$$

where

$$\xi_{x_{s}^{*}}^{(s)i}\left(s, x_{s}^{*}\right) = \frac{[A_{i}(x_{s}^{*})^{1/2}+C_{i}]A(x_{s}^{*})^{-1/2}+[A(x_{s}^{*})^{1/2}+C]A_{i}(x_{s}^{*})^{-1/2}}{2\sum_{j=1}^{2}[A_{j}(x_{s}^{*})^{1/2}+C_{j}]}$$

$$- \frac{[A_{i}(x_{s}^{*})^{1/2}+C_{i}][A(x_{s}^{*})^{1/2}+C]}{(\sum_{j=1}^{2}[A_{j}(x_{s}^{*})^{1/2}+C_{j}])^{2}}\left(\frac{1}{2}\sum_{j=1}^{2}A_{j}\left(x_{s}^{*}\right)^{-1/2}\right);$$

and

$$
\xi_{x_s^* x_s^*}^{(\tau)i}\left(s, x_s^*\right) = -\frac{C_i A(x_s^*)^{-3/2} + C A_i(x_s^*)^{-3/2}}{4\sum_{j=1}^{2}[A_j(x_s^*)^{1/2} + C_j]}
$$

$$
-\frac{[A_i(x_s^*)^{1/2} + C_i]A(x_s^*)^{-1/2} + [A(x_s^*)^{1/2} + C]A_i(x_s^*)^{-1/2}}{(2\sum_{j=1}^{2}[A_j(x_s^*)^{1/2} + C_j])^2}
$$

$$
\times \sum_{j=1}^{2}[A_j(x_s^*)^{-1/2}]
$$

$$
+\frac{[A_i(x_s^*)^{1/2} + C_i][A(x_s^*)^{1/2} + C]}{(\sum_{j=1}^{2}[A_j(x_s^*)^{1/2} + C_j])^2}\left(\frac{1}{4}\sum_{j=1}^{2}A_j(x_s^*)^{-3/2}\right)
$$

$$
-\left(\frac{1}{2}\sum_{j=1}^{2}A_j(x_s^*)^{-1/2}\right)
$$

$$
\times \left[\frac{A_i A + \frac{1}{2}[A_i C + AC_i](x_s^*)^{-1/2}}{(\sum_{j=1}^{2}[A_j(x_s^*)^{1/2} + C_j])^2}\right.
$$

$$
\left.-\frac{[A_i(x_\tau^*)^{1/2} + C_i][A(x_\tau^*)^{1/2} + C]}{(\sum_{j=1}^{2}[A_j(x_\tau^*)^{1/2} + C_j])^3}\sum_{j=1}^{2}A_j(x_\tau^*)^{-1/2}\right]. \quad (8.81)
$$

With extractors using the cooperative strategies in (8.79), the instantaneous receipt of agent i at time instant $\upsilon \in [\tau, \infty)$ is

$$
\zeta_i\left(\upsilon, x_\upsilon^*\right) = \frac{(x_\upsilon^*)^{1/2}}{2[c_i + A/2]} - \frac{c_i(x_\upsilon^*)^{1/2}}{4[c_i + A/2]^2}, \quad \text{for } i \in \{1, 2\}, \quad (8.82)
$$

if $x_\upsilon^* \in X_\upsilon^*$ occurs.

Under the cooperative agreement, the instantaneous payment that agent $i \in \{1, 2\}$ should receive under the agreed-upon optimality principle is $B_i(\upsilon, x_\upsilon^*)$ in (8.81). Hence an instantaneous transfer payment

$$
\chi^i\left(\upsilon, x_\upsilon^*\right) = B_i\left(\upsilon, x_\upsilon^*\right) - \zeta_i\left(\upsilon, x_\upsilon^*\right), \quad (8.83)
$$

has to be given to agent i at time υ, for $i \in \{1, 2\}$ and $x_\upsilon^* \in X_\upsilon^*$.

8.7 Exercises

8.1 Consider the case of two nations harvesting fish in common waters. The growth rate of the fish biomass is subject to stochastic shocks and follows the differential equation

$$dx(s) = \left[8x(s)^{1/2} - x(s) - u_1(s) - u_2(s)\right]ds + 0.05x(s)\,dz(s), \quad x(0) = 100,$$

where $z(s)$ is a Wiener process, $x(s)$ is the fish stock, and $u_i(s)$ is the amount of fish harvested by nation i, for $i \in \{1, 2\}$. The horizon of the game is $[0, 3]$.

The harvesting cost for nation $i \in \{1, 2\}$ depends on the quantity of the resource extracted $u_i(s)$ and the resource stock size $x(s)$. In particular, nation 1's extraction cost is $1.5u_1(s)x(s)^{-1/2}$ and nation 2's is $u_2(s)x(s)^{-1/2}$. The fish harvested by nation i at time s will generate a net benefit of the amount $[u_i(s)]^{1/2}$. At terminal time 4, nations 1 and 2 will receive termination bonuses $8x(3)^{1/2}$ and $6x(3)^{1/2}$ while the interest rate is 0.05.

At time 0 the expected payoffs of nations 1 and 2 are, respectively,

$$E\left\{\int_0^3 \left[[u_1(s)]^{1/2} - \frac{2.5}{x(s)^{1/2}}u_i(s)\right]\exp(-0.05s)\,ds + \exp\left[-r(3)\right]8x(3)^{\frac{1}{2}}\right\}, \quad \text{and}$$

$$E\left\{\int_0^3 \left[[u_2(s)]^{1/2} - \frac{3}{x(s)^{1/2}}u_i(s)\right]\exp(-0.05s)\,ds + \exp\left[-r(3)\right]6x(3)^{\frac{1}{2}}\right\}.$$

Obtain a Nash equilibrium solution for this stochastic transnational market activity.

8.2 If these nations agree to cooperate and maximize their expected joint payoff, compute the optimal cooperative strategies and optimal stock path of the fish biomass.

8.3 Furthermore, if these nations agree to share the excess of their expected gain equally, obtain a subgame consistent solution.

8.4 Consider the case when the game horizon in exercise 1 is extended to infinity.

(i) Obtain a Nash equilibrium solution for this stochastic dynamic transnational market activity.

(ii) If these nations agree to cooperate and maximize their expected joint payoff and share the excess of their expected gain equally, obtain a subgame consistent solution.

Chapter 9
Cost-Saving Joint Venture Under Uncertainty

In this chapter, we consider a cost-saving joint venture in the presence of stochastic elements. Section 9.1 formulates a dynamic cost-saving corporate joint venture in a stochastic environment and characterizes its subgame consistent solutions. An explicitly solvable illustration is given in Sect. 9.2. A characterization of the Shapley Value solution to a stochastic cost-saving joint venture is presented in Sect. 9.3 and a payoff distribution procedure leading to a subgame consistent solution is computed. Extensions to infinite-horizon ventures are formulated with explicit illustrations in the subsequent two sections.

9.1 Dynamic Corporate Joint Venture Under Uncertainty

To incorporate uncertainty in the corporate joint venture models in Chap. 5 we formulate the technology state dynamics of the ith firm as a set of stochastic differential equations

$$dx^i(s) = f^i\big[s, x^i(s), u_i(s)\big] ds + \sigma_i\big[s, x^i(s)\big] dz_i(s), \quad x^i(t_0) = x_0^i,$$

$$\text{for } i \in N, \tag{9.1}$$

where $x^i(s) \in X^i \subset R^{m_i}$ denotes the technology state of firm i, $u_i \in U_i \subset \text{comp } R^\ell$ is the control vector of firm i, $\sigma_i[s, x^i(s)]$ is a $m_i \times \Theta_i$, and $z_i(s)$ is a Θ_i-dimensional Wiener process and the initial state x_0^i is given. Let $\Omega_i[s, x^i(s)] = \sigma_i[s, x^i(s)]\sigma_i[s, x^i(s)]^T$ denote the covariance matrix with its element in row h and column ζ denoted by $\Omega_i^{h\zeta}[s, x^i(s)]$. For $i \neq j, x^i \cap x^j = \varnothing$, and $z_i(s)$ and $z_j(s)$ are independent Wiener processes. We also used $x^N(s)$ to denote the vector $[x^1(s), x^2(s), \dots, x^n(s)]$ and x_0^N the vector $[x_0^1, x_0^2, \dots, x_0^n]$.

D.W.K. Yeung, L.A. Petrosyan, *Subgame Consistent Economic Optimization*,   239
Static & Dynamic Game Theory: Foundations & Applications,
DOI 10.1007/978-0-8176-8262-0_9, © Springer Science+Business Media, LLC 2012

The expected profit of firm i is

$$E_{t_0}\left\{\int_{t_0}^{T}\left(g^i[s,x^i(s)]-c_i^{\{i\}}u_i(s)]\right)\exp\left[-\int_{t_0}^{s}r(y)\,dy\right]ds\right.$$

$$\left.+\exp\left[-\int_{t_0}^{T}r(y)\,dy\right]q^i\left(x^i(T)\right)\right\},\quad\text{for }i\in[1,2,\ldots,n]\equiv N,\quad(9.2)$$

where $\exp[-\int_{t_0}^{t}r(y)\,dy]$ is the discount factor and $q^i(x_i(T))$ the terminal payoff. In particular, $g^i[s,x_i,u_i]$ and $q^i(x_i)$ are positively related to x_i, reflecting the earning potent of the technology.

Since the expected payoffs and state dynamics in a noncooperative equilibrium are independent across firms, the market outcome is represented by an n stochastic neoclassical theory of the firm problems. Let $V^{(t_0)i}(t,x^i)$ and $\phi_i^*(t,x^i)$ denote the expected payoff and investment strategies of firm i, for $i\in N$, by which a firm's equilibrium is characterized (see Theorem A.5 in the Technical Appendixes). In particular, a set of investment strategies $\phi_i^*(t,x^i)$ for firm i constitutes an optimal solution to the stochastic investment problem in (9.1) and (9.2) if there exists a continuously twice differentiable function $V^{(t_0)i}(t,x^i)$ defined by $[t_0,T]\times R^{m_i}\to R$ and satisfying the following Bellman equation:

$$-V_t^{(t_0)i}\left(t,x^i\right)-\frac{1}{2}\sum_{h,\zeta=1}^{m_i}\Omega_i^{h\zeta}\left(t,x^i\right)V_{x^{i(h)}x^{i(\zeta)}}^{(t_0)i}\left(t,x^i\right)$$

$$=\max_{u_i}\left\{[g(t,x^i)-c_i^{\{i\}}(u_i)]\exp\left[-\int_{t_0}^{t}r(y)\,dy\right]+V_x^{(t_0)i}(t,x)f^i\left[t,x^i,u_i\right]\right\},$$

$$V^{(t_0)i}\left(T,x^i\right)=q^i\left(x^i\right)\exp\left[-\int_{t_0}^{T}r(y)\,dy\right].$$

Let $V^{(\tau)i}(t,x^i)$ denote the payoff function of firm i in a game with the dynamics in (9.1) and payoff in (9.2), which starts at time τ for $\tau\in[t_0,T)$. Note that the equilibrium feedback strategies are Markovian in the sense that they depend on the current time and current state. Invoking Remark 2.2 of Chap. 2, one can obtain

$$\exp\left[\int_{t_0}^{\tau}r(y)\,dy\right]V^{(t_0)i}\left(t,x^i\right)=V^{(\tau)i}\left(t,x^i\right),$$

for $\tau\in[t_0,T]$ and $i\in N$.

For the sake of clarity in exposition, we consider the case where $m_i=1$, for $i\in N$.

9.1.1 Joint Venture and Expected Profit Maximization

Consider a joint venture consisting of all these n companies. The participating firms can gain core skills and technology that would be impossible for them to obtain on their own individually. Cost-saving opportunities are created under joint venture, for instance, savings in joint R&D, administration, marketing, customer services, purchasing, financing, and economy of scales and scope. The cost of control of firm j under the joint venture becomes $c_j^N[u_j(s)]$. With absolute joint venture cost advantage we have

$$c_j^N(u_j) \leq c_j^{\{j\}}(u_j), \quad \text{for } j \in N. \tag{9.3}$$

Moreover, marginal cost advantages lead to

$$\partial c_j^N(u_j)/\partial u_j \leq \partial c_j^{\{j\}}(u_j)/\partial u_j, \quad \text{for } j \in N.$$

At time t_0, the joint venture would maximize the expected joint venture profit

$$E_{t_0} \left\{ \int_{t_0}^T \sum_{j=1}^n (g^j[s, x^j(s)] - c_j^N[u_j(s)]) \exp\left[-\int_{t_0}^s r(y)\,dy\right] ds \right.$$
$$\left. + \sum_{j=1}^n \exp\left[-\int_{t_0}^T r(y)\,dy\right] q^j(x^j(T)) \right\}, \tag{9.4}$$

subject to (9.3).

Invoking Fleming's techniques of stochastic optimal control in Theorem A.5 of the Technical Appendixes, the solution to the problem in (9.3) and (9.4) can be characterized as follows.

Corollary 9.1 *A set of controls $\{\psi_i^*(t, x), \text{for } i \in N \text{ and } t \in [t_0, T]\}$, provides an optimal solution to the control problem in (9.3) and (9.4) if there exists a continuously twice differentiable function $W^{(t_0)}(t, x) : [t_0, T] \times R^n \to R$ satisfying the following Bellman equation:*

$$-W_t^{(t_0)}(t, x) - \frac{1}{2} \sum_{h,\zeta=1}^n \Omega^{h\zeta}(t, x) W_{x^h x^\zeta}^{(t_0)}(t, x)$$
$$= \max_{u_1, u_2, \ldots, u_n} \left\{ \sum_{j=1}^n [g^j(t, x^j) - c_j^N(u_j)] \exp\left[-\int_{t_0}^t r(y)\,dy\right] \right.$$
$$\left. + \sum_{j=1}^n W_{x^j}^{(t_0)}(t, x) f^j(t, x^j, u_j) \right\}, \tag{9.5}$$
$$W^{(t_0)}(T, x) = \exp\left[-\int_{t_0}^T r(y)\,dy\right] \sum_{j=1}^n q^j(x^j),$$

where $x = \{x^1, x^2, \ldots, x^n\}$.

Hence the firms will adopt the cooperative control $\{\psi_i^*(t, x), \text{for } i \in N \text{ and } t \in [t_0, T]\}$, to obtain the maximized level of expected joint profit. Substituting this set of controls into (9.3) yields the dynamics of technology advancement under cooperation as

$$dx^i(s) = f^i\left[s, x^i(s), \psi_i^*(s, x(s))\right] ds + \sigma_i\left[s, x^i(s)\right] dz_i(s),$$

$$x^i(t_0) = x_0^i, \quad \text{for } i \in N. \tag{9.6}$$

Let $x^*(t) = \{x^{1*}(t), x^{2*}(t), \ldots, x^{n*}(t)\}$ denote the solution to (9.6). The optimal cooperative trajectory can be expressed as

$$x^{i*}(t) = x_0^i + \int_{t_0}^t f^i\left[s, x^{i*}(s), \psi_i^*(s, x^*(s))\right] ds + \int_{t_0}^t \sigma_i\left[s, x^{i*}(s)\right] dz_i(s),$$

$$\text{for } i \in N. \tag{9.7}$$

We use X_t^* to denote the set of realizable values of $x^*(t)$ at time t generated by (9.6). The term $x_t^* \in X_t^*$ is used to denote an element in X_t^*.

The cooperative investment strategies for the joint venture with the dynamics of (9.3) and the expected joint venture profit in (9.4) over the time interval $[t_0, T]$ can be expressed more precisely as

$$\{\psi_i^*(t, x^*(t)), \text{ for } i \in N \text{ and } t \in [t_0, T]\}. \tag{9.8}$$

Note that for group optimality to be achievable, the cooperative investment strategies $\{\psi_i^*(t, x^*(t)), \text{ for } i \in N \text{ and } t \in [t_0, T]\}$, must be exercised throughout time interval $[t_0, T]$.

Along the cooperative investment path $\{x^*(t)\}_{t=t_0}^T$, the present value of the total expected joint venture profit over the interval $[t, T]$, for $t \in [t_0, T)$, can be expressed as

$$W^{(t_0)}(t, x_t^*) = E_{t_0}\left\{\int_t^T \sum_{j=1}^n (g^j[s, x^{j*}(s)] - c_j^N[\psi_j^*(s, x^*(s))])\right.$$

$$\times \exp\left[-\int_{t_0}^s r(y) dy\right] ds$$

$$\left. + \exp\left[-\int_{t_0}^T r(y) dy\right] \sum_{j=1}^n q^j(x^{j*}(T)) \,|\, x^*(t) = x_t^* \in X_t^*\right\}. \tag{9.9}$$

Let $W^{(\tau)}(t, x_t^*)$ denote the total venture profit from the control problem with the dynamics in (9.3) and payoff in (9.4), which begins at time $\tau \in [t_0, T]$ with initial state x_τ^*. Invoking Remark 8.1 of Chap. 8, one can readily obtain

$$\exp\left[\int_{t_0}^\tau r(y) dy\right] W^{(t_0)}(t, x_t^*) = W^{(\tau)}(t, x_t^*),$$

for $\tau \in [t_0, T]$ and $t \in [\tau, T)$.

9.1.2 Subgame Consistent Joint Venture

Since the sizes and earning potentials of the firms in a corporate joint venture may vary significantly, we consider the case when the venture agrees to share the excess of the expected total cooperative payoff over the sum of expected individual noncooperative payoffs proportionally to the firms' expected noncooperative payoffs.

The imputation scheme has to fulfill the following condition.

Condition 9.1 An imputation

$$\xi^{(t_0)i}(t_0, x_0) = \frac{V^{(t_0)i}(t_0, x_0^i)}{\sum_{j=1}^{n} V^{(t_0)j}(t_0, x_0^j)} W^{(t_0)}(t_0, x_0),$$

is assigned to firm i, for $i \in N$ at the outset; and an imputation

$$\xi^{(\tau)i}\left(\tau, x_\tau^*\right) = \frac{V^{(\tau)i}(\tau, x_\tau^{i*})}{\sum_{j=1}^{n} V^{(\tau)j}(\tau, x_\tau^{j*})} W^{(\tau)}\left(\tau, x_\tau^*\right) \tag{9.10}$$

is assigned to firm i, for $i \in N$ at time $\tau \in (t_0, T]$.

The imputation in (9.10) satisfies

(i) $\xi^{(\tau)i}(\tau, x_\tau^*) \geq V^{(\tau)i}(\tau, x_\tau^{i*})$, for $i \in N$ and $\tau \in [t_0, T]$; and
(ii) $\sum_{j=1}^{n} \xi^{(\tau)j}(\tau, x_\tau^*) = W^{(\tau)}(\tau, x_\tau^*)$, for $\tau \in [t_0, T]$.

Hence the imputation vector $\xi^{(\tau)}(\tau, x_\tau^*)$ in (9.10) satisfies individual rationality and group optimality throughout the game horizon $[t_0, T]$.

The optimality principle guiding the joint venture can then be stated as follows:

Optimality Principle PI

(i) the maximization of the venture's expected payoffs and
(ii) the sharing of the expected venture cooperative profit proportionally to individual firms' expected noncooperative payoffs.

All the participating firms in the joint venture will have no incentive to exit the venture if the agreed-upon optimality principle is maintained at every instant $t \in [t_0, T]$ along the cooperative state trajectory. Hence a subgame consistent solution has to be sought.

As in Chap. 4, we let the solution under this optimality principle be expressed as

$$P(x_0, T - t_0) = \left\{u\left(s, x_s^*\right) \text{ and } B^{t_0}\left(s, x_s^*\right) \text{ for } s \in [t_0, T]; \xi^{(t_0)}(t_0, x_0)\right\}.$$

Invoking Theorem 8.2 of Chap. 8, a subgame consistent solution for the joint venture under the above optimality principle can be obtained as follows.

Corollary 9.2 *For the joint venture characterized by (9.3) and (9.4) under optimality Principle* PI, *the solution* $P(x_0, T - t_0) = \{u(s, x_s^*)$ *and* $B(s, x_s^*)$ *for* $s \in [t_0, T]$ *and* $\xi^{(t_0)}(t_0, x_0)\}$ *in which*

(i) $u(s, x_s^*)$ *for* $s \in [t_0, T]$ *is the set of group optimal strategies* $\psi^*(s, x_s^*)$ *for the game* $\Gamma_c(x_0, T - t_0)$, *and*

(ii) $B(s, x_s^*) = \{B_1(s, x_s^*), B_2(s, x_s^*), \ldots, B_n(s, x_s^*)\}$ *for* $s \in [t_0, T]$ *where*

$$B_i(s, x_s^*) = -\left[\xi_t^{(s)i}(t, x_t^*)|_{t=s}\right] - \frac{1}{2}\sum_{h,\zeta=1}^{n}\Omega^{h\zeta}(s, x_s^*)\left[\xi_{x_t^h x_t^\zeta}^{(s)i}(t, x_t^*)|_{t=s}\right]$$

$$-\sum_{h=1}^{n}\left[\xi_{x_t^h}^{(s)i}(t, x_t^*)|_{t=s}\right]f^h\left[s, x_s^{h*}, \psi_h^*(s, x_s^*)\right]$$

$$= -\frac{\partial}{\partial t}\left[\frac{V^{(s)i}(t, x_t^{i*})}{\sum_{j=1}^{n}V^{(s)j}(t, x_t^{j*})}W^{(s)}(t, x_t^*)|_{t=s}\right]$$

$$-\frac{1}{2}\sum_{h,\zeta=1}^{n}\Omega^{h\zeta}(s, x_s^*)\frac{\partial^2}{\partial x_t^{h*}\partial x_t^{\zeta*}}$$

$$\times\left[\frac{V^{(s)i}(t, x_t^{i*})}{\sum_{j=1}^{n}V^{(s)j}(t, x_t^{j*})}W^{(s)}(t, x_t^*)|_{t=s}\right]$$

$$-\sum_{h=1}^{n}\frac{\partial}{\partial x_t^{h*}}\left[\frac{V^{(s)i}(t, x_t^{i*})}{\sum_{j=1}^{n}V^{(t)j}(t, x_t^{j*})}W^{(s)}(t, x_t^*)|_{t=s}\right]$$

$$\times f^h\left[s, x_s^{h*}, \psi_h^*(s, x_s^*)\right], \quad \text{for } i \in N \text{ and } x_s^* \in X_s^*, \qquad (9.11)$$

and $\xi^{(s)i}(s, x_s^*) = \dfrac{V^{(s)i}(s, x_s^{i*})}{\sum_{j=1}^{n}V^{(s)j}(s, x_s^{j*})}W^{(s)}(s, x_s^*)$ *is subgame consistent.*

With firms using the cooperative investment strategies $\{\psi_i^*(\tau, x_\tau^*)$, for $\tau \in [t_0, T]$ and $i \in N\}$, the instantaneous receipt of firm i at time instant τ is

$$\zeta_i(\tau, x_\tau^*) = g^i(\tau, x_\tau^{i*}) - c_i^N[\psi_i^*(\tau, x_\tau^*)],$$

$$\text{for } \tau \in [t_0, T] \text{ and } i \in N. \qquad (9.12)$$

According to Corollary 9.2, the instantaneous payment that firm i should receive under the agreed-upon optimality principle is $B_i(\tau, x_\tau^*)$ as stated in (9.11). Hence an instantaneous transfer payment

$$\chi^i(\tau, x_\tau^*) = B_i(\tau, x_\tau^*) - \zeta_i(\tau, x_\tau^*) \qquad (9.13)$$

has to be given or charged to firm i at time τ, for $i \in N$ and $\tau \in [t_0, T]$.

9.2 A Cost-Saving Joint Venture with Stochasticity

Consider the case when there are three companies involved in a joint venture. The planning period is $[t_0, T]$. Company i's expected profit is

$$
E_{t_0}\left\{\int_{t_0}^{T}\left[P_i\left[x^i(s)\right]^{1/2}-c_i^{\{i\}}u_i(s)\right]\exp\left[-r(s-t_0)\right]\mathrm{d}s\right.
$$

$$
\left.+\exp\left[-r(T-t_0)\right]q_i\left[x^i(T)\right]^{1/2}\right\},\quad\text{for }i\in\{1,2,3\},\qquad(9.14)
$$

where $P_i, c_i^{\{i\}}$, and q_i are positive constants, r is the discount rate, $x_i(s)\subset R^+$ is the level of technology of company i at time s, and $u_i(s)\subset R^+$ is its physical investment in technological advancement. The term $P_i[x^i(s)]^{1/2}$ reflects the net operating revenue of company i at technology level $x_i(s)$, and $c_i^{\{i\}}u_i$ is the cost of investment if firm i operates on its own. The term $q_i[x^i(T)]^{1/2}$ gives the salvage value of company i's technology at time T.

The dynamics of the technology level of company i follows the stochastic differential equation

$$
\mathrm{d}x^i(s)=\left[\alpha_i\left[u_i(s)x^i(s)\right]^{1/2}-\delta x^i(s)\right]\mathrm{d}s+\sigma_i x^i(s)\,\mathrm{d}z_i(s),\quad x^i(t_0)=x_0^i\in X^i,
$$

$$
\text{for }i\in\{1,2,3\},\qquad(9.15)
$$

where $\alpha_i[u_i(s)x^i(s)]^{1/2}$ is the addition to the technology brought about by $u_i(s)$ amount of physical investment, δ is the rate of obsolescence, and $z_1(s), z_2(s)$, and $z_3(s)$ are independent Wiener processes.

In the case when each of these three firms acts independently, using Theorem A.5 in the Technical Appendixes, we obtain the corresponding partial differential equations as

$$
-V_t^{(t_0)i}\left(t,x^i\right)-\frac{(\sigma_i x^i)^2}{2}V_{x^i x^i}^{(t_0)i}\left(t,x^i\right)
$$

$$
=\max_{u_i}\left\{\left[P_i\left(x^i\right)^{1/2}-c_i^{\{i\}}u_i\right]\exp\left[-r(t-t_0)\right]\right.
$$

$$
\left.+V_{x^i}^{(t_0)i}\left(t,x^i\right)\left[\alpha_i\left(u^i x^i\right)^{1/2}-\delta x^i\right]\right\},
$$

$$
V^{(t_0)i}\left(T,x^i\right)=\exp\left[-r(T-t_0)\right]q_i\left(x^i\right)^{1/2},\quad\text{for }i\in\{1,2,3\}.
$$

Performing the indicated maximization yields

$$
u_i=\frac{\alpha_i^2}{4(c_i^{\{i\}})^2}\left[V_{x^i}^{(t_0)i}\left(t,x^i\right)\exp\left[r(t-t_0)\right]\right]^2 x^i,\quad\text{for }i\in\{1,2,3\}.
$$

Substituting u_i into the above partial differential equations yields

$$-V_t^{(t_0)i}(t, x^i) - \frac{(\sigma_i x^i)^2}{2} V_{x^i x^i}^{(t_0)i}(t, x^i)$$

$$= P_i(x^i)^{1/2} \exp[-r(t-t_0)] - \frac{\alpha_i^2}{4c_i^{\{i\}}} [V_{x^i}^{(t_0)i}(t, x^i)]^2 \exp[r(t-t_0)]x^i$$

$$+ \frac{\alpha_i^2}{2c_i^{\{i\}}} [V_{x^i}^{(t_0)i}(t, x^i)]^2 \exp[r(t-\tau)]x^i - \delta V_{x^i}^{(t_0)i}(t, x^i)x_i, \quad \text{for } i \in [1, 2, 3].$$

Solving the above system of partial differential equations yields

$$V^{(t_0)i}(t, x^i) = [A_i^{\{i\}}(t)(x^i)^{1/2} + C_i^{\{i\}}(t)] \exp[-r(\tau - t_0)], \quad \text{for } i \in \{1, 2, 3\},$$

$$(9.16)$$

where

$$\dot{A}_i^{\{i\}}(t) = \left(r + \frac{\delta}{2} + \frac{\sigma_i^2}{8}\right) A_i^{\{i\}}(t) - P_i, \qquad \dot{C}_i^{\{i\}}(t) = r C_i^{\{i\}}(t) - \frac{\alpha_i^2}{16c_i^{\{i\}}} [A_i^{\{i\}}(t)]^2,$$

$$(9.17)$$

$$A_i^{\{i\}}(T) = q_i, \quad \text{and} \quad C_i^{\{i\}}(T) = 0.$$

The first equation in the block-recursive system in (9.17) is a first-order linear differential equation in $A_i^{\{i\}}(t)$ that can be solved independently by standard techniques. Substituting the solution of $A_i^{\{i\}}(t)$ into the second equation of (9.17) yields a first-order linear differential equation in $C_i^{\{i\}}(t)$. The solution of $C_i^{\{i\}}(t)$ can be readily obtained by standard techniques.

Moreover, one can easily derive for $\tau \in [t_0, T]$

$$V^{(\tau)i}(t, x^i) = [A_i^{\{i\}}(t)(x^i)^{1/2} + C_i^{\{i\}}(t)] \exp[-r(t-\tau)],$$

for $i \in \{1, 2, 3\}$ and $\tau \in [t_0, T]$.

9.2.1 Expected Venture Profit and Cost Savings

Consider the case when all three firms agree to form a joint venture and share their expected joint profit proportionally to their expected noncooperative profits. Cost-saving opportunities are created under joint venture from joint R&D, administration, purchasing, financing, and economy of scales and scope. The cost of control of firm j under the joint venture becomes $c_j^{\{1,2,3\}}[u_j(s)]$, with joint venture cost advantage

$$c_j^{\{1,2,3\}} \leq c_j^{\{j\}}, \quad \text{for } j \in N. \tag{9.18}$$

The expected profit of the joint venture is the sum of the participating firms' expected profits

$$
E_{t_0} \left\{ \int_{t_0}^{T} \sum_{j=1}^{3} \left[P_j \left[x^j(s) \right]^{1/2} - c_j^{\{1,2,3\}} u_j(s) \right] \exp\left[-r(s - t_0) \right] ds \right.
$$

$$
\left. + \sum_{j=1}^{3} \exp\left[-r(T - t_0) \right] q_j \left[x^j(T) \right]^{1/2} \right\}. \tag{9.19}
$$

The firms in the joint venture then act cooperatively to maximize (9.19) subject to (9.18). In particular, (9.18)–(9.19) becomes an optimization problem under a joint venture involving all three firms with technology spillover. Using Theorem A.5 in the Technical Appendixes, we obtain the equation

$$
- W_t^{(t_0)\{1,2,3\}} \left(t, x^1, x^2, x^3 \right) - \sum_{h,\zeta=1}^{3} \frac{(\sigma_h x^h)(\sigma_\zeta x^\zeta)}{2} W_{x^h x^\zeta}^{(t_0)\{1,2,3\}} \left(t, x^1, x^2, x^3 \right)
$$

$$
= \max_{u_i} \left\{ \sum_{i=1}^{3} \left[P_i \left(x^i \right)^{1/2} - c_i^{\{1,2,3\}} u_i \right] \exp\left[-r(t - t_0) \right] \right.
$$

$$
\left. + \sum_{i=1}^{3} W_{x^i}^{(t_0)\{1,2,3\}} \left(t, x^1, x^2, x^3 \right) \left[\alpha_i \left(u_i x^i \right)^{1/2} - \delta x^i \right] \right\}, \tag{9.20}
$$

$$
W^{(t_0)\{1,2,3\}} \left(T, x^1, x^2, x^3 \right) = \sum_{j=1}^{3} \exp\left[-r(T - t_0) \right] q_j \left(x^j \right)^{1/2},
$$

for $i, j, h \in \{1, 2, 3\}$ and $i \neq j \neq h$.

Performing the indicated maximization yields

$$
u_i = \frac{\alpha_i^2}{4(c_i^{\{1,2,3\}})^2} \left[W_{x^i}^{(t_0)\{1,2,3\}} \left(t, x^1, x^2, x^3 \right) \exp\left[r(t - t_0) \right] \right]^2 x^i,
$$

for $i \in \{1, 2, 3\}$. \tag{9.21}

Substituting (9.21) into (9.20) yields

$$
- W_t^{(t_0)\{1,2,3\}} \left(t, x^1, x^2, x^3 \right) - \sum_{j=1}^{3} \frac{(\sigma_j x^j)^2}{2} W_{x^j x^j}^{(t_0)\{1,2,3\}} \left(t, x^1, x^2, x^3 \right)
$$

$$
= \sum_{i=1}^{3} \left[P_i \left(x^i \right)^{1/2} \exp\left[-r(t - t_0) \right] - \frac{\alpha_i^2 x^i}{4 c_i^{\{1,2,3\}}} \left[W_{x^i}^{(t_0)\{1,2,3\}} \left(t, x^1, x^2, x^3 \right) \right]^2 \right.
$$

$$\times \exp\big[r(t - t_0)\big]\Big] + \sum_{i=1}^{3} W_{x_i}^{(t_0)\{1,2,3\}}(t, x_1, x_2, x_3)$$

$$\times \left[\frac{\alpha_i^2}{2c_i^2} \big[W_{x_i}^{(t_0)i}(t, x_1, x_2, x_3) \exp[r(t - t_0)]\big] x^i - \delta x^i \right], \quad \text{and} \qquad (9.22)$$

$$W^{(t_0)\{1,2,3\}}(T, x^1, x^2, x^3) = \sum_{j=1}^{3} \exp[-r(T - t_0)]q_j(x^j)^{1/2},$$

for $i, j, h \in \{1, 2, 3\}$ and $i \neq j \neq h$.

Solving (9.23) yields

$$W^{(t_0)\{1,2,3\}}(t, x^1, x^2, x^3)$$

$$= \big[A_1^{\{1,2,3\}}(t)(x^1)^{1/2} + A_2^{\{1,2,3\}}(t)(x^2)^{1/2} + A_3^{\{1,2,3\}}(t)(x^3)^{1/2} + C^{\{1,2,3\}}(t) \big]$$

$$\times \exp[-r(t - t_0)], \qquad (9.23)$$

where $A_1^{\{1,2,3\}}(t)$, $A_2^{\{1,2,3\}}(t)$, $A_3^{\{1,2,3\}}(t)$, and x_3, $C^{\{1,2,3\}}(t)$ satisfy

$$\dot{A}_i^{\{1,2,3\}}(t) = \left(r + \frac{\delta}{2} + \frac{\sigma_i^2}{8} \right) A_i^{\{1,2,3\}}(t) - \frac{b_i^{[i,j]}}{2} A_j^{\{1,2,3\}}(t) - \frac{b_i^{[i,h]}}{2} A_h^{\{1,2,3\}}(t) - P_i$$

for $i, j, h \in \{1, 2, 3\}$ and $i \neq j \neq h$,

(9.24)

$$\dot{C}^{\{1,2,3\}}(t) = rC^{\{1,2,3\}}(t) - \sum_{i=1}^{3} \frac{\alpha_i^2}{16c_i^{\{1,2,3\}}} \big[A_i^{\{1,2,3\}}(t) \big]^2,$$

$$A_i^{\{1,2,3\}}(T) = q_i, \quad \text{for } i \in \{1, 2, 3\}, \text{ and } C^{\{1,2,3\}}(T) = 0.$$

The first three equations in the block recursive system in (9.24) is a system of three linear differential equations that can be solved explicitly by standard techniques. Upon solving $A_i^{\{1,2,3\}}(t)$ for $i \in \{1, 2, 3\}$, and substituting them into the fourth equation of (9.24), one has a linear differential equation in $C^{\{1,2,3\}}(t)$.

The investment strategies of the grand coalition joint venture can be derived as

$$\psi_i^{\{1,2,3\}}(t, x) = \frac{\alpha_i^2}{16(c_i^{\{1,2,3\}})^2} \big[A_i^{\{1,2,3\}}(t) \big]^2, \quad \text{for } i \in \{1, 2, 3\}. \qquad (9.25)$$

The dynamics of technological progress of the joint venture over the time interval $s \in [t_0, T]$ can be expressed as

$$dx^i(s) = \left(\frac{\alpha_i^2}{4c_i^{\{1,2,3\}}} A_i^{\{1,2,3\}}(t)[x^i(s)]^{1/2} - \delta x^i(s) \right) ds + \sigma_i x^i(s) \, dz_i(s),$$

$$x^i(t_0) = x_0^i, \quad \text{for } i \in \{1, 2, 3\}. \qquad (9.26)$$

Taking the transforming $y^i(s) = x^i(s)^{1/2}$, for $i \in \{1, 2, 3\}$, the equation system in (9.26) can be expressed as

$$dy^i(s) = \left(\frac{\alpha_i^2}{8c_i} A_i^{\{1,2,3\}}(t) - \frac{\delta}{2} y^i(s) - \frac{\sigma_i^2}{8} A_i^{\{1,2,3\}}(s) y^i(s) \right) ds + \frac{1}{2} \sigma_i y^i(s) \, dz_i(s),$$

$$y^i(t_0) = \left(x_0^i \right)^{1/2}, \quad \text{for } i \in \{1, 2, 3\}. \tag{9.27}$$

Equation (9.27) is a system of linear stochastic differential equations that can be solved by standard techniques. Solving (9.27) yields the joint venture's state trajectory. Let $\{y^{1*}(t), y^{2*}(t), y^{3*}(t)\}$ denote the solution to (9.27). Transforming $x^i = (y^i)^2$, we obtain the state trajectories of the joint venture over the time interval $s \in [t_0, T]$ as

$$x^*(s) = \left\{ x^{1*}(t), x^{2*}(t), x^{3*}(t) \right\}_{t=t_0}^{T} = \left\{ \left[y^{1*}(t) \right]^2, \left[y^{2*}(t) \right]^2, \left[y^{3*}(t) \right]^2 \right\}_{t=t_0}^{T}. \tag{9.28}$$

We use X_t^* to denote the set of realizable values of $x^*(t)$ at time t and the term $x_t^* \in X_t^*$ is used to denote an element in X_t^*.

Remark 9.1 One can readily verify that

$$W^{(t_0)\{1,2,3\}}\left(t, x^{1*}, x^{2*}, x^{3*} \right) = W^{(t)\{1,2,3\}}\left(t, x^{1*}, x^{2*}, x^{3*} \right) \exp\left[-r(t - t_0) \right],$$

for $i \in \{1, 2, 3\}$.

9.2.2 Subgame Consistent Venture Profit Sharing

Since the firms agree to share their expected joint profit proportionally to their expected noncooperative profits, the imputation scheme has to fulfill the following condition.

Condition 9.2 In the game $\Gamma_c(x_0, T - t_0)$, an imputation

$$\xi^{(t_0)i}(t_0, x_0) = \frac{V^{(t_0)i}(t_0, x_0^i)}{\sum_{j=1}^{n} V^{(t_0)j}(t_0, x_0^i)} W^{(t_0)\{1,2,3\}}\left(t_0, x_0^1, x_0^2, x_0^3 \right),$$

is assigned to firm i, for $i \in \{1, 2, 3\}$, and in the subgame $\Gamma_c(x_\tau^*, T - \tau)$, for $\tau \in (t_0, T]$, an imputation

$$\xi^{(\tau)i}(\tau, x_\tau^*) = \frac{V^{(\tau)i}(\tau, x_\tau^{i*})}{\sum_{j=1}^{n} V^{(\tau)j}(\tau, x_\tau^{i*})} W^{(\tau)\{1,2,3\}}\left(\tau, x_\tau^{1*}, x_\tau^{2*}, x_\tau^{3*} \right) \tag{9.29}$$

is assigned to firm i, for $i \in \{1, 2, 3\}$.

To formulate a payoff distribution procedure over time so that the agreed imputations in Condition 9.2 are satisfied we invoke Corollary 9.2 to obtain

$$
B_i\left(\tau, x_\tau^{1*}, x_\tau^{2*}, x_\tau^{3*}\right)
$$

$$
= -\frac{\partial}{\partial t}\left[\frac{V^{(\tau)i}\left(t, x_t^{i*}\right)}{\sum_{j=1}^3 V^{(\tau)j}\left(t, x_t^{j*}\right)} W^{(\tau)\{1,2,3\}}\left(t, x_t^{1*}, x_t^{2*}, x_t^{3*}\right)\big|_{t=\tau}\right]
$$

$$
- \frac{1}{2}\sum_{h,\zeta=1}^3 \sigma_h x^h \sigma_\zeta x^\zeta \frac{\partial^2}{\partial x_\tau^{h*}\partial x_\tau^{\zeta*}}
$$

$$
\times\left[\frac{V^{(\tau)i}\left(\tau, x_\tau^{i*}\right)}{\sum_{j=1}^3 V^{(\tau)j}\left(\tau, x_\tau^{j*}\right)} W^{(\tau)\{1,2,3\}}\left(\tau, x_\tau^{1*}, x_\tau^{2*}, x_\tau^{3*}\right)\right]
$$

$$
- \sum_{h=1}^3 \frac{\partial}{\partial x_\tau^{h*}}\left[\frac{V^{(\tau)i}\left(\tau, x_\tau^{i*}\right)}{\sum_{j=1}^3 V^{(\tau)j}\left(\tau, x_\tau^{j*}\right)} W^{(\tau)\{1,2,3\}}\left(\tau, x_\tau^{1*}, x_\tau^{2*}, x_\tau^{3*}\right)\right]
$$

$$
\times\left[\frac{\alpha_h^2}{4c_h^{\{1,2,3\}}} A_h^{\{1,2,3\}}(\tau)\left(x_\tau^{h*}\right)^{1/2} - \delta x_\tau^{h*}(s)\right],
$$

for $i \in \{1, 2, 3\}$, $x_\tau^* \in X_\tau^*$, $\tau \in [t_0, T]$.   (9.30)

Finally, with firms using the cooperative investment strategies the instantaneous
receipt of firm i at time instant τ is

$$
\zeta_i\left(\tau, x_\tau^*\right) = P_i\left(x_\tau^{i*}\right)^{1/2} - \frac{\alpha_i^2}{16(c_i^{\{1,2,3\}})}\left[A_i^{\{1,2,3\}}(\tau)\right]^2,
$$

for $i \in \{1, 2, 3\}$, $x_\tau^* \in X_\tau^*$, and $\tau \in [t_0, T]$.   (9.31)

Under cooperation, the instantaneous payment that firm i should receive under
the agreed-upon optimality principle is $B_i(\tau, x_\tau^*)$ as stated in (9.30). Hence an instantaneous transfer payment

$$
\chi^i\left(\tau, x_\tau^*\right) = B_i\left(\tau, x_\tau^*\right) - \zeta_i\left(\tau, x_\tau^*\right),
$$ (9.32)

has to be given or charged to firm i at time τ, for $i \in \{1, 2, 3\}$, $x_\tau^* \in X_\tau^*$, and $\tau \in [t_0, T]$.

9.3 A Shapley Value Solution to a Joint Venture Under Uncertainty

Consider again the stochastic dynamic venture model in (9.1) and (9.2). If firms
are allowed to form different coalitions consisting of a subset of companies $K \subseteq N$.

There are k firms in the subset K. The participating firms in a coalition can gain core skills and technology from each other. In particular, they can obtain cost reduction and with joint venture cost advantage as

$$c_j^K[u_j(s)] \leq c_j^L[u_j(s)], \quad \text{for } j \in L \subseteq K, \tag{9.33}$$

where $c_j^K[u_j(s)]$ represents the costs of the controls of the firm j in the subset K and $c_j^L[u_j(s)]$ represents the costs of the controls of the firm j in the subset L.

Moreover, marginal cost advantages lead to

$$\partial c_j^K[u_j(s)]/\partial u_j(s) \leq \partial c_j^L[u_j(s)]/\partial u_j(s), \quad \text{for } j \in L \subseteq K.$$

At time t_0, the expected profit to the joint venture K becomes

$$E_{t_0}\left\{\int_{t_0}^T \sum_{j \in K}\left(g^j[s, x^j(s)] - c_j^K[u_j(s)]\right)\exp\left[-\int_{t_0}^s r(y)\,dy\right]ds\right.$$

$$\left. + \sum_{j \in K}\exp\left[-\int_{t_0}^T r(y)\,dy\right]q^j\left(x^j(T)\right)\right\}, \quad \text{for } K \subseteq N. \tag{9.34}$$

9.3.1 Expected Joint Venture Profits and Optimal Trajectory

To compute the expected profit of the joint venture K we have to consider the stochastic control problem $\varpi[K; t_0, x_0^K]$, which maximizes the expected joint venture profit in (9.34) subject to the technology accumulation dynamics in (9.33).

Invoking Theorem A.5 in the Technical Appendixes, the solution to the control problem $\varpi[K; t_0, x_0^K]$ can be characterized as follows.

Corollary 9.3 *A set of controls* $\{\psi_i^{K*}(t, x^K), \text{for } i \in K \text{ and } t \in [t_0, T]\}$, *provides an optimal solution to the stochastic control problem* $\varpi[K; t_0, x_0^K]$ *if there exists a continuously twice differentiable function* $W^{(t_0)K}(t, x) : [t_0, T] \times R^k \to R$ *satisfying the following partial differential equation:*

$$-W_t^{(t_0)K}\left(t, x^K\right) - \frac{1}{2}\sum_{h, \zeta \in K}\Omega^{h\zeta}\left(t, x^K\right)W_{x^h x^\zeta}^{(t_0)K}\left(t, x^K\right)$$

$$= \max_{u_K}\left\{\sum_{j \in K}\left[g^j\left(t, x^j\right) - c_j^K(u_j)\right]\exp\left[-\int_{t_0}^t r(y)\,dy\right]\right.$$

$$\left. + \sum_{j \in K}W_{x^j}^{(t_0)K}\left(t, x^K\right)f^j\left(s, x^j, u_j\right)\right\}, \tag{9.35}$$

$$W^{(t_0)K}\left(T, x^K\right) = \exp\left[-\int_{t_0}^T r(y)\,dy\right]\sum_{j \in K}q^j\left(x^j\right).$$

Following Corollary 9.3, one can characterize the maximized expected payoff $W^{(\tau)K}(t, x^K)$ to the optimal control problem $\varpi[K; \tau, x_\tau^K]$ which maximizes

$$E_\tau\left\{\int_\tau^T \sum_{j \in K} \left(g^j[(s, x^j(s)] - c_j^K[u_j(s)]\right) \exp\left[-\int_\tau^s r(y)\,dy\right] ds\right.$$

$$\left. + \sum_{j \in K} \exp\left[-\int_\tau^T r(y)\,dy\right] q^j(x^j(T))\right\},$$

subject to $\dot{x}^j(s) = f^j[s, x^j(s), u_j(s)],$ $x^j(\tau) = x_\tau^j,$ for $j \in K.$

The superadditivity of the expected coalition payoff can be demonstrated.

Proposition 9.1 *The expected coalition profits $W^{(\tau)K}(t, x^K)$ is superadditivity, that is,*

$$W^{(\tau)K}\left(\tau, x^K\right) \geq W^{(\tau)L}\left(\tau, x^L\right) + W^{(\tau)K\backslash L}\left(\tau, x^{K\backslash L}\right), \quad \text{for } L \subset K \subseteq N,$$

where $K\backslash L$ is the relative complement of L in K.

Proof Follow the Proof of Proposition 5.2 in the Appendix of Chap. 5.   □

Now consider the case of a grand coalition N in which all the n firms are in the coalition. Following Corollary 9.3, the solution to the stochastic control problem $\varpi[N; t_0, x_0^N]$ can be characterized as in Corollary 9.1. The cooperative state dynamics is (9.6) and the optimal stochastic trajectory is (9.7). The optimal cooperative strategies are in (9.8). Along the cooperative investment path $\{x^*(t)\}_{t=t_0}^T$ the expected total venture profit over the interval $[t, T]$, for $t \in [t_0, T)$, can be expressed as (9.9).

9.3.2 The PDP for Shapley Value

Consider the case where the participating firms agree to share their expected cooperative profits according to the Shapley Value (1953). The imputation has to satisfy the following condition.

Condition 9.3 In the game $\Gamma_c(x_0, T - t_0)$, an imputation

$$\xi^{(t_0)i}\left(t_0, x_0^N\right) = \sum_{K \subseteq N} \frac{(k-1)!(n-k)!}{n!}\left[W^{(t_0)K}\left(t_0, x_0^K\right) - W^{(t_0)K\backslash i}\left(t_0, x_0^{K\backslash i}\right)\right],$$

is assigned to firm i, for $i \in N$, and in the subgame $\Gamma_c(x^*_\tau, T - \tau)$, for $\tau \in (t_0, T]$, an imputation

$$\xi^{(\tau)i}\left(\tau, x_\tau^{N*}\right) = \sum_{K \subseteq N} \frac{(k-1)!(n-k)!}{n!}\left[W^{(\tau)K}\left(\tau, x_\tau^{K*}\right) - W^{(\tau)K \setminus i}\left(\tau, x_\tau^{K \setminus i*}\right)\right],$$

(9.36)

is assigned to firm i, for $i \in N$.

To formulate a payoff distribution procedure over time so that the agreed imputations satisfy the Shapley Value in Condition 9.3 we invoke Theorem 8.1 in Chap. 8 and obtain the following.

Corollary 9.4 *A PDP with a terminal payment $q^i(x^*_T))$ at time T and an instantaneous payment at time $\tau \in [t_0, T]$ when $x^*(\tau) = x^*_\tau \in X^*_\tau$*

$$B_i\left(\tau, x^*_\tau\right) = -\sum_{K \subseteq N} \frac{(k-1)!(n-k)!}{n!}\left\{\left[W_t^{(\tau)K}\left(t, x_t^{K*}\right)\big|_{t=\tau}\right]\right.$$

$$-\left[W_t^{(\tau)K \setminus i}\left(t, x_t^{K \setminus i*}\right)\big|_{t=\tau}\right]$$

$$+\sum_{h \in K}\left[\frac{\partial}{\partial x_\tau^{h*}} W^{(\tau)K}\left(\tau, x_\tau^{K*}\right)\right]f^h\left[\tau, x_\tau^{h*}, \psi_h^*\left(\tau, x^*_\tau\right)\right]$$

$$-\sum_{h \in K \setminus i}\left[\frac{\partial}{\partial x_\tau^{h*}} W^{(\tau)K \setminus i}\left(\tau, x_\tau^{K \setminus i*}\right)\right]f^h\left[\tau, x_\tau^{h*}, \psi_h^*\left(\tau, x^*_\tau\right)\right]$$

$$+\frac{1}{2}\sum_{h,\zeta \in K}\Omega^{h\zeta}\left(\tau, x^*_\tau\right)\frac{\partial^2}{\partial x_\tau^{h*}\partial x_\tau^{\zeta*}}\left[W^{(\tau)K}\left(\tau, x_\tau^{K*}\right)\right]$$

$$-\frac{1}{2}\sum_{h,\zeta \in K \setminus i}\Omega^{h\zeta}\left(\tau, x^*_\tau\right)\frac{\partial^2}{\partial x_\tau^{h*}\partial x_\tau^{\zeta*}}\left[W^{(\tau)K \setminus i}\left(\tau, x_\tau^{K \setminus i*}\right)\right]\right\},$$

for $i \in N$, (9.37)

would lead to the realization of the Shapley Value imputations $\xi^{(\tau)i}(\tau, x_\tau^{N*})$ in Condition 9.3.

Invoking Theorem 8.2 in Chap. 8, a subgame consistent Shapley Value solution for the joint venture can be obtained as

$$P(x_0, T - t_0) = \left\{u\left(s, x^*_s\right) \text{ and } B\left(s, x^*_s\right) \text{ for } s \in [t_0, T] \text{ and } \xi^{(t_0)}(t_0, x_0)\right\},$$

where $u(s, x^*_s) = \psi^{N*}(s, x^*(s))$ is the set of group optimal strategies in the grand coalition, $B(s, x^*_s)$ is the PDP given in (9.37), and $\xi^{(t_0)}(t_0, x_0)$ is the Shapley Value imputation in Condition 9.3.

Finally, with firms using the cooperative investment strategies $\{\psi_i^*(\tau, x_\tau^*), \text{ for } \tau \in [t_0, T] \text{ and } i \in N\}$, the instantaneous receipt of firm i at time instant τ when $x^*(\tau) = x_\tau^* \in X_\tau^*$ is

$$\zeta_i\left(\tau, x_\tau^*\right) = g^i\left(\tau, x_\tau^{i*}\right) - c_i^N\left[\psi_i^*\left(\tau, x_\tau^*\right)\right], \quad \text{for } \tau \in [t_0, T] \text{ and } i \in N. \tag{9.38}$$

According to Corollary 9.4, the instantaneous payment that firm i should receive under the agreed-upon optimality principle is $B_i(\tau, x_\tau^*)$ as stated in (9.37). Hence an instantaneous transfer payment

$$\chi^i\left(\tau, x_\tau^*\right) = B_i\left(\tau, x_\tau^*\right) - \zeta_i\left(\tau, x_\tau^*\right) \tag{9.39}$$

would be given or charged to firm i at time τ, for $i \in N, x_\tau^* \in X_\tau^*$, and $\tau \in [t_0, T]$.

9.4 A Stochastic Joint Venture with Shapley Value Profit Sharing

Consider a similar venture as that in Sect. 9.2. When the firms act independently, their expected profits and state dynamics are, respectively, (9.14) and (9.15). The expected profits of firm $i \in \{1, 2, 3\}$ are given in (9.16). However, the participating firms would like to share their expected cooperative profits according to the Shapley Value.

9.4.1 Expected Coalition Payoffs

Cost-saving opportunities are created under joint venture. In particular, the cost savings in joint venture is depicted as follows:

$$c_i^{\{i\}} \leq c_i^{\{i,j\}}, \quad \text{for } i, j \in \{1, 2, 3\} \text{ and } i \neq j,$$
$$c_i^{\{i,j\}} \leq c_i^{\{i,j,k\}}, \quad \text{for } i, j, k \in \{1, 2, 3\} \text{ and } i \neq j \neq k. \tag{9.40}$$

The firms in the joint venture maximize the sum of their expected profits

$$E_{t_0}\left\{\int_{t_0}^T \sum_{j=1}^3 \left[P_j\left[x^j(s)\right]^{1/2} - c_j u_j^{\{1,2,3\}}(s)\right]\exp\left[-r(s - t_0)\right]ds\right.$$

$$\left.+ \sum_{j=1}^3 \exp\left[-r(T - t_0)\right]q_j\left[x^j(T)\right]^{1/2}\right\}, \tag{9.41}$$

subject to (9.15).

Following the analysis in Sect. 9.2, one can obtain the maximized expected venture profit $W^{(t_0)\{1,2,3\}}(t, x^1, x^2, x^3)$ as in (9.23), and the investment strategies

$$\psi_i^{\{1,2,3\}}(t, x) = \frac{\alpha_i^2}{16(c_i)^2}\left[A_i^{\{1,2,3\}}(t)\right]^2, \quad \text{for } i \in \{1, 2, 3\}. \tag{9.42}$$

The cooperative state dynamics of the joint venture over the time interval $s \in [t_0, T]$ is in (9.26).

For the computation of the dynamic the Shapley Value, we consider cases when two of the firms form a coalition $\{i, j\} \subset \{1, 2, 3\}$ to maximize the expected joint profit

$$E_{t_0}\left\{\int_{t_0}^T \left[P_i\left[x^i(s)\right]^{1/2} - c_i^{\{i,j\}}u_i(s) + P_j\left[x^j(s)\right]^{1/2} - c_j^{\{i,j\}}u_j(s)\right]\right.$$

$$\left. \times \exp\left[-r(s - t_0)\right]ds + \exp\left[-r(T - t_0)\right]\left\{q_i\left[x^i(T)\right]^{1/2} + q_j\left[x^j(T)\right]^{1/2}\right\}\right\}, \tag{9.43}$$

subject to

$$dx^i(s) = \left[\alpha_i\left[u_i(s)x^i(s)\right]^{1/2} - \delta x_i(s)\right]ds + \sigma_i x^i(s)\,dz_i(s),$$

$$x^i(t_0) = x_0^i \in X^i, \quad \text{for } i, j, \in \{1, 2, 3\} \text{ and } i \neq j. \tag{9.44}$$

Following the analysis in Sect. 9.2, we obtain the following value functions:

$$W^{(t_0)\{i,j\}}(t, x^i, x^j) = \left[A_i^{\{i,j\}}(t)(x^i)^{1/2} + A_j^{\{i,j\}}(t)(x^j)^{1/2} + C^{\{i,j\}}(t)\right]$$

$$\times \exp\left[-r(t - t_0)\right], \tag{9.45}$$

for $i, j, \in \{1, 2, 3\}$ and $i \neq j$, where $A_i^{\{i,j\}}(t)$, $A_j^{\{i,j\}}(t)$, and $C^{\{i,j\}}(t)$ satisfy

$$\dot{A}_i^{\{i,j\}}(t) = \left(r + \frac{\delta}{2} + \frac{\sigma_i^2}{8}\right)A_i^{\{i,j\}}(t) - P_i, \quad \text{and} \quad A_i^{\{i,j\}}(T) = q_i$$

for $i, j, \in \{1, 2, 3\}$ and $i \neq j$;

$$\dot{C}^{\{i,j\}}(t) = rC^{\{i,j\}}(t) - \sum_{h \in \{i,j\}} \frac{\alpha_h^2}{16c_h^{\{i,j\}}}\left[A_h^{\{i,j\}}(t)\right]^2,$$

$$C^{\{i,j\}}(T) = 0.$$

Moreover, one can easily derive, for $\tau \in [t_0, T]$,

$$W^{(t_0)\{i,j\}}(t, x^i, x^j) = \exp\left[-r(\tau - t_0)\right]W^{(\tau)\{i,j\}}(t, x^i, x^j),$$

for $i, j, \in \{1, 2, 3\}$ and $i \neq j$.

9.4.2 Subgame Consistent Shapley Value Solution

To formulate a payoff distribution procedure over time so that the agreed imputations satisfy the Shapley Value we invoke Corollary 9.4 and obtain the following.

A PDP with a terminal payment $q^i(x_T^*)$ at time T and an instantaneous payment at time $\tau \in [t_0, T]$

$$B_i(\tau, x_\tau^*)$$

$$= -\sum_{K \subseteq \{1,2,3\}} \frac{(k-1)!(3-k)!}{3!} \left\{ \left[W_t^{(\tau)K}(t, x_t^{K*}) \big|_{t=\tau} \right] - \left[W_t^{(\tau)K \setminus i}(t, x_t^{K \setminus i*}) \big|_{t=\tau} \right] \right.$$

$$+ \sum_{h \in K} \left[\frac{\partial}{\partial x_\tau^{h*}} W^{(\tau)K}(\tau, x_\tau^{K*}) \right] \left[\frac{\alpha_h^2}{4c_h^{\{1,2,3\}}} A_h^{\{1,2,3\}}(\tau)(x_\tau^{i*})^{1/2} - \delta x_\tau^{h*} \right]$$

$$- \sum_{h \in K \setminus i} \left[\frac{\partial}{\partial x_\tau^{h*}} W^{(\tau)K \setminus i}(\tau, x_\tau^{K \setminus i*}) \right] \left[\frac{\alpha_h^2}{4c_h^{\{1,2,3\}}} A_h^{\{1,2,3\}}(\tau)(x_\tau^{i*})^{1/2} - \delta x_\tau^{h*} \right]$$

$$+ \frac{1}{2} \sum_{h,\zeta \in K} (\sigma_h x_\tau^{h*})(\sigma_\zeta x_\tau^{\zeta*}) \frac{\partial^2}{\partial x_\tau^{h*} \partial x_\tau^{\zeta*}} [W^{(\tau)K}(\tau, x_\tau^{K*})]$$

$$- \frac{1}{2} \sum_{h,\zeta \in K \setminus i} (\sigma_h x_\tau^{h*})(\sigma_\zeta x_\tau^{\zeta*}) \frac{\partial^2}{\partial x_\tau^{h*} \partial x_\tau^{\zeta*}} [W^{(\tau)K \setminus i}(\tau, x_\tau^{K \setminus i*})] \right\},$$

for $i \in \{1, 2, 3\}$,   (9.46)

would lead to the realization of the Shapley Value imputations in Condition 9.3.

Using (9.23) and (9.45),

$$\left[W_t^{(\tau)i}(t, x_t^{i*}) \big|_{t=\tau} \right] = r \left[A_i^{\{i\}}(\tau)(x_\tau^{i*})^{1/2} + C_i^{\{i\}}(\tau) \right] + \left[\dot{A}_i^{\{i\}}(\tau)(x_\tau^{i*})^{1/2} + \dot{C}_i^{\{i\}}(\tau) \right],$$

for $i \in \{1, 2, 3\}$;

$$\left[W_t^{(\tau)\{i,j\}}(t, x_t^{i*}) \big|_{t=\tau} \right] = r \left[A_i^{\{i,j\}}(\tau)(x_\tau^{i*})^{1/2} + A_j^{\{i,j\}}(\tau)(x_\tau^{j*})^{1/2} + C^{\{i,j\}}(\tau) \right]$$

$$+ \left[\dot{A}_i^{\{i,j\}}(\tau)(x_\tau^{i*})^{1/2} + \dot{A}_j^{\{i,j\}}(\tau)(x_\tau^{j*})^{1/2} + \dot{C}^{\{i,j\}}(\tau) \right],$$

for $i, j \in \{1, 2, 3\}$ and $i \neq j$;

$$\left[W_t^{(\tau)\{1,2,3\}}(t, x_t^{i*}) \big|_{t=\tau} \right]$$

$$= r \left[A_1^{\{1,2,3\}}(\tau)(x_\tau^{1*})^{1/2} + A_2^{\{1,2,3\}}(\tau)(x_\tau^{2*})^{1/2} + A_3^{\{1,2,3\}}(\tau)(x_\tau^{3*})^{1/2} \right.$$

$$+ C^{\{1,2,3\}}(\tau) \right] + \left[\dot{A}_1^{\{1,2,3\}}(\tau)(x_\tau^{1*})^{1/2} \right.$$

$$+ \dot{A}_2^{\{1,2,3\}}(\tau)(x_\tau^{2*})^{1/2} + \dot{A}_3^{\{1,2,3\}}(\tau)(x_\tau^{3*})^{1/2} + \dot{C}^{\{1,2,3\}}(\tau) \right];$$

$$\left[\frac{\partial}{\partial x_\tau^{h*}} W^{(\tau)K}(\tau, x_\tau^{K*}) \right] = \frac{1}{2} A_h^K(\tau)(x_\tau^{h*})^{-1/2}, \quad \text{for } h \in K \subseteq \{1, 2, 3\},$$

$$\frac{\partial^2}{\partial (x_\tau^{h*})^2}\left[W^{(\tau)K}\left(\tau, x_\tau^{K*}\right)\right] = \frac{-1}{4} A_h^K(\tau)\left(x_\tau^{h*}\right)^{-3/2}, \quad \text{and}$$

$$\frac{\partial^2}{\partial x_\tau^{h*}\partial x_\tau^{\zeta*}}\left[W^{(\tau)K}\left(\tau, x_\tau^{K*}\right)\right] = 0, \quad \text{for } h \neq \zeta.$$

Invoking Theorem 8.2 in Chap. 8 a subgame consistent solution for the joint venture can be obtained as

$$P(x_0, T - t_0) = \left\{u\left(s, x_s^*\right) \text{ and } B\left(s, x_s^*\right) \text{ for } s \in [t_0, T] \text{ and } \xi^{(t_0)}(t_0, x_0)\right\},$$

where $u(s, x_s^*) = \psi^{N*}(s, x^*(s))$ is the set of group optimal strategies in the grand coalition, $B(s, x_s^*)$ is the PDP given in (9.46), and $\xi^{(t_0)}(t_0, x_0)$ is the Shapley Value imputation in Condition 9.3.

Finally, with firms using the cooperative investment strategies the instantaneous receipt of firm i at time instant τ is

$$\zeta_i\left(\tau, x_\tau^*\right) = P_i\left(x_\tau^{i*}\right)^{1/2} - \frac{\alpha_i^2}{16(c_i^{\{1,2,3\}})}\left[A_i^{\{1,2,3\}}(\tau)\right]^2,$$

$$\text{for } i \in \{1, 2, 3\}, x_\tau^* \in X_\tau^*, \text{ and } \tau \in [t_0, T]. \tag{9.47}$$

According to (9.46), the instantaneous payment that firm i should receive under the agreed-upon optimality principle is $B_i(\tau, x_\tau^*)$. Hence an instantaneous transfer payment

$$\chi^i\left(\tau, x_\tau^*\right) = B_i\left(\tau, x_\tau^*\right) - \zeta_i\left(\tau, x_\tau^*\right), \tag{9.48}$$

has to be given or charged to firm i at time τ, for $i \in \{1, 2, 3\}, x_\tau^* \in X_\tau^*$, and $\tau \in [t_0, T]$.

9.5 Infinite-Horizon Analysis

Consider the case when the horizon of the analysis approaches infinity. The state dynamics of the ith firm is characterized by the set of vector-valued differential equations

$$dx^i(s) = f^i\left[x^i(s), u_i(s)\right] ds + \sigma_i\left[x^i(s)\right] dz_i(s), \quad x^i(t_0) = x_0^i, \quad \text{for } i \in N, \tag{9.49}$$

where $\sigma_i[x^i(s)]$ is a $m_i \times \Theta_i$ and $z_i(s)$ is a Θ_i-dimensional Wiener process and the initial state x_0^i is given. Let $\Omega_i[x^i(s)] = \sigma_i[x^i(s)]\sigma_i[x^i(s)]^T$ denote the covariance matrix with its element in row h and column ζ denoted by $\Omega_i^{h\zeta}[x^i(s)]$.

The objective of firm i to be maximized is

$$E_{t_0}\left\{\int_{t_0}^\infty \left(g^i\left[x^i(s)\right] - c_i^{\{i\}}[u_i(s)]\right) \exp[-r(s - t_0)] ds\right\}, \quad \text{for } i \in N. \tag{9.50}$$

Consider the alternative formulation of (9.49) and (9.50) as

$$\max_{u_i} E_t \left\{ \int_t^{\infty} \left(g^i [x^i(s)] - c_i^{\{i\}} [u_i(s)] \right) \exp[-r(s-t)] \, ds \right\}, \quad \text{for } i \in N, \quad (9.51)$$

subject to

$$dx^i(s) = f_i [x^i(s), u_i(s)] \, ds + \sigma_i [x^i(s)] \, dz_i(s), \quad x^i(t) = x^i, \quad \text{for } i \in N. \quad (9.52)$$

The infinite-horizon problem in (9.51) and (9.52) is independent of the choice of t and dependent only upon the state at the starting time.

Invoking Theorem A.6 in the Technical Appendixes, a noncooperative equilibrium can be characterized by a set of strategies $\{\phi_i^*(x) \text{ for } i \in N\}$ constitutes a firm's equilibrium solution to the problem in (9.51) and (9.52), if there exist functionals $\hat{V}^i(x^i) : R^m \to R$ for $i \in N$, satisfying the following set of partial differential equations:

$$r\hat{V}^i(x^i) - \frac{1}{2} \sum_{h,\zeta=1}^{m_i} \Omega^{h\zeta}(x^i) \hat{V}_{x^{i(h)} x^{i(\zeta)}}^i (x^i)$$

$$= \max_{u_i} \{ g(x^i) - c_i^{\{i\}}(u_i) + \hat{V}_x^i(x) f[x^i, u_i] \}. \quad (9.53)$$

Once again, for the sake of clarity in exposition, we consider the case where $m_i = 1$, for $i \in N$.

9.5.1 Infinite-Horizon Dynamic Joint Venture

Consider the case when all these n companies form a joint venture. Cost-saving opportunities are created under joint venture. The cost of control of firm j under the joint venture becomes $c_j^{\{1,2,3\}} [u_j(s)]$. With joint venture cost advantage

$$c_j^{\{1,2,3\}}(u_j) \le c_j^{\{j\}}(u_j), \quad \text{for } j \in N, \quad (9.54)$$

the joint venture would maximize the expected joint venture profit

$$E_t \left\{ \int_t^{\infty} \sum_{j=1}^n (g^j [x^j(s)] - c_j^N [u_j(s)]) \exp[-r(s-t)] \, ds \right\}, \quad (9.55)$$

subject to (9.52).

An optimal solution of the control problem in (9.52) and (9.55) can be characterized using Theorem A.6 in the Technical Appendixes as follows.

Corollary 9.5 *A set of control strategies* $\{\psi_i^*(x) \text{ for } i \in N_1\}$ *provides a solution to the control problem in (9.52) and (9.55), if there exist continuously twice differentiable functions* $W(x) : R^n \to R$, *satisfying the following partial differential equation:*

$$rW(x) - \frac{1}{2} \sum_{h,\zeta=1}^{n} \Omega^{h\zeta}(x) W_{x^h x^\zeta}(x)$$

$$= \max_{u_1, u_2, \ldots, u_n} \left\{ \sum_{j=1}^{n} [g^j(x^j) - c_j^N(u_j)] + \sum_{j=1}^{n} W_{x_j}(x) f^j(x^j, u_j) \right\}, \quad (9.56)$$

where $x = \{x^1, x^2, \ldots, x^n\}$.

Hence the firms will adopt the cooperative control $\{\psi_i^*(x), \text{ for } i \in N\}$, to obtain the maximized level of expected joint profit. Substituting this set of control into (9.52) yields the dynamics of technology advancement under cooperation as

$$dx^i(s) = f^i[x^i(s), \psi_i^*(x(s))] \, ds + \sigma_i[x^i(s)] \, dz_i(s),$$

$$x^i(t_0) = x_0^i, \quad \text{for } i \in N. \quad (9.57)$$

Let $x^*(t) = \{x^{1*}(t), x^{2*}(t), \ldots, x^{n*}(t)\}$ denote the solution to (9.57). The optimal trajectory $\{x^*(t)\}_{t=t_0}^{\infty}$ can be expressed as

$$x^{i*}(t) = x_0^i + \int_{t_0}^{t} f^i[x^{i*}(s), \psi_i^*(x^*(s))] \, ds + \int_{t_0}^{t} \sigma_i[s, x^{i*}(s)] \, dz_i(s),$$

$$\text{for } i \in N. \quad (9.58)$$

We use X_t^* to denote the set of realizable values of $x^*(t)$ at time t generated by (9.58). The term $x_t^* \in X_t^*$ is used to denote an element in X_t^*.

Substituting the optimal extraction strategies in $\{\psi_i^*(x), \text{ for } i \in N\}$ into (9.55) yields the expected venture profit as

$$W(x_t^*) = E_t \left\{ \int_t^{\infty} \sum_{j=1}^{n} (g^j[x^{j*}(s)] - c_j^N[\psi_j^*(x^*(s))]) \exp[-r(s-t)] \, ds \right\}. \quad (9.59)$$

9.5.2 Subgame Consistent Venture Profit Sharing

Consider the case when the firms in the venture share the excess of the total expected cooperative payoff over the sum of the expected individual noncooperative payoffs proportionally to the firms' expected noncooperative payoffs.

The imputation scheme has to fulfill the following condition.

Condition 9.4 An imputation

$$\xi^{(\tau)i}\left(\tau, x_\tau^*\right) = \frac{\hat{V}^i(x_\tau^*)}{\sum_{i=1}^n \hat{V}^i(x_\tau^*)} W\left(x_\tau^*\right), \tag{9.60}$$

is assigned to firm i, for $i \in N$ at time $\tau \in [t_0, \infty)$ if $x^*(\tau) = x_\tau^* \in X_\tau^*$.

To formulate a payoff distribution procedure over time so that the agreed imputations satisfy Condition 9.4 we invoke Theorem 8.3 in Chap. 8 and obtain the following.

Corollary 9.6 *A PDP with an instantaneous payment at time $\tau \in [t_0, \infty)$*

$$B_i\left(\tau, x_\tau^*\right) = r\left[\frac{\hat{V}^i(x_\tau^{i*})}{\sum_{j=1}^n \hat{V}^j(x_\tau^{j*})} W\left(x_\tau^*\right)\right]$$

$$- \frac{1}{2} \sum_{h,\zeta=1}^n \Omega^{h\zeta}\left(x_\tau^*\right) \frac{\partial^2}{\partial x_\tau^{h*}\partial x_\tau^{\zeta*}}\left[\frac{\hat{V}^i(x_\tau^{i*})}{\sum_{j=1}^n \hat{V}^j(x_\tau^{j*})} W\left(x_\tau^*\right)\right]$$

$$- \sum_{h=1}^n \frac{\partial}{\partial x_\tau^{h*}}\left[\frac{\hat{V}^i(x_\tau^{i*})}{\sum_{j=1}^n \hat{V}^j(x_\tau^{j*})} W\left(x_\tau^*\right)\right] f^h\left[x_\tau^{h*}, \psi_h^*(x_\tau^*)\right],$$

for $i \in N$ and $x^*(\tau) = x_\tau^* \in X_\tau^*$, \qquad (9.61)

would lead to the realization of the solution imputations in Condition 9.4.

With (9.61) a subgame consistent solution can be obtained. Note that while the firms use the cooperative investment strategies $\{\psi_i^*(x_\tau^*), \text{ for } i \in N\}$, the instantaneous receipt of firm i at time instant τ is

$$\zeta_i\left(\tau, x_\tau^*\right) = g^i\left(x_\tau^{i*}\right) - c_i^N\left[\psi_i^*(x_\tau^*)\right],$$

for $i \in N, x^*(\tau) = x_\tau^* \in X_\tau^*$ and $\tau \in [t_0, \infty)$.

According to Corollary 9.6, the instantaneous payment that firm i should receive under the agreed-upon optimality principle is $B_i(\tau, x_\tau^*)$, for $i \in N$, as stated in (9.61). Hence an instantaneous transfer payment

$$\chi^i\left(\tau, x_\tau^*\right) = B_i\left(\tau, x_\tau^*\right) - \zeta_i\left(\tau, x_\tau^*\right),$$

has to be given or charged to firm i at time τ, for $i \in N$ if $x^*(\tau) = x_\tau^* \in X_\tau^*$.

9.5.3 Shapley Value Profit Sharing

Consider again the infinite-horizon dynamic venture model in (9.51) and (9.52). The member firms would maximize their expected joint profit and share their expected cooperative profits according to the Shapley Value. If firms are allowed to form different coalitions consisting of a subset of companies $K \subseteq N$. There are k firms in the subset K. In particular, they can obtain cost reduction and with absolute joint venture cost advantage

$$c_j^K[u_j(s)] \leq c_j^L[u_j(s)], \quad \text{for } j \in L \subseteq K, \tag{9.62}$$

where $c_j^K[u_j(s)]$ represents the costs of the controls of the firm j in the subset K and $c_j^L[u_j(s)]$ represents the costs of the controls of the firm j in the subset L.

Moreover, marginal cost advantages lead to

$$\partial c_j^K[u_j(s)]/\partial u_j(s) \leq \partial c_j^L[u_j(s)]/\partial u_j(s), \quad \text{for } j \in L \subseteq K.$$

The expected profit to the joint venture K becomes

$$E_t\left\{\int_t^\infty \sum_{j \in K}(g^j[s, x^j(s)] - c_j^K[u_j(s)])\exp[-r(s-t)]\,ds\right\}, \quad \text{for } K \subseteq N. \tag{9.63}$$

To compute the profit of the joint venture K we have to consider the optimal control problem in (9.62) and (9.63). Invoking Theorem A.6 of the Technical Appendixes, the solution to the stochastic control problem can be characterized as follows.

Corollary 9.7 *A set of controls* $\{\psi_i^{K*}(x^K), \text{for } i \in K \text{ and } t \in [t_0, \infty)\}$ *provides an optimal solution to the stochastic control problem in (9.62) and (9.63) if there exists a continuously twice differentiable function* $W^K(x^K) : R^k \to R$ *satisfying the following equation:*

$$rW^K(x^K) - \frac{1}{2}\sum_{h, \zeta \in K}\Omega^{h\zeta}(x^K)W_{x^h x^\zeta}^K(x^K)$$

$$= \max_{u_K}\left\{\sum_{j \in K}[g^j(x^j) - c_j^K(u_j)] + \sum_{j \in K}W_{x_j}^K(x^K)f^j(x^j, u_j)\right\}. \tag{9.64}$$

Now consider the case of a grand coalition N in which all the n firms are in the coalition. Using the result in Corollary 9.5, the cooperative state trajectory can be obtained as in (9.58).

To share the venture profit among participating firms according to the Shapley Value, the imputation has to satisfy the following.

Condition 9.5 An imputation

$$\xi^{(\tau)i}\left(\tau, x_\tau^{N*}\right) = \sum_{K \subseteq N} \frac{(k-1)!(n-k)!}{n!} \left[W^K\left(x_\tau^{K*}\right) - W^{K \setminus i}\left(x_\tau^{K \setminus i*}\right)\right], \quad (9.65)$$

is assigned to firm i, for $i \in N$ at time τ when the state is x_τ^*.

To formulate a payoff distribution procedure over time so that the agreed imputations satisfy the Shapley Value in Condition 9.5 we invoke Theorem 8.3 in Chap. 8 and obtain the following.

Corollary 9.8 *A PDP with an instantaneous payment at time $\tau \in [t_0, \infty)$*

$$B_i\left(\tau, x_\tau^*\right) = -\sum_{K \subseteq N} \frac{(k-1)!(n-k)!}{n!} \left\{ r W^{K \setminus i}\left(x_\tau^{K \setminus i*}\right) - r W^K\left(x_\tau^{K*}\right) \right.$$

$$+ \sum_{h \in K} \left[\frac{\partial}{\partial x_\tau^{h*}} W^K\left(x_\tau^{K*}\right)\right] f^h\left[x_\tau^{h*}, \psi_h^*\left(x_\tau^*\right)\right]$$

$$- \sum_{h \in K \setminus i} \left[\frac{\partial}{\partial x_\tau^{h*}} W^{K \setminus i}\left(x_\tau^{K \setminus i*}\right)\right] f^h\left[x_\tau^{h*}, \psi_h^*\left(x_\tau^*\right)\right]$$

$$+ \frac{1}{2} \sum_{h,\zeta \in K} \Omega^{h\zeta}\left(x_\tau^*\right) \frac{\partial^2}{\partial x_\tau^{h*} \partial x_\tau^{\zeta*}} \left[W^K\left(x_\tau^{K*}\right)\right]$$

$$\left. - \frac{1}{2} \sum_{h,\zeta \in K \setminus i} \Omega^{h\zeta}\left(x_\tau^*\right) \frac{\partial^2}{\partial x_\tau^{h*} \partial x_\tau^{\zeta*}} \left[W^{K \setminus i}\left(x_\tau^{K \setminus i*}\right)\right] \right\}, \quad \text{for } i \in N, \quad (9.66)$$

would lead to the realization of the Shapley Value in Condition 9.5.

A subgame consistent solution (as that in Theorem 8.3 of Chap. 8) can be constructed with the optimal cooperative strategies and the PDP in (9.66).

9.6 An Infinite-Horizon Stochastic Joint Venture

Consider the infinite-horizon version of the three-company joint venture in Sect. 9.2. The planning period is $[t_0, \infty)$. Company i's expected profit is

$$E_{t_0}\left\{ \int_{t_0}^{\infty} \left[P_i\left[x^i(s)\right]^{1/2} - c_i^{\{i\}} u_i(s)\right] \exp\left[-r(s - t_0)\right] ds \right\}, \quad (9.67)$$

for $i \in \{1, 2, 3\}$.

The evolution of the technology level of company i follows the dynamics

$$dx^i(s) = \left[\alpha_i\left[u_i(s)x^i(s)\right]^{1/2} - \delta x^i(s)\right]ds + \sigma_i x^i(s)\,dz_i(s),$$

$$x^i(t_0) = x_0^i \in X^i, \quad \text{for } i \in \{1, 2, 3\}, \tag{9.68}$$

in the case when each of these three firms acts independently. Using Theorem A.6 in the Technical Appendixes, we obtain the Bellman equation as

$$rW^i(x^i) - \frac{(\sigma_i x^i)^2}{2}W^i_{x^i x^i}(x^i)$$

$$= \max_{u_i}\left\{\left[P_i(x^i)^{1/2} - c_i^{\{i\}}u_i\right] + W^i_{x^i}(x^i)\left[\alpha_i\left(u^i x_i\right)^{1/2} - \delta x^i\right]\right\}, \tag{9.69}$$

for $i \in \{1, 2, 3\}$.

Performing the indicated maximization yields

$$u_i = \frac{\alpha_i^2}{4(c_i^{\{i\}})^2}\left[W^i_{x^i}(x^i)\right]^2 x^i, \quad \text{for } i \in \{1, 2, 3\}.$$

Substituting u_i into (9.69) yields

$$rW^i(x^i) - \frac{(\sigma_i x^i)^2}{2}W^i_{x^i x^i}(x^i)$$

$$= P_i(x^i)^{1/2} - \frac{\alpha_i^2}{4c_i^{\{i\}}}\left[W^i_{x^i}(x^i)\right]^2 x^i$$

$$+ \frac{\alpha_i^2}{2c_i^{\{i\}}}\left[W^i_{x^i}(x^i)\right]^2 x^i - \delta W^i_{x^i}(x^i)x_i, \quad \text{for } i \in \{1, 2, 3\}.$$

Solving the above system of partial differential equations yields

$$W^i(x^i) = \left[A_i^{\{i\}}(x^i)^{1/2} + C_i^{\{i\}}\right], \quad \text{for } i \in \{1, 2, 3\}, \tag{9.70}$$

where

$$0 = \left(r + \frac{\delta}{2} + \frac{\sigma_i^2}{8}\right)A_i^{\{i\}} - P_i, \qquad rC_i^{\{i\}} = \frac{\alpha_i^2}{16c_i^{\{i\}}}(A_i^{\{i\}})^2.$$

9.6.1 Cost-Saving Joint Venture

Consider the case when all three firms agree to form a joint venture and share their expected joint profit proportionally to their expected noncooperative profits. The cost of control of firm j under the joint venture becomes $c_j^{\{1,2,3\}} u_j(s)$, with joint venture cost advantage

$$c_j^{\{1,2,3\}} \le c_j^{\{j\}}, \quad \text{for } j \in N. \tag{9.71}$$

The expected profit of the joint venture is the sum of the participating firms' profits

$$E_t\left\{ \int_t^\infty \sum_{j=1}^3 \left[P_j\left[x^j(s)\right]^{1/2} - c_j^{\{1,2,3\}} u_j(s) \right] \exp\left[-r(s-t)\right] ds \right\}. \tag{9.72}$$

The firms in the joint venture then act cooperatively to maximize (9.72) subject to (9.71). Using Theorem A.6 in the Technical Appendixes, we obtain

$$r W^{\{1,2,3\}}\left(x^1, x^2, x^3\right) - \sum_{h,\zeta=1}^3 \frac{(\sigma_h x^h)(\sigma_\zeta x^\zeta)}{2} W_{x^h x^\zeta}^{\{1,2,3\}}\left(x^1, x^2, x^3\right)$$

$$= \max_{u_1, u_2, u_3}\left\{ \sum_{i=1}^3 \left[P_i\left(x^i\right)^{1/2} - c_i^{\{1,2,3\}} u_i \right] \right.$$

$$\left. + \sum_{i=1}^3 W_{x^i}^{\{1,2,3\}}\left(x^1, x^2, x^3\right)\left[\alpha_i\left[u_i x^i\right]^{1/2} - \delta x^i\right] \right\}. \tag{9.73}$$

Performing the indicated maximization yields

$$u_i = \frac{\alpha_i^2}{4(c_i^{\{1,2,3\}})^2}\left[W_{x^i}^{\{1,2,3\}}\left(x^1, x^2, x^3\right)\right]^2 x^i, \quad \text{for } i \in \{1, 2, 3\}. \tag{9.74}$$

Substituting (9.74) into (9.73) yields

$$r W^{\{1,2,3\}}\left(x^1, x^2, x^3\right) - \sum_{h,\zeta=1}^3 \frac{(\sigma_h x^h)(\sigma_\zeta x^\zeta)}{2} W_{x^h x^\zeta}^{\{1,2,3\}}\left(x^1, x^2, x^3\right)$$

$$= \sum_{i=1}^3 \left[P_i\left(x^i\right)^{1/2} - \frac{\alpha_i^2 x^i}{4 c_i^{\{1,2,3\}}}\left[W_{x^i}^{\{1,2,3\}}\left(x^1, x^2, x^3\right)\right]^2 \right]$$

$$+ \sum_{i=1}^3 W_{x^i}^{\{1,2,3\}}(x_1, x_2, x_3)\left[\frac{\alpha_i^2}{2 c_i^2}\left[W_{x^i}^{\{1,2,3\}}(x_1, x_2, x_3)\right]x^i - \delta x_i\right]. \tag{9.75}$$

Solving (9.75) yields

$$W^{\{1,2,3\}}\left(x^1, x^2, x^3\right) = \left[A_1^{\{1,2,3\}}\left(x^1\right)^{1/2} + A_2^{\{1,2,3\}}\left(x^2\right)^{1/2} + A_3^{\{1,2,3\}}\left(x^3\right)^{1/2} \right.$$

$$\left. + C^{\{1,2,3\}}\right], \tag{9.76}$$

where $A_1^{\{1,2,3\}}$, $A_2^{\{1,2,3\}}$, $A_3^{\{1,2,3\}}$, and $C^{\{1,2,3\}}$ satisfy

$$0 = \left(r + \frac{\delta}{2} + \frac{\sigma_i^2}{8} \right) A_i^{\{1,2,3\}} - P_i,$$

for $i, j, h \in \{1, 2, 3\}$ and $i \neq j \neq h$,

$$rC^{\{1,2,3\}} = \sum_{i=1}^{3} \frac{\alpha_i^2}{16c_i^{\{1,2,3\}}} \left(A_i^{\{1,2,3\}} \right)^2. \tag{9.77}$$

The investment strategies of the grand coalition joint venture can be derived as

$$\psi_i^{\{1,2,3\}}(x) = \frac{\alpha_i^2}{16(c_i^{\{1,2,3\}})^2} \left[A_i^{\{1,2,3\}} \right]^2, \quad \text{for } i \in \{1, 2, 3\}. \tag{9.78}$$

The dynamics of the technological progress of the joint venture over the time interval $s \in [t_0, \infty)$ can be expressed as

$$dx^i(s) = \left(\frac{\alpha_i^2}{4c_i^{\{1,2,3\}}} A_i^{\{1,2,3\}} [x^i(s)]^{1/2} - \delta x^i(s) \right) ds + \sigma_i x^i(s) \, dz_i(s),$$

$$x^i(t_0) = x_0^i, \tag{9.79}$$

for $i \in \{1, 2, 3\}$.

Taking the transforming $y^i(s) = x^i(s)^{1/2}$, for $i \in \{1, 2, 3\}$, equation system (9.79) can be expressed as

$$dy^i(s) = \left(\frac{\alpha_i^2}{8c_i^{\{1,2,3\}}} A_i^{\{1,2,3\}} - \frac{\delta}{2} y^i(s) - \frac{\sigma_i^2}{8} A_i^{\{1,2,3\}} y^i(s) \right) ds + \frac{1}{2} \sigma_i y^i(s) \, dz_i(s),$$

$$y^i(t_0) = \left(x_0^i \right)^{1/2}, \tag{9.80}$$

for $i \in \{1, 2, 3\}$.

Equation (9.80) is a system of linear stochastic differential equations that can be solved by standard techniques. Solving (9.80) yields the joint venture's state trajectory. Let $\{y^{1*}(t), y^{2*}(t), y^{3*}(t)\}$ denote the solution to (9.80). Transforming $x^i = (y^i)^2$, we obtain the state trajectories of the joint venture over the time interval $s \in [t_0, \infty)$ as

$$\{x^*(t)\}_{t=t_0}^{\infty} \equiv \{x^{1*}(t), x^{2*}(t), x^{3*}(t)\}_{t=t_0}^{T}$$

$$= \{[y^{1*}(t)]^2, [y^{2*}(t)]^2, [y^{3*}(t)]^2\}_{t=t_0}^{T}. \tag{9.81}$$

We use X_t^* to denote the set of realizable values of $x^*(t)$ at time t and the term $x_t^* \in X_t^*$ is used to denote an element in X_t^*.

9.6.2 Subgame Consistent Venture Profit Sharing

If the firms agree to share their expected joint profit proportionally to their expected noncooperative profits, the imputation scheme has to fulfill the following condition.

Condition 9.6 An imputation

$$\xi^{(\tau)i}\left(\tau, x_\tau^*\right) = \frac{\hat{V}^i(x_\tau^{i*})}{\sum_{j=1}^3 \hat{V}^j(x_\tau^{i*})} W^{\{1,2,3\}}\left(x_\tau^{1*}, x_\tau^{2*}, x_\tau^{3*}\right), \tag{9.82}$$

is assigned to firm i, for $i \in \{1, 2, 3\}$ at time τ when the state is $x_\tau^* \in X_\tau^*$.

To formulate a payoff distribution procedure over time so that the agreed imputations in Condition 9.6 are satisfied we invoke Corollary 9.6 and obtain the following.

Corollary 9.9 *A PDP with and an instantaneous payment at time $\tau \in [t_0, \infty)$ when $x^*(\tau) = x_\tau^* \in X_\tau^*$:*

$$B_i\left(\tau, x_\tau^*\right) = r\frac{\hat{V}^i(x_\tau^{i*})}{\sum_{j=1}^3 \hat{V}^j(x_\tau^{i*})} W^{\{1,2,3\}}\left(x_\tau^{1*}, x_\tau^{2*}, x_\tau^{3*}\right)$$

$$-\frac{1}{2}\sum_{h,\zeta=1}^3 \sigma_h x^h \sigma_\zeta x^\zeta \frac{\partial^2}{\partial x_\tau^{h*}\partial x_\tau^{\zeta*}}$$

$$\times \left[\frac{\hat{V}^i(x_\tau^{i*})}{\sum_{j=1}^3 \hat{V}^j(x_\tau^{j*})} W^{\{1,2,3\}}\left(x_\tau^{1*}, x_\tau^{2*}, x_\tau^{3*}\right)\right]$$

$$-\sum_{h=1}^3 \frac{\partial}{\partial x_\tau^{h*}}\frac{\hat{V}^i(x_\tau^{i*})}{\sum_{j=1}^3 \hat{V}^j(x_\tau^{i*})} W^{\{1,2,3\}}\left(x_\tau^{1*}, x_\tau^{2*}, x_\tau^{3*}\right)$$

$$\times \left[\frac{\alpha_h^2}{4c_h^{\{1,2,3\}}} A_h^{\{1,2,3\}}\left(x_\tau^{h*}\right)^{1/2} - \delta x_\tau^{h*}(s)\right],$$

for $i \in \{1, 2, 3\}$, $\tag{9.83}$

will lead to the realization of the imputation in Condition 9.6.

A subgame consistent solution can be readily obtained using (9.78) and (9.83). Using the cooperative strategies, the instantaneous receipt of firm i at time instant τ given $x^*(\tau) = x_\tau^* \in X_\tau^*$ is

$$\zeta_i\left(\tau, x_\tau^*\right) = P_i\left(x_\tau^{i*}\right)^{1/2} - \frac{\alpha_i^2}{16(c_i^{\{1,2,3\}})}\left[A_i^{\{1,2,3\}}\right]^2, \tag{9.84}$$

for $i \in \{1, 2, 3\}$ along the cooperative path $\{x^*(t)\}_{t=t_0}^\infty$.

According to (9.83), the instantaneous payment that firm i should receive under the agreed-upon optimality principle is $B_i(\tau, x_\tau^*)$, for $i \in \{1, 2, 3\}$. Hence an instantaneous transfer payment

$$\chi^i(\tau, x_\tau^*) = B_i(\tau, x_\tau^*) - \zeta_i(\tau, x_\tau^*), \tag{9.85}$$

has to be given or charged to firm i at time τ given $x^*(\tau) = x_\tau^* \in X_\tau^*$, for $i \in \{1, 2, 3\}$ along the cooperative path $\{x^*(t)\}_{t=t_0}^\infty$.

9.6.3 Shapley Value Solution

Consider the case when the participating firms agree to share their expected cooperative profits according to the Shapley Value. For the computation of the dynamic of the Shapley Value, we consider cases when two of the firms form a coalition $\{i, j\} \subset \{1, 2, 3\}$. The cost savings in the joint venture are depicted as follows:

$$\begin{aligned}
c_i^{\{i\}} &\leq c_i^{\{i,j\}}, && \text{for } i, j \in \{1, 2, 3\} \text{ and } i \neq j, \\
c_i^{\{i,j\}} &\leq c_i^{\{i,j,k\}}, && \text{for } i, j, k \in \{1, 2, 3\} \text{ and } i \neq j \neq k.
\end{aligned} \tag{9.86}$$

Coalition $\{i, j\}$ would maximize the expected joint profit

$$E_t \Bigg\{ \int_t^\infty \Big[P_i [x^i(s)]^{1/2} - c_i u_i(s) + P_j [x^j(s)]^{1/2} - c_j u_j(s) \Big]$$

$$\times \exp[-r(s - t)] ds \Bigg\}, \tag{9.87}$$

subject to (9.68).

Following the above analysis, we obtain the following value functions:

$$W^{\{i,j\}}(x^i, x^j) = \big[A_i^{\{i,j\}}(x^i)^{1/2} + A_j^{\{i,j\}}(x^j)^{1/2} + C^{\{i,j\}} \big], \tag{9.88}$$

for $i, j, \in \{1, 2, 3\}$ and $i \neq j$, where $A_i^{\{i,j\}}$, $A_j^{\{i,j\}}$, and $C^{\{i,j\}}$ satisfy

$$0 = \left(r + \frac{\delta}{2} + \frac{\sigma_i^2}{8} \right) A_i^{\{1,2\}} - P_i, \quad \text{for } i, j, \in \{1, 2, 3\} \text{ and } i \neq j, \quad \text{and}$$

$$rC^{\{i,j\}} = \sum_{h \in \{i,j\}} \frac{\alpha_h^2}{16 c_h^{\{i,j\}}} \big(A_h^{\{i,j\}} \big)^2.$$

To formulate a payoff distribution procedure over time so that the agreed imputations satisfy the Shapley Value in Condition 9.5 we invoke Corollary 9.8 and obtain the following.

Corollary 9.10 *A PDP with an instantaneous payment at time $\tau \in [t_0, \infty)$:*

$$
B_i\left(\tau, x_\tau^*\right) = - \sum_{K \subseteq \{1,2,3\}} \frac{(k-1)!(3-k)!}{3!} \left\{ r W^{K \backslash i}\left(x_\tau^{K \backslash i*}\right) - r W^K\left(x_\tau^{K*}\right) \right.
$$

$$
+ \sum_{h \in K} \left[\frac{\partial}{\partial x_\tau^{h*}} W^K\left(x_\tau^{K*}\right)\right]\left[\frac{\alpha_h^2}{4c_h^{\{i,j\}}} A_h^{\{1,2,3\}}\left(x_\tau^{i*}\right)^{1/2} - \delta x_\tau^{h*}\right]
$$

$$
- \sum_{h \in K \backslash i} \left[\frac{\partial}{\partial x_\tau^{h*}} W^{K \backslash i}\left(x_\tau^{K \backslash i*}\right)\right]\left[\frac{\alpha_h^2}{4c_h^{\{i,j\}}} A_h^{\{1,2,3\}}\left(x_\tau^{i*}\right)^{1/2} - \delta x_\tau^{h*}\right]
$$

$$
+ \frac{1}{2} \sum_{h,\zeta \in K} \left(\sigma_h x_\tau^{h*}\right)\left(\sigma_\zeta x_\tau^{\zeta*}\right) \frac{\partial^2}{\partial x_\tau^{h*} \partial x_\tau^{\zeta*}}\left[W^K\left(x_\tau^{K*}\right)\right]
$$

$$
\left. - \frac{1}{2} \sum_{h,\zeta \in K \backslash i} \left(\sigma_h x_\tau^{h*}\right)\left(\sigma_\zeta x_\tau^{\zeta*}\right) \frac{\partial^2}{\partial x_\tau^{h*} \partial x_\tau^{\zeta*}}\left[W^{K \backslash i}\left(x_\tau^{K \backslash i*}\right)\right] \right\}, \qquad (9.89)
$$

for $i \in \{1, 2, 3\}$, would lead to the realization of the Shapley Value in Condition 9.5.

In particular, $W^K\left(x_\tau^{K*}\right)$ is given in (9.70), (9.76), and (9.88) and

$$
\left[\frac{\partial}{\partial x_\tau^{h*}} W^K\left(x_\tau^{K*}\right)\right] = \frac{1}{2} A_h^K\left(x_\tau^{h*}\right)^{-1/2}, \quad \text{for } h \in K \subseteq \{1, 2, 3\},
$$

$$
\frac{\partial^2}{\partial (x_\tau^{h*})^2}\left[W^K\left(x_\tau^{K*}\right)\right] = \frac{-1}{4} A_h^K\left(x_\tau^{h*}\right)^{-3/2}, \quad \text{and}
$$

$$
\frac{\partial^2}{\partial x_\tau^{h*} \partial x_\tau^{\zeta*}}\left[W^K\left(x_\tau^{K*}\right)\right] = 0, \quad \text{for } h \neq \zeta.
$$

A subgame consistent solution can be obtained using (9.78) and (9.89).

9.7 Exercises

9.1 Consider the case when there are three companies involved in a joint venture. The planning period is $[0, 2]$. We use $x_i(s)$ to denote the level of technology of company i at time $s \in [0, 2]$, and $u_i(s) \subset R^+$ is its physical investment in technological advancement. The increments of the levels of technology are subject to stochastic disturbances. The discount rate is 0.05. The salvage values of the firms' technologies are $4[x^1(2)]^{1/2}$, $3[x^2(2)]^{1/2}$, and $1.5[x^3(2)]^{1/2}$. If the companies act independently, the costs of the physical investment of these three firms are, respectively, $2u_1(s)$, $3u_2(s)$, and $1.5u_3(s)$.

The expected profits for companies 1, 2, and 3 are, respectively,

$$E\left\{\int_0^2 \left[10[x^1(s)]^{1/2} - 2u_1(s)\right]\exp(-0.05s)\,ds + \exp[-0.05(4)]4[x^1(2)]^{1/2}\right\},$$

$$E\left\{\int_0^2 \left[8[x^2(s)]^{1/2} - 3u_2(s)\right]\exp(-0.05s)\,ds\right.$$

$$\left. + \exp[-0.05(4)]3[x^2(2)]^{1/2}\right\}, \quad \text{and}$$

$$E\left\{\int_0^2 \left[12[x^3(s)]^{1/2} - 1.5u_3(s)\right]\exp(-0.05s)\,ds + \exp[-0.05(4)]1.5[x^3(2)]^{1/2}\right\}.$$

The evolution of the technology level of company $i \in \{1, 2, 3\}$ follows a system of stochastic dynamics

$$dx^1(s) = \left[4[u_1(s)x^1(s)]^{1/2} - 0.1x^1(s)\right]ds + 0.5x^1(s)\,dz_1(s), \quad x^1(0) = 30,$$

$$dx^2(s) = \left[2[u_2(s)x^2(s)]^{1/2} - 0.08x^2(s)\right]ds$$

$$+ 0.8x^2(s)\,dz_2(s), \quad x^2(0) = 20, \quad \text{and}$$

$$dx^3(s) = \left[3[u_3(s)x^3(s)]^{1/2} - 0.05x^3(s)\right]ds + 0.75x^i(s)\,dz_3(s), \quad x^3(0) = 25,$$

where $z_1(s)$, $z_2(s)$, and $z_3(s)$ are independent Wiener processes.

Compute a Nash equilibrium solution when these three firms act independently.

9.2 Consider the case when these three companies form a joint venture. The participating firms in a coalition can gain core skills and technology from each other. In particular, they can obtain cost reduction and with absolute joint venture cost advantage.

With joint venture cost advantage, the cost of the investment of firm $j \in \{1, 2, 3\}$ under the joint venture becomes $c_j^{\{1,2,3\}}u_j(s)$, where $c_1^{\{1,2,3\}} = 1$, $c_2^{\{1,2,3\}} = 1.5$, and $c_3^{\{1,2,3\}} = 0.8$.

If the joint venture firms agree to maximize their expected joint profit and share the excess gain equally, characterize a subgame consistent solution.

9.3 Consider the joint venture in Exercise 9.2. In particular, the firms would like to share the expected venture profit according to the Shapley Value. The costs under joint ventures in different coalitions $K \subseteq \{1, 2, 3\}$ are

$$c_1^{\{1,2,3\}} = 1, \qquad c_2^{\{1,2,3\}} = 1.5, \quad \text{and} \quad c_3^{\{1,2,3\}} = 0.8;$$

$$c_1^{\{1,2\}} = 1.5 \quad \text{and} \quad c_2^{\{1,2\}} = 2.5;$$

$$c_1^{\{1,3\}} = 1.4 \quad \text{and} \quad c_3^{\{1,3\}} = 1;$$

$$c_2^{\{2,3\}} = 2 \quad \text{and} \quad c_3^{\{2,3\}} = 1.1;$$

$$c_1^{\{1\}} = 2, \qquad c_2^{\{2\}} = 3, \quad \text{and} \quad c_3^{\{3\}} = 1.5.$$

Characterize a subgame consistent solution.

Chapter 10
Collaborative Environmental Management Under Uncertainty

In this chapter, we introduce stochastic elements in collaborative environmental management. Similar to the deterministic analysis in Chap. 6, the industrial sector is characterized by an international trading zone involving n nations or regions. Each government adopts its own abatement policy and tax scheme to reduce pollution. The governments have to promote business interests and at the same time have to handle the financing of the costs brought about by pollution. The industrial sectors remain competitive among themselves while the governments cooperate in pollution abatement. Industrial production creates two types of negative environmental externalities. First, pollutants emitted via industrial production cause short-term local impacts on neighboring areas of the origin of production. Examples of these short-term local impacts include passing-by waste in waterways, wind-driven suspended particles in the air, unpleasant odor, noise, dust, and heat generated in the production processes. Second, the emitted pollutants will add to the existing pollution stock in the environment and produce long-term impacts to extensive and far-away areas. Greenhouse-gases, CFC, and atmospheric particulates are examples of this form of negative environmental externality. This specification permits the proximity of the origin of industrial production to receive heavier environmental damages as production increases. Given these neighboring impacts, the individual government tax policy has to take into consideration the tax policies of other nations and these policies' intricate effects on outputs and environmental effects. In particular, while designing tax policies to curtail their outputs, governments have to consider the inducement to neighboring nations' output that can cause local negative environmental impacts to themselves.

To incorporate the widely observed uncertainty in nature's capability to replenish the environment this article adopts a stochastic pollution stock dynamics and formulates a cooperative stochastic differential game of transboundary industrial pollution. The number of solvable cooperative stochastic differential games so far remains low because of the difficulties in deriving tractable solutions (like Haurie et al. 1994; Yeung and Petrosyan 2004, 2005, 2006a). The stringent condition of subgame consistency is required for a dynamically stable cooperative solution in stochastic differential games.

D.W.K. Yeung, L.A. Petrosyan, *Subgame Consistent Economic Optimization*,
Static & Dynamic Game Theory: Foundations & Applications,
DOI 10.1007/978-0-8176-8262-0_10, © Springer Science+Business Media, LLC 2012

An analytical framework is constructed in Sect. 10.1 and its noncooperative outcome is characterized in Sect. 10.2. Cooperative arrangement and subgame consistent collaborative environmental schemes are presented in the next two sections. A stochastic industrial pollution model is presented and a subgame consistent collaboration management scheme is explicitly characterized in Sects. 10.5 and 10.6.

10.1 An Analytical Framework

In this section we present an analytical framework to study transboundary industrial pollution management under uncertainty.

10.1.1 The Industrial Sector

Following Chap. 6 we consider a global economy that is comprised of n nations. At time instant s the demand system of the outputs of the nations is

$$P_i(s) = f^i[q_1(s), q_2(s), \ldots, q_n(s), s], \quad i \in N \equiv \{1, 2, \ldots, n\}, \qquad (10.1)$$

where $P_i(s)$ is the price vector of the output vector of nation i and $q_j(s)$ is the output of nation j. The demand system in (10.1) shows that the world economy is a form of generalized differentiated products oligopoly.

Industrial profits of nation i at time s can be expressed as

$$f^i[q_1(s), q_2(s), \ldots, q_n(s), s]q_i(s) - c^i[q_i(s), v_i(s)], \quad \text{for } i \in N, \qquad (10.2)$$

where $v_i(s)$ is the set of environmental policy instruments of government i. Policy instruments may include tools like taxes, subsidies, technology choices, and pollution legislations. The cost of producing $q_i(s)$ under policy $v_i(s)$ is $c^i[q_i(s), v_i(s)]$.

Profit maximization by the industrial sectors yields

$$f^i[q_1(s), q_2(s), \ldots, q_n(s), s] + f^i_{q_i}[q_1(s), q_2(s), \ldots, q_n(s), s]q_i(s)$$
$$- c^i_{q_i}[q_i(s), v_i(s)] = 0, \quad \text{for } i \in N. \qquad (10.3)$$

Nation i's instantaneous market equilibrium output can be expressed as

$$q_i^*(s) = \hat{q}^i[v_1(s), v_2(s), \ldots, v_n(s), s] \equiv \hat{q}^i[v(s), s], \quad \text{for } i \in N. \qquad (10.4)$$

The fact that each nation's output decision depends on government environmental policies is reflected in (10.4).

10.1.2 Impacts of Pollution and Stochastic Accumulation Dynamics

Industrial production emits pollutants into the environment and the amount of pollution created by different nations' outputs may be different. For an output of $q_i(s)$ produced by nation i, there will be an instantaneous damaging environmental impact of $\varepsilon_i^i[q_i(s)]$ on nation i itself and a damaging impact of $\varepsilon_j^i[q_i(s)]$ on its adjacent nation j for $j \in \underline{K}^i$. On the other hand, nation i will receive instantaneous damaging environmental impacts from its adjacent nations measured as $\varepsilon_i^j[q_j(s)]$ for $j \in \bar{K}^i$. This type of externality is typical in spatial environmental impacts.

Moreover, the pollutant will then add to the stock of existing pollution. Each government adopts its own pollution abatement policy to reduce the pollution stock. Let $x(s) \subset R^m$ denote the level of pollution at time s. The dynamics of the pollution stock is governed by the stochastic differential equation

$$
dx(s) = \left(\sum_{j=1}^{n} a_j\big[q_j(s), v_j(s)\big] - \sum_{j=1}^{n} b_j\big[u_j(s), x(s)\big] - \delta\big[x(s)\big]x(s) \right) ds
$$
$$
+ \sigma\big[s, x(s)\big] dz(s),
$$
$$
x(t_0) = x_{t_0}, \tag{10.5}
$$

where $\sigma[s, x(s)]$ is a scaling function, $z(s)$ is a Wiener process, $\mathbf{\Omega}[s, x(s)] = \sigma[s, x(s)]\sigma[s, x(s)]'$ is the covariance matrix with its element in row h and column ζ denoted by $\Omega^{h\zeta}[s, x(s)]$, $a_j[q_j(s), v_j(s)]$ is the amount of pollution created by the $q_j(s)$ amount of output produced under policy $v_j(s)$, $u_j(s)$ is the pollution abatement effort of nation j, $b_j[u_j(s), x(s)]$ is the amount of pollution removed by the $u_j(s)$ unit of abatement effort of nation j, and $\delta[x(s)]$ is the natural rate of decay of the pollutants.

10.1.3 The Governments' Objectives

The governments have to promote business interests and at the same time handle the financing of the costs brought about by pollution. In particular, each government maximizes the net gains in the industrial sector plus tax revenue minus expenditures on pollution abatement and damages from pollution. A lump-sum income tax is levied on the industrial sector to balance the government budget. The last item turns out to be a net transfer between the government and the public (with no effect on industrial output). The instantaneous objective of government i at time s can be expressed as

$$
f^i\big[q_1(s), q_2(s), \ldots, q_n(s), s\big]q_i(s) - c^i\big[q_i(s), v_i(s)\big] - c_i^P\big[v_i(s)\big] - c_i^a\big[u_i(s)\big]
$$
$$
- \varepsilon_i^i\big[q_i(s)\big] - \sum_{j \in \bar{K}^i} \varepsilon_i^j\big[q_j(s)\big] - h_i\big[x(s)\big], \quad i \in N, \tag{10.6}
$$

where $c_i^P[v_i(s)]$ is the cost of implementing the vector policy instrument $v_i(s)$, $c_i^a[u_i(s)]$ is the cost of employing a u_i amount of pollution abatement effort, and $h_i[x(s)]$ is the value of damage to country i from an $x(s)$ amount of pollution.

The governments' planning horizon is $[t_0, T]$. It is possible that T may be very large. The discount rate is r. At time T, the terminal appraisal of pollution damage is $g^i[x(T)]$ where $\partial g^i/\partial x < 0$. Each one of the n governments seeks to maximize the integral of its expected instantaneous objective in (10.6) over the planning horizon subject to the pollution dynamics of (10.5) with controls on the level of abatement effort and output tax.

Substitute $q_i(s)$, for $i \in N$, from (10.4) into (10.5) and (10.6) one obtains a stochastic differential game in which government $i \in N$ seeks to

$$
\max_{v_i(s), u_i(s)} E_{t_0} \left\{ \int_{t_0}^{T} \left[f^i \{ \hat{q}^1[v(s), s], \hat{q}^2[v(s), s], \ldots, \hat{q}^n[v(s), s], s \} \hat{q}^i[v(s), s] \right. \right.
$$

$$
- c^i \{ \hat{q}^i[v(s), s], v_i(s) \} - c_i^P[v_i(s)] - c_i^a[u_i(s)] - \varepsilon_i^i \{ \hat{q}^i[v(s), s] \}
$$

$$
\left. \left. - \sum_{j \in \bar{K}^i} \varepsilon_i^j \{ \hat{q}^j[v(s), s] \} - h_i[x(s)] \right] e^{-r(s-t_0)} \, ds + g^i[x(T)] e^{-r(T-t_0)} \right\}, \quad (10.7)
$$

subject to

$$
dx(s) = \left(\sum_{j=1}^{n} a_j \{ \hat{q}^j[v(s), s], v_j(s) \} - \sum_{j=1}^{n} b_j[u_j(s), x(s)] - \delta[x(s)]x(s) \right) ds
$$

$$
+ \sigma[s, x(s)] \, dz(s),
$$

$$
x(t_0) = x_{t_0}. \quad (10.8)
$$

Thus the economic interactions among nations in industrial production, pollution emission, and abatement are characterized as a stochastic differential game with the expected the payoffs in (10.7) and stochastic pollution dynamics of (10.8).

10.2 Noncooperative Outcomes

In this section we discuss the solution to the stochastic differential game in (10.7) and (10.8). Since the payoffs of nations are measured in monetary terms, the game is a transferable payoff game. Under a noncooperative framework, a Nash equilibrium solution can be characterized as the following (see Theorem 2.5 in Chap. 2).

Corollary 10.1 *A set of feedback strategies* $\{u_i^*(t) = \mu_i(t, x), v_i^*(t) = \phi_i(t, x),$ *for* $i \in N\}$, *provides a feedback Nash equilibrium solution to the game in* (10.7) *and* (10.8) *if there exist suitably smooth functions* $V^{(t_0)i}(t, x) : [t_0, T] \times R \to R, i \in N$, *satisfying the following partial differential equations:*

$$
-V_t^{(t_0)i}(t, x) - \frac{1}{2} \sum_{h, \zeta = 1}^m \Omega^{h\zeta}(t, x) V_{x^h x^\zeta}^{(t_0)i}(t, x)
$$

$$
= \max_{v_i, u_i} \left\{ \left[f^i \left\{ \hat{q}^1 [v_i, \phi_{\neq i}(t, x), t], \hat{q}^2 [v_i, \phi_{\neq i}(t, x), t], \ldots, \hat{q}^n [v_i, \phi_{\neq i}(t, x), t] \right\} \right. \right.
$$

$$
\times \hat{q}^i [v_i, \phi_{\neq i}(t, x), t] - c \{\hat{q}^i [v_i, \phi_{\neq i}(t, x), t], v_i\} - c_i^P [v_i] - c_i^a [u_i]
$$

$$
\left. - \varepsilon_i^i \{\varphi^i [v_i, \phi_{\neq i}(t, x), t]\} - \sum_{j \in \bar{K}^i} \varepsilon_i^j \{\hat{q}^j [v_i, \phi_{\neq i}(t, x), t]\} - h_i(x) \right] e^{-r(t - t_0)}
$$

$$
+ V_x^{(t_0)i}(t, x) \left[\sum_{j=1}^n a_j \{\hat{q}^j [v_i, \phi_{\neq i}(t, x), t], v_j\} - b_i(u_i, x) \right.
$$

$$
\left. \left. - \sum_{\substack{j=1 \\ j \neq i}}^n b_j [\mu_j(t, x), x] - \delta(x)x \right] \right\}, \tag{10.9}
$$

$$
V^{(t_0)i}(T, x) = g^i [x] e^{-r(T - t_0)}, \tag{10.10}
$$

where

$$
\phi_{\neq i}(t, x) = \left[\phi^1(t, x), \phi^2(t, x), \ldots, \phi^{i-1}(t, x), \phi^{i+1}(t, x), \ldots, \phi^n(t, x) \right].
$$

In a prevailing Nash equilibrium the function $V^{(t_0)i}(t, x)$ is then the integral

$$
E_{t_0} \left\{ \int_t^T \left[f^i \{ \hat{q}^1 [\phi(s, x(s)), s], \hat{q}^2 [\phi(s, x(s)), s], \ldots, \hat{q}^n [\phi(s, x(s)), s], s \} \right. \right.
$$

$$
\times \hat{q}^i [\phi(s, x(s)), s] - c^i \{ \hat{q}^i [\phi(s, x(s)), s], \phi_i(s, x(s)) \}
$$

$$
- c_i^P [\phi_i(s, x(s))] - c_i^a [\mu_i(s, x(s))] - \varepsilon_i^i \{ \hat{q}^i [\phi(s, x(s)), s] \}
$$

$$
\left. - \sum_{j \in \bar{K}^i} \varepsilon_i^j \{ \hat{q}^j [\phi(s, x(s)), s] \} - h_i [x(s)] \right] e^{-r(s - t_0)} \, ds
$$

$$
\left. + g^i [x(T)] e^{-r(T - t_0)} \, \middle| \, x(t) = x \in X \right\}, \quad \text{for } i \in N. \tag{10.11}
$$

The game equilibrium dynamics then becomes

$$
dx(s) = \left(\sum_{j=1}^{n} a_j \{ \hat{q}^j [\phi(s, x(s)), s], \phi_j(s, x(s)) \} - \sum_{j=1}^{n} b_j [\mu_j(s, x(s)), x(s)] \right.
$$

$$
\left. - \delta[x(s)]x(s) \right) ds + \sigma[s, x(s)] dz(s),
$$

$$
x(t_0) = x_{t_0}. \tag{10.12}
$$

Remark 10.1 One can readily verify that $V^{(\tau)i}(t, x_t) = V^i(t, x_t)e^{r(\tau - t_0)}$, for $\tau \in [t_0, T]$, is the value function to nation i at time $t \in [\tau, T]$ when the state $x(t) = x_t$ in the game in (10.7) and (10.8), which starts at time τ.

10.3   Cooperative Arrangement

Now consider the case when all the nations agree to cooperate and adhere to an optimality principle. Since the nations are asymmetric and the number of nations may be large, a reasonable solution optimality principle for gain distribution is to share the expected gain from cooperation proportional to the nations' relative sizes of expected noncooperative payoffs. Group optimality requires the nations to maximize their joint payoff. Hence the optimality principle entails (i) group optimality, and (ii) sharing the expected gain from cooperation proportional to the nations' relative sizes of expected noncooperative payoffs.

For the cooperative scheme to be sustainable the agreed-upon optimality has to be upheld throughout the game horizon.

10.3.1   Group Optimality and Cooperative State Trajectory

Consider the collaborative environmental scheme with the participating nations' expected payoff structure in (10.7) and pollution dynamics in (10.8). To secure group optimality the participating nations seek to maximize their expected joint payoff by solving the following stochastic control problem:

$$
\max_{v_1, v_2, \dots, v_n; u_1, u_2, \dots, u_n} E_{t_0} \left\{ \int_{t_0}^{T} \left[\sum_{i=1}^{n} f^i \{ [\hat{q}^1[v(s), s], \hat{q}^2[v(s), s], \dots, \hat{q}^n[v(s), s], s] \} \right. \right.
$$

$$
\times \hat{q}^i[v(s), s] - c^i\{\hat{q}^i[v(s), s], v_i(s)\}
$$

$$
\left. - c_i^P[v_i(s)] - c_i^a[u_i(s)] - \varepsilon_i^i\{\hat{q}^i[v(s), s]\} \right.
$$

$$- \sum_{j \in \bar{K}^i} \varepsilon_i^j \left\{ \hat{q}^j [v(s), s] \right\} - h_i [x(s)] \right] e^{-r(t-t_0)} \, ds$$

$$+ \sum_{i=1}^{n} g^i [x(T)] e^{-r(T-t_0)} \Bigg\},$$ (10.13)

subject to (10.8).

Invoking Theorem A.5 in the Technical Appendixes, a set of controls $\{[v_i^{**}(t), u_i^{**}(t)] = [\psi_i(t, x), \varpi_i(t, x)], \text{ for } i \in N\}$ constitutes an optimal solution to the stochastic control problem in (10.13) and (10.8) if there exists a continuously twice differentiable function $W^{(t_0)}(t, x) : [t_0, T] \times R^m \to R, i \in N$, satisfying the following partial differential equations:

$$- W_t^{(t_0)}(t, x) - \frac{1}{2} \sum_{h, \zeta=1}^{m} \Omega^{h\zeta}(t, x) W_{x^h x^\zeta}^{(t_0)}(t, x)$$

$$= \max_{v_1, v_2, \ldots, v_n; u_1, u_2, \ldots, u_n} \left\{ \sum_{i=1}^{n} f^i [\hat{q}^1(v, t), \hat{q}^2(v, t), \ldots, \hat{q}^n(v, t), t] \hat{q}^i(v, t) \right.$$

$$- c^i [\hat{q}^i(v, t), v_i] - c_i^P(v_i) - c_i^a(u_i) - \varepsilon_i^i [\hat{q}^i(v, t)]$$

$$- \sum_{j \in \bar{K}^i} \varepsilon_i^j [\hat{q}^j(v, t)] - h_i(x)] e^{-r(t-t_0)}$$ (10.14)

$$+ W_x^{(t_0)}(t, x) \left[\sum_{j=1}^{n} a_j [\hat{q}^j(v, t), v_j] - \sum_{j=1}^{n} b_j(u_j, x) - \delta(x)x \right] \right\}, \quad \text{and}$$

$$W^{(t_0)}(T, x) = \sum_{i=1}^{n} g^i(x) e^{-r(T-t_0)}.$$

Hence the nations will adopt the cooperative control $\{[\psi_i(t, x), \varpi_i(t, x)], \text{ for } i \in N \text{ and } t \in [t_0, T]\}$. The optimal trajectory under cooperation becomes

$$dx(s) = \left(\sum_{j=1}^{n} a_j \{\hat{q}^j [\psi(s, x(s)), s], \psi_j(s, x(s))\} - \sum_{j=1}^{n} b_j [\varpi_j(s, x(s)), x(s)] \right.$$

$$- \delta[x(s)] x(s) \right) ds + \sigma[s, x(s)] dz(s), \quad x(t_0) = x_{t_0}.$$ (10.15)

The solution to (10.15) can be expressed as

$$
x^*(t) = x_0 + \int_{t_0}^{t} \left\{ \sum_{j=1}^{n} a_j \{ \hat{q}^j [\psi(s, x^*(s)), s], \psi_j(s, x^*(s)) \} \right.
$$
$$
\left. - \sum_{j=1}^{n} b_j [\varpi_j(s, x^*(s)), x^*(s)] - \delta[x^*(s)] x^*(s) \right\} ds
$$
$$
+ \int_{t_0}^{t} \sigma[s, x(s)] dz(s). \tag{10.16}
$$

We use X_t^* to denote the set of realizable values of $x^*(t)$ at time t generated by (10.16). The term x_t^* is used to denote an element in the set X_t^*.

The cooperative control for the game $\Gamma_c(x_0, T - t_0)$ over the time interval $[t_0, T]$ can be expressed more precisely as

$$
\psi_i(t, x^*(t)), \qquad \varpi_i(t, x^*(t)) \quad \text{for } t \in [t_0, T] \text{ and } i \in N. \tag{10.17}
$$

Note that for group optimality to be achievable, the cooperative controls in (10.17) must be exercised throughout time interval $[t_0, T]$.

The value function $W^{(t_0)i}(t, x)$ reflects

$$
E_{t_0} \left\{ \int_{t}^{T} \left[\sum_{i=1}^{n} f^i \{ \hat{q}^1 [\psi(s, x^*(s)), s], \hat{q}^2 [\psi(s, x^*(s)), s], \dots, \right. \right.
$$
$$
\hat{q}^n [\psi(s, x^*(s)), s], s \}
$$
$$
\times \hat{q}^i [\psi(s, x^*(s)), s] - c^i \{ \hat{q}^i [\psi(s, x^*(s)), s], \psi_i(s, x^*(s)) \}
$$
$$
- c_i^P [\psi_i(s, x^*(s))] - c_i^a [\varpi_i(s, x^*(s))] - \varepsilon_i^i \{ \hat{q}^i [\psi(s, x^*(s)), s] \}
$$
$$
\left. - \sum_{j \in \bar{K}^i} \varepsilon_i^j \{ \hat{q}^j [\psi(s, x^*(s)), s] \} - h_i[x^*(s)] \right] e^{-r(s-t_0)} ds
$$
$$
\left. + \sum_{i=1}^{n} g^i [x^*(T)] e^{-r(T-t_0)} \right\}, \quad \text{for } i \in N, \tag{10.18}
$$

where $\psi^*(s, x^*(s)) = \{ \psi_1^*(s, x^*(s)), \psi_2^*(s, x^*(s)), \dots, \psi_n^*(s, x^*(s)) \}$.

Remark 10.2 One can readily verify that $W^{(\tau)}(t, x_t^*) = W^{(t_0)}(t, x_t^*) e^{r(\tau - t_0)}$, for $\tau \in [t_0, T]$, is the value function at time $t \in [\tau, T]$ of the control problem in (10.8) and (10.13) that starts at time τ with $x(t) = x_t^* \in X_t^*$.

10.3.2 Imputation Scheme

The agreed-upon optimality principle must be maintained at every instant of time within the cooperative duration $[t_0, T]$ given any realizable state $x_\tau^* \in X_\tau^*$ generated by the cooperative trajectory in (10.16). For $\tau \in [t_0, T]$, let $\xi^{(\tau)i}(\tau, x_\tau^*)$ denote the solution imputation (payoff under cooperation) over the period $[\tau, T]$ to nation $i \in N$ given that the state is $x_\tau^* \in X_\tau^*$.

Hence the imputation scheme $\{\xi^{(\tau)i}(\tau, x_\tau^*); \text{ for } i \in N\}$ has to satisfy the following condition:

Condition 10.1

$$\xi^{(\tau)i}(\tau, x_\tau^*) = \frac{V^{(\tau)i}(\tau, x_\tau^*)}{\sum_{j=1}^n V^{(\tau)j}(\tau, x_\tau^*)} W^{(\tau)}(\tau, x_\tau^*), \tag{10.19}$$

for $i \in N$, $x_\tau^* \in X_\tau^*$ and $\tau \in [t_0, T]$.

The imputation scheme in Condition 10.1 satisfies individual rationality. Crucial to the analysis is the formulation of a payment distribution mechanism that would lead to the realization of Condition 10.1. This will be done in the next section.

10.4 Subgame Consistent Collaborative Environmental Management

To formulate a payoff distribution procedure over time so that the agreed imputations satisfy Condition 10.1 we invoke Theorem 8.1 in Chap. 8 and obtain the following.

Corollary 10.2 *A distribution scheme with a terminal payment* $-g^i[x_T^* - \bar{x}^i]$ *at time T and an instantaneous payment at time* $\tau \in [t_0, T]$

$$B_i(\tau, x_\tau^*) = -\frac{\partial}{\partial t}\left[\frac{V^{(\tau)i}(t, x_t^*)}{\sum_{j=1}^n V^{(\tau)j}(t, x_t^*)} W^{(\tau)}(t, x_t^*)\Big|_{t=\tau}\right]$$

$$-\frac{\partial}{\partial x_\tau^*}\left[\frac{V^{(\tau)i}(\tau, x_\tau^{i*})}{\sum_{j=1}^n V^{(\tau)j}(\tau, x_\tau^{j*})} W^{(\tau)}(\tau, x_\tau^*)\right]$$

$$\times \left[\sum_{j=1}^n a_j\{\hat{q}^j[\psi(\tau, x_\tau^*), \tau], \psi_j(\tau, x_\tau^*)\}\right]$$

$$-\sum_{j=1}^{n} b_j \big[\varpi_j\big(\tau, x_\tau^*\big), x_\tau^*\big] - \delta\big(x_\tau^*\big)x_\tau^*\Bigg]$$

$$-\frac{1}{2}\sum_{h,\zeta=1}^{m} \Omega^{h\zeta}\big(\tau, x_\tau^*\big) \frac{\partial^2}{\partial x_\tau^{h*}\partial x_\tau^{\zeta*}}\Bigg[\frac{V^{(\tau)i}\big(\tau, x_\tau^{i*}\big)}{\sum_{j=1}^{n} V^{(\tau)j}\big(\tau, x_\tau^{j*}\big)} W^{(\tau)}\big(\tau, x_\tau^*\big)\Bigg],$$

for $i \in N$,   (10.20)

will lead to the realization of the solution imputations $\xi^{(\tau)i}(\tau, x_\tau^*)$, *for* $i \in N$ *and* $\tau \in [t_0, T]$, *satisfying Condition* 10.1.

Invoking Theorem 8.2 in Chap. 8 a subgame consistent solution is constructed with the group optimal strategies $[\psi(\tau, x_\tau^*), \varpi(\tau, x_\tau^*)]$, the imputation in Condition 10.1, and the distribution scheme $B(\tau, x_\tau^*)$ in (10.20).

With nations using the cooperative environmental policy instruments $\psi(s, x^*(s))$ and pollution abatement efforts $\varpi(s, x^*(s))$, the instantaneous receipt of nation i at time instant τ given $x(\tau) = x_\tau^* \in X_\tau^*$ is

$$\zeta_i\big(\tau, x_\tau^*\big) = f^i\big\{\hat{q}^1\big[\psi\big(\tau, x_\tau^*\big), \tau\big], \hat{q}^2\big[\psi\big(\tau, x_\tau^*\big), \tau\big], \dots, \hat{q}^n\big[\psi\big(\tau, x_\tau^*\big), \tau\big], s\big\}$$

$$\times \hat{q}^i\big[\psi\big(\tau, x_\tau^*\big), \tau\big] - c^i\big\{\hat{q}^i\big[\psi\big(\tau, x_\tau^*\big), \tau\big], \psi_i\big(\tau, x_\tau^*\big)\big\}$$

$$- c_i^P\big[\psi_i\big(\tau, x_\tau^*\big)\big] - c_i^a\big[\varpi_i\big(s, x^*(s)\big)\big] - \varepsilon_i^i\big\{\hat{q}^i\big[\psi\big(s, x^*(s)\big), s\big]\big\}$$

$$- \sum_{j\in\bar{K}^i} \varepsilon_i^j\big\{\hat{q}^j\big[\psi\big(\tau, x_\tau^*\big), \tau\big]\big\} - h_i\big(x_\tau^*\big),$$ (10.21)

for $\tau \in [t_0, T]$ and $i \in N$.

According to Corollary 10.2, the instantaneous payment that firm i should receive under the solution to the agreed-upon optimality principle is $B_i(\tau, x_\tau^*)$, for $\tau \in [t_0, T]$ and $i \in N$, as stated in (10.20). Hence an instantaneous transfer payment

$$\chi^i\big(\tau, x_\tau^*\big) = B_i\big(\tau, x_\tau^*\big) - \zeta_i\big(\tau, x_\tau^*\big),$$ (10.22)

will be given or charged to firm i at time τ, for $i \in N, x(\tau) = x_\tau^* \in X_\tau^*$, and $\tau \in [t_0, T]$.

10.5 A Model of Stochastic Industrial Pollution Management

This section presents a model of the collaborative industrial pollution management under uncertainty.

10.5.1 A Multinational Economy with Industrial Pollution

We follow the multinational economy with n asymmetric nations or regions in Chap. 6. Industrial pollution is generated via the production process.

10.5.1.1 The Industrial Economy

The inverse demand function of the output of nation $i \in N$ at time instant s is

$$P_i(s) = \alpha^i - \sum_{j=1}^{n} \beta_j^i q_j(s), \tag{10.23}$$

where $P_i(s)$ is the price of the output of nation i, $q_j(s)$ is the output of nation j, and α^i and β_j^i for $i \in N$ and $j \in N$ are positive constants. The output choice $q_j(s) \in [0, \bar{q}_j]$ is nonnegative and bounded by a maximum output constraint $\bar{q}_j$. The output price equals zero if the right-hand side of (10.23) becomes negative.

Industrial profits of nation i at time s can be expressed as

$$\pi_i(s) = \left[\alpha^i - \sum_{j=1}^{n} \beta_j^i q_j(s) \right] q_i(s) - c_i q_i(s) - v_i(s) q_i(s), \quad \text{for } i \in N, \tag{10.24}$$

where $v_i(s) \geq 0$ is the tax rate imposed by government i on its industrial output at time s and c_i is the unit cost of production. At each time instant s, the industrial sector of nation $i \in N$ seeks to maximize (10.24). The first-order condition for a Nash equilibrium for the n nations economy yields

$$\sum_{j=1}^{n} \beta_j^i q_j(s) + \beta_i^i q_i(s) = \alpha^i - c_i - v_i(s), \quad \text{for } i \in N. \tag{10.25}$$

With output tax rates $v(s) = \{v_1(s), v_2(s), \ldots, v_n(s)\}$ being regarded as parameters, (10.25) becomes a system of equations linear in $q(s) = \{q_1(s), q_2(s), \ldots, q_n(s)\}$. Solving (10.25) yields an industry equilibrium

$$q_i(s) = \phi_i\big(v(s)\big) = \bar{\alpha}^i + \sum_{j \in N} \bar{\beta}_j^i v_j(s), \tag{10.26}$$

where $\bar{\alpha}^i$ and $\bar{\beta}_j^i$, for $i \in N$ and $j \in N$, are constants involving the model parameters $\{\beta_1^1, \beta_2^1, \ldots, \beta_n^1; \beta_1^2, \beta_2^2, \ldots, \beta_n^2; \ldots; \beta_1^n, \beta_2^n, \ldots, \beta_n^n\}$, $\{\alpha^1, \alpha^2, \ldots, \alpha^n\}$, and $\{c_1, c_2, \ldots, c_n\}$.

The industry equilibrium generated by this oligopoly model is computable and fully tractable.

10.5.1.2 Local and Global Environmental Impacts

Industrial production emits pollutants into the environment. The emitted pollutants cause short-term local impacts on neighboring areas of the origin of production in forms like passing-by waste in waterways, wind-driven suspended particles in the air, unpleasant odor, noise, dust, and heat. For an output of $q_i(s)$ produced by nation

i, there will be a short-term local environmental impact (cost) of $\varepsilon_i^i q_i(s)$ on nation i itself and a local impact of $\varepsilon_j^i q_i(s)$ on its neighbor nation j. In particular, ε_j^i is a positive constant. Nation i will receive short-term local environmental impacts from its adjacent nations measured as $\varepsilon_i^j q_j(s)$ for $j \in \bar{K}^i$. Thus $\bar{K}^i$ is the subset of nations whose outputs produce local environmental impacts to nation i. Moreover, industrial production will also create long-term global environmental impacts by building up existing pollution stocks like greenhouse gases, CFC, and atmospheric particulates.

Each government adopts its own pollution abatement policy to reduce the pollution stock. Let $x(s) \subset R^+$ denote the level of pollution at time s, the dynamics of pollution stock is governed by the stochastic differential equation

$$dx(s) = \left[\sum_{j=1}^{n} a_j q_j(s) - \sum_{j=1}^{n} b_j u_j(s) [x(s)]^{1/2} - \delta x(s) \right] ds + \sigma x(s) \, dz(s),$$

$$x(t_0) = x_{t_0}, \tag{10.27}$$

where σ is a noise parameter and $z(s)$ is a Wiener process, $a_j q_j$ is the amount added to the pollution stock by a unit of nation j's output, $u_j(s)$ is the pollution abatement effort of nation j, $b_j u_j(s)[x(s)]^{1/2}$ is the amount of pollution removed by a $u_j(s)$ unit of abatement effort of nation j, and δ is the natural rate of decay of the pollutants.

Short-term local impacts are closely related to the level of production activities and hence are characterized by a deterministic scheme. However, the accumulation of pollution stock like greenhouse gases often involves the interactions between the natural environment and the pollutants emitted and hence stochastic elements will appear. For instance, nature's capability to replenish the environment, the rate of pollution degradation, and climate change are subject to certain degrees of uncertainty. Hence a stochastic dynamic is used to model the evolution of the pollution stock in (10.27). Finally, the damage (cost) of the pollution stock in the environment to nation i at time s is $h_i x(s)$.

10.5.1.3 The Governments' Objectives

The governments have to promote business interests and at the same time handle the financing of the costs brought about by pollution. The instantaneous objective of government i at time s can be expressed as

$$\left[\alpha^i - \sum_{j=1}^{n} \beta_j^i q_j(s) \right] q_i(s) - c_i q_i(s) - c_i^a [u_i(s)]^2$$

$$- \sum_{j \in \bar{K}^i} \varepsilon_i^j [q_j(s)] - h_i x(s), \quad i \in N, \tag{10.28}$$

where $c_i^a > 0$ and $h_i > 0$ are constants, $c_i^a[u_i(s)]^2$ is the cost of employing a u_i amount of pollution abatement effort, and $h_i x(s)$ is the value of damage to country i from an $x(s)$ amount of pollution.

The governments' planning horizon is $[t_0, T]$. It is possible that T may be very large. At time T, the terminal appraisal associated with the state of pollution is $g^i[\bar{x}^i - x(T)]$, where $g^i \geq 0$ and $\bar{x}^i \geq 0$. The discount rate is r. Each one of the n governments seeks to maximize the expected value of the integral of its instantaneous objective in (10.28) over the planning horizon subject to the pollution dynamics in (10.27) with controls on the level of the abatement effort and output tax.

By substituting $q_i(s)$, for $i \in N$, from (10.26) into (10.27) and (10.28), one obtains a stochastic differential game in which government $i \in N$ seeks to

$$
\max_{v_i(s),u_i(s)} E_{t_0} \left\{ \int_{t_0}^{T} \left[\left(\alpha^i - \sum_{j=1}^{n} \beta_j^i \left[\bar{\alpha}^j + \sum_{h \in N} \bar{\beta}_h^j v_h(s) \right] \right) \left[\bar{\alpha}^i + \sum_{h \in N} \bar{\beta}_h^i v_h(s) \right] \right. \right.
$$

$$
- c_i \left[\bar{\alpha}^i + \sum_{j \in N} \bar{\beta}_j^i v_j(s) \right] - c_i^a[u_i(s)]^2 - \sum_{j \in \bar{K}^i} \varepsilon_i^j \left[\bar{\alpha}^j + \sum_{\ell \in N} \bar{\beta}_\ell^j v_\ell(s) \right]
$$

$$
\left. \left. - h_i x(s) \right] e^{-r(s-t_0)} \, ds - g^i \left[x(T) - \bar{x}^i \right] e^{-r(T-t_0)} \right\},
\tag{10.29}
$$

subject to

$$
dx(s) = \left(\sum_{j=1}^{n} a_j \left[\bar{\alpha}^j + \sum_{h \in N} \bar{\beta}_h^j v_h(s) \right] - \sum_{j=1}^{n} b_j u_j(s)[x(s)]^{1/2} - \delta x(s) \right) ds
$$

$$
+ \sigma x(s) \, dz(s),
$$

$$
x(t_0) = x_{t_0}.
\tag{10.30}
$$

In the game in (10.29) and (10.30) one can readily observe that government i's tax policy $v_i(s)$ is not only explicitly reflected in its own output, but also on the outputs of other nations. As mentioned before, this modeling formulation allows some intriguing scenarios to arise. For instance, an increase of $v_i(s)$ may just cause a minor drop in nation i's industrial profit, but may cause significant increases in its neighbors' outputs that produce large, local, negative environmental impacts to nation i. This results in the nations' reluctance to increase or impose taxes on industrial outputs.

10.5.2 Noncooperative Outcomes

In this section we discuss the solution to the noncooperative game in (10.29) and (10.30). Under a noncooperative framework, a Nash equilibrium solution can be characterized with Theorem 2.5 in Chap. 2 as follows.

Corollary 10.3 *A set of strategies* $\{u_i^*(t) = \mu_i(t, x), v_i^*(t) = \phi_i(t, x), \text{for } i \in N\}$, *provides a Nash equilibrium solution to the stochastic differential game in* (10.29) *and* (10.30) *if there exist suitably smooth functions* $V^{(t_0)i}(t, x) : [t_0, T] \times R \to R, i \in N$, *satisfying the following partial differential equations:*

$$
-V_t^{(t_0)i}(t, x) - \frac{\sigma^2 x^2}{2} V_{xx}^{(t_0)i}(t, x)
$$

$$
= \max_{v_i, u_i} \left\{ \left[\left(\alpha^i - \sum_{j=1}^n \beta_j^i \left[\bar{\alpha}^j + \sum_{\substack{h \in N \\ h \neq i}}^n \bar{\beta}_h^j \phi_h(t, x) + \bar{\beta}_i^j v_i \right] \right) \right. \right.
$$

$$
\times \left[\bar{\alpha}^i + \sum_{\substack{h \in N \\ h \neq i}} \bar{\beta}_h^i \phi_h(t, x) + \bar{\beta}_i^i v_i \right]
$$

$$
- c_i \left[\bar{\alpha}^i + \sum_{\substack{j \in N \\ j \neq i}} \bar{\beta}_j^i \phi_j(t, x) + \bar{\beta}_i^i v_i \right] - c_i^a [u_i]^2
$$

$$
- \sum_{j \in \bar{K}^i} \varepsilon_i^j \left[\bar{\alpha}^j + \sum_{\substack{\ell \in N \\ \ell \neq i}} \bar{\beta}_\ell^j \phi_\ell(t, x) + \bar{\beta}_i^j v_i \right] - h_i x \right] e^{-r(t-t_0)}
$$

$$
+ V_x^{(t_0)i}(t, x) \left[\sum_{j=1}^n a_j \left[\bar{\alpha}^j + \sum_{\substack{h \in N \\ h \neq i}} \bar{\beta}_h^j \phi_h(t, x) + \bar{\beta}_i^j v_i \right] \right.
$$

$$
\left. \left. - \sum_{\substack{j=1 \\ j \neq i}}^n b_j \mu_j(t, x) x^{1/2} - b_i u_i x^{1/2} - \delta x \right] \right\}, \tag{10.31}
$$

$$
V^{(t_0)i}(T, x) = -g^i [x - \bar{x}^i] e^{-r(T-t_0)}. \tag{10.32}
$$

Performing the indicated maximization in (10.31) *yields*

$$
\mu_i(t, x) = -\frac{b_i}{2c_i^a} V_x^{(t_0)i}(t, x) e^{r(t-t_0)} x^{1/2}, \tag{10.33}
$$

$$
\left(\alpha^i - \sum_{j=1}^n \beta_j^i \left[\bar{\alpha}^j + \sum_{h \in N} \bar{\beta}_h^j \phi_h(t, x) \right] \right) \bar{\beta}_i^i - \left[\sum_{j=1}^n \beta_j^i \bar{\beta}_i^j \right] \left[\bar{\alpha}^i + \sum_{h \in N} \bar{\beta}_h^i \phi_h(t, x) \right]
$$

$$
- c_i \bar{\beta}_i^i - \sum_{j \in \bar{K}^i} \varepsilon_i^j \bar{\beta}_i^j + V_x^{(t_0)i}(t, x) \sum_{j=1}^n a_j \bar{\beta}_i^j e^{r(t-t_0)} = 0, \tag{10.34}
$$

for $t \in [t_0, T]$ *and* $i \in N$.

　　The system in (10.34) *forms a set of equations that are linear in* $\{\phi_1(t, x), \phi_2(t, x), \dots, \phi_n(t, x)\}$, *with* $\{V_x^{(t_0)1}(t, x) e^{r(t-t_0)}, V_x^{(t_0)2}(t, x) e^{r(t-t_0)}, \dots, V_x^{(t_0)n}(t, x) e^{r(t-t_0)}\}$

being taken as a set of parameters. Solving (10.34) *yields*

$$\phi_i(t,x) = \hat{\alpha}^i + \sum_{j \in N} \hat{\beta}^i_j V^{(t_0)j}_x(t,x) e^{r(t-t_0)}, \quad i \in N, \tag{10.35}$$

where $\hat{\alpha}^i$ *and* $\hat{\beta}^i_j$, *for* $i \in N$ *and* $j \in N$, *are constants involving the constant coefficients in* (10.34). *Substituting the results in* (10.33) *and* (10.35) *into* (10.31) *and* (10.32) *we obtain the following proposition.*

Proposition 10.1 *The system in* (10.31) *and* (10.32) *admits a solution*

$$V^{(t_0)i}(t,x) = \left[A_i(t)x + C_i(t) \right] e^{-r(t-t_0)}, \quad for\ i \in N, \tag{10.36}$$

where $\{A_1(t), A_2(t), \ldots, A_n(t)\}$ *satisfies the following set of constant coefficient quadratic ordinary differential equations:*

$$\dot{A}_i(t) = (r+\delta)A_i(t) - \frac{b_i^2}{4c_i^a}\left[A_i(t)\right]^2 - A_i(t) \sum_{\substack{j=1 \\ j \neq i}}^{n} \frac{b_j^2}{2c_j^a} A_j(t) + h_i,$$

$$A_i(T) = -g^i;\ for\ i \in N, \tag{10.37}$$

and

$$\{C_i(t); i \in N\} \quad is\ given\ by\ C_i(t) = e^{r(t-t_0)}\left[\int_{t_0}^{t} F_i(y)e^{-r(y-t_0)}\,dy + C_i^0 \right],$$
$$\tag{10.38}$$

where

$$C_i^0 = g^i \bar{x}^i e^{-r(T-t_0)} - \int_{t_0}^{T} F_i(y)e^{-r(y-t_0)}\,dy,$$

$$F_i(t) = -\left(\alpha^i - \sum_{j=1}^{n} \beta^i_j \left\{ \bar{\alpha}^j + \sum_{h \in N} \bar{\beta}^j_h \left[\hat{\alpha}^h + \sum_{k \in N} \hat{\beta}^h_k A_k(t) \right] \right\} \right)$$

$$\times \left(\bar{\alpha}^i + \sum_{h \in N} \bar{\beta}^i_h \left[\hat{\alpha}^h + \sum_{k \in N} \hat{\beta}^h_k A_k(t) \right] \right)$$

$$+ c_i \left\{ \bar{\alpha}^i - \sum_{j \in N} \bar{\beta}^i_j \left[\hat{\alpha}^j + \sum_{k \in N} \hat{\beta}^j_k A_k(t) \right] \right\}$$

$$+ \sum_{j \in \bar{K}^i} \varepsilon^j_i \left\{ \bar{\alpha}^j + \sum_{\ell \in N} \bar{\beta}^j_\ell \left[\hat{\alpha}^\ell + \sum_{k \in N} \hat{\beta}^\ell_k A_k(t) \right] \right\}$$

$$- A_i(t) \left[\sum_{j=1}^{n} a_j \left\{ \bar{\alpha}^j + \sum_{h \in N} \bar{\beta}^j_h \left[\hat{\alpha}^h + \sum_{k \in N} \hat{\beta}^h_k A_k(t) \right] \right\} \right].$$

Proof Follow the proof of Proposition 6.2 in the Appendixes of Chap. 6.   □

The corresponding feedback Nash equilibrium strategies of the game in (10.29) and (10.30) can be obtained as

$$\mu_i(t,x) = -\frac{b_i}{2c_i^a} A_i(t) x^{1/2} \quad \text{and} \quad \phi_i(t,x) = \hat{\alpha}^i + \sum_{j \in N} \hat{\beta}_j^i A_j(t) \quad (10.39)$$

for $i \in N$ and $t \in [t_0, T]$.

A remark that will be utilized in the subsequent analysis is given below.

Remark 10.3 Let $V^{(\tau)i}(t, x_t)$ denote the value function of nation i in a game with the payoffs in (10.29) and dynamics in (10.30) that starts at time τ. One can readily verify that $V^{(\tau)i}(t, x_t) = V^{(t_0)i}(t, x_t) e^{r(\tau - t_0)}$, for $\tau \in [t_0, T]$.

10.6 Collaborative Scheme in Stochastic Industrial Pollution Management

Now consider the case when all the nations agree to cooperate and adhere to an optimality principle that entails (i) group optimality and (ii) sharing the expected gain from cooperation proportional to the nations' relative sizes of the expected noncooperative payoffs.

10.6.1 Cooperative Optimization and State Trajectory

To secure group optimality the participating nations seek to maximize their expected joint payoff by solving the following optimal control problem:

$$\max_{v_1, v_2, \ldots, v_n; u_1, u_2, \ldots, u_n} E_{t_0} \left\{ \int_{t_0}^{T} \sum_{\ell=1}^{n} \left[\left(\alpha^{\ell} - \sum_{j=1}^{n} \beta_j^{\ell} \left[\bar{\alpha}^j + \sum_{h \in N} \bar{\beta}_h^j v_h(s) \right] \right) \right.\right.$$

$$\times \left[\bar{\alpha}^{\ell} + \sum_{h \in N} \bar{\beta}_h^{\ell} v_h(s) \right] - c_{\ell} \left[\bar{\alpha}^{\ell} + \sum_{j \in N} \bar{\beta}_j^{\ell} v_j(s) \right] - c_{\ell}^a [u_{\ell}(s)]^2$$

$$- \sum_{j \in \bar{K}^{\ell}} \varepsilon_{\ell}^j \left[\bar{\alpha}^j + \sum_{k \in N} \bar{\beta}_k^j v_k(s) \right] - h_{\ell} x(s) \right] e^{-r(s-t_0)} ds$$

$$- \sum_{\ell=1}^{n} g^{\ell} [x(T) - \bar{x}^{\ell}] e^{-r(T-t_0)} \right\}, \quad (10.40)$$

subject to (10.30).

Invoking Theorem A.5 in the Technical Appendixes, a set of controls $\{[v_i^{**}(t), u_i^{**}(t)] = [\psi_i(t, x), \varpi_i(t, x)], \text{ for } i \in N\}$ constitutes an optimal solution to the stochastic control problem in (10.30) and (10.40) if there exists a continuously twice differentiable function $W^{(t_0)}(t, x) : [t_0, T] \times R \to R, i \in N$, satisfying the following partial differential equations:

$$-W_t^{(t_0)}(t, x) - \frac{\sigma^2 x^2}{2} W_{xx}^{(t_0)}(t, x)$$

$$= \max_{v_1, v_2, \dots, v_n; u_1, u_2, \dots, u_n} \left\{ \sum_{\ell=1}^n \left[\left(\alpha^\ell - \sum_{j=1}^n \beta_j^\ell \left[\bar{\alpha}^j + \sum_{h \in N} \bar{\beta}_h^j v_h \right] \right) \left[\bar{\alpha}^\ell + \sum_{h \in N} \bar{\beta}_h^\ell v_h \right] \right. $$

$$- c_\ell \left[\bar{\alpha}^\ell + \sum_{j \in N} \bar{\beta}_j^\ell v_j \right] - c_\ell^a [u_\ell]^2 - \sum_{j \in \bar{K}^\ell} \varepsilon_\ell^j \left[\bar{\alpha}^j + \sum_{k \in N} \bar{\beta}_k^j v_k \right] - h_\ell x \right] e^{-r(s-t_0)}$$

$$+ W_x^{(t_0)}(t, x) \left[\sum_{j=1}^n a_j \left[\bar{\alpha}^j + \sum_{h \in N} \bar{\beta}_h^j v_h \right] - \sum_{j=1}^n b_j u_j x^{1/2} - \delta x \right] \right\}, \qquad (10.41)$$

$$W^{(t_0)}(T, x) = -\sum_{i=1}^n g^i [x(T) - \bar{x}^i] e^{-r(T-t_0)}. \qquad (10.42)$$

Performing the indicated maximization in (10.41) yields the optimal controls under cooperation as

$$\varpi_i(t, x) = -\frac{b_i}{2c_i^a} W_x^{(t_0)}(t, x) e^{r(t-t_0)} x^{1/2}, \quad \text{for } i \in N; \qquad (10.43)$$

$$\sum_{\ell=1}^n \left[\left(\alpha^\ell - \sum_{j=1}^n \beta_j^\ell \left[\bar{\alpha}^j + \sum_{h \in N} \bar{\beta}_h^j \psi_h(t, x) \right] \right) \bar{\beta}_i^\ell \right.$$

$$- \left[\sum_{j=1}^n \beta_j^\ell \bar{\beta}_i^j \right] \left[\bar{\alpha}^\ell + \sum_{h \in N} \bar{\beta}_h^\ell \psi_h(t, x) \right] \right]$$

$$- \sum_{\ell=1}^n \left[c_\ell \bar{\beta}_i^\ell + \sum_{j \in \bar{K}^i} \varepsilon_\ell^j \bar{\beta}_i^j \right] + W_x^{(t_0)} \sum_{j=1}^n a_j \bar{\beta}_i^j e^{r(t-t_0)} = 0,$$

for $i \in N$. $\qquad (10.44)$

The system in (10.44) can be viewed as a set of equations that are linear in $\{\psi_1(t, x), \psi_2(t, x), \dots, \psi_n(t, x)\}$, with $W_x(t, x) e^{r(t-t_0)}$ being taken as a parameter. Solving (10.44) yields

$$\psi_i(t, x) = \hat{\alpha}^i + \hat{\beta}^i W_x^{(t_0)}(t, x) e^{r(t-t_0)}, \qquad (10.45)$$

where $\hat{\alpha}^i$ and $\hat{\beta}^i$, for $i \in N$, are constants involving the model parameters.

Proposition 10.2 *The system in* (10.41) *and* (10.42) *admits a solution*

$$W^{(t_0)}(t, x) = \left[A^*(t)x + C^*(t)\right]e^{-r(t-t_0)}, \qquad (10.46)$$

with

$$A^*(t) = A_*^P + \Phi^*(t)\left[\bar{C}^* - \int_{t_0}^t \sum_{j=1}^n \frac{b_j^2}{2c_j^a}\Phi^*(y)\,dy\right]^{-1}, \quad and$$

$$C^*(t) = e^{r(t-t_0)}\left[\int_{t_0}^t F^*(y)e^{-r(y-t_0)}\,dy + C_*^0\right],$$

where

$$\Phi^*(t) = \exp\left\{\int_{t_0}^t \left[\sum_{j=1}^n \frac{b_j^2}{2c_j^a}A_*^P + (r+\delta)\right]dy\right\},$$

$$\bar{C}^* = \frac{-\Phi^*(T)}{(A_*^P + \sum_{j=1}^n g^j)} + \int_{t_0}^T \sum_{j=1}^n \frac{b_j^2}{2c_j^a}\Phi^*(y)\,dy,$$

$$A_*^P = \left\{(r+\delta) - \left[(r+\delta)^2 + 4\sum_{j=1}^n \frac{b_j^2}{2c_j^a}\sum_{j=1}^n h_j\right]^{1/2}\right\}\Big/ \sum_{j=1}^n \frac{b_j^2}{c_j^a},$$

$$F^*(t) = -\sum_{\ell=1}^n \left[\left(\alpha^\ell - \sum_{j=1}^n \beta_j^\ell \left\{\bar{\alpha}^j + \sum_{h\in N}\bar{\beta}_h^j[\hat{\alpha}^h + \hat{\beta}^h A^*(t)]\right\}\right)\right.$$

$$\times \left\{\bar{\alpha}^\ell + \sum_{h\in N}\bar{\beta}_h^\ell[\hat{\alpha}^h + \hat{\beta}^h A^*(t)]\right\} - c_\ell\left\{\bar{\alpha}^\ell + \sum_{j\in N}\bar{\beta}_j^\ell[\hat{\alpha}^j + \hat{\beta}^j A^*(t)]\right\}$$

$$- \sum_{j\in\bar{K}^\ell}\varepsilon_\ell^j\left\{\bar{\alpha}^j + \sum_{k\in N}\bar{\beta}_k^j[\hat{\alpha}^k + \hat{\beta}^{kj} A^*(t)]\right\}\Bigg]$$

$$- A_x^*(t)\left[\sum_{j=1}^n a_j\left\{\bar{\alpha}^j + \sum_{h\in N}\bar{\beta}_h^j[\hat{\alpha}^h + \hat{\beta}^h A^*(t)]\right\}\right], \quad and$$

$$C_*^0 = \sum_{j=1}^n g^j \bar{x}^j e^{-r(T-t_0)} - \int_{t_0}^T F^*(y)e^{-r(y-t_0)}\,dy.$$

Proof Follow the Proof of Proposition 6.3 in the Appendixes of Chap. 6.   □

Using (10.43), (10.45), and (10.46), the control strategy under cooperation can be obtained as

$$\psi_i(t, x) = \hat{\alpha}^i + \hat{\beta}^i A^*(t) \quad \text{and} \quad \varpi_i(t, x) = -\frac{b_i}{2c_i^a} A^*(t) x^{1/2}, \qquad (10.47)$$

for $t \in [t_0 < T]$ and $i = 1, 2, \ldots, n$.

Substituting the control strategy from (10.47) into (10.25) yields the dynamics of pollution accumulation under cooperation. Solving the stochastic cooperative pollution dynamics yields the cooperative state trajectory

$$x^*(t) = e^{[\int_{t_0}^{t}[\sum_{j=1}^{n}\frac{b_j^2}{2c_j^a}A^*(s)-\delta-\frac{\sigma^2}{2}]ds+\int_{t_0}^{t}\sigma\,dz(s)]}$$
$$\times \left[x_{t_0} + \int_{t_0}^{t} \sum_{j=1}^{n} a_j \left\{ \bar{\alpha}^j + \sum_{h\in N} \bar{\beta}_h^j[\hat{\alpha}^h + \hat{\beta}^h A^*(s)] \right\} \right.$$
$$\left. \times e^{[\int_{t_0}^{s}[\frac{\sigma^2}{2}+\delta-\sum_{j=1}^{n}\frac{b_j^2}{2c_j^a}A^*(\tau)]d\tau-\int_{t_0}^{s}\sigma\,dz(\tau)]} \, ds \right], \qquad (10.48)$$

for $t \in [t_0, T]$.

We use X_t^* to denote the set of realizable values of $x^*(t)$ at time t generated by (10.48). The term x_t^* is used to denote an element in the set X_t^*.

A remark that will be utilized in the subsequent analysis is given below.

Remark 10.4 Let $W^{(\tau)}(t, x_t)$ denote the value function of the stochastic control problem with the objective in (10.40) and dynamics in (10.30), which starts at time τ. One can readily verify that $W^{(\tau)}(t, x_t^*) = W^{(t_0)}(t, x_t^*)e^{r(\tau - t_0)}$, for $\tau \in [t_0, T]$.

10.6.2 Subgame Consistent Solution and Benefit Distribution

According to the agreed-upon optimality principle, the imputation scheme $\{\xi^{(\tau)i}(\tau, x_\tau^*); \text{ for } i \in N\}$ has to satisfy the following condition:

Condition 10.2

$$\xi^{(\tau)i}(\tau, x_\tau^*) = \frac{V^{(\tau)i}(\tau, x_\tau^*)}{\sum_{j=1}^{n} V^{(\tau)j}(\tau, x_\tau^*)} W^{(\tau)}(\tau, x_\tau^*), \qquad (10.49)$$

for $i \in N, x_\tau^* \in X_\tau^*$, and $\tau \in [t_0, T]$.

Invoking Theorem 8.2 in Chap. 8, a subgame consistent solution can then be expressed as

$$P(x_0, T - t_0) = \left\{ [\psi(s, x^*(s))]_{s=t_0}^{T}, [\varpi(s, x^*(s))]_{s=t_0}^{T}, [B(s, x_s^*)]_{s=t_0}^{T}, s\xi^{(t_0)}(t_0, x_{t_0}) \right\},$$

where $\psi_i(s, x^*(s)) = \hat{\alpha}^i + \hat{\beta}^i A^*(s)$ for $i \in N$ is a set of environmental policy instruments in the collaborative environmental scheme,

$$\varpi\left(s, x^*(s)\right) = -\frac{b_i}{2c_i^a} A^*(s)\left[x^*(s)\right]^{1/2},$$

is the set of pollution abatement efforts in the collaborative environmental scheme, $\xi^{(t_0)}(t_0, x_{t_0})$ is an imputation scheme satisfying Condition 10.2, and

$$
\begin{aligned}
B_i\left(s, x_s^*\right) = &-\frac{\partial}{\partial t}\left[\frac{V^{(s)i}(t, x_t^*)}{\sum_{j=1}^n V^{(s)j}(t, x_t^*)} W^{(s)}(t, x_t^*)\Big|_{t=s}\right] \\
&-\frac{\partial}{\partial x_s^*}\left[\frac{V^{(s)i}(s, x_s^{i*})}{\sum_{j=1}^n V^{(s)j}(s, x_s^{j*})} W^{(s)}(s, x_s^*)\right] \\
&\times\left[\sum_{j=1}^n a_j\left[\bar{\alpha}^j + \sum_{h\in N}\bar{\beta}_h^j \psi_h(s, x_s^*)\right] - \sum_{j=1}^n b_j \varpi_j(s, x_s^*)(x_s^*)^{1/2} - \delta x_s^*\right] \\
&-\frac{\sigma^2 x_s^*}{2}\frac{\partial^2}{\partial(x_s^*)^2}\left[\frac{V^{(s)i}(s, x_s^{i*})}{\sum_{j=1}^n V^{(s)j}(s, x_s^{j*})} W^{(s)}(s, x_s^*)\right]
\end{aligned}
$$

for $i \in N$ and $x_s^* \in X_s^*$.

Invoking Propositions 10.1, 10.2, and (10.47), one can express $B_i(s, x_s^*)$ as

$$
\begin{aligned}
B_i\left(s, x_s^*\right) = &\frac{-[A_i(s)x_s^* + C_i(s)]}{(\sum_{j=1}^2 [A_j(s)x_s^* + C_j(s)])}\left\{[\dot{A}(s)x_s^* + \dot{C}(s)] - r[A(s)x_s^* + C(s)]\right\} \\
&-\frac{[A(s)x_s^* + C(s)]}{(\sum_{j=1}^2 [A_j(s)x_s^* + C_j(s)])} \\
&\times\left\{[\dot{A}_i(s)x_s^* + \dot{C}_i(s)] - r[A_i(s)x_s^* + C_i(s)]\right\} \\
&+\frac{[A_i(s)x_s^* + C_i(s)][A(s)x_s^* + C(s)]}{(\sum_{j=1}^2 [A_j(s)x_s^* + C_j(s)])^2} \\
&\times\sum_{j=1}^2\left\{[\dot{A}_j(s)x_s^* + \dot{C}_j(s)] - r[A_j(s)x_s^* + C_j(s)]\right\} \\
&+\left[\frac{[A_i(s)x_s^* + C_i(s)][A(\tau)x_s^* + C(s)]}{(\sum_{j=1}^2 [A_j(s)x_s^* + C_j(s)])^2}\left(\sum_{j=1}^2 A_j(s)\right)\right]
\end{aligned}
$$

$$\times \left[\sum_{j=1}^{n} a_j \left[\bar{\alpha}^j + \sum_{h \in N} \bar{\beta}_h^j (\hat{\alpha}^h + \hat{\beta}^h A^*(s)) \right] \right.$$

$$\left. + \sum_{j=1}^{n} \frac{b_j^2}{2c_j^a} A^*(s) x_s^* - \delta x_s^* \right], \tag{10.50}$$

for $i \in N$ and $x_s^* \in X_s^*$.

When all nations are adopting the cooperative strategies, the rate of instantaneous payment that nation $\ell \in N$ will realize at time t with the state being x_t^* can be expressed as

$$\Re_\ell (t, x_t^*) = \left(\alpha^\ell - \sum_{j=1}^{n} \beta_j^\ell \left\{ \bar{\alpha}^j + \sum_{h \in N} \bar{\beta}_h^j [\hat{\alpha}^h + \hat{\beta}^h A^*(t)] \right\} \right)$$

$$\times \left\{ \bar{\alpha}^\ell + \sum_{h \in N} \bar{\beta}_h^\ell [\hat{\alpha}^h + \hat{\beta}^h A^*(t)] \right\}$$

$$- c_\ell \left\{ \bar{\alpha}^\ell + \sum_{j \in N} \bar{\beta}_j^\ell [\hat{\alpha}^j + \hat{\beta}^j A^*(t)] \right\} - c_\ell^a \left[\frac{b_\ell}{2c_\ell^a} A^*(t) \right]^2 x_t^*$$

$$- \sum_{j \in \bar{K}^\ell} \varepsilon_\ell^j \left\{ \bar{\alpha}^j + \sum_{k \in N} \bar{\beta}_k^j [\hat{\alpha}^k + \hat{\beta}^{kj} A^*(t)] \right\} - h_\ell x_t^*. \tag{10.51}$$

Since, according to (10.50), under the cooperative scheme an instantaneous payment to nation ℓ equaling $B_\ell(t, x_t^*)$ at time t when $x^*(\tau) = x_\tau^* \in X_\tau^*$, a side payment of the value $B_\ell(t, x_t^*) - \Re_\ell(t, x_t^*)$ will be offered to nation ℓ.

The model analyzed in Sects. 10.5 and 10.6 comes mainly from Yeung and Petrosyan (2008). Another version of stochastic differential games on cooperative pollution management can be found in Yeung (2007).

10.7 Exercises

10.1 Consider a two-nation international economy in which transboundary pollution is generated by the production process. The planning horizon is $[0, 2]$. The inverse demand function of the outputs of nations 1 and 2 at time instant $s \in [0, 2]$ are, respectively,

$$P_1(s) = 150 - 2q_1(s) - q_2(s) \quad \text{and} \quad P_2(s) = 130 - q_1(s) - 0.5q_2(s),$$

where $P_i(s)$ is the price of the output and $q_i(s)$ is the output of nation. The unit costs of production in these nations are $c_1 = 0.5$ and $c_2 = 1$. The instantaneous industrial profits of nations 1 and 2 at time s can be expressed as

$$\pi_1(s) = \big[150 - 2q_1(s) - q_2(s)\big]q_1(s) - 0.5q_1(s) - v_1(s)q_1(s), \quad \text{and}$$

$$\pi_2(s) = \big[130 - 0.5q_1(s) - q_2(s)\big]q_2(s) - q_2(s) - v_2(s)q_2(s),$$

where $v_i(s) \geq 0$ is the tax rate imposed by government i on its industrial output at time s. At each time instant s, the industrial sector of nation $i \in \{1, 2\}$ seeks to maximize its instantaneous profit.

Derive the market equilibrium at time instant s.

10.2 Industrial production emits pollutants into the environment. The emitted pollutants cause short-term local impacts on neighboring areas of the origin of production in forms like passing-by waste in waterways, wind-driven suspended particles in the air, unpleasant odor, noise, dust, and heat. For an output of $q_1(s)$ produced by nation 1, there will be a short-term local environmental impact (cost) of $0.2q_1(s)$ on nation 1 itself and a local impact of $0.15q_1(s)$ on its neighbor nation 2. On the other hand, for an output of $q_2(s)$ produced by nation 2, there will be a short-term local environmental impact (cost) of $0.35q_2(s)$ on nation 2 itself and a local impact of $0.3q_2(s)$ on its neighbor nation 1.

Moreover, industrial production will also create long-term global environmental impacts by building up existing pollution stocks like greenhouse gases, CFC, and atmospheric particulates. Each government adopts its own pollution abatement policy to reduce the pollution stock. Let $x(s)$ denote the level of pollution and $u_j(s)$ the pollution abatement effort of nation j at time s, the dynamics of pollution stock is governed by the stochastic differential equation

$$dx(s) = \big[4q_1(s) + 3q_2(s) - 0.05u_1(s)\big[x(s)\big]^{1/2} - 0.02u_2(s)\big[x(s)\big]^{1/2}$$
$$- 0.04x(s)\big]ds + 0.2\,dz(s), \quad x(0) = 100,$$

where $z(s)$ is a Wiener process.

The governments have to promote business interests and at the same time handle the financing of the costs brought about by pollution. The damages to countries 1 and 2 from an $x(s)$ amount of pollution are $0.25x(s)$ and $0.2x(s)$. In particular, each government maximizes the net gains in the industrial sector minus the sum of expenditures on pollution abatement and damages from pollution. The instantaneous objective of governments 1 and 2 at time s are, respectively,

$$\big[150 - 2q_1(s) - q_2(s)\big]q_1(s) - 0.5q_1(s) - 0.2q_1(s) - 0.3q_2(s) - 0.25x(s), \quad \text{and}$$

$$\big[130 - 0.5q_1(s) - q_2(s)\big]q_2(s) - q_2(s) - 0.15q_1(s) - 0.35q_2(s) - 0.2x(s).$$

The governments' planning horizon is $[0, 2]$. At terminal time 2, the terminal appraisal associated with the state of pollution is $3[60 - x(2)]$ for nation 1 and $2[40 - x(2)]$ for nation 2. The discount rate is 0.05. Each government seeks to maximize the integral of its instantaneous objective over the planning horizon subject to pollution stock dynamics.

Construct a stochastic differential game of noncooperative pollution management by these two nations. Obtain a Nash equilibrium solution for the game.

10.3 Consider the case when both nations want to cooperate and agree to act so that an international optimum can be achieved.

Obtain the optimal cooperative levels of outputs and abatement efforts.

10.4 These cooperating nations adopt an optimality principle that distributes the expected gain from cooperation proportional to the relative sizes of the nations' expected noncooperative payoffs. Characterize a subgame consistent solution.

Chapter 11
Subgame Consistent Dormant Firm Cartel

In this chapter, we introduce uncertainty into the dormant-firm cartel discussed in Chap. 7.

Section 11.1 presents a stochastic dynamic oligopoly in which there are cost differentials among firms. The optimal cartel output trajectory, subgame consistent imputation schemes, and profit sharing arrangement are derived in Sect. 11.2. An illustration is shown in the following section. The case when the planning horizon becomes infinite is analyzed in Sect. 11.4; an illustration with an explicit subgame consistent solution in a stochastic framework is given in Sect. 11.5.

11.1 A Stochastic Dynamic Oligopoly

In this section, we extend the dynamic model of oligopoly in Chap. 7 to a stochastic environment.

11.1.1 Basic Settings

Consider an oligopoly in which n firms are allowed to extract a renewable resource within the duration $[t_0, T]$. Among the n firms, n_1 of them have absolute and marginal cost disadvantages over the other $n_2 = n - n_1$ firms. For notational convenience, the firms with cost advantages are numbered from 1 to n_1 and the firms with cost disadvantages are numbered from $n_1 + 1$ to n. The subset of firms with cost advantages is denoted by N_1 and that of firms with cost disadvantages is denoted by N_2. The firms with cost advantages are identical and so are the firms with cost disadvantages.

D.W.K. Yeung, L.A. Petrosyan, *Subgame Consistent Economic Optimization*,
Static & Dynamic Game Theory: Foundations & Applications,
DOI 10.1007/978-0-8176-8262-0_11, © Springer Science+Business Media, LLC 2012

The dynamics of the resource is characterized by the stochastic differential equations

$$dx(s) = \left(f\left[s, x(s), \sum_{j^i \in N_1} u_{j^1}(s) + \sum_{j^2 \in N_2} u_{j^2}(s) \right] \right) ds + \sigma[s, x(s)] dz(s),$$

$$x(t_0) = x_0 \in X, \qquad\qquad\qquad\qquad (11.1)$$

where $u_j \in U_j$ is the (nonnegative) amount of the resource extracted by firm i, for $i \in N$, and $x(s)$ is the resource stock, $\sigma[s, x(s)]$ is a $m \times \Theta$ matrix, and $z(s)$ is a Θ-dimensional Wiener process and the initial state x_0 is given. Let $\Omega[s, x(s)] = \sigma[s, x(s)]\sigma[s, x(s)]'$ denote the covariance matrix with its element in row h and column ζ denoted by $\Omega^{h\zeta}[s, x(s)]$.

The extraction cost depends on the quantity of the resource extracted $u^i(s)$ and the resource stock size $x(s)$. In particular, the extraction cost for the n_1 firms with cost advantages is

$$c^{j^1}\left[u_{j^1}(s), x(s) \right], \quad \text{for } j^1 \in N_1,$$

and the extraction cost for the n_1 firms with cost disadvantages is

$$c^{j^2}\left[u_{j^2}(s), x(s) \right], \quad \text{for } j^2 \in N_2.$$

This formulation of unit cost follows from two assumptions: (i) the cost of extraction is positively related to extraction effort and (ii) the amount of resource extracted, seen as the output of a production function of two inputs (effort and stock level) is increasing in both inputs (see Clark 1976). In particular, firm $j^1 \in N_1$ has cost advantage so that

$$\partial c^{j^1}\left[u_{j^1}(s), x(s) \right]/\partial u_{j^1}(s) < \partial c^{j^2}\left[u_{j^2}(s), x(s) \right]/\partial u_{j^2}(s),$$

for all levels of $u_{j^1} \in U_{j^1}$ and $u_{j^2} \in U_{j^2}$ at any $x \in X$.

The market price of the resource depends on the total amount extracted and supplied to the market. The price-output relationship at time s is given by the following downward-sloping inverse demand curve $P(s) = g[Q(s)]$, where $Q(s) = \sum_{j^1 \in N_1} u_{j^1}(s) + \sum_{j^2 \in N_2} u_{j^2}(s)$ is the total amount of the resource extracted and marketed at time s. At time T, firm $j^1 \in N_1$ will receive a termination bonus $q^{j^1}[x(T)]$ and firm $j^2 \in N_2$ will receive a termination bonus $q^{j^2}[x(T)]$. There exists a discount rate r, and the profits received at time t have to be discounted by the factor $\exp[-r(t - t_0)]$.

At time t_0, firm $j^1 \in N_1$, which has cost advantages, seeks to maximize its expected profit

$$E_{t_0}\left\{ \int_{t_0}^{T} \left(g\left[\sum_{h \in N_i} u_h(s) + \sum_{\ell \in N_2} u_\ell(s) \right] u_{j^1}(s) - c^{j^1}\left[u_{j^1}(s), x(s) \right] \right) \right.$$
$$\left. \times \exp[-r(s - t_0)] ds + \exp[-r(T - t_0)] q^{j^1}[x(T)] \right\}, \qquad (11.2)$$

subject to (11.1).

At time t_0, firm $j^2 \in N_2$, which has cost disadvantages, seeks to maximize the expected profit

$$E_{t_0}\left\{\int_{t_0}^T \left(g\left[\sum_{h\in N_i} u_h(s) + \sum_{\ell\in N_2} u_\ell(s)\right]u_{j^2}(s) - c^{j^2}\left[u_{j^2}(s), x(s)\right]\right)\right.$$

$$\left. \times \exp\left[-r(s-t_0)\right]ds + \exp\left[-r(T-t_0)\right]q^{j^2}\left[x(T)\right]\right\}, \qquad (11.3)$$

subject to (11.1).

11.1.2 Market Outcome

We use $\Gamma(x_0, T - t_0)$ to denote the game in (11.1)–(11.3) and $\Gamma(x_\tau, T - \tau)$ to denote an alternative game with the state dynamics in (11.1) and payoff structures in (11.2) and (11.3), which starts at time $\tau \in [t_0, T]$ with the initial state $x_\tau \in X$. Invoking Theorem 2.5 in Chap. 2, a noncooperative Nash equilibrium solution of the game $\Gamma(x_\tau, T - \tau)$ can be characterized as follows.

Corollary 11.1 *A set of feedback strategies* $\{\phi^*_{j^1}(t, x)$ *for* $j^1 \in N_1$ *and* $\phi^*_{j^2}(t, x)$ *for* $j^2 \in N_2\}$ *provides a Nash equilibrium solution to the game* $\Gamma(x_\tau, T - \tau)$ *if there exist continuously twice differentiable functions* $V^{(\tau)j^1}(t, x) : [\tau, T] \times R \to R$ *for* $j^1 \in N_1$ *and* $V^{(\tau)j^2}(t, x) : [\tau, T] \times R \to R$ *for* $j^2 \in N_2$, *satisfying the following partial differential equations:*

$$-V_t^{(\tau)j^1}(t, x) - \frac{1}{2}\sum_{h,\zeta=1}^m \Omega^{h\zeta}(t, x)V_{x^h x^\zeta}^{(\tau)j^1}(t, x)$$

$$= \max_{u_{j^1}}\left\{\left(g\left[\sum_{\substack{h\in N_i \\ h\neq j^1}} \phi^*_h(t, x) + u_{j^1} + \sum_{\ell\in N_2} \phi^*_\ell(t, x)\right]u_{j^1}, -c^{j^1}(u_{j^1}, x)\right)\right.$$

$$\times \exp\left[-r(t-\tau)\right]$$

$$\left. + V_x^{(\tau)j^1}(t, x)f\left[t, x, \sum_{\substack{h\in N_i \\ h\neq j^1}} \phi^*_h(t, x) + u_{j^1} + \sum_{\ell\in N_2} \phi^*_\ell(t, x)\right]\right\}, \quad and$$

$$V^{(\tau)j^1}(T, x) = \exp\left[-r(T - t_0)\right]q^{j^1}(x), \quad for\ j^1 \in N_1; \qquad (11.4)$$

$$-V_t^{(\tau)j^2}(t, x) - \frac{1}{2}\sum_{h,\zeta=1}^m \Omega^{h\zeta}(t, x)V_{x^h x^\zeta}^{(\tau)j^2}(t, x)$$

$$= \max_{u_{j^2}} \left\{ \left(g \left[\sum_{h \in N_i} \phi_h^*(t,x) + \sum_{\substack{\ell \in N_2 \\ \ell \neq j^2}} \phi_\ell^*(t,x) + u_{j^2} \right] u_{j^2} - c^{j^2}(u_{j^2},x) \right) \right.$$

$$\times \exp[-r(t-\tau)]$$

$$\left. + V_x^{(\tau)j^2}(t,x) f \left[t,x, \sum_{h \in N_i} \phi_h^*(t,x) + \sum_{\substack{\ell \in N_2 \\ \ell \neq j^2}} \phi_\ell^*(t,x) + u_{j^2} \right] \right\}, \quad \text{and}$$

$$V^{(\tau)j^2}(T,x) = \exp[-r(T-t_0)] q^{j^2}(x), \quad \text{for } j^2 \in N_2.$$

Conditions satisfying the indicated maximization in (11.4) yield

$$\left\{ g \left(\sum_{\substack{h \in N_i \\ h \neq j^1}} \phi_h^*(t,x) + u_{j^1} + \sum_{\ell \in N_2} \phi_\ell^*(t,x) \right) \right.$$

$$+ g' \left(\sum_{\substack{h \in N_i \\ h \neq j^1}} \phi_h^*(t,x) + u_{j^1} + \sum_{\ell \in N_2} \phi_\ell^*(t,x) \right) u_{j^1}$$

$$\left. - \frac{\partial}{\partial u_{j^1}} c^{j^1}(u_{j^1},x) \right\} \exp[-r(t-\tau)]$$

$$+ V_x^{(\tau)j^1}(t,x) \frac{\partial}{\partial u_{j^1}} f \left[t,x, \sum_{\substack{h \in N_i \\ h \neq j^1}} \phi_h^*(t,x) + u_{j^1} + \sum_{\ell \in N_2} \phi_\ell^*(t,x) \right] = 0,$$

for $j^1 \in N_1$;

$$\left\{ g \left(\sum_{h \in N_i} \phi_h^*(t,x) + \sum_{\substack{\ell \in N_2 \\ \ell \neq j^2}} \phi_\ell^*(t,x) + u_{j^2} \right) \right.$$

$$+ g' \left(\sum_{h \in N_i} \phi_h^*(t,x) + \sum_{\substack{\ell \in N_2 \\ \ell \neq j^2}} \phi_\ell^*(t,x) + u_{j^2} \right) u_{j^2}$$

$$\left. - \frac{\partial}{\partial u_{j^2}} c^{j^2}(u_{j^1},x) \right\} \exp[-r(t-\tau)]$$

$$+ V_x^{(\tau)j^2}(t,x) \frac{\partial}{\partial u_{j^2}} f \left[t,x, \sum_{h \in N_i} \phi_h^*(t,x) + \sum_{\substack{\ell \in N_2 \\ \ell \neq j^2}} \phi_\ell^*(t,x) + u_{j^2} \right] = 0,$$

for $j^2 \in N_2$.

(11.5)

The expected profits of firm $j^1 \in N_1$, which has cost advantages, can be expressed as

$$
\begin{aligned}
V^{(\tau)j^1}(t, x_\tau) = E_\tau \Bigg\{ & \int_\tau^T \bigg(g\bigg[\sum_{h \in N_i} \phi_h^*[s, x(s)] + \sum_{\ell \in N_2} \phi_\ell^*[s, x(s)] \bigg] \phi_{j1}^*[s, x(s)] \\
& - c^{j^1}\big[\phi_{j1}^*\big(s, x(s)\big), x(s)\big] \bigg) \exp\big[-r(s - \tau)\big] ds \\
& + \exp\big[-r(T - \tau)\big] q^{j^1}\big[x(T)\big] \Bigg\},
\end{aligned}
$$

for $j^1 \in N_1$. The expected profits of firm $j^2 \in N_2$, which has cost disadvantages, can be expressed as

$$
\begin{aligned}
V^{(\tau)j^2}(t, x_\tau) = E_\tau \Bigg\{ & \int_\tau^T \bigg(g\bigg[\sum_{h \in N_i} \phi_h^*[s, x(s)] + \sum_{\ell \in N_2} \phi_\ell^*[s, x(s)] \bigg] \phi_{j2}^*[s, x(s)] \\
& - c^{j^2}\big[\phi_{j2}^*\big(s, x(s)\big), x(s)\big] \bigg) \exp\big[-r(s - \tau)\big] ds \\
& + \exp\big[-r(T - \tau)\big] q^{j^2}\big[x(T)\big] \Bigg\},
\end{aligned}
$$

for $j^2 \in N_2$, where

$$
dx(s) = f\bigg[s, x(s), \sum_{j^i \in N_1} \phi_{j1}^*\big(s, x(s)\big) + \sum_{j^2 \in N_2} \phi_{j2}^*\big(s, x(s)\big) \bigg] ds + \sigma\big[s, x(s)\big] dz(s),
$$

$$
x(\tau) = x_\tau \in X.
$$

The dynamic oligopoly model presented above is an extension of the dormant-firm duopoly model in Yeung (2005).

11.2 Subgame Consistent Cartel

Assume that the firms in the oligopoly agree to form a cartel to restrain output and enhance their expected profits.

11.2.1 Pareto Optimal Output Path

To achieve a group optimum, these firms are required to solve the following expected joint profit maximization problem:

$$
\max_{u_1,u_2,\ldots,u_n} E_{t_0}\left\{ \int_{t_0}^{T} \left(g\left[\sum_{h\in N_i} u_h(s) + \sum_{\ell\in N_2} u_\ell(s) \right]\left[\sum_{h\in N_i} u_h(s) + \sum_{\ell\in N_2} u_\ell(s) \right] \right. \right.
$$

$$
\left. - \left[\sum_{h\in N_i} c^h[u_h(s), x(s)] + \sum_{\ell\in N_2} c^\ell[u_\ell(s), x(s)] \right] \right) \exp[-r(s-t_0)]\, ds
$$

$$
\left. + \exp[-r(T-t_0)] \left[\sum_{h\in N_i} q^h[x(T)] + \sum_{\ell\in N_2} q^\ell[x(T)] \right] \right\}, \tag{11.6}
$$

subject to (11.1).

An optimal solution of the stochastic control problem in (11.1) and (11.6) can be characterized using Theorem A.5 in the Technical Appendixes as follows.

Corollary 11.2 *A set of control strategies* $\{\psi^*_{j^1}(t,x)$ *for* $j^1 \in N_1$ *and* $\psi^*_{j^2}(t,x)$ *for* $j^2 \in N_2\}$, *provides a solution to the control problem in* (11.1) *and* (11.6) *if there exist continuously twice differentiable functions* $W^{(t_0)}(t,x) : [\tau, T] \times R^m \to R$, *satisfying the following partial differential equation:*

$$
- W_t^{(t_0)}(t,x) - \frac{1}{2}\sum_{h,\zeta=1}^{m} \Omega^{h\zeta}(t,x) W_{x^h x^\zeta}^{(t_0)}(t,x)
$$

$$
= \max_{u_1,u_2,\ldots,u_n}\left\{ \left(g\left[\sum_{h\in N_i} u_h + \sum_{\ell\in N_2} u_\ell \right]\left[\sum_{h\in N_i} u_h + \sum_{\ell\in N_2} u_\ell \right] \right. \right.
$$

$$
\left. - \left[\sum_{h\in N_i} c^h[u_h, x] + \sum_{\ell\in N_2} c^\ell[u_\ell, x] \right] \right) \exp[-r(t-t_0)] \tag{11.7}
$$

$$
\left. + W_x^{(t_0)}(t,x) f\left[t, x, \sum_{h\in N_1} u_h + \sum_{\ell\in N_2} u_\ell \right] \right\}, \quad and
$$

$$
W^{(t_0)}(T,x) = \exp[-r(T-t_0)] \left[\sum_{h\in N_i} q^h x + \sum_{\ell\in N_2} q^\ell x \right].
$$

Conditions satisfying the indicated maximization in (11.7) include

$$
\left\{ g\left[\sum_{h\in N_i} u_h + \sum_{\ell\in N_2} u_\ell \right] + g'\left[\sum_{h\in N_i} u_h + \sum_{\ell\in N_2} u_\ell \right]\left[\sum_{h\in N_i} u_h + \sum_{\ell\in N_2} u_\ell \right] \right.
$$

$$
\left. - \frac{\partial}{\partial u_{j^1}} c^{j^1}(u_{j^1}, x) \right\} \exp[-r(t-t_0)]
$$

$$+ W_x^{(t_0)}(t, x) \frac{\partial}{\partial u_{j^1}} f\left[t, x, \sum_{h \in N_1} u_h + \sum_{\ell \in N_2} u_\ell\right] \leq 0,$$

$$u_{j^1} \geq 0,$$

and if $u_{j^1} > 0$, the equality sign must hold, for $j^1 \in N_1$;

$$\left\{ g\left[\sum_{h \in N_i} u_h + \sum_{\ell \in N_2} u_\ell\right] + g'\left[\sum_{h \in N_i} u_h + \sum_{\ell \in N_2} u_\ell\right]\left[\sum_{h \in N_i} u_h + \sum_{\ell \in N_2} u_\ell\right] \right.$$

$$\left. - \frac{\partial}{\partial u_{j^2}} c^{j^2}(u_{j^1}, x) \right\} \exp[-r(t - t_0)]$$

$$+ W_x^{(t_0)}(t, x) \frac{\partial}{\partial u_{j^2}} f\left[t, x, \sum_{h \in N_1} u_h + \sum_{\ell \in N_2} u_\ell\right] \leq 0,$$

$$u_{j^2} \geq 0. \tag{11.8}$$

If $u_{j^2} > 0$, the equality sign must hold, for $j^2 \in N_2$.

Since $\frac{\partial}{\partial u_{j^1}} c^{j^1}(u_{j^1}, x) < \frac{\partial}{\partial u_{j^2}} c^{j^2}(u_{j^1}, x)$, all the firms that have cost disadvantages will refrain from extraction. The optimal extraction strategies under cooperation become

$$u_{j^1}^*(t) = \psi_{j^1}^*(t, x), \quad \text{for } j^1 \in N_1, \quad \text{and} \quad u_{j^2}^*(t) = 0, \quad \text{for } j^2 \in N_2. \tag{11.9}$$

The optimal cooperative state dynamics follows:

$$dx(s) = f\left[s, x(s), \sum_{j^i \in N_1} \psi_{j^1}^*(s, x(s))\right] ds + \sigma[s, x(s)] dz(s),$$

$$x(t_0) = x_0. \tag{11.10}$$

The solution to (11.10) yields a group optimal trajectory, which can be expressed as

$$x^*(t) = x_0 + \int_{t_0}^t f\left[s, x^*(s), \sum_{j^i \in N_1} \psi_{j^1}^*(s, x^*(s))\right] ds$$

$$+ \int_{t_0}^t \sigma[s, x(s)] dz(s). \tag{11.11}$$

We use X_t^* to denote the set of realizable values of $x^*(t)$ at time t generated by (11.11). The term $x_t^* \in X_t^*$ is used to denote an element in X_t^*.

Substituting the optimal extraction strategies in (11.9) into (11.6) yields the expected cartel profit as

$$
W^{(t_0)}(t_0, x_0) = E_{t_0} \left\{ \int_{t_0}^{T} \left(g \left[\sum_{h \in N_i} \psi_h^*[s, x^*(s)] \right] \left[\sum_{h \in N_i} \psi_h^*[s, x^*(s)] \right] \right. \right.
$$
$$
- \left[\sum_{h \in N_i} c^h [\psi_h^*(s, x^*(s)), x^*(s)] \right] \right) \exp[-r(s - t_0)]
$$
$$
\left. + \exp[-r(T - t_0)] \left[\sum_{h \in N_i} q^h [x^*(T)] + \sum_{\ell \in N_2} q^\ell [x^*(T)] \right] \right\}. \quad (11.12)
$$

Let $W^{(\tau)}(t, x_t^*)$ denote the expected total venture profit from the control problem with the dynamics in (11.1) and payoff in (11.6), which begins at time $\tau \in [t_0, T]$ with initial state $x_\tau^* \in X_\tau^*$. One can readily obtain

$$
\exp \left[\int_{t_0}^{\tau} r(y) \, dy \right] W^{(t_0)}(t, x_t^*) = W^{(\tau)}(t, x_t^*),
$$

for $\tau \in [t_0, T]$ and $t \in [\tau, T)$, and $x_t^* \in X_t^*$.

Next we consider subgame consistent profit sharing schemes for the cartel along the optimal output path.

11.2.2 Subgame Consistent Cartel Profit Sharing

In a dormant-firm cartel, firms having cost disadvantages will refrain from extraction to enhance the cartel's expected profit to a group optimum. Compensation must be made to the dormant firms for stopping their production activities. Since there are cost differentials among the firms in the cartel, the sizes and earning potentials of the firms cannot be identical. Consider the case when the firms in the cartel agree to share the expected total cooperative payoff proportional to the firms' expected noncooperative payoffs.

The imputation scheme has to fulfill the following condition.

Condition 11.1 An imputation

$$
\xi^{(t_0)j^1}(t_0, x_0) = \frac{V^{(t_0)j^1}(t_0, x_0)}{\sum_{h \in N_1} V^{(t_0)h}(t_0, x_0) + \sum_{\ell \in N_2} V^{(t_0)\ell}(t_0, x_0)} W^{(t_0)}(t_0, x_0),
$$

is assigned to firm j^1, for $j^1 \in N_1$ at the outset, an imputation

$$
\xi^{(t_0)j^2}(t_0, x_0) = \frac{V^{(t_0)j^2}(t_0, x_0)}{\sum_{h \in N_1} V^{(t_0)h}(t_0, x_0) + \sum_{\ell \in N_2} V^{(t_0)\ell}(t_0, x_0)} W^{(t_0)}(t_0, x_0),
$$

is assigned to firm j^2, for $j^2 \in N_2$ at the outset, an imputation

$$\xi^{(\tau)j^1}(\tau, x_\tau^*) = \frac{V^{(\tau)j^1}(\tau, x_\tau^*)}{\sum_{h \in N_1} V^{(\tau)h}(\tau, x_\tau^*) + \sum_{\ell \in N_2} V^{(\tau)\ell}(\tau, x_\tau^*)} W^{(\tau)}(\tau, x_\tau^*)$$

is assigned to firm j^1, for $j^1 \in N_1$ at time $\tau \in (t_0, T]$, and an imputation

$$\xi^{(\tau)j^2}(\tau, x_\tau^*) = \frac{V^{(\tau)j^2}(\tau, x_\tau^*)}{\sum_{h \in N_1} V^{(\tau)h}(\tau, x_\tau^*) + \sum_{\ell \in N_2} V^{(\tau)\ell}(\tau, x_\tau^*)} W^{(\tau)}(\tau, x_\tau^*) \quad (11.13)$$

is assigned to firm j^2, for $j^2 \in N_2$ at time $\tau \in (t_0, T]$.

Invoking Theorem 8.2 in Chap. 8, a subgame consistent solution for the cartel can then be expressed as

$$P(x_0, T - t_0) = \left\{ [\psi^*(s, x^*(s))]_{s=t}^T, [B_{j^1}(s, x_s^*)]_{s=t_0}^T, [B_{j^2}(s, x_s^*)]_{s=t_0}^T, \xi^{(t_0)}(t_0, x_0) \right\},$$

where $\psi^*(s, x^*(s)) = \{\psi_{j^1}^*(s, x)$ for $j^1 \in N_1\}$, $\xi^{(t_0)}(t_0, x_0)$ is an imputation scheme satisfying Condition 11.1, and

$$B_{j^1}(s, x_s^*) = -\frac{\partial}{\partial t}\left[\frac{V^{(s)j^1}(t, x_t^*)}{\sum_{h \in N_1} V^{(s)h}(t, x_t^*) + \sum_{\ell \in N_2} V^{(s)\ell}(t, x_t^*)} W^{(s)}(t, x_t^*)|_{t-s} \right]$$

$$- \frac{\partial}{\partial x_s^*}\left[\frac{V^{(s)j^1}(s, x_s^*)}{\sum_{h \in N_1} V^{(s)h}(s, x_s^*) + \sum_{\ell \in N_2} V^{(s)\ell}(s, x_s^*)} W^{(s)}(s, x_s^*) \right]$$

$$\times f\left[s, x^*(s), \sum_{j^i \in N_1} \psi_{j^1}^*(s, x^*(s)) \right]$$

$$- \frac{1}{2} \sum_{h, \zeta = 1}^m \Omega^{h\zeta}(s, x_s^*) \frac{\partial^2}{\partial x_s^{h*} \partial x_s^{\zeta*}}$$

$$\times \left[\frac{V^{(s)j^1}(s, x_s^*)}{\sum_{h \in N_1} V^{(s)h}(s, x_s^*) + \sum_{\ell \in N_2} V^{(s)\ell}(s, x_s^*)} W^{(s)}(s, x_s^*) \right],$$

for $j^1 \in N_1;$ (11.14)

$$B_{j^2}(s, x_s^*) = -\frac{\partial}{\partial t}\left[\frac{V^{(s)j^2}(t, x_t^*)}{\sum_{h \in N_1} V^{(s)h}(t, x_t^*) + \sum_{\ell \in N_2} V^{(s)\ell}(t, x_t^*)} W^{(s)}(t, x_t^*)|_{t=s} \right]$$

$$- \frac{\partial}{\partial x_s^*}\left[\frac{V^{(s)j^2}(s, x_s^*)}{\sum_{h \in N_1} V^{(s)h}(s, x_s^*) + \sum_{\ell \in N_2} V^{(s)\ell}(s, x_s^*)} W^{(s)}(s, x_s^*) \right]$$

$$\times f\left[s, x^*(s), \sum_{j^i \in N_1} \psi_{j1}^*(s, x^*(s))\right]$$

$$-\frac{1}{2} \sum_{h,\zeta=1}^{m} \Omega^{h\zeta}(s, x_s^*) \frac{\partial^2}{\partial x_s^{h*} \partial x_s^{\zeta*}}$$

$$\times\left[\frac{V^{(s)j^2}(s, x_s^*)}{\sum_{h \in N_1} V^{(s)h}(s, x_s^*) + \sum_{\ell \in N_2} V^{(s)\ell}(s, x_s^*)} W^{(s)}(s, x_s^*)\right],$$

for $j^2 \in N_1$.

With firms having cost advantages producing an output $\psi_{j1}^*(t, x)$ for $j^1 \in N_1$ and firms having cost disadvantages refraining from production, the instantaneous receipt of firm i at time instant τ when $x_\tau^*(\tau) = x_\tau^* \in X_\tau^*$ is

$$\zeta_{j1}(\tau, x_\tau^*) = g\left[\sum_{h \in N_i} \psi_h^*(\tau, x_\tau^*)\right] \psi_{j1}^*(\tau, x_\tau^*) - c^{j^1}\left[\psi_{j1}^*(\tau, x_\tau^*), x_\tau^*(s)\right],$$

for $\tau \in [t_0, T]$ and $j^1 \in N_1$, and

$$\zeta_{j2}(\tau, x_\tau^*) = 0, \quad \text{for } \tau \in [t_0, T] \text{ and } j^2 \in N_2.$$

According to (11.15), the instantaneous payment that firm i should receive under the agreed-upon optimality principle is $B_{j1}(\tau, x_\tau^*)$ for $j^1 \in N_1$ and $B_{j2}(\tau, x_\tau^*)$ for $j^2 \in N_2$. Hence an instantaneous transfer payment

$$\chi^{j^1}(\tau, x_\tau^*) = \zeta_{j1}(\tau, x_\tau^*) - B_{j1}(\tau, x_\tau^*), \quad \text{for firm } j^1 \in N_1 \text{ and } \tau \in [t_0, T];$$

$$(11.15)$$

$$\chi^{j^2}(\tau, x_\tau^*) = B_{j1}(\tau, x_\tau^*), \quad \text{for firm } j^2 \in N_2 \text{ and } \tau \in [t_0, T];$$

would have to be arranged.

11.3 A Dormant-Firm Cartel

Consider the dormant-firm duopoly game example in Yeung (2005) in which two firms are allowed to extract a renewable resource within the duration $[t_0, T]$. The dynamics of the resource is characterized by the differential equations

$$dx(s) = \left[ax(s)^{1/2} - bx(s) - u_1(s) - u_2(s)\right] ds + \sigma x(s) dz(s),$$

$$x(t_0) = x_0 \in X, \tag{11.16}$$

where $u_i \in U_i$ is the (nonnegative) amount of the resource extracted by firm i, for $i \in \{1, 2\}$, a and b are positive constants.

The extraction cost for firm $i \in \{1, 2\}$ depends on the quantity of the resource extracted $u_i(s)$, the resource stock size $x(s)$, and a parameter c_i. In particular, firm i's extraction cost can be specified as $c_i u^i(s) x(s)^{-1/2}$. In particular, firm 1 has absolute and marginal cost advantages with $c_1 < c_2$.

The market price of the resource depends on the total amount extracted and supplied to the market. The price-output relationship at time s is given by the following downward-sloping inverse demand curve $P(s) = Q(s)^{-1/2}$, where $Q(s) = u_1(s) + u_2(s)$ is the total amount of the resource extracted and marketed at time s. At time T, firm i will receive a termination bonus with satisfaction $q_i x(T)^{1/2}$, where q_i is nonnegative. There exists a discount rate r, and the profits received at time t have to be discounted by the factor $\exp[-r(t - t_0)]$.

At time t_0 the expected profit of firm $i \in \{1, 2\}$ is

$$
E_{t_0} \left\{ \int_{t_0}^T \left[\frac{u_i(s)}{[u_1(s) + u_2(s)]^{1/2}} - \frac{c_i}{x(s)^{1/2}} u_i(s) \right] \exp[-r(s - t_0)] \, ds \right.
$$

$$
\left. + \exp[-r(T - t_0)] q_i x(T)^{\frac{1}{2}} \right\}. \tag{11.17}
$$

A set of strategies $\{u_i^*(t) = \phi_i^*(t, x), \text{ for } i \in \{1, 2\}\}$, provides a Nash equilibrium solution to the stochastic differential game in (11.16) and (11.17) if there exist continuously twice differentiable functions $V^{(t_0)i}(t, x) : [t_0, T] \times R \rightarrow R, i \in \{1, 2\}$, satisfying the following partial differential equations:

$$
-V_t^{(t_0)i}(t, x) - \frac{1}{2}\sigma^2 x^2 V_{xx}^{(t_0)i}(t, x)
$$

$$
= \max_{u_i} \left\{ \left[\frac{u_i}{(u_i + \phi_j^*(t, x))^{1/2}} - \frac{c_i}{x^{1/2}} u_i \right] \exp[-r(t - t_0)] \right.
$$

$$
\left. + V_x^{(t_0)i}(t, x)\left[ax^{1/2} - bx - u_i - \phi_j^*(t, x)\right] \right\}, \quad \text{and} \tag{11.18}
$$

$$
V^{(t_0)i}(T, x) = q_i x^{1/2} \exp[-r(T - t_0)], \quad \text{for } i \in \{1, 2\}, j \in \{1, 2\}, \text{ and } j \neq i.
$$

Performing the indicated maximization yields

$$
\phi_1^*(t, x) = \frac{x}{4[c_1 + V_x^{(t_0)1} \exp[r(t - t_0)]x^{1/2}]^2}, \quad \text{and}
$$

$$
\phi_2^*(t, x) = \frac{x}{4[c_2 + V_x^{(t_0)2} \exp[r(t - t_0)]x^{1/2}]^2}. \tag{11.19}
$$

Proposition 11.1 *The value function of firm i in the game in (11.16) and (11.17) is*

$$
V^{(t_0)i}(t, x) = \exp[-r(t - t_0)]\left[A_i(t)x^{1/2} + C_i(t)\right],
$$

$$
for \ i \in \{1, 2\} \ and \ t \in [t_0, T], \tag{11.20}
$$

where $A_i(t), C_i(t), A_j(t),$ and $C_j(t),$ for $i \in \{1,2\}, j \in \{1,2\}$ and $i \neq j$, satisfy

$$\dot{A}_i(t) = \left[r + \frac{b}{2} + \frac{\sigma^2}{8} \right] A_i(t) - \left(\frac{3}{2} \right) \frac{[2c_j - c_i + A_j(t) - A_i(t)/2]}{[c_1 + c_2 + A_1(t)/2 + A_2(t)/2]^2}$$

$$+ \left(\frac{3}{2} \right)^2 \frac{c_i[2c_j - c_i + A_j(t) - A_i(t)/2]}{[c_1 + c_2 + A_1(t)/2 + A_2(t)/2]^3}$$

$$+ \left(\frac{9}{8} \right) \frac{A_i(t)}{[c_1 + c_2 + A_1(t)/2 + A_2(t)/2]^2},$$

$$A_i(T) = q_i,$$

$$\dot{C}_i(t) = rC_i(t) - \frac{a}{2} A_i(t), \quad \text{and} \quad C_i(T) = 0.$$

Proof First substitute the results in (11.19) and $V^{(t_0)1}(t,x), V_x^{(t_0)1}(t,x), V^{(t_0)2}(t,x),$ and $V_x^{(t_0)2}(t,x)$ obtained via (11.20) into the set of partial differential equations in (11.18). One can readily show that for this set of equations to be satisfied, Proposition 11.1 has to hold.   □

One can readily verify that

$$V^{(\tau)i}(t,x) = \exp\left[-r(t-\tau)\right]\left[A_i(t)x^{1/2} + C_i(t)\right], \quad \text{for } i \in \{1,2\} \text{ and } t \in [\tau, T].$$

Assume that the firms agree to form a cartel and seek to solve the following expected joint profit maximization problem to achieve a group optimum

$$\max_{u_1, u_2} E_{t_0}\left\{ \int_{t_0}^{T} \left[[u_1(s) + u_2(s)]^{1/2} - \frac{c_1 u_1(s) + c_2 u_2(s)}{x(s)^{1/2}} \right] \exp\left[-r(s-t_0)\right] ds \right.$$

$$\left. + \exp\left[-r(T-t_0)\right][q_1 + q_2]x(T)^{1/2} \right\}, \tag{11.21}$$

subject to the dynamics in (11.16).

A set of strategies $[\psi_1^*(s,x), \psi_2^*(s,x)]$, for $s \in [t_0, T]$, provides an optimal solution to the stochastic control problem in (11.16) and (11.21) if there exist a continuously twice differentiable function $W^{(t_0)}(t,x) : [t_0, T] \times R \to R$, satisfying the following partial differential equations:

$$- W_t^{(t_0)}(t,x) - \frac{1}{2}\sigma^2 x^2 W_{xx}^{(t_0)}(t,x)$$

$$= \max_{u_1, u_2}\left\{ \left[(u_1 + u_2)^{1/2} - (c_1 u_1 + c_2 u_2)x^{-1/2} \right] \exp\left[-r(t-t_0)\right] \right.$$

$$\left. + W_x^{(t_0)}(t,x)\left[ax^{1/2} - bx - u_1 - u_2 \right] \right\}, \quad \text{and} \tag{11.22}$$

$$W^{(t_0)}(T,x) = (q_1 + q_2)x^{1/2} \exp\left[-r(T-t_0)\right].$$

Performing the indicated maximization operation in (11.22) yields

$$\psi_1^*(t, x) = \frac{x}{4[c_1 + W_x \exp[r(t - t_0)]x^{1/2}]^2} \quad \text{and} \quad \psi_2^*(t, x) = 0. \quad (11.23)$$

Firm 2 has to refrain from extraction. The more efficient firm (firm 1) would buy the less efficient firm (firm 2) out from the resource extraction process. Firm 2 becomes a dormant firm under cooperation.

Proposition 11.2 *The value function of the control problem in* (11.16) *and* (11.21) *can be obtained as*

$$W^{(t_0)}(t, x) = \exp[-r(t - t_0)][A(t)x^{1/2} + C(t)], \quad (11.24)$$

where $A(t)$ and $B(t)$ satisfy

$$\dot{A}(t) = \left[r + \frac{b}{2} + \frac{\sigma^2}{8}\right]A(t) - \frac{1}{4[c_1 + A(t)/2]},$$

$$A(T) = q_1 + q_2,$$

$$\dot{C}(t) = rC(t) - \frac{a}{2}A(t), \quad \text{and} \quad B(T) = 0.$$

Proof First, substitute the results in (11.23) and $W^{(t_0)}(t, x)$ and $W_x^{(t_0)}(t, x)$, obtained via (11.24), into the set of partial differential equations in (11.22). One can readily show that for this set of equations to be satisfied, Proposition 11.2 has to hold.   □

Again, one can readily verify that

$$W^{(\tau)}(t, x) = \exp[-r(t - \tau)][A(t)x^{1/2} + B(t)].$$

Substituting $\psi_1^*(t, x)$ and $\psi_2^*(t, x)$ into (11.16) yields the optimal cooperative state dynamics as

$$dx(s) = \left[ax(s)^{1/2} - bx(s) - \frac{x(s)}{4[c_1 + A(s)/2]^2}\right]ds + \sigma x(s)\,dz(s),$$

$$x(t_0) = x_0 \in X. \quad (11.25)$$

The solution to (11.25) yields a Pareto optimal trajectory, which can be expressed as

$$x^*(t) = \left\{\Phi(t, t_0)\left[x_0^{1/2} + \int_{t_0}^{t} \Phi^{-1}(s, t_0)\frac{a}{2}\,ds\right]\right\}^2, \quad (11.26)$$

where

$$\Phi(t, t_0) = \exp\left[\int_{t_0}^{t}\left(\frac{-b}{2} - \frac{1}{8[c_1 + A(s)/2]^2} - \frac{3\sigma^2}{8}\right)ds + \int_{t_0}^{t}\frac{\sigma}{2}\,dz(s)\right].$$

We denote the set containing realizable values of $x^*(t)$ by X_t, for $t \in (t_0, T]$.

Consider the case when the firms in the cartel agree to share the total expected cooperative payoff proportional to the firms' expected noncooperative payoffs.

The imputation scheme has to fulfill the following condition:

Condition 11.2 $\xi^{(\tau)i}(\tau, x_\tau^*) = \dfrac{V^{(\tau)i}(\tau, x_\tau^*)}{\sum_{j=1}^2 V^{(\tau)j}(\tau, x_\tau^*)} W^{(\tau)}(\tau, x_\tau^*)$, for $i \in \{1, 2\}$ and $\tau \in$ $[t_0, T]$.

Invoking the results in (11.15), a subgame consistent solution for the cartel can then be expressed as

$$P(x_0, T - t_0) = \left\{ [\psi^*(s, x^*(s))]_{s=t}^T, [B_1(s, x_s^*)]_{s=t_0}^T, [B_2(s, x_s^*)]_{s=t_0}^T, \xi^{(t_0)}(t_0, x_0) \right\}.$$

In particular,

$$\psi_1^*(s, x_s^*) = \frac{x}{4[c_1 + A(s)/2]^2} \quad \text{and} \quad \psi_2^*(s, x_s^*) = 0;$$

$$\xi^{(t_0)i}(t_0, x_0) = \left[\frac{V^{(t_0)i}(t_0, x_0)}{\sum_{j=1}^2 V^{(t_0)j}(t_0, x_0)} W^{(t_0)}(t_0, x_0) \right], \quad \text{for } i \in \{1, 2\}; \quad \text{and}$$

$$
\begin{aligned}
B_i(s, x_s^*) = &-\frac{\partial}{\partial t} \left[\frac{V^{(s)i}(t, x_t^*)}{\sum_{j=1}^2 V^{(s)j}(t, x_t^*)} W^{(s)}(t, x_t^*) \Big|_{t=s} \right] \\
&- \frac{\partial}{\partial x_s^*} \left[\frac{V^{(s)i}(s, x_s^*)}{\sum_{j=1}^2 V^{(s)j}(s, x_s^*)} W^{(s)}(s, x_s^*) \right] \\
&\times \left[a(x_s^*)^{1/2} - bx_s^* - \frac{x_s^*}{4[c_1 + A(s)/2]^2} \right] \\
&- \frac{\sigma^2 (x_s^*)^2}{2} \frac{\partial^2}{\partial (x_s^*)^2} \left[\frac{V^{(s)i}(s, x_s^*)}{\sum_{j=1}^2 V^{(s)j}(s, x_s^*)} W^{(s)}(s, x_s^*) \right], \quad \text{for } i \in \{1, 2\}.
\end{aligned}
$$

(11.27)

Under cooperation, firm 1 would derive an expected payoff

$$
\begin{aligned}
W^{(t_0)1}(t_0, x_0) = E_{t_0} \Bigg\{ \int_{t_0}^T &\left[[\psi_1^*(s, x^*(s))]^{1/2} - \frac{c_1}{x^*(s)^{1/2}} \psi_1^*(s, x^*(s)) \right] \\
&\times \exp[-r(s - t_0)] \, ds + \exp[-r(T - t_0)] q_1 x^*(T)^{\frac{1}{2}} \Bigg\},
\end{aligned}
$$

where $\psi_1^*(s, x^*(s)) = \frac{x^*(s)}{4[c_1 + A(s)/2]^2}$, and firm 2 would derive an expected payoff

$$W^{(t_0)2}(t_0, x_0) = 0 \quad \text{for being dormant.} \tag{11.28}$$

The instantaneous receipt of firm 1 at time instant τ is

$$\zeta_1\left(\tau, x_\tau^*\right) = \frac{(x_\tau^*)^{1/2}}{2[c_1 + A(\tau)/2]} - \frac{c_1(x_\tau^*)^{1/2}}{4[c_1 + A(\tau)/2]^2},$$

for $\tau \in [t_0, T]$ when $x^*(\tau) = x_\tau^* \in X_\tau^*$.

The instantaneous receipt of firm 2 at time instant τ is

$$\zeta_2\left(\tau, x_\tau^*\right) = 0, \quad \text{for } \tau \in [t_0, T].$$

According to (11.27), the instantaneous payment that firm i should receive under the agreed-upon optimality principle is $B_i(\tau, x_\tau^*)$. Hence an instantaneous transfer payment

$$\chi^1\left(\tau, x_\tau^*\right) = \zeta_1\left(\tau, x_\tau^*\right) - B_1\left(\tau, x_\tau^*\right), \quad \text{for firm 1,} \quad \text{and}$$

$$\chi^2\left(\tau, x_\tau^*\right) = B_2\left(\tau, x_\tau^*\right), \quad \text{for firm 2 at time } \tau \in [t_0, T] \qquad (11.29)$$

$$\text{when } x^*(\tau) = x_\tau^* \in X_\tau^*,$$

would be arranged.

11.4 Infinite-Horizon Cartel

In this section we consider the Dormant-Firm Cartel in Sect. 11.1 with an infinite horizon. Consider an oligopoly in which n firms are given extraction rights of a renewable resource.

The dynamics of the resource is characterized by the differential equations

$$dx(s) = f\left[x(s), \sum_{j^i \in N_1} u_{j^1}(s) + \sum_{j^2 \in N_2} u_{j^2}(s)\right] ds + \sigma\left[x(s)\right] dz(s),$$

$$x(t_0) = x_0 \in X, \qquad (11.30)$$

where $u_j \in U_j$ is the (nonnegative) amount of the resource extracted by firm i, for $i \in N$, and $x(s)$ is the resource stock, $\sigma[s, x(s)]$ is a $m \times \Theta$ matrix, $z(s)$ is a Θ-dimensional Wiener process, and the initial state x_0 is given. Let $\Omega[s, x(s)] = \sigma[s, x(s)]\sigma[s, x(s)]'$ denote the covariance matrix with its element in row h and column ζ denoted by $\Omega^{h\zeta}[s, x(s)]$.

The extraction cost for the n_1 firms with cost advantages is

$$c^{j^1}\left[u_{j^1}(s), x(s)\right], \quad \text{for } j^1 \in N_1,$$

and the extraction cost for the n_1 firms with cost advantages is

$$c^{j^2}\left[u_{j^2}(s), x(s)\right], \quad \text{for } j^2 \in N_2.$$

In particular, firm $j^1 \in N_1$ has absolute and marginal cost advantage so that $\partial c^{j^1}[u_{j^1}(s), x(s)]/\partial u_{j^1}(s) < \partial c^{j^2}[u_{j^2}(s), x(s)]/\partial u_{j^2}(s)$, for any levels of $u_{j^1} \in U_{j^1}, u_{j^2} \in U_{j^2}$, and $x \in X$.

The market price of the resource is governed by the following downward-sloping inverse demand curve

$$P(s) = g[Q(s)],$$

where $Q(s) = \sum_{j^i \in N_1} u_{j^1}(s) + \sum_{j^2 \in N_2} u_{j^2}(s)$ is the total amount of the resource extracted and marketed at time s. There exists a discount rate r, and the profits received at time t have to be discounted by the factor $\exp[-r(t - t_0)]$.

At time t_0, firm $j^1 \in N_1$, which has cost advantages, seeks to maximize its expected profit

$$E_{t_0}\left\{\int_{t_0}^{\infty}\left(g\left[\sum_{h \in N_i} u_h(s) + \sum_{\ell \in N_2} u_\ell(s)\right] u_{j^1}(s) - c^{j^1}[u_{j^1}(s), x(s)]\right) \right.$$
$$\left. \times \exp[-r(s - t_0)]\,ds\right\}, \tag{11.31}$$

subject to (11.30).

At time t_0, firm $j^2 \in N_2$, which has cost disadvantages, seeks to maximize its expected profit

$$E_{t_0}\left\{\int_{t_0}^{\infty}\left(g\left[\sum_{h \in N_i} u_h(s) + \sum_{\ell \in N_2} u_\ell(s)\right] u_{j^2}(s) - c^{j^2}[u_{j^2}(s), x(s)]\right) \right.$$
$$\left. \times \exp[-r(s - t_0)]\,ds\right\}, \tag{11.32}$$

subject to (11.30).

Consider the alternative formulation of (11.30)–(11.32) as

$$\max_{u_{j^1}} E_t\left\{\int_t^{\infty}\left(g\left[\sum_{h \in N_i} u_h(s) + \sum_{\ell \in N_2} u_\ell(s)\right] u_{j^1}(s) - c^{j^1}[u_{j^1}(s), x(s)]\right) \right.$$
$$\left. \times \exp[-r(s - t)]\,ds\right\}, \tag{11.33}$$

for $j^1 \in N_1$, and

$$\max_{u_{j^2}} E_t\left\{\int_t^{\infty}\left(g\left[\sum_{h \in N_i} u_h(s) + \sum_{\ell \in N_2} u_\ell(s)\right] u_{j^2}(s) - c^{j^2}[u_{j^2}(s), x(s)]\right) \right.$$
$$\left. \times \exp[-r(s - t)]\,ds\right\}, \tag{11.34}$$

subject to the state dynamics

$$dx(s) = f\left[x(s), \sum_{j^i \in N_1} u_{j^1}(s) + \sum_{j^2 \in N_2} u_{j^2}(s)\right] ds + \sigma\left[x(s)\right] dz(s),$$

$$x(t) = x.$$

(11.35)

The infinite-horizon autonomous game in (11.33)–(11.35) is independent of the choice of t and dependent only upon the state at the starting time, that is, x.

Invoking Theorem 2.6 in Chap. 2, a noncooperative feedback Nash equilibrium solution can be characterized by a set of strategies $\{\phi_{j^1}^*(x)$ for $j^1 \in N_1$ and $\phi_{j^2}^*(x)$ for $j^2 \in N_2\}$ constitutes a Nash equilibrium solution to the game in (11.33)–(11.35) if there exist functionals $\hat{V}^{j^1}(x) : R^m \to R$ for $j^1 \in N_1$ and $\hat{V}^{j^2}(x) : R^m \to R$ for $j^2 \in N_2$, satisfying the following set of partial differential equations:

$$r\hat{V}^{j^1}(x) - \frac{1}{2}\sum_{h,\zeta=1}^{m} \Omega^{h\zeta}(x)\hat{V}_{x^h x^\zeta}^{j^1}(x)$$

$$= \max_{u_{j^1}}\left\{\left(g\left[\sum_{\substack{h \in N_i \\ h \neq j^1}} \phi_h^*(x) + u_{j^1} + \sum_{\ell \in N_2} \phi_\ell^*(x)\right] u_{j^1} - c^{j^1}(u_{j^1}, x)\right)\right.$$

$$\left. + \hat{V}_x^{j^1}(x)f\left[x, \sum_{\substack{h \in N_i \\ h \neq j^1}} \phi_h^*(x) + u_{j^1} + \sum_{\ell \in N_2} \phi_\ell^*(x)\right]\right\}, \quad \text{for } j^1 \in N_1; \quad \text{and}$$

(11.36)

$$r\hat{V}^{j^2}(x) - \frac{1}{2}\sum_{h,\zeta=1}^{m} \Omega^{h\zeta}(x)\hat{V}_{x^h x^\zeta}^{j^2}(x)$$

$$= \max_{u_{j^2}}\left\{\left(g\left[\sum_{h \in N_i} \phi_h^*(x) + \sum_{\substack{\ell \in N_2 \\ \ell \neq j^2}} \phi_\ell^*(x) + u_{j^2}\right] u_{j^2} - c^{j^2}(u_{j^2}, x)\right)\right.$$

$$\left. + \hat{V}_x^{j^2}(x)f\left[x, \sum_{h \in N_i} \phi_h^*(x) + \sum_{\substack{\ell \in N_2 \\ \ell \neq j^2}} \phi_\ell^*(x) + u_{j^2}\right]\right\}, \quad \text{for } j^2 \in N_2.$$

In particular, the expected profits of firm $j^1 \in N_1$, which has cost advantages, can be expressed as

$$\hat{V}^{j^1}(x) = E_t\left\{\int_t^\infty \left(g\left[\sum_{h \in N_i} \phi_h^*[x(s)] + \sum_{\ell \in N_2} \phi_\ell^*[x(s)]\right]\phi_{j^1}^*[x(s)]\right.\right.$$

$$\left.\left. - c^{j^1}\left[\phi_{j^1}^*(x(s)), x(s)\right]\right) \exp[-r(s-\tau)] ds\right\},$$

for $j^1 \in N_1$. The expected profits of firm $j^2 \in N_2$, which has cost disadvantages, can be expressed as

$$\hat{V}^{j^2}(x_\tau) = E_t \left\{ \int_t^\infty \left(g \left[\sum_{h \in N_i} \phi_h^*[x(s)] + \sum_{\ell \in N_2} \phi_\ell^*[x(s)] \right] \phi_{j^2}^*[x(s)] \right. \right.$$

$$\left. \left. - c^{j^2} \left[\phi_{j^2}^*(x(s)), x(s) \right] \right) \exp[-r(s - \tau)] \, ds \right\},$$

for $j^2 \in N_2$,

where

$$dx(s) = f \left[x(s), \sum_{h \in N_1} \phi_h^*(s) + \sum_{\ell \in N_2} \phi_\ell^*(s) \right] ds + \sigma \left[x(s) \right] dz(s), \quad x(t) = x \in X.$$

After characterizing the noncooperative market we proceed to consider the Pareto optimal output trajectory if a cartel of these firms is formed.

11.4.1 Pareto Optimal Trajectory

Assume that the firms agree to form a cartel to restrain output and enhance their expected profits. To achieve a group optimum, these firms are required to solve the following expected joint profit maximization problem:

$$\max_{u_1, u_2, \dots, u_n} E_t \left\{ \int_t^\infty \left(g \left[\sum_{h \in N_i} u_h(s) + \sum_{\ell \in N_2} u_\ell(s) \right] \left[\sum_{h \in N_i} u_h(s) + \sum_{\ell \in N_2} u_\ell(s) \right] \right. \right.$$

$$\left. - \left[\sum_{h \in N_i} c^h \left[u_h(s), x(s) \right] + \sum_{\ell \in N_2} c^\ell \left[u_\ell(s), x(s) \right] \right] \right)$$

$$\times \exp[-r(s - t)] \, ds \right\}, \tag{11.37}$$

subject to (11.35).

An optimal solution of the stochastic control problem in (11.35) and (11.37) can be characterized using Theorem A.6 in the Technical Appendixes as follows.

Corollary 11.3 *A set of control strategies* $\{\psi_{j^1}^*(x)$ *for* $j^1 \in N_1$ *and* $\psi_{j^2}^*(x)$ *for* $j^2 \in N_2\}$ *provides a solution to the stochastic control problem in (11.35) and (11.37) if there exist continuously twice differentiable functions* $W(x) : R^m \to R$, *satisfying* *the following partial differential equation:*

$$
rW(x) - \frac{1}{2} \sum_{h,\zeta=1}^{m} \Omega^{h\zeta}(x) W_{x^h x^\zeta}(x)
$$

$$
= \max_{u_1, u_2, \dots, u_n} \left\{ \left(g\left[\sum_{h \in N_i} u_h + \sum_{\ell \in N_2} u_\ell \right]\left[\sum_{h \in N_i} u_h + \sum_{\ell \in N_2} u_\ell \right] \right.\right.
$$

$$
\left. - \left[\sum_{h \in N_i} c^h[u_h, x] + \sum_{\ell \in N_2} c^\ell[u_\ell, x] \right] \right)
$$

$$
\left. + W_x(x) f\left[t, x, \sum_{h \in N_1} u_h + \sum_{\ell \in N_2} u_\ell \right] \right\}. \tag{11.38}
$$

Conditions satisfying the indicated maximization in (11.38) include

$$
\left\{ g\left[\sum_{h \in N_i} u_h + \sum_{\ell \in N_2} u_\ell \right] + g'\left[\sum_{h \in N_i} u_h + \sum_{\ell \in N_2} u_\ell \right]\left[\sum_{h \in N_i} u_h + \sum_{\ell \in N_2} u_\ell \right] \right.
$$

$$
\left. - \frac{\partial}{\partial u_{j^1}} c^{j^1}(u_{j^1}, x) \right\} + W_x(x) \frac{\partial}{\partial u_{j^1}} f\left[x, \sum_{h \in N_1} u_h + \sum_{\ell \in N_2} u_\ell \right] \le 0,
$$

$$
u_{j^1} \ge 0,
$$

and if $u_{j^1} > 0$, the equality sign must hold, for $j^1 \in N_1$;

$$
\left\{ g\left[\sum_{h \in N_i} u_h + \sum_{\ell \in N_2} u_\ell \right] + g'\left[\sum_{h \in N_i} u_h + \sum_{\ell \in N_2} u_\ell \right]\left[\sum_{h \in N_i} u_h + \sum_{\ell \in N_2} u_\ell \right] \right.
$$

$$
\left. - \frac{\partial}{\partial u_{j^2}} c^{j^2}(u_{j^1}, x) \right\} + W_x(x) \frac{\partial}{\partial u_{j^2}} f\left[x, \sum_{h \in N_1} u_h + \sum_{\ell \in N_2} u_\ell \right] \le 0, \tag{11.39}
$$

$$
u_{j^2} \ge 0.
$$

If $u_{j^2} > 0$, the equality sign must hold, for $j^2 \in N_2$.

Since $\frac{\partial}{\partial u_{j^1}} c^{j^1}(u_{j^1}, x) < \frac{\partial}{\partial u_{j^2}} c^{j^2}(u_{j^1}, x)$, all the firms that have cost disadvantages would refrain from extraction. The optimal extraction strategies under cooperation become

$$
u^*_{j^1}(t) = \psi^*_{j^1}(x), \quad \text{for } j^1 \in N_1, \quad \text{and} \quad u^*_{j^2}(t) = 0, \quad \text{for } j^2 \in N_2. \tag{11.40}
$$

The optimal cooperative state dynamics follows

$$dx(s) = f\left[x(s), \sum_{j^i \in N_1} \psi_{j^1}^*(x(s))\right] ds + \sigma[x(s)] dz(s), \quad x(t_0) = x_0. \quad (11.41)$$

The solution to (11.41) yields a group optimal trajectory, which can be expressed as

$$x^*(t) = x_0 + \int_{t_0}^t f\left[x^*(s), \sum_{j^i \in N_1} \psi_{j^1}^*(x^*(s))\right] ds + \int_{t_0}^t \sigma[x(s)] dz(s). \quad (11.42)$$

Substituting the optimal extraction strategies in (11.40) into (11.35) yields the expected cartel profit as

$$W(x) = E_t\left\{ \int_t^\infty \left(g\left[\sum_{h \in N_i} \psi_h^*[x^*(s)]\right]\left[\sum_{h \in N_i} \psi_h^*[x^*(s)]\right] \right. \right.$$
$$\left. \left. - \left[\sum_{h \in N_i} c^h[\psi_h^*(x^*(s)), x^*(s)]\right] \right) \exp[-r(s-t)] \,\middle|\, x(t) = x \right\}. \quad (11.43)$$

We then examine the subgame consistent cartel profit sharing mechanisms in the next section.

11.4.2 Subgame Consistent Cartel Profit Sharing

In a dormant-firm cartel, firms having cost disadvantages will refrain from extraction to enhance the cartel's expected profit to a group optimum. Consider the case when the firms in the cartel agree to share the excess of the total expected cooperative payoff proportional to the firms' expected noncooperative payoffs.

The imputation scheme has to fulfill the condition.

Condition 11.3 An imputation

$$\xi^{(\tau)j^1}(\tau, x_\tau^*) = \frac{\hat{V}^{j^1}(x_\tau^*)}{\sum_{h \in N_1} \hat{V}^h(x_\tau^*) + \sum_{\ell \in N_2} \hat{V}^\ell(x_\tau^*)} W(x_\tau^*)$$

is assigned to firm j^1, for $j^1 \in N_1$ at time $\tau \in [t_0, \infty)$ when $x^*(\tau) = x_\tau^* \in X_\tau^*$; an imputation

$$\xi^{(\tau)j^2}(\tau, x_\tau^*) = \frac{\hat{V}^{j^2}(x_\tau^*)}{\sum_{h \in N_1} \hat{V}^h(x_\tau^*) + \sum_{\ell \in N_2} \hat{V}^\ell(x_\tau^*)} W(x_\tau^*), \quad (11.44)$$

is assigned to firm j^2, for $j^2 \in N_2$ at time $\tau \in [t_0, \infty)$ when $x^*(\tau) = x_\tau^* \in X_\tau^*$.

To formulate a set of subgame consistent payoff distribution procedures we invoke Theorem 8.3 in Chap. 8 and obtain the following.

Corollary 11.4 *A PDP with instantaneous payments at time* $\tau \in [t_0, \infty)$ *when* $x^*(\tau) = x_\tau^* \in X_\tau^*$ *equaling*

$$
B_{j^1}(\tau, x_\tau^*) = \frac{r\hat{V}^{j^1}(x_\tau^*)}{\sum_{h \in N_1} \hat{V}^h(x_\tau^*) + \sum_{\ell \in N_2} \hat{V}^\ell(x_\tau^*)} W(x_\tau^*)
$$

$$
- \frac{\partial}{\partial x_\tau^*} \left[\frac{\hat{V}^{j^1}(x_\tau^*)}{\sum_{h \in N_1} \hat{V}^h(x_\tau^*) + \sum_{\ell \in N_2} \hat{V}^\ell(x_\tau^*)} W(x_\tau^*) \right]
$$

$$
\times f\left[x_\tau^*, \sum_{j^i \in N_1} \psi_{j^1}^*(x_\tau^*) \right] - \frac{1}{2} \sum_{h,\zeta=1}^{m} \Omega^{h\zeta}(x_\tau^*) \frac{\partial^2}{\partial x_\tau^{h*} \partial x_\tau^{\zeta*}}
$$

$$
\times \left[\frac{\hat{V}^{j^1}(x_\tau^*)}{\sum_{h \in N_1} \hat{V}^h(x_\tau^*) + \sum_{\ell \in N_2} \hat{V}^\ell(x_\tau^*)} W(x_\tau^*) \right], \quad \text{for } j^1 \in N_1;
$$

$$
\tag{11.45}
$$

$$
B_{j^2}(\tau, x_\tau^*) = \frac{r\hat{V}^{j^2}(x_\tau^*)}{\sum_{h \in N_1} \hat{V}^h(x_\tau^*) + \sum_{\ell \in N_2} \hat{V}^\ell(x_\tau^*)} W(x_\tau^*)
$$

$$
- \frac{\partial}{\partial x_\tau^*} \left[\frac{\hat{V}^{j^2}(x_\tau^*)}{\sum_{h \in N_1} \hat{V}^h(x_\tau^*) + \sum_{\ell \in N_2} \hat{V}^\ell(x_\tau^*)} W(x_\tau^*) \right]
$$

$$
\times f\left[x_\tau^*, \sum_{j^i \in N_1} \psi_{j^1}^*(x_\tau^*) \right] - \frac{1}{2} \sum_{h,\zeta=1}^{m} \Omega^{h\zeta}(x_\tau^*) \frac{\partial^2}{\partial x_\tau^{h*} \partial x_\tau^{\zeta*}}
$$

$$
\times \left[\frac{\hat{V}^{j^2}(x_\tau^*)}{\sum_{h \in N_1} \hat{V}^h(x_\tau^*) + \sum_{\ell \in N_2} \hat{V}^\ell(x_\tau^*)} W(x_\tau^*) \right], \quad \text{for } j^2 \in N_2;
$$

yields a subgame consistent payoff distribution procedure to the cooperative game $\Gamma_c(x_0)$ *with the imputation as specified in Condition* 11.3.

With firms having cost advantages producing an output $\psi_{j^1}^*(x)$, for $j^1 \in N_1$, and firms having cost disadvantages refraining from production, the instantaneous receipt of firm i at time instant τ when $x^*(\tau) = x_\tau^* \in X_\tau^*$ is

$$
\zeta_{j^1}(\tau, x_\tau^*) = g\left[\sum_{h \in N_i} \psi_h^*(x_\tau^*) \right] \psi_{j^1}^*(x_\tau^*) - c^{j^1}\left[\psi_{j^1}^*(x_\tau^*), x_\tau^*(s) \right],
$$

for $\tau \in [t_0, \infty)$ and $j^1 \in N_1$, and $\zeta_{j^2}(\tau, x_\tau^*) = 0$, for $\tau \in [t_0, \infty)$ and $j^2 \in N_2$.

According to Corollary 11.4, the instantaneous payment that firm i should receive under the agreed-upon optimality principle is $B_{j^1}(\tau, x_\tau^*)$, for $j^1 \in N_1$, and

$B_{j^2}(\tau, x_\tau^*)$, for $j^2 \in N_2$, as stated in (11.45). Hence an instantaneous transfer payment

$$\chi^{j^1}(\tau, x_\tau^*) = \zeta_{j^1}(\tau, x_\tau^*) - B_{j^1}(\tau, x_\tau^*), \quad \text{for firm } j^1 \in N_1 \text{ and } \tau \in [t_0, T], \tag{11.46}$$

$$\chi^{j^2}(\tau, x_\tau^*) = B_{j^2}(\tau, x_\tau^*), \quad \text{for firm } j^2 \in N_2 \text{ and } \tau \in [t_0, T]$$

would have to be arranged.

11.5 An Infinite-Horizon Dormant-Firm Cartel

Consider an infinite-horizon version of the game in Sect. 11.3. The dynamics of the resource is characterized by the stochastic differential equations

$$dx(s) = \left[ax(s)^{1/2} - bx(s) - u_1(s) - u_2(s) \right] ds + \sigma x(s) dz(s),$$
$$x(t_0) = x_0 \in X, \tag{11.47}$$

where $u_i \in U_i$ is the (nonnegative) amount of the resource extracted by firm i, for $i \in \{1, 2\}, a$ and b are positive constants.

At time t_0 the expected profit of firm $i \in \{1, 2\}$ is

$$E_{t_0} \left\{ \int_{t_0}^{\infty} \left[\frac{u_i(s)}{[u_1(s) + u_2(s)]^{1/2}} - \frac{c_i}{x(s)^{1/2}} u_i(s) \right] \exp[-r(s - t_0)] ds \right\},$$
$$\text{where } c_1 < c_2. \tag{11.48}$$

Consider the alternative game problem that starts at time $t \in [t_0, \infty)$ with initial state $x(t) = x$

$$\max_{u_i} E_{t_0} \left\{ \int_t^{\infty} \left[\frac{u_i(s)}{[u_1(s) + u_2(s)]^{1/2}} - \frac{c_i}{x(s)^{1/2}} u_i(s) \right] \exp[-r(s - t)] ds \right\}, \tag{11.49}$$

subject to

$$dx(s) = \left[ax(s)^{1/2} - bx(s) - u_1(s) - u_2(s) \right] ds + \sigma x(s) dz(s),$$
$$x(t) = x \in X. \tag{11.50}$$

A set of strategies $\{\phi_i^*(x), \text{ for } i \in \{1, 2\}\}$ provides a Nash equilibrium solution to the game in (11.49) and (11.50) if there exist continuously twice differentiable functions $\hat{V}^i(x) : R \to R, i \in \{1, 2\}$, satisfying the following partial differential equations:

$$r\hat{V}^i(x) - \frac{1}{2}\sigma^2 x^2 \hat{V}_{xx}^i(x)$$
$$= \max_{u_i} \left\{ \left[\frac{u_i}{(u_i + \phi_j^*(x))^{1/2}} - \frac{c_i}{x^{1/2}} u_i \right] + \hat{V}_x^i(t, x)\left[ax^{1/2} - bx - u_i - \phi_j^*(x)\right] \right\},$$
$$\text{for } i \in \{1, 2\}, j \in \{1, 2\} \text{ and } j \neq i. \tag{11.51}$$

Performing the indicated maximization yields

$$\phi_1^*(x) = \frac{x}{4[c_1 + \hat{V}_x^1 x^{1/2}]^2}, \quad \text{and}$$

$$\phi_2^*(t, x) = \frac{x}{4[c_2 + \hat{V}_x^2 x^{1/2}]^2}. \tag{11.52}$$

Proposition 11.3 *The value function of firm i in the game in* (11.49) *and* (11.50) *is*

$$\hat{V}^i(x) = [A_i x^{1/2} + C_i], \quad \text{for } i \in \{1, 2\}, \tag{11.53}$$

where $A_i, C_i, A_j,$ and $C_j,$ for $i \in \{1, 2\}$ and $j \in \{1, 2\}$ and $i \neq j,$ satisfy

$$0 = \left[r + \frac{b}{2} + \frac{\sigma^2}{8}\right] A_i - \left(\frac{3}{2}\right) \frac{[2c_j - c_i + A_j - A_i/2]}{[c_1 + c_2 + A_1/2 + A_2/2]^2}$$

$$+ \left(\frac{3}{2}\right)^2 \frac{c_i[2c_j - c_i + A_j - A_i/2]}{[c_1 + c_2 + A_1/2 + A_2/2]^3} + \left(\frac{9}{8}\right) \frac{A_i}{[c_1 + c_2 + A_1/2 + A_2/2]^2},$$

$$rC_i = \frac{a}{2} A_i.$$

Proof First substitute the results in (11.52), and $\hat{V}^1(x)$, $\hat{V}_x^1(x)$, $\hat{V}^2(x)$, and $\hat{V}_x^2(x)$, obtained via (11.53), into the set of partial differential equations in (11.51). One can readily show that for this set of equations to be satisfied, Proposition 11.3 has to hold. □

A noncooperative market equilibrium can be explicitly obtained from (11.52) and (11.53).

11.5.1 Cartel Output

Assume that the firms agree to form a cartel and seek to solve the following expected joint profit maximization problem to achieve a group optimum:

$$\max_{u_1, u_2} E_{t_0} \left\{ \int_{t_0}^{\infty} \left[[u_1(s) + u_2(s)]^{1/2} - \frac{c_1 u_1(s) + c_2 u_2(s)}{x(s)^{1/2}} \right] \right.$$

$$\left. \times \exp[-r(s - t_0)] \, ds \right\} \tag{11.54}$$

subject to the dynamics in (11.47).

Consider the alternative stochastic control problem that starts at time $t \in [t_0, \infty)$ with initial state $x(t) = x$

$$\max_{u_1, u_2} E_t \left\{ \int_t^\infty \left[[u_1(s) + u_2(s)]^{1/2} - \frac{c_1 u_1(s) + c_2 u_2(s)}{x(s)^{1/2}} \right] \right.$$
$$\left. \times \exp[-r(s-t)] \, ds \right\} \tag{11.55}$$

subject to (11.50).

A set of strategies $[\psi_1^*(x), \psi_2^*(x)]$ provides an optimal solution to the problem in (11.55) and (11.50) if there exists a continuously twice differentiable function $W(x) : R \to R$ satisfying the following partial differential equations:

$$rW(x) - \frac{1}{2}\sigma^2 x^2 W_{xx}(x) = \max_{u_1, u_2} \{ [(u_1 + u_2)^{1/2} - (c_1 u_1 + c_2 u_2)x^{-1/2}]$$
$$+ W_x(x)[ax^{1/2} - bx - u_1 - u_2] \}. \tag{11.56}$$

Performing the indicated maximization operation in (11.56) yields

$$\psi_1^*(x) = \frac{x}{4[c_1 + W_x x^{1/2}]^2} \quad \text{and} \quad \psi_2^*(x) = 0. \tag{11.57}$$

Firm 2 has to refrain from extraction. The more efficient firm (firm 1) would buy the less efficient firm (firm 2) out from the resource extraction process. Firm 2 becomes a dormant firm under cooperation.

Proposition 11.4 *The value function of the control problem in* (11.55) *and* (11.50) *can be obtained as*

$$W(x) = [Ax^{1/2} + C], \tag{11.58}$$

where A and B satisfy

$$0 = \left[r + \frac{b}{2} + \frac{\sigma^2}{8} \right] A - \frac{1}{4[c_1 + A/2]} \quad \text{and} \quad rC = \frac{a}{2}A.$$

Proof First substitute the results in (11.57) and $W(x)$ and $W_x(x)$, obtained via (11.58), into the partial differential equation in (11.56). One can readily show that for this equation to be satisfied, Proposition 11.4 has to hold.   □

Substituting $\psi_1^*(x)$ and $\psi_2^*(x)$ into (11.45) yields the optimal cooperative state dynamics as

$$dx(s) = \left[ax(s)^{1/2} - bx(s) - \frac{x(s)}{4[c_1 + A/2]^2} \right] + \sigma x(s) \, dz(s),$$
$$x(t_0) = x_0 \in X. \tag{11.59}$$

The solution to (11.25) yields a Pareto optimal trajectory, which can be expressed as

$$x^*(t) = \left\{ \Phi(t, t_0) \left[x_0^{1/2} + \int_{t_0}^{t} \Phi^{-1}(s, t_0) \frac{a}{2} \, ds \right] \right\}^2, \tag{11.60}$$

where

$$\Phi(t, t_0) = \exp \left[\int_{t_0}^{t} \left(\frac{-b}{2} - \frac{1}{8[c_1 + A/2]^2} - \frac{3\sigma^2}{8} \right) ds + \int_{t_0}^{t} \frac{\sigma}{2} \, dz(s) \right].$$

We denote the set containing realizable values of $x^*(t)$ by X_t, for $t \in (t_0, T]$.

11.5.2 Subgame Consistent Cartel Profits Sharing

Consider the case when the firms in the cartel agree to share the expected cooperative payoff proportional to the firms' expected noncooperative payoffs.

The imputation scheme has to fulfill Condition 11.3, that is,

$$\xi^{(\tau)i} \left(\tau, x_\tau^* \right) = \frac{\hat{V}^i(x_\tau^*)}{\sum_{j=1}^{2} \hat{V}^j(x_\tau^*)},$$

for $i \in \{1, 2\}$ along the cooperative path $\{x_\tau^*\}_{\tau=t_0}^{\infty}$.

To formulate a set of subgame consistent payoff distribution procedures we invoke Corollary 11.4 and obtain the following.

Corollary 11.5 *A PDP with an instantaneous payment at time $\tau \in [t_0, \infty)$ when $x^*(\tau) = x_\tau^* \in X_\tau^*$*

$$B_i \left(\tau, x_\tau^* \right) = r \frac{\hat{V}^i(x_\tau^*)}{\sum_{j=1}^{2} \hat{V}^j(x_\tau^*)} W\left(x_\tau^* \right) - \frac{\sigma^2(x_\tau^*)^2}{2} \frac{\partial^2}{\partial(x_\tau^*)^2} \left[\frac{\hat{V}^i(x_\tau^*)}{\sum_{j=1}^{2} \hat{V}^j(x_\tau^*)} W\left(x_\tau^* \right) \right],$$

$$- \frac{\partial}{\partial x_\tau^*} \left[\frac{\hat{V}^i(x_\tau^*)}{\sum_{j=1}^{2} \hat{V}^j(x_\tau^*)} W\left(x_\tau^* \right) \right]$$

$$\times \left[a\left(x_\tau^* \right)^{1/2} - b x_\tau^* - \frac{x_\tau^*}{4[c_1 + A/2]^2} \right], \tag{11.61}$$

for $i \in \{1, 2\}$, yields a subgame consistent PDP to the cooperative game with the payoff (11.48), dynamics in (11.47), and imputation as specified in Condition 11.3.

The instantaneous receipt of firm 1 at time instant τ with $x^*(\tau) = x_\tau^* \in X_\tau^*$ is

$$\zeta_1 \left(\tau, x_\tau^* \right) = \frac{(x_\tau^*)^{1/2}}{2[c_1 + A/2]} - \frac{c_1(x_\tau^*)^{1/2}}{4[c_1 + A/2]^2}, \quad \text{for } \tau \in [t_0, \infty).$$

The instantaneous receipt of firm 2 at time instant τ is

$$\zeta_2(\tau, x_\tau^*) = 0, \quad \text{for } \tau \in [t_0, \infty) \text{ with } x^*(\tau) = x_\tau^* \in X_\tau^*.$$

According to Corollary 11.5, the instantaneous payments that firm i should receive under the agreed-upon optimality principle are $B_1(\tau, x_\tau^*)$ and $B_2(\tau, x_\tau^*)$, as stated in (11.61). Hence when $x^*(\tau) = x_\tau^* \in X_\tau^*$ an instantaneous transfer payment

$$\chi^1(\tau, x_\tau^*) = \zeta_1(\tau, x_\tau^*) - B_1(\tau, x_\tau^*), \quad \text{for firm 1 \quad and}$$

$$\chi^2(\tau, x_\tau^*) = B_2(\tau, x_\tau^*), \quad \text{for firm 2}$$

would be arranged.

11.6 Exercises

11.1 Consider a duopoly in which two firms are allowed to extract a renewable resource within the duration $[0, 3]$. The dynamics of the resource is characterized by the stochastic dynamics

$$dx(s) = \left[5x(s)^{1/2} - 0.4x(s) - u_1(s) - u_2(s)\right] ds + 0.1x(s) \, dz(s), \quad x(0) = 120,$$

where $z(s)$ is a Wiener process, $x(s)$ is the resource biomass, and $u_i(s)$ is the amount of the resource extracted by firm i at time $s \in [0, 3]$, for $i \in \{1, 2\}$.

The extraction cost for firms 1 and 2 are, respectively, $u_1(s)x(s)^{-1/2}$ and $2.5u_1(s)x(s)^{-1/2}$. The market price of the resource depends on the total amount extracted and supplied to the market. The price-output relationship at time s is given by the following downward-sloping inverse demand curve $P(s) = Q(s)^{-1/2}$, where $Q(s) = u_1(s) + u_2(s)$ is the total amount of the resource extracted and marketed at time s. At time 3, firm 1 will receive a termination bonus $2x(3)^{1/2}$ and firm 2 a bonus $x(3)^{1/2}$. The discount factor is 0.05.

At time 0, the expected profits of firms 1 and 2 then are

$$E\left\{\int_0^3 \left[\frac{u_1(s)}{[u_1(s) + u_2(s)]^{1/2}} - \frac{u_1(s)}{x(s)^{1/2}}\right] \exp(-0.05s) \, ds + \exp\left[-0.05(3)\right] 2x(3)^{\frac{1}{2}}\right\},$$

and

$$E\left\{\int_0^3 \left[\frac{u_2(s)}{[u_1(s) + u_2(s)]^{1/2}} - \frac{2.5u_2(s)}{x(s)^{1/2}}\right] \exp(-0.05s) \, ds + \exp\left[-0.05(3)\right] x(3)^{\frac{1}{2}}\right\}.$$

Obtain a Nash equilibrium solution when these firms act independently.

11.2 If these two firms form a cartel, show that firm 2 has to be dormant. Derive the optimal output strategies of the cartel.

11.3 Consider the case when the firms in the cartel agree to share the excess of the total expected cooperative profits proportional to the firms' expected noncooperative profits. Characterize a subgame consistent solution.

Chapter 12
Dynamic Consistency in Discrete-Time Cooperative Games

In some economic situations, the economic process is in discrete time rather than in continuous time. The discrete-time counterpart of differential games are known as dynamic games. Bylka et al. (2000) analyzed oligopolistic price competition in a dynamic game model. Wie and Choi (2000) examined discrete-time traffic network. Beard and McDonald (2007) investigated water sharing agreements and Amir and Nannerup (2006) considered resource extraction problems in a discrete-time dynamic framework. Yeung (2011b) examined dynamically consistent collaborative environmental management with technology selection in a discrete-time dynamic game framework. The properties of Nash equilibria in dynamic games are examined in Basar (1974, 1976). The solution algorithm for solving dynamic games can be found in Basar (1977a, 1977b). Petrosyan and Zenkevich (1996) presented an analysis on cooperative dynamic games in a discrete time framework. The SIAM Classics on Dynamic Noncooperative Game Theory by Basar and Olsder (1995) gave a comprehensive treatment of discrete-time noncooperative dynamic games.

An extension of the analysis in a differential game framework to a discrete-time dynamic framework is provided in this chapter. We first present a general formulation of dynamic economic games in discrete time and the noncooperative market outcomes in Sect. 12.1. Group optimality and individual rationality under dynamic cooperation are discussed in Sect. 12.2. Time (optimal-trajectory-subgame) consistent cooperative solutions with the corresponding payoff distribution procedures are derived in Sect. 12.3. An illustration of cooperative resource extraction in discrete time is given in Sect. 12.4.

12.1 Dynamic Games

In this section we present dynamic games in economics and their corresponding noncooperative outcomes.

D.W.K. Yeung, L.A. Petrosyan, *Subgame Consistent Economic Optimization*, Static & Dynamic Game Theory: Foundations & Applications, DOI 10.1007/978-0-8176-8262-0_12, © Springer Science+Business Media, LLC 2012

12.1.1 Game Formulation

We begin with the presentation of a general formation of cooperative dynamic economic games. Consider the general T-stage n-person nonzero-sum discrete-time dynamic game with initial state x^0. The state space of the game is $X \in R^m$ and the state dynamics of the game is characterized by the difference equation

$$x_{k+1} = f_k\big(x_k, u_k^1, u_k^2, \ldots, u_k^n\big), \tag{12.1}$$

for $k \in \{1, 2, \ldots, T\} \equiv \kappa$ and $x_1 = x^0$, where $u_k^i \in U^i \subset R^{m_i}$ is the control vector of agent i at stage k, $x_k \in X$ is the state.

The objective of economic agent i is

$$\sum_{\zeta=1}^{T} g_\zeta^i\big[x_\zeta, u_\zeta^1, u_\zeta^2, \ldots, u_\zeta^n, x_{\zeta+1}\big]\bigg(\frac{1}{1+r}\bigg)^{\zeta-1}, \quad \text{for } i \in \{1, 2, \ldots, n\} \equiv N, \tag{12.2}$$

where r is the discount rate.

The payoffs of the agents are transferable.

12.1.2 Noncooperative Outcome

In this section, we characterize the noncooperative outcome of the discrete-time economic game in (12.1) and (12.2). Let $\{\phi_k^i(x),$ for $k \in \kappa$ and $i \in N\}$ denote a set of strategies that provides a feedback Nash equilibrium solution (if it exists) to the game in (12.1) and (12.2), and

$$V^i(k, x) = \sum_{\zeta=k}^{T} g_\zeta^i\big[x_\zeta, \phi_\zeta^1(x_\zeta), \phi_\zeta^2(x_\zeta), \ldots, \phi_\zeta^n(x_\zeta), x_{\zeta+1}\big]\bigg(\frac{1}{1+r}\bigg)^{\zeta-1},$$
$$\text{where } x_k = x,$$

for $k \in K$ and $i \in N$, denote the value functions indicating the game equilibrium payoff to agent i over the stages from k to T. A frequently used way to characterize and derive a feedback Nash equilibrium of the game is as follows.

Theorem 12.1 *A set of strategies $\{\phi_k^i(x),$ for $k \in \kappa$ and $i \in N\}$, provides a feedback Nash equilibrium solution to the game in (12.1) and (12.2) if there exist functions $V^i(k, x),$ for $k \in K$ and $i \in N$, such that the following recursive relations are satisfied:*

$$V^i(k, x)$$
$$= \max_{u_k^i}\Big\{ g_k^i\big[x, \phi_k^1(x), \phi_k^2(x), \ldots, \phi_k^{i-1}(x), u_k^i, \phi_k^{i+1}(x), \ldots, \phi_k^n(x), \tilde{f}_k^i\big(x, u_k^i\big)\big]$$

$$\times \left(\frac{1}{1+r}\right)^{k-1} + V^i\big[k+1, \tilde{f}_k^i\big(x, u_k^i\big)\big]\bigg\}$$

$$= g_k^i\big[x, \phi_k^1(x), \phi_k^2(x), \ldots, \phi_k^n(x), f_k\big(x, \phi_k^1(x), \phi_k^2(x), \ldots, \phi_k^n(x)\big)\big]$$

$$\times \left(\frac{1}{1+r}\right)^{k-1} + V^i\big[k+1, f_k\big(x, \phi_k^1(x), \phi_k^2(x), \ldots, \phi_k^n(x)\big)\big],$$

$$\hspace{11cm}(12.3)$$

$$V^i(T+1, x) = 0, \hspace{7cm}(12.4)$$

for $i \in N$ *and* $k \in \kappa$, *where* $\tilde{f}_k^i(x, u_k^i) = f_k[x, \phi_k^1(x), \phi_k^2(x), \ldots, \phi_k^{i-1}(x), u_k^i, \phi_k^{i+1}(x), \ldots, \phi_k^n(x)]$.

Proof The game in (12.1) and (12.2) is an n-person multistage game with T stages of play. We first consider the last stage, that is, stage T, and let the set containing all possible values of states be denoted by X_T. One has a single-stage game in which economic agent i maximizes

$$g_T^i\big[x, u_T^1, u_T^2, \ldots, u_T^n, f_T\big(x, u_T^1, u_T^2, \ldots, u_T^n\big)\big]\left(\frac{1}{1+r}\right)^{T-1}, \quad \text{for } i \in N, \quad (12.5)$$

where $x_T = x \in X_T$.

The solution to the problem in (12.5) is the Nash equilibria of an n-person static game, which yields a set of strategies $\{\phi_T^i(x), \text{for } i \in N\}$. Substituting the game equilibrium strategies into agent $i's$ objective yields his game equilibrium payoff in stage T

$$V^i(T, x) = g_k^i\big[x, \phi_k^1(x), \phi_k^2(x), \ldots, \phi_k^n(x), f_k\big(x, \phi_k^1(x), \phi_k^2(x), \ldots, \phi_k^n(x)\big)\big]$$

$$\times \left(\frac{1}{1+r}\right)^{k-1} + V^i\big[k+1, f_k\big(x, \phi_k^1(x), \phi_k^2(x), \ldots, \phi_k^n(x)\big)\big]. \quad (12.6)$$

Using the relation for $k = T$ in (12.3) and the condition in (12.4), we have the problem

$$V^i(T, x) = \max_{u_T^i}\bigg\{g_T^i\big[x, \phi_T^1(x), \phi_T^2(x), \ldots, \phi_T^{i-1}(x), u_T^i, \phi_T^{i+1}(x), \ldots,$$

$$\phi_T^n(x), \tilde{f}_T^i(x, u_T^i)\big]\left(\frac{1}{1+r}\right)^{k-1}\bigg\}, \quad \text{for } i \in N, \quad (12.7)$$

with $x_T = x \in X_T$.

The problem in (12.7) is equivalent to the problem in (12.5).

Now consider the problem in stage $T - 1$. The problem becomes a game in which agent i

$$\max_{u_{T-1}^i, u_T^i}\bigg\{\sum_{\zeta=T-1}^{T} g_\zeta^i\big[x_\zeta, u_\zeta^1, u_\zeta^2, \ldots, u_\zeta^n, x_{\zeta+1}\big]\left(\frac{1}{1+r}\right)^{\zeta-1}\bigg\}, \quad \text{for } i \in N,$$

subject to

$$x_T = f_{T-1}\left(x_{T-1}, u_{T-1}^1, u_{T-1}^2, \ldots, u_{T-1}^n\right), x_{T-1} = x \in X_{T-1}.$$

Using the analysis in stage T, the stage $T-1$ problem can be expressed as a single-stage game in which agent i seeks to maximize

$$g_{T-1}^i\left[x_{T-1}, u_{T-1}^1, u_{T-1}^2, \ldots, u_{T-1}^n, x_T\right]\left(\frac{1}{1+r}\right)^{T-2} + V^i(T, x_T)$$

$$= g_{T-1}^i\left[x_{T-1}, u_{T-1}^1, u_{T-1}^2, \ldots, u_{T-1}^n, f_{T-1}\left(x, u_{T-1}^1, u_{T-1}^2, \ldots, u_{T-1}^n\right)\right]$$

$$\times \left(\frac{1}{1+r}\right)^{T-2} + V^i\left[T, f_{T-1}\left(x, u_{T-1}^1, u_{T-1}^2, \ldots, u_{T-1}^n\right)\right]. \tag{12.8}$$

The game equilibrium for (12.8) can be expressed as a set of strategies $\{\phi_{T-1}^i(x), i \in N\}$ for its explicit dependence on $x_{T-1} = x \in X_{T-1}$. Note that the relation for $k = T-1$ in (12.3) characterizes the game equilibrium of (12.8).

Proceeding recursively onward for stage $k \in \{T-2, T-1, \ldots, 1\}$, the stage k problem can be expressed as a single-stage game in which agent i seeks to maximize

$$g_k^i\left[x, u_k^1, u_k^2, \ldots, u_k^n, f_k\left(x, u_k^1, u_k^2, \ldots, u_k^n\right)\right]\left(\frac{1}{1+r}\right)^{k-1}$$

$$+ V^i\left[k+1, f_k\left(x, u_k^1, u_k^2, \ldots, u_k^n\right)\right], \tag{12.9}$$

where $x_k = x \in X_k$.

The game equilibrium for the problem in (12.9) can be expressed as a set of strategies $\{\phi_k^i(x), i \in N\}$ because of its explicit dependence on x. Note that the relation for $k \in \{1, 2, \ldots, T-2\}$ in (12.3) characterizes the game equilibrium of (12.9). Hence Theorem 12.1 follows. $\square$

For the sake of exposition, we sidestep the issue of multiple equilibria and focus on solvable games in which a particular noncooperative Nash equilibrium is chosen by the agents in the entire subgame.

12.2 Dynamic Cooperation

Now consider the case when the agents agree to cooperate and distribute the payoff among themselves according to an optimality principle. Two essential properties that a cooperative scheme has to satisfy are group optimality and individual rationality. Let the agreed-upon optimality principle be as follows.

Principle PI It is an optimality principle that entails (i) group optimality and (ii) the distribution of the total cooperative payoff according to an imputation that equals $\xi(k, x_k^*) = [\xi^1(k, x_k^*), \xi^2(k, x_k^*), \ldots, \xi^n(k, x_k^*)]$, for $k \in \kappa$.

We first examine the conditions under which group optimality and individual rationality will be maintained.

12.2.1 Group Optimality

Maximizing the agents' joint payoff guarantees group optimality in a game where payoffs are transferable. To maximize their joint payoff the agents have to solve the discrete-time dynamic programming problem of maximizing

$$\sum_{j=1}^{n}\sum_{k=1}^{T}\left[g_k^j\left[x_k, u_k^1, u_k^2, \ldots, u_k^n, x_{k+1}\right]\left(\frac{1}{1+r}\right)^{k-1}\right], \tag{12.10}$$

subject to (12.1).

A frequently used method to characterize and derive an optimal solution to the problem in (12.10) and (12.1) is as follows.

Theorem 12.2 *A set of strategies* $\{\psi_k^i(x), \text{for } k \in \kappa \text{ and } i \in N\}$, *provides an optimal solution to the problem in* (12.10) *and* (12.1) *if there exist functions* $W(k, x)$, *for* $k \in K$, *such that the following recursive relations are satisfied*:

$$W(k, x) = \max_{u_k^1, u_k^2, \ldots, u_k^n}\left\{\sum_{j=1}^{n} g_k^j\left[x_k, u_k^1, u_k^2, \ldots, u_k^n, f_k\left(x_k, u_k^1, u_k^2, \ldots, u_k^n\right)\right]\right.$$

$$\times \left(\frac{1}{1+r}\right)^{k-1} + W\left[k+1, f_k\left(x_k, u_k^1, u_k^2, \ldots, u_k^n\right)\right]\Big\}$$

$$= \sum_{j=1}^{n} g_k^j\left[x, \psi_k^1(x), \psi_k^2(x), \ldots, \psi_k^n(x), f_k\left(x, \psi_k^1(x), \psi_k^2(x), \ldots, \psi_k^n(x)\right)\right]$$

$$\times \left(\frac{1}{1+r}\right)^{k-1} + W\left[k+1, f_k\left(x, \psi_k^1(x), \psi_k^2(x), \ldots, \psi_k^n(x)\right)\right], \tag{12.11}$$

$$W(T+1, x) = 0. \tag{12.12}$$

Proof The method of dynamic programming is based on the principle of optimality, which states that an optimal strategy has the property that, whatever the initial state and time are, all remaining decisions (from that particular initial state and particular initial time onward) must also constitute an optimal strategy. To exploit this principle we work backward in time, starting at all possible final states with the corresponding final times.

Consider the joint maximization problem that maximizes the objective in (12.10) subject to the dynamics in (12.1). To determine the optimal control strategy we shall need the expression for the maximized payoff from any starting point at any initial time. This is the value function

$$W(k, x) = \max_{u_k, u_{k+1}, \dots, u_T} \left\{ \sum_{\zeta=k}^{T} \sum_{j=1}^{n} g_\zeta^j \left[x_\xi, u_\zeta^1, u_\zeta^2, \dots, u_\zeta^n, x_{\xi+1} \right] \left(\frac{1}{1+r} \right)^{\zeta-1} \right\},$$

for $k \in \kappa$,

with feedback control $u_\zeta^i = \gamma_\zeta^i(x_\zeta) \in U_i$ for $i \in N$ and $x_k = x$. A direct application of the principle of optimality readily leads to the recursive relation

$$W(k, x) = \max_{u_k^1, u_k^2, \dots, u_k^n} \left\{ \sum_{j=1}^{n} g_k^j \left[x_k, u_k^1, u_k^2, \dots, u_k^n, f_k \left(x_k, u_k^1, u_k^2, \dots, u_k^n \right) \right] \right. $$
$$\times \left(\frac{1}{1+r} \right)^{k-1} + W \left[k+1, f_k \left(x_k, u_k^1, u_k^2, \dots, u_k^n \right) \right] \right\}.$$

Hence Theorem 12.2 follows.   □

Substituting the optimal control $\{ \psi_k^i(x), \text{for } k \in \kappa \text{ and } i \in N \}$ into the state dynamics in (12.1), one can obtain the dynamics of the cooperative trajectory as

$$x_{k+1} = f_k \left(x_k, \psi_k^1(x_k), \psi_k^2(x_k), \dots, \psi_k^n(x_k) \right), \qquad (12.13)$$

for $k \in \kappa$ and $x_1 = x^0$.

Let $\{x_k^*\}_{k=1}^{T}$ denote the solution to (12.13) and hence the optimal cooperative path. The total cooperative payoff over the stages from k to T can be expressed as

$$W \left(k, x_k^* \right) = \sum_{\zeta=k}^{T} \sum_{j=1}^{n} g_\zeta^j \left[x_\xi^*, \psi_\zeta^1(x_\zeta^*), \psi_\zeta^2(x_\zeta^*), \dots, \psi_\zeta^n(x_\zeta^*), f_\zeta \left(x_\zeta^*, \psi_\zeta^1(x_\zeta^*), \right. \right.$$
$$\left. \left. \psi_\zeta^2(x_\zeta^*), \dots, \psi_\zeta^n(x_\zeta^*) \right) \right] \left(\frac{1}{1+r} \right)^{\zeta-1}, \qquad \text{for } k \in \kappa.$$

We then proceed to consider individual rationality.

12.2.2 Individual Rationality

The agents have to agree on an optimality principle in distributing the total cooperative payoff among themselves. For individual rationality to be upheld, the payoffs an agent receives under cooperation have to be no less than his noncooperative payoff along the cooperative state trajectory. For instance, (i) the agents may share the excess of the total cooperative payoff over the sum of individual noncooperative payoffs equally, or (ii) they may share the total cooperative payoff proportionally to their noncooperative payoffs.

Let $\xi(\cdot, \cdot)$ denote the imputation vector guiding the distribution of the total co-operative payoff under the agreed-upon optimality principle along the cooperative trajectory $\{x_k^*\}_{k=1}^T$. At stage k, the imputation vector according to $\xi(\cdot, \cdot)$ is

$$\xi(k, x_k^*) = \left[\xi^1(k, x_k^*), \xi^2(k, x_k^*), \ldots, \xi^n(k, x_k^*)\right], \quad \text{for } k \in \kappa.$$

If, for example, the optimality principle specifies that the agents share the excess of the total cooperative payoff over the sum of individual noncooperative payoffs equally, then the imputation to agent i becomes

$$\xi^i(k, x_k^*) = V^i(k, x_k^*) + \frac{1}{n}\left[W(k, x_k^*) - \sum_{j=1}^n V^j(k, x_k^*)\right], \quad (12.14)$$

for $i \in N$ and $k \in \kappa$.

For individual rationality to be maintained throughout all the stages $k \in \kappa$, it is required that

$$\xi^i(k, x_k^*) \geq V^i(k, x_k^*), \quad \text{for } i \in N \text{ and } k \in \kappa.$$

In particular, the above condition guarantees that the payoff allocated to an economic agent under cooperation will be no less than its noncooperative payoff.

To satisfy group optimality, the imputation vector has to satisfy

$$W(k, x_k^*) = \sum_{j=1}^n \xi^j(k, x_k^*), \quad \text{for } k \in \kappa.$$

This condition guarantees the highest joint payoffs for the participating agents.

12.3 Time Consistent Solutions and Payment Mechanism

To guarantee dynamical stability in a dynamic cooperation scheme, the solution has to satisfy the property of time consistency. Similar to the case of continuous-time analysis, the stringent condition of time consistency is required for a dynamically stable solution. Petrosyan and Zenkevich (1996) were the first to provide the analysis of time consistent solutions in dynamic cooperative games. In particular, the specific agreed-upon optimality principle must remain effective at any stage of the game along the optimal state trajectory. Since at any stage of the game the agents are guided by the same optimality principles, they do not have any grounds for deviation from the previously adopted optimal behavior throughout the game. Therefore, for time consistency to be satisfied, the imputation $\xi(\cdot, \cdot)$ according to the original optimality principle has to be maintained at all the T stages along the cooperative trajectory $\{x_k^*\}_{k=1}^T$. In other words, the imputation

$$\xi(k, x_k^*) = \left[\xi^1(k, x_k^*), \xi^2(k, x_k^*), \ldots, \xi^n(k, x_k^*)\right] \quad \text{at stage } k, \text{ for } k \in \kappa, \quad (12.15)$$

has to be upheld.

Crucial to the analysis is the formulation of a payment mechanism so that the imputation in (12.15) can be realized.

12.3.1 Payoff Distribution Procedure

Following the continuous-time analysis of Yeung and Petrosyan (2006a) for cooperative differential games, we formulate a discrete-time version of the Payoff Distribution Procedure (PDP) so that the agreed imputations of (12.15) can be realized. Let $B_k^i(x_k^*)$ denote the payment that agent i will receive at stage k under the cooperative agreement along the cooperative trajectory $\{x_k^*\}_{k=1}^T$.

The payment scheme involving $B_k^i(x_k^*)$ constitutes a PDP in the sense that the imputation to agent i over the stages from k to T can be expressed as

$$\xi^i(k, x_k^*) = B_k^i(x_k^*)\left(\frac{1}{1+r}\right)^{k-1} + \sum_{\zeta=k+1}^T \left\{ B_\zeta^i(x_\zeta^*)\left(\frac{1}{1+r}\right)^{\zeta-1} \right\}, \quad (12.16)$$

for $i \in N$ and $k \in \kappa$.

Using (12.16) one can obtain

$$\xi^i(k+1, x_{k+1}^*) = B_{k+1}^i(x_{k+1}^*)\left(\frac{1}{1+r}\right)^k + \sum_{\zeta=k+2}^T \left\{ B_\zeta^i(x_\zeta^*)\left(\frac{1}{1+r}\right)^{\zeta-1} \right\}. \quad (12.17)$$

Substituting (12.17) into (12.16) yields

$$\xi^i(k, x_k^*) = B_k^i(x_k^*)\left(\frac{1}{1+r}\right)^{k-1} + \xi^i[k+1, f_k(x_k^*, \psi_k(x_k^*))], \quad (12.18)$$

for $i \in N$ and $k \in \kappa$.

Theorem 12.3 *A payment equaling*

$$B_k^i(x_k^*) = (1+r)^{k-1}\left[\xi^i(k, x_k^*) - \xi^i[k+1, f_k(x_k^*, \psi_k(x_k^*))]\right], \quad \text{for } i \in N, \quad (12.19)$$

given to agent i at stage $k \in \kappa$ along the cooperative trajectory $\{x_k^*\}_{k=1}^T$ will lead to the realization of the imputation $\{\xi(k, x_k^*), \text{for } k \in \kappa\}$.

Proof From (12.18), one can readily obtain (12.19). Theorem 12.3 can also be verified alternatively by showing that from (12.16)

$$\xi^i(k, x_k^*) = B_k^i(x_k^*)\left(\frac{1}{1+r}\right)^{k-1} + \sum_{\zeta=k+1}^T \left\{ B_\zeta^i(x_\zeta^*)\left(\frac{1}{1+r}\right)^{\zeta-1} \right\}$$

$$= \left\{ \xi^i(k, x_k^*) - \left(\xi^i[k+1, f_k(x_k^*, \psi_k(x_k^*))]\right) \right\}$$

$$+ \sum_{\zeta=k+1}^{T} \left\{ \xi^i\left(\zeta, x_\zeta^*\right) - \left(\xi^i\left[\zeta+1, f_\zeta\left(x_\zeta^*, \psi_\zeta\left(x_\zeta^*\right)\right)\right]\right)\right\}$$

$$= \xi^i\left(k, x_k^*\right),$$

given that $\xi^i(T+1, x_{T+1}^*) = 0$.

Hence Theorem 12.3 follows.   □

The payment scheme in Theorem 12.3 gives rise to the realization of the imputation guided by the agreed-upon optimal principle and will be used to derive time
(optimal-trajectory-subgame) consistent solutions in the next section.

12.3.2 Time (Optimal-Trajectory-Subgame) Consistent Solution

We denote the discrete-time cooperative game with the dynamics in (12.1) and
payoff in (12.2) by $\Gamma_c(1, x_0)$, and the game with the dynamics in (12.1) and pay-
off in (12.2), which starts at stage υ and initial state x_υ by $\Gamma_c(\upsilon, x_\upsilon^*)$. Moreover,
we let $P(1, x_0) = \{u_h^i$ and B_h^i for $h \in \kappa$ and $i \in N, \xi(1, x_0)\}$ denote the solution
to the cooperative game $\Gamma_c(1, x_0)$ under optimality Principle PI. Let $P(x_\upsilon^*, \upsilon) = \{u_h^i$ and B_h^i for $h \in \{\upsilon, \upsilon+1, \ldots, T\}$ and $i \in N, \xi(\upsilon, x_\upsilon^*)\}$ denote the solution to the
cooperative game $\Gamma_c(\upsilon, x_\upsilon^*)$ under optimality Principle PI.

A theorem characterizing a time (optimal-trajectory-subgame) consistent solution for the discrete-time cooperative game $\Gamma_c(1, x_0)$ under optimality Principle PI
is presented below.

Theorem 12.4 *For the cooperative game* $\Gamma_c(1, x_0)$ *with optimality Principle* PI *the
solution* $P(1, x_0) = \{u_h^i$ *and* B_h^i *for* $h \in \kappa$ *and* $i \in N, \xi(1, x_0)\}$—*in which*

(i) $u_h^i = \psi_h^i(x_h^*),$ *for* $h \in \kappa$ *and* $i \in N,$ *is the set of group optimal strategies for the
game* $\Gamma_c(1, x_0),$ *and*

(ii) $B_h^i = B_h^i(x_h^*),$ *for* $h \in \kappa$ *and* $i \in N,$ *where*

$$B_h^i\left(x_h^*\right) = (1+r)^{h-1}\left[\xi^i\left(h, x_h^*\right) - \xi^i\left[k+1, f_h\left(x_h^*, \psi_h\left(x_h^*\right)\right)\right]\right], \quad (12.20)$$

and $[\xi^1(h, x_h^*), \xi^2(h, x_h^*), \ldots, \xi^i(h, x_h^*)] \in P(h, x_h^*)$ *is the imputation according to optimality Principle* PI—*is time (optimal-trajectory-subgame) consistent.*

Proof Follow the proof of the continuous-time analog in Theorem 4.2 of Chap. 4. □

When all agents use the cooperative strategies, the payoff that agent i will directly
receive at stage k given that along the cooperative trajectory $\{x_k^*\}_{k=1}^{T}$ is

$$g_k^i\left[x_k^*, \psi_k^1\left(x_k^*\right), \psi_k^2\left(x_k^*\right), \ldots, \psi_k^n\left(x_k^*\right), x_{k+1}^*\right].$$

However, according to the agreed-upon imputation, agent i will receive $B_k^i(x_k^*)$ at stage k. Therefore, a side payment

$$\varpi_k^i(x_k^*) = B_k^i(x_k^*) - g_k^i\left[x_k^*, \psi_k^1(x_k^*), \psi_k^2(x_k^*), \ldots, \psi_k^n(x_k^*), x_k^*\right], \quad (12.21)$$

for $k \in \kappa$ and $i \in N$, will be given to agent i to yield the cooperative imputation $\xi^i(k, x_k^*)$.

12.4 An Illustration in Cooperative Resource Extraction

Consider an economy endowed with a renewable resource and with two resource extractors (firms). The lease for resource extraction begins at stage 1 and ends at stage 3 for these two firms. Let u_k^i denote the rate of resource extraction of firm i at stage k, for $i \in \{1, 2\}$. Let U^i be the set of admissible extraction rates and $x_k \in X \subset R^+$ the size of the resource stock at stage k. In particular, we have $U^i \in R^+$ and $u_k^1 + u_k^2 \leq x_k$. The extraction cost for firm $i \in \{1, 2\}$ depends on the quantity of the resource extracted u_k^i, the resource stock size x_k, and cost parameters c_1 and c_2. In particular, the extraction cost for firm i at stage k is specified as $c_i(u_k^i)^2/x_k$. The price of the resource is P.

The profits that firms 1 and 2 will obtain at stage k are, respectively,

$$\left[Pu_k^1 - \frac{c_1}{x_k}(u_k^1)^2\right] \quad \text{and} \quad \left[Pu_k^2 - \frac{c_2}{x_k}(u_k^2)^2\right]. \quad (12.22)$$

The growth dynamics of the resource is governed by the difference equation

$$x_{k+1} = x_k + a - bx_k - \sum_{j=1}^{2} u_k^j, \quad (12.23)$$

for $k \in \{1, 2, 3\}$ and $x_1 = x^0$.

With no human harvesting, the natural growth of the resource stock is $x_{k+1} - x_k = a - bx_k$. The objective of extractor $i \in \{1, 2\}$ is to maximize the present value of the stream of future profits

$$\sum_{k=1}^{3}\left[Pu_k^i - \frac{c_i}{x_k}(u_k^i)^2\right]\left(\frac{1}{1+r}\right)^{k-1}, \quad \text{for } i \in \{1, 2\}, \quad (12.24)$$

subject to (12.23).

Invoking Theorem 12.1, one can characterize the noncooperative equilibrium strategies in a feedback solution for the game in (12.23) and (12.24). In particular, a set of strategies $\{\phi_k^i(x), \text{ for } k \in \{1, 2, 3\} \text{ and } i \in \{1, 2\}\}$ provides a Nash equilibrium solution to the game in (12.23) and (12.24) if there exist functions $V^i(k, x)$, for

$i \in \{1, 2\}$ and $k \in \{1, 2, 3\}$, such that the following recursive relations are satisfied:

$$V^i(k, x) = \max_{u_k^i} \left\{ \left[Pu_k^i - \frac{c_i}{x}(u_k^i)^2 \right] \left(\frac{1}{1+r} \right)^{k-1} \right.$$

$$\left. + V^i \left[k + 1, x + a - bx - u_k^i - \phi_k^j(x) \right] \right\}, \qquad (12.25)$$

for $k \in \{1, 2, 3\}$,

$$V^i(4, x) = 0.$$

Performing the indicated maximization in (12.25) yields

$$\left(P - \frac{2c_i u_k^i}{x} \right) \left(\frac{1}{1+r} \right)^{k-1} - V_{x_{k+1}}^i \left[k + 1, x + a - bx - u_k^i - \phi_k^j(x) \right] = 0, \quad (12.26)$$

for $i \in \{1, 2\}$ and $k \in \{1, 2, 3\}$.

From (12.26), the game equilibrium strategies can be expressed as

$$\phi_k^i(x) = \left(P - V_{x_{k+1}}^i \left[k + 1, x + a - bx - \sum_{\ell=1}^{2} \phi_k^\ell(x) \right] (1+r)^{k-1} \right) \frac{x}{2c_i}, \quad (12.27)$$

for $i \in \{1, 2\}$ and $k \in \{1, 2, 3\}$.

Proposition 12.1 *The value function*

$$V^i(k, x) = \left[A_k^i x + C_k^i \right], \quad for \ i \in \{1, 2\} \ and \ k \in \{1, 2, 3\}, \qquad (12.28)$$

where A_k^i and C_k^i, for $i \in \{1, 2\}$ and $k \in \{1, 2, 3\}$, are constants in terms of the parameters of the game in (12.23) and (12.24).

Proof For the proof see Appendix 1 of this chapter. □

Substituting the relevant derivatives of the value functions in Proposition 12.1 into the game equilibrium strategies in (12.27) yields a noncooperative feedback equilibrium solution of the game in (12.23) and (12.24).

Now consider the case when the extractors agree to maximize their joint profit and share the excess of cooperative gains over their noncooperative payoffs equally. To maximize their joint payoff, they solve the problem of maximizing

$$\sum_{j=1}^{2} \sum_{k=1}^{3} \left[Pu_k^j - \frac{c_j}{x_k}(u_k^j)^2 \right] \left(\frac{1}{1+r} \right)^{k-1}, \qquad (12.29)$$

subject to (12.23).

Invoking Theorem 12.2, one can characterize the optimal controls in the dynamic programming problem in (12.23) and (12.29). In particular, a set of control strategies $\{\psi_k^i(x)$, for $k \in \{1, 2, 3\}$ and $i \in \{1, 2\}\}$ provides an optimal solution to the problem in (12.23) and (12.29) if there exist functions $W(k, x) : R \to R$, for $k \in \{1, 2, 3\}$, such that the following recursive relations are satisfied:

$$W(k, x) = \max_{u_k^1, u_k^2} \left\{ \sum_{j=1}^{2} \left[Pu_k^j - \frac{c_j}{x}(u_k^j)^2 \right] \left(\frac{1}{1+r} \right)^{k-1} \right.$$

$$\left. + W\left[k+1, x + a - bx - \sum_{j=1}^{2} u_k^j \right] \right\}, \quad \text{for } k \in \{1, 2, 3\} \qquad (12.30)$$

$$W(4, x) = 0.$$

Performing the indicated maximization in (12.31) yields

$$u_k^i = \left(P - \frac{2c_i u_k^i}{x} \right) \left(\frac{1}{1+r} \right)^{k-1} - W_{x_{k+1}} \left[k+1, x + a - bx - \sum_{j=1}^{2} u_k^j \right]$$

$$= 0, \qquad (12.31)$$

for $i \in \{1, 2\}$ and $k \in \{1, 2, 3\}$.

In particular, the optimal cooperative strategies can be obtained from (12.31) as

$$\left(P - W_{x_{k+1}} \left[k+1, x + a - bx - \sum_{j=1}^{2} u_k^j \right] (1+r)^{k-1} \right) \frac{x}{2c_i}, \qquad (12.32)$$

for $i \in \{1, 2\}$ and $k \in \{1, 2, 3\}$.

Proposition 12.2 *The value function*

$$W(k, x) = [A_k x + C_k], \quad \text{for } k \in \{1, 2, 3\}, \qquad (12.33)$$

where A_k and C_k, for $k \in \{1, 2, 3\}$, are constants in terms of the parameters of the problem in (12.29) and (12.23).

Proof For the proof see Appendix 2 of this chapter.   □

Using (12.32) and Proposition 12.2, the optimal cooperative strategies of the agents can be expressed as

$$\psi_k^i(x) = \left[P - A_{k+1}(1+r)^{k-1} \right] \frac{x}{2c_i}, \quad \text{for } i \in \{1, 2\} \text{ and } k \in \{1, 2, 3\}. \qquad (12.34)$$

Substituting $\psi_k^i(x)$ from (12.34) into (12.23) yields the optimal cooperative state trajectory

$$x_{k+1} = x_k + a - bx_k - \sum_{j=1}^{2}[P - A_{k+1}(1+r)^{k-1}]\frac{x_k}{2c_j}, \qquad (12.35)$$

for $k \in \{1, 2, 3\}$ and $x_1 = x^0$.

The dynamics in (12.35) is a linear difference equation readily solvable by standard techniques. Let $\{x_k^*, \text{ for } k \in \{1, 2, 3\}\}$ denote the solution to (12.35).

Since the extractors agree to share the excess of the cooperative gains over their noncooperative payoffs equally, an imputation

$$\xi^i(k, x_k^*) = V^i(k, x_k^*) + \frac{1}{2}\left[W(k, x_k^*) - \sum_{j=1}^{2}V^j(k, x_k^*)\right]$$

$$= (A_k^i x_k^* + C_k^i) + \frac{1}{2}\left[(A_k x_k^* + C_k) - \sum_{j=1}^{2}(A_k^j x_k^* + C_k^j)\right], \qquad (12.36)$$

for $k \in \{1, 2, 3\}$ and $i \in \{1, 2\}$, has to be maintained.

Invoking Theorem 12.4, if $x_k^* \in X$ is realized at stage k a payment equaling

$$B_k^i(x_k^*) - (1+r)^{k-1}[\xi^i(k, x_k^*) - \xi^i[k+1, f_k(x_k^*, \psi_k(x_k^*)) + \theta_{k+1}]]$$

$$= (1+r)^{k-1}\left\{(A_k^i x_k^* + C_k^i) + \frac{1}{2}\left((A_k x_k^* + C_k) - \sum_{j=1}^{2}(A_k^j x_k^* + C_k^j)\right)\right.$$

$$- \left[(A_{k+1}^i x_{k+1}^* + C_{k+1}^i) + \frac{1}{2}\left((A_{k+1} x_{k+1}^* + C_{k+1})\right.\right.$$

$$\left.\left.\left. - \sum_{j=1}^{2}(A_{k+1}^j x_{k+1}^* + C_{k+1}^j)\right)\right]\right\}, \quad \text{for } i \in \{1, 2\}, \qquad (12.37)$$

given to agent i at stage $k \in \kappa$ will lead to the realization of the imputation in (12.36).

A time (optimal-trajectory-subgame) consistent solution can be readily obtained from (12.34), (12.36), and (12.37).

12.5 Exercises

12.1 Consider an economy endowed with a renewable resource and with two resource extractors (firms). The lease for resource extraction begins at stage 1 and ends at stage 3 for these two firms. Let u_k^i denote the rate of resource extraction of firm i at stage k, for $i \in \{1, 2\}$. Let U^i be the set of admissible extraction rates and $x_k \in X \subset R^+$ the size of the resource stock at stage k. In particular, we have $U^i \in R^+$ and $u_k^1 + u_k^2 \leq x_k$. The extraction cost for firms 1 and 2 are, respectively, $(u_k^1)^2 / x_k$ and $1.5(u_k^1)^2 / x_k$. The price of the resource is 20.

The profits that firms 1 and 2 will obtain at stage k are, respectively,

$$\left[20u_k^1 - \frac{1}{x_k}(u_k^1)^2 \right] \quad \text{and} \quad \left[20u_k^2 - \frac{1.5}{x_k}(u_k^2)^2 \right].$$

The growth dynamics of the resource is governed by the difference equation

$$x_{k+1} = x_k + 10 - 0.2x_k - \sum_{j=1}^{2} u_k^j, \quad \text{for } k \in \{1, 2, 3\} \text{ and } x_1 = 45.$$

The objectives of extractors 1 and 2 are, respectively,

$$\sum_{k=1}^{3} \left[20u_k^1 - \frac{1}{x_k}(u_k^1)^2 \right] \left(\frac{1}{1+r} \right)^{k-1}, \quad \text{and}$$

$$\sum_{k=1}^{3} \left[20u_k^2 - \frac{1.5}{x_k}(u_k^2)^2 \right] \left(\frac{1}{1+r} \right)^{k-1}.$$

Characterize the noncooperative equilibrium strategies in a feedback Nash equilibrium solution for the above resource economy.

12.2 If the extractors agree to cooperate and maximize their joint payoff, derive the optimal cooperative strategies and the optimal resource trajectory.

12.3 Consider the case when the extractors agree to share the excess of the cooperative gains over their noncooperative payoffs equally. Characterize a time (optimal-trajectory-subgame) consistent solution.

Appendix 1: Proof of Proposition 12.1

Consider first the last stage, that is, stage 3. Remembering that $V^i(3, x) = [A_3^i x + C_3^i]$ from Proposition 12.1 and $V^i(4, x) = 0$, the conditions in (12.25) become

$$V^i(3, x) = [A_3^i x + C_3^i] = \max_{u_3^i} \left\{ \left[Pu_3^i - \frac{c_i}{x}(u_3^i)^2 \right] \left(\frac{1}{1+r} \right)^2 \right\}, \quad \text{for } i \in \{1, 2\}.$$
$$(12.38)$$

Performing the indicated maximization in (12.38) yields

$$\left(P - \frac{2c_i u_3^i}{x} \right) \left(\frac{1}{1+r} \right)^2 = 0, \quad \text{for } i \in \{1, 2\}. \tag{12.39}$$

The game equilibrium strategies in stage 3 can then be expressed as

$$\phi_3^i(x) = \frac{Px}{2c_i}, \quad \text{for } i \in \{1, 2\}. \tag{12.40}$$

Substituting (12.40) into (12.38) yields

$$V^i(3, x) = \left[A_3^i x + C_3^i \right] = \left[\frac{P^2 x}{2c_i} - \frac{P^2 x}{4c_i} \right] \left(\frac{1}{1+r} \right)^2 = \left(\frac{1}{1+r} \right)^2 \frac{P^2}{4c_i} x,$$

$$\text{for } i \in \{1, 2\}. \tag{12.41}$$

Using (12.41), we obtain

$$V^i(3, x) = \left[A_3^i x + C_3^i \right], \quad \text{where } A_3^i = \left(\frac{1}{1+r} \right)^2 \frac{P^2}{4c_i} \text{ and } C_3^i = 0, \text{ for } i \in \{1, 2\}. \tag{12.42}$$

Now we proceed to stage 2. The conditions in (12.25) become

$$V^i(2, x) = \left[A_2^i x + C_2^i \right]$$

$$= \max_{u_2^i} \left\{ \left[P u_2^i - \frac{c_i}{x} (u_2^i)^2 \right] \left(\frac{1}{1+r} \right) \right.$$

$$\left. + V^i \left[3, x + a - bx - u_2^i - \phi_2^j(x) \right] \right\}, \quad i, j \in \{1, 2\} \text{ and } i \neq j. \tag{12.43}$$

Invoking (12.41), the condition in (12.43) can be expressed as

$$\left[A_2^i x + C_2^i \right] = \max_{u_2^i} \left\{ \left[P u_2^i - \frac{c_i}{x} (u_2^i)^2 \right] \left(\frac{1}{1+r} \right) \right.$$

$$\left. + A_3^i \left[x + a - bx - u_2^i - \phi_2^j(x) \right] \right\}, \quad \text{for } i, j \in \{1, 2\} \text{ and } i \neq j. \tag{12.44}$$

Performing the indicated maximization in (12.44) yields

$$\left(P - \frac{2c_i u_2^i}{x} \right) \left(\frac{1}{1+r} \right) - A_3^i = 0, \quad \text{for } i \in \{1, 2\}. \tag{12.45}$$

The game equilibrium strategies in stage 2 can then be expressed as

$$\phi_2^i(x) = \left[P - (1+r) A_3^i \right] \frac{x}{2c_i}, \quad \text{for } i \in \{1, 2\}. \tag{12.46}$$

Substituting (12.46) into (12.44) yields

$$
\begin{aligned}
[A_2^i x + C_2^i] = \left(\frac{1}{1+r}\right) & \left([P - (1+r)A_3^i]\frac{P}{2c_i}x - [P - (1+r)A_3^i]^2\frac{1}{4c_i}x\right) \\
& + A_3^i\left(x + a - bx - [P - (1+r)A_3^i]\frac{1}{2c_i}x\right. \\
& \left. - [P - (1+r)A_3^j]\frac{1}{2c_j}x\right), \quad \text{for } i, j \in \{1, 2\} \text{ and } i \neq j. \quad (12.47)
\end{aligned}
$$

Collecting the terms together, (12.47) can be expressed as

$$
\begin{aligned}
V^i(2, x) = & [A_2^i x + C_2^i] \\
= & \left\{\left(\frac{1}{1+r}\right)[P - (1+r)A_3^i]\frac{P + (1+r)A_3^i}{4c_i}\right. \\
& \left. + A_3^i(1-b) - [P - (1+r)A_3^i]\frac{A_3^i}{2c_i} - [P - (1+r)A_3^j]\frac{A_3^i}{2c_j}\right\}x \\
& + aA_3^i, \quad \text{for } i, j \in \{1, 2\} \text{ and } i \neq j. \quad (12.48)
\end{aligned}
$$

Substituting $A_3^i = (\frac{1}{1+r})^2\frac{P^2}{4c_i}$ for $i \in \{1, 2\}$ into (12.48), A_2^i and C_2^i for $i \in \{1, 2\}$ are obtained in explicit terms.

Finally, we proceed to the first stage. The conditions in (12.25) become

$$
\begin{aligned}
V^i(1, x) = & [A_1^i x + C_1^i] \\
= & \max_{u_1^i}\left\{\left[Pu_1^i - \frac{c_i}{x}(u_1^i)^2\right]\right. \\
& \left. + V^i[2, x + a - bx - u_1^i - \phi_1^j(x)]\right\}, \\
& \text{for } i, j \in \{1, 2\} \text{ and } i \neq j. \quad (12.49)
\end{aligned}
$$

Invoking (12.48), the condition in (12.49) can be expressed as

$$
[A_1^i x + C_1^i] = \max_{u_1^i}\left\{\left[Pu_1^i - \frac{c_i}{x}(u_1^i)^2\right] + (A_2^i[x + a - bx - u_1^i - \phi_1^j(x)] + C_2^i)\right\},
$$

for $i, j \in \{1, 2\}$ and $i \neq j$. $\quad (12.50)$

Performing the indicated maximization in (12.50) yields

$$
\left(P - \frac{2c_i u_1^i}{x}\right) - A_2^i = 0, \quad \text{for } i \in \{1, 2\}. \quad (12.51)
$$

The game equilibrium strategies in stage 1 can then be expressed as

$$\phi_1^i(x) = \left[P - A_2^i\right]\frac{x}{2c_i}, \quad \text{for } i \in \{1, 2\}. \tag{12.52}$$

Substituting (12.52) into (12.50) yields

$$\left[A_1^i x + C_1^i\right] = \left((P - A_2^i)\frac{P}{2c_i}x - (P - A_2^i)^2\frac{1}{4c_i}x\right)$$

$$+ \left[A_2^i\left(x + a - bx - (P - A_2^i)\frac{1}{2c_i}x - (P - A_2^j)\frac{1}{2c_j}x\right) + C_2^i\right],$$

$$\tag{12.53}$$

for $i, j \in \{1, 2\}$ and $i \neq j$.

Collecting the terms together, (12.53) can be expressed as

$$V^i(3, x) = \left[A_1^i x + C_1^i\right]$$

$$= \left[(P - A_2^i)\frac{P + A_2^i}{4c_i} + A_2^i(1 - b) - (P - A_2^i)\frac{A_2^i}{2c_i} - (P - A_2^j)\frac{A_2^i}{2c_j}\right]x$$

$$+ aA_2^i + C_2^i, \quad \text{for } i, j \in \{1, 2\} \text{ and } i \neq j. \tag{12.54}$$

Substituting the explicit terms for A_2^i, A_2^j, C_2^i, and C_2^j from (12.48) into (12.54), A_1^i and C_1^i for $i \in \{1, 2\}$ are obtained in explicit terms.

Appendix 2: Proof of Proposition 12.2

Consider first the last stage, that is, stage 3. Remembering that $W(3, x) = [A_3x + C_3]$ from Proposition 12.2 and $W(4, x) = 0$. The conditions in (12.31) become

$$W(3, x) = [A_3x + C_3] = \max_{u_3^1, u_3^2}\left\{\sum_{j=1}^{2}\left[Pu_3^j - \frac{c_j}{x}(u_3^j)^2\right]\left(\frac{1}{1+r}\right)^2\right\}. \tag{12.55}$$

Performing the indicated maximization in (12.38) yields

$$\left(P - \frac{2c_i u_3^i}{x}\right)\left(\frac{1}{1+r}\right)^2 = 0, \quad \text{for } i \in \{1, 2\}. \tag{12.56}$$

The optimal cooperative strategies in stage 3 can then be expressed as

$$\psi_3^i(x) = \frac{Px}{2c_i}, \quad \text{for } i \in \{1, 2\}. \tag{12.57}$$

Substituting (12.57) into (12.55) yields

$$[A_3x + C_3] = \left(\frac{1}{1+r}\right)^2 \sum_{j=1}^{2} \frac{P^2}{4c_j} x, \quad \text{for } i \in \{1, 2\}. \tag{12.58}$$

Using (12.58), we obtain

$$W(3, x) = [A_3x + C_3], \quad \text{where } A_3 = \left(\frac{1}{1+r}\right)^2 \sum_{j=1}^{2} \frac{P^2}{4c_j} \text{ and } C_3 = 0. \tag{12.59}$$

Now we proceed to stage 2. The conditions in (12.31) become

$$W(2, x) = [A_2x + C_2]$$
$$= \max_{u_2^1, u_2^2} \left\{ \sum_{j=1}^{2} \left[Pu_2^j - \frac{c_j}{x}(u_2^j)^2 \right] \left(\frac{1}{1+r}\right) \right.$$
$$\left. + W\left[3, x + a - bx - \sum_{j=1}^{2} u_2^j \right] \right\}. \tag{12.60}$$

Invoking (12.58), the condition in (12.60) can be expressed as

$$[A_2x + C_2] = \max_{u_2^1, u_2^2} \left\{ \sum_{j=1}^{2} \left[Pu_2^j - \frac{c_j}{x}(u_2^j)^2 \right] \left(\frac{1}{1+r}\right) \right.$$
$$\left. + A_3\left[x + a - bx - \sum_{j=1}^{2} u_2^j \right] \right\}. \tag{12.61}$$

Performing the indicated maximization in (12.61) yields

$$\left(P - \frac{2c_i u_2^i}{x}\right)\left(\frac{1}{1+r}\right) - A_3 = 0, \quad \text{for } i \in \{1, 2\}. \tag{12.62}$$

The optimal cooperative strategies in stage 2 can then be expressed as

$$\psi_2^i(x) = [P - (1+r)A_3]\frac{x}{2c_i}, \quad \text{for } i \in \{1, 2\}. \tag{12.63}$$

Substituting (12.63) into (12.61) yields

$$[A_2x + C_2] = \left(\frac{1}{1+r}\right) \sum_{j=1}^{2} [P - (1+r)A_3]\frac{P + (1+r)A_3}{4c_j} x$$
$$+ A_3\left(x + a - bx - [P - (1+r)A_3]\frac{1}{2c_1}x \right.$$
$$\left. - [P - (1+r)A_3]\frac{1}{2c_2}x \right). \tag{12.64}$$

Collecting the terms together, (12.64) can be expressed as

$$
\begin{aligned}
W(2, x) &= [A_2 x + C_2] \\
&= \left[\left(\frac{1}{1+r} \right) \sum_{j=1}^{2} [P - (1+r)A_3] \frac{P + (1+r)A_3}{4c_j} + A_3(1-b) \right. \\
&\quad \left. - [P - (1+r)A_3] \frac{A_3}{2c_1} - [P - (1+r)A_3] \frac{A_3^i}{2c_2} \right] x + a A_3 \right).
\end{aligned} \quad (12.65)
$$

Substituting $A_3 = (\frac{1}{1+r})^2 \sum_{j=1}^{2} \frac{P^2}{4c_j}$ into (12.65), A_2 and C_2 are obtained in explicit terms.

Finally, we proceed to the first stage, the conditions in (12.31) become

$$
\begin{aligned}
W(1, x) &= [A_1 x + C_1] \\
&= \max_{u_1^1, u_1^2} \left\{ \sum_{j=1}^{2} \left[P u_1^j - \frac{c_j}{x} (u_1^j)^2 \right] \right. \\
&\quad \left. + W\left[2, x + a - bx - \sum_{j=1}^{2} u_1^j \right] \right\}.
\end{aligned} \quad (12.66)
$$

Invoking (12.65), the condition in (12.66) can be expressed as

$$
\begin{aligned}
[A_1 x + C_1] &= \max_{u_1^1, u_1^2} \left\{ \sum_{j=1}^{2} \left[P u_1^j - \frac{c_j}{x} (u_1^j)^2 \right] \right. \\
&\quad \left. + \left(A_2 \left[x + a - bx - \sum_{j=1}^{2} u_1^j \right] + C_2 \right) \right\}.
\end{aligned} \quad (12.67)
$$

Performing the indicated maximization in (12.67) yields

$$
\left(P - \frac{2c_i u_1^i}{x} \right) - A_2 = 0, \quad \text{for } i \in \{1, 2\}. \quad (12.68)
$$

The optimal cooperative strategies in stage 1 can then be expressed as

$$
\psi_1^i(x) = (P - A_2) \frac{x}{2c_i}, \quad \text{for } i \in \{1, 2\}. \quad (12.69)
$$

Substituting (12.69) into (12.67) yields

$$
\begin{aligned}
[A_1 x + C_1] &= \sum_{j=1}^{2} (P - A_2) \frac{P + A_2}{4c_j} x \\
&\quad + \left[A_2 \left(x + a - bx - (P - A_2) \frac{1}{2c_1} x - (P - A_2) \frac{1}{2c_2} x \right) + C_2 \right].
\end{aligned} \quad (12.70)
$$

Collecting the terms together, (12.70) can be expressed as

$$W(2, x) = [A_1 x + C_1]$$

$$= \left[\sum_{j=1}^{2} (P - A_2) \frac{P + A_2}{4c_j} + A_2(1 - b) - (P - A_2) \frac{A_2}{2c_1} - (P - A_2) \frac{A_2}{2c_2} \right] x$$

$$+ a A_2 + C_2. \tag{12.71}$$

Substituting the explicit terms for A_2 and C_2 from (12.65) into (12.71), A_1 and C_1 are obtained in explicit terms.

Chapter 13
Discrete-Time Cooperative Games
Under Uncertainty

In some economic processes in discrete-time, uncertainty may also arise. For instance, Smith and Zenou (2003) considered a discrete-time stochastic job search model. Esteban-Bravo and Nogales (2008) analyzed mathematical programming for stochastic discrete-time dynamics arising in economic systems including examples in a stochastic national growth model and international growth model with uncertainty. The discrete-time counterpart of stochastic differential games is known as stochastic dynamic games. Basar and Ho (1974) examined informational properties of the Nash solutions of stochastic nonzero-sum games. The elimination of the informational nonuniqueness in a Nash equilibrium through a stochastic formulation was first discussed in Basar (1976) and further examined in Basar (1975, 1979, 1989). Basar and Mintz (1972, 1973) and Basar (1978) developed an equilibrium solution of linear-quadratic stochastic dynamic games with noisy observation. Again, the SIAM Classics on Dynamic Noncooperative Game Theory by Basar and Olsder (1995) gave a comprehensive treatment of noncooperative stochastic dynamic games. Yeung and Petrosyan (2010) provided the techniques in characterizing subgame consistent solutions for stochastic dynamic. Furthermore, they also presented a stochastic dynamic game in resource extraction. Analyses of noncooperative and cooperative discrete-time dynamic games with random game horizons were presented in Yeung and Petrosyan (2011). The recently emerging robust control techniques in discrete time along the lines of Hansen and Sargent (2008) should prove to be fruitful in developing into stochastic dynamic interactive economic models.

In the first section of this chapter, we present a general formulation of stochastic dynamic economic games in discrete time and noncooperative market outcomes. Group optimality and individual rationality under stochastic dynamic cooperation are discussed in Sect. 13.2. Subgame consistent cooperative solutions with corresponding payoff distribution procedures are derived in Sect. 13.3. An illustration of cooperative resource extraction under uncertainty in discrete time is given in Sect. 13.4.

D.W.K. Yeung, L.A. Petrosyan, *Subgame Consistent Economic Optimization*,
Static & Dynamic Game Theory: Foundations & Applications,
DOI 10.1007/978-0-8176-8262-0_13, © Springer Science+Business Media, LLC 2012

13.1 Stochastic Dynamic Games

In this section we consider the formulation of stochastic dynamic games in economics and their solutions.

13.1.1 Game Formulation

Consider the general T-stage n-person nonzero-sum discrete-time stochastic dynamic game with initial state x^0. The state space of the game is $X \in R^m$ and the state dynamics of the game is characterized by the stochastic difference equation

$$x_{k+1} = f_k(x_k, u_k^1, u_k^2, \ldots, u_k^n) + \theta_k, \tag{13.1}$$

for $k \in \{1, 2, \ldots, T\} \equiv \kappa$ and $x_1 = x^0$, where $u_k^i \in R^{m_i}$ is the control vector of agent i at stage k, $x_k \in X$ is the state, and θ_k is a set of statistically independent random variables.

The objective of agent i is

$$E_{\theta_1, \theta_2, \ldots, \theta_T} \left\{ \sum_{\zeta=1}^{T} g_\zeta^i \left[x_\zeta, u_\zeta^1, u_\zeta^2, \ldots, u_\zeta^n \right] \left(\frac{1}{1+r} \right)^{\zeta-1} \right\},$$

$$\text{for } i \in \{1, 2, \ldots, n\} \equiv N, \tag{13.2}$$

where r is the discount rate and $E_{\theta_1, \theta_2, \ldots, \theta_T}$ is the expectation operation with respect to the statistics of $\theta_1, \theta_2, \ldots, \theta_T$.

The payoffs of the agents are transferable.

13.1.2 Noncooperative Solution

In this section, we characterize the noncooperative outcome of the discrete-time stochastic economic game in (13.1) and (13.2). Let $\{\phi_k^i(x), \text{ for } k \in \kappa \text{ and } i \in N\}$ denote a set of strategies that provides a feedback Nash equilibrium solution (if it exists) to the game in (13.1) and (13.2), and

$$V^i(k, x)$$

$$= E_{\theta_k, \theta_{k+1}, \ldots, \theta_T} \left\{ \sum_{\zeta=k}^{T} g_\zeta^i \left[x_\zeta, \phi_\zeta^1(x_\zeta), \phi_\zeta^2(x_\zeta), \ldots, \phi_\zeta^n(x_\zeta) \right] \left(\frac{1}{1+r} \right)^{\zeta-1} \right\},$$

where $x_k = x$, for $k \in K$ and $i \in N$, denote the value functions indicating the expected game equilibrium payoff to agent i over the stages from k to T.

A frequently used way to characterize and derive a feedback Nash equilibrium of the game is as follows.

Theorem 13.1 *A set of strategies* $\{\phi_k^i(x), \text{for } k \in \kappa \text{ and } i \in N\}$ *provides a feedback Nash equilibrium solution to the game in* (13.1) *and* (13.2) *if there exist functions* $V^i(k, x)$, *for* $k \in K$ *and* $i \in N$, *such that the following recursive relations are satisfied:*

$$V^i(k, x) = \max_{u_k^i} E_{\theta_k} \left\{ g_k^i \left[x, \phi_k^1(x), \phi_k^2(x), \ldots, \phi_k^{i-1}(x), u_k^i, \phi_k^{i+1}(x), \ldots, \phi_k^n(x) \right] \right.$$

$$\times \left(\frac{1}{1+r} \right)^{k-1} + V^i \left[k+1, \tilde{f}_k^i \left(x, u_k^i \right) + \theta_k \right] \right\}$$

$$= E_{\theta_k} \left\{ g_k^i \left[x, \phi_k^1(x), \phi_k^2(x), \ldots, \phi_k^n(x) \right] \right.$$

$$\times \left(\frac{1}{1+r} \right)^{k-1} + V^i \left[k+1, f_k \left(x, \phi_k^1(x), \phi_k^2(x), \ldots, \phi_k^n(x) \right) + \theta_k \right] \right\},$$

$$\tag{13.3}$$

$$V^i(T+1, x) = 0; \tag{13.4}$$

for $i \in N$ *and* $k \in \kappa$, *where* $\tilde{f}_k^i(x, u_k^i) = f_k[x, \phi_k^1(x), \phi_k^2(x), \ldots, \phi_k^{i-1}(x), u_k^i, \phi_k^{i+1}(x), \ldots, \phi_k^n(x)]$ *and* E_{θ_k} *is the expectation operation with respect to the statistics of* θ_k.

Proof See Appendix 1 of this chapter. □

Again, for the sake of exposition, we sidestep the issue of multiple equilibria and focus on solvable games in which a particular noncooperative Nash equilibrium is chosen by the agents in the entire subgame.

13.2 Dynamic Cooperation Under Uncertainty

Now consider the case when the agents agree to cooperate and distribute the payoff among themselves according to an optimality principle. Once again, the essential properties of group optimality and individual rationality have to be satisfied.

Let the agreed-upon optimality principle be as follows.

Principle PI It is an optimality principle that entails (i) group optimality and (ii) the distribution of the total cooperative payoff according to an imputation that equals $\xi(k, x_k^*) = [\xi^1(k, x_k^*), \xi^2(k, x_k^*), \ldots, \xi^n(k, x_k^*)]$, for $k \in \kappa$.

Again, we first examine the conditions under which group optimality and individual rationality will be maintained.

13.2.1 Group Optimality

Maximizing the agents' expected joint payoff guarantees group optimality in a game where payoffs are transferable. To maximize their expected joint payoff the agents have to solve the discrete-time stochastic dynamic programming problem of maximizing

$$E_{\theta_1,\theta_2,\dots,\theta_T}\left\{\sum_{j=1}^{n}\sum_{k=1}^{T}\left[g_k^j\left[x_k,u_k^1,u_k^2,\dots,u_k^n\right]\left(\frac{1}{1+r}\right)^{k-1}\right]\right\}, \tag{13.5}$$

subject to (13.1).

A frequently used method to characterize and derive an optimal solution to the problem in (13.5) and (13.1) is as follows.

Theorem 13.2 *A set of strategies $\{\psi_k^i(x), for\ k \in \kappa\ and\ i \in N\}$ provides an optimal solution to the problem in (13.5) and (13.1) if there exist functions $W(k,x)$, for $k \in K$, such that the following recursive relations are satisfied:*

$$W(k,x) = \max_{u_k^1,u_k^2,\dots,u_k^n} E_{\theta_k}\left\{\sum_{j=1}^{n}g_k^j\left[x_k,u_k^1,u_k^2,\dots,u_k^n\right]\right.$$

$$\times \left(\frac{1}{1+r}\right)^{k-1} + W\left[k+1,f_k\left(x_k,u_k^1,u_k^2,\dots,u_k^n\right)+\theta_k\right]\right\}$$

$$= E_{\theta_k}\left\{\sum_{j=1}^{n}g_k^j\left[x,\psi_k^1(x),\psi_k^2(x),\dots,\psi_k^n(x)\right]\right.$$

$$\times \left(\frac{1}{1+r}\right)^{k-1} + W\left[k+1,f_k\left(x,\psi_k^1(x),\psi_k^2(x),\dots,\psi_k^n(x)\right)+\theta_k\right]\right\}, \tag{13.6}$$

$$W(T+1,x) = 0. \tag{13.7}$$

Proof Again, invoking the principle of optimality, we work backward in time, starting at all possible final states with the corresponding final times. Consider the joint maximization problem that maximizes the objective in (13.5) subject to the dynamics in (13.1). To determine the optimal control strategy, we shall need the expression for the maximized payoff from any starting point at any initial time with any given initial value of the state. This is the value function

$$W(k,x)$$

$$= \max_{u_k,u_{k+1},\dots,u_T} E_{\theta_k\theta_{k+1}\dots\theta_T} \left\{ \sum_{\varsigma=k}^{T} \sum_{j=1}^{n} g_\varsigma^j \big[x_\xi, u_\varsigma^1, u_\varsigma^2, \dots, u_\varsigma^n \big] \left(\frac{1}{1+r} \right)^{\varsigma-1} \right\},$$

$$x_k = x, \quad \text{for } k \in \kappa,$$

with $u_\varsigma^i = \gamma_\varsigma^i(x_\varsigma) \in U_i$ for $i \in N$. Utilizing the property of independence of the random vectors involved, a direct application of the principle of optimality readily shows that the value function satisfies the recursive relation

$$W(k,x) = \max_{u_k^1,u_k^2,\dots,u_k^n} E_{\theta_k} \left\{ \sum_{j=1}^{n} g_k^j \big[x_k, u_k^1, u_k^2, \dots, u_k^n \big] \right.$$

$$\left. \times \left(\frac{1}{1+r} \right)^{k-1} + W \big[k+1, f_k \big(x_k, u_k^1, u_k^2, \dots, u_k^n \big) + \theta_k \big] \right\},$$

which also leads to the conclusion that the optimal strategies ϕ_k^i, for $i \in N$, depend only on the current value of the state x_k. Hence Theorem 13.2 follows.   □

Substituting the optimal control $\{\psi_k^i(x), \text{for } k \in \kappa \text{ and } i \in N\}$ into the state dynamics in (13.1), one can obtain the dynamics of the cooperative trajectory as

$$x_{k+1} = f_k \big(x_k, \psi_k^1(x_k), \psi_k^2(x_k), \dots, \psi_k^n(x_k) \big) + \theta_k, \tag{13.8}$$

for $k \in \kappa$ and $x_1 = x^0$.

We use X_k^* to denote the set of realizable values of x_k^* at stage k generated by (13.8). The term $x_k^* \in X_k^*$ is used to denote an element in X_k^*.

The term $W(k, x_k^*)$ gives the expected total cooperative payoff over the stages from k to T if $x_k^* \in X_k^*$ is realized at stage $k \in \kappa$. We then proceed to consider individual rationality.

13.2.2 Individual Rationality

The agents have to agree to an optimality principle in distributing the total cooperative payoff among themselves. For individual rationality to be upheld the expected payoffs an agent receives under cooperation have to be no less than his expected noncooperative payoff along the cooperative state trajectory. Let $\xi(\cdot, \cdot)$ denote the imputation vector guiding the distribution of the total cooperative payoff under the agreed-upon optimality principle along the cooperative trajectory $\{x_k^*\}_{k=1}^T$. At stage k, the imputation vector according to $\xi(\cdot, \cdot)$ is $\xi(k, x_k^*) = [\xi^1(k, x_k^*), \xi^2(k, x_k^*), \ldots, \xi^n(k, x_k^*)]$, for $k \in \kappa$.

For individual rationality to be maintained throughout all the stages $k \in \kappa$, it is required that

$$\xi^i(k, x_k^*) \geq V^i(k, x_k^*), \quad \text{for } i \in N \text{ and } k \in \kappa.$$

In particular, the above condition guarantees that the expected payoff allocated to an economic agent under cooperation will be no less than its expected noncooperative payoff.

To satisfy group optimality the imputation vector has to satisfy

$$W(k, x_k^*) = \sum_{j=1}^n \xi^j(k, x_k^*), \quad \text{for } k \in \kappa.$$

This condition guarantees the highest expected joint payoffs for the participating agents.

13.3 Subgame Consistent Solutions and Payment Mechanism

Now we proceed to consider dynamically stable solutions in cooperative stochastic dynamic games. To guarantee dynamical stability in a stochastic dynamic cooperation scheme, the solution has to satisfy the property of subgame consistency. A cooperative solution is subgame consistent if an extension of the solution policy to a subgame starting at a later time with any realizable state brought about by prior optimal behavior would remain optimal under the agreed-upon optimality principle. In particular, subgame consistency ensures that, as the game proceeds, agents are guided by the same optimality principle at each stage of the game, and hence do not possess incentives to deviate from the previously adopted optimal behavior. Yeung and Petrosyan (2010) developed conditions leading to subgame consistent solutions in stochastic differential games.

For subgame consistency to be satisfied, the imputation $\xi(\cdot, \cdot)$ according to the original optimality principle has to be maintained at all the T stages along the cooperative trajectory $\{x_k^*\}_{k=1}^T$. In other words, the imputation

$$\xi(k, x_k^*) = \left[\xi^1(k, x_k^*), \xi^2(k, x_k^*), \ldots, \xi^n(k, x_k^*)\right] \quad \text{at stage } k, \text{ for } k \in \kappa, \quad (13.9)$$

has to be upheld.

Crucial to the analysis is the formulation of a payment mechanism so that the imputation in (13.9) can be realized.

13.3.1 Payoff Distribution Procedure

Following the analysis of Yeung and Petrosyan (2010), we formulate a discrete-time Payoff Distribution Procedure (PDP) so that the agreed imputations in (13.9) can be realized. Let $B_k^i(x_k^*)$ denote the payment that agent i will receive at stage k under the cooperative agreement if $x_k^* \in X_k^*$ is realized at stage $k \in \kappa$.

The payment scheme involving $B_k^i(x_k^*)$ constitutes a PDP in the sense that, if $x_k^* \in X_k^*$ is realized at stage k, the imputation to agent i over the stages from k to T can be expressed as

$$\xi^i\left(k, x_k^*\right) = B_k^i\left(x_k^*\right)\left(\frac{1}{1+r}\right)^{k-1} + \sum_{\zeta=k+1}^{T} E_{\theta_{k+1}, \theta_{k+2}, \dots, \theta_\zeta} \left\{ B_\zeta^i\left(x_\zeta^*\right)\left(\frac{1}{1+r}\right)^{\zeta-1} \right\},$$

(13.10)

for $i \in N$ and $k \in \kappa$.

Using (13.10) one can obtain

$$\xi^i\left(k+1, x_{k+1}^*\right) = B_{k+1}^i\left(x_{k+1}^*\right)\left(\frac{1}{1+r}\right)^{k}$$

$$+ \sum_{\zeta=k+2}^{T} E_{\theta_{k+2}, \theta_{k+3}, \dots, \theta_\zeta} \left\{ B_\zeta^i\left(x_\zeta^*\right)\left(\frac{1}{1+r}\right)^{\zeta-1} \right\}. \quad (13.11)$$

Substituting (13.11) into (13.10) yields

$$\xi^i\left(k, x_k^*\right) = B_k^i\left(x_k^*\right)\left(\frac{1}{1+r}\right)^{k-1}$$

$$+ E_{\theta_{k+1}}\left(\xi^i\left[k+1, f_k\left(x_k^*, \psi_k\left(x_k^*\right)\right) + \theta_k\right]\right), \quad (13.12)$$

for $i \in N$ and $k \in \kappa$.

Theorem 13.3 *A payment equaling*

$$B_k^i\left(x_k^*\right) = (1+r)^{k-1}\left\{\xi^i\left(k, x_k^*\right) - E_{\theta_k}\left(\xi^i\left[k+1, f_k\left(x_k^*, \psi_k\left(x_k^*\right)\right) + \theta_k\right]\right)\right\},$$

for $i \in N$,

(13.13)

given to agent i at stage $k \in \kappa$, if $x_k^* \in X_k^*$ will lead to the realization of the imputation $\{\xi(k, x_k^*), \text{for } k \in \kappa\}$.

Proof From (13.12) one can readily obtain (13.13). Theorem 13.3 can also be verified alternatively by showing that from (13.10)

$$
\xi^i\left(k, x_k^*\right) = B_k^i\left(x_k^*\right)\left(\frac{1}{1+r}\right)^{k-1} + \sum_{\zeta=k+1}^{T} E_{\theta_{k+1},\theta_{k+2},...,\theta_\zeta}\left\{B_\zeta^i\left(x_\zeta^*\right)\left(\frac{1}{1+r}\right)^{\zeta-1}\right\}
$$

$$
= \left\{\xi^i\left(k, x_k^*\right) - E_{\theta_k}\left(\xi^i\left[k+1, f_k\left(x_k^*, \psi_k\left(x_k^*\right)\right) + \theta_k\right]\right)\right\}
$$

$$
+ \sum_{\zeta=k+1}^{T} E_{\theta_{k+1},\theta_{k+2},...,\theta_\zeta}\left\{\xi^i\left(\zeta, x_\zeta^*\right)\right.
$$

$$
\left. - E_{\theta_\zeta}\left(\xi^i\left[\zeta+1, f_\zeta\left(x_\zeta^*, \psi_\zeta\left(x_\zeta^*\right)\right) + \theta_\zeta\right]\right)\right\} = \xi^i\left(k, x_k^*\right),
$$

given that $\xi^i\left(T+1, x_{T+1}^*\right) = 0$. □

The payment scheme in Theorem 13.3 gives rise to the realization of the imputation guided by the agreed-upon optimal principle and will be used to derive subgame consistent solutions in the next section.

13.3.2 Subgame Consistent Solution

We denote the discrete-time cooperative game with the dynamics in (13.1) and payoff in (13.2) by $\Gamma_c(1, x_0)$, and the game with the dynamics in (13.1) and payoff in (13.2), which starts at stage υ and initial state $x_\upsilon^* \in X_\upsilon^*$ by $\Gamma_c(\upsilon, x_\upsilon^*)$. Moreover, we let $P(1, x_0) = \{u_h^i(x_h^*)$ and $B_h^i(x_h^*)$ for $h \in \kappa$ and $i \in N$ and $x_h^* \in X_h^*, \xi(1, x_0)\}$ denote the solution to the cooperative game $\Gamma_c(1, x_0)$ under optimality Principle PI. Let $P(x_\upsilon^*, \upsilon) = \{u_h^i(x_h^*)$ and $B_h^i(x_h^*)$ for $h \in \{\upsilon, \upsilon+1, \ldots, T\}$ and $i \in N$ and $x_h^* \in X_h^*, \xi(\upsilon, x_\upsilon^*)\}$ denote the solution to the cooperative game $\Gamma_c(\upsilon, x_\upsilon^*)$ under optimality Principle PI.

A theorem characterizing a subgame consistent solution for the discrete-time cooperative game $\Gamma_c(1, x_0)$ under optimality Principle PI is presented below.

Theorem 13.4 *For the cooperative game $\Gamma_c(1, x_0)$ with optimality Principle PI the solution $P(1, x_0) = \{u_h^i(x_h^*)$ and $B_h^i(x_h^*)$ for $h \in \kappa$ and $i \in N$ and $x_h^* \in X_h^*, \xi(1, x_0)\}$ in which (i) $u_h^i(x_h^*) = \psi_h^i(x_h^*)$, for $h \in \kappa$ and $i \in N$ and $x_h^* \in X_h^*$, is the set of group optimal strategies for the game $\Gamma_c(1, x_0)$, and (ii) $B_h^i(x_h^*) = B_h^i(x_h^*)$, for $h \in \kappa$ and $i \in N$ and $x_h^* \in X_h^*$, where*

$$
B_h^i\left(x_h^*\right) = (1+r)^{h-1}\left\{\xi^i\left(h, x_h^*\right) - E_{\theta_h}\left(\xi^i\left[h+1, f_h\left(x_h^*, \psi_h\left(x_h^*\right)\right) + \theta_h\right]\right)\right\}, \quad (13.14)
$$

and $[\xi^1(h, x_h^*), \xi^2(h, x_h^*), \ldots, \xi^i(h, x_h^*)] \in P(h, x_h^*)$ is the imputation according to optimality Principle PI, is subgame consistent.

Proof Follow the proof of the continuous-time analog in Theorem 8.2 of Chap. 8. $\square$

When all agents use the cooperative strategies, the payoff that agent i will directly receive at stage k given that $x_k^* \in X_k^*$ is

$$g_k^i[x_k^*, \psi_k^1(x_k^*), \psi_k^2(x_k^*), \ldots, \psi_k^n(x_k^*)].$$

However, according to the agreed-upon imputation, agent i will receive $B_k^i(x_k^*)$ at stage k. Therefore, a side payment

$$\varpi_k^i(x_k^*) = B_k^i(x_k^*) - g_k^i[x_k^*, \psi_k^1(x_k^*), \psi_k^2(x_k^*), \ldots, \psi_k^n(x_k^*), x_k^*], \quad (13.15)$$

for $k \in \kappa$ and $i \in N$, will be given to agent i to yield the cooperative imputation $\xi^i(k, x_k^*)$.

13.4 Cooperative Resource Extraction Under Uncertainty

Consider an economy endowed with a renewable resource and with two resource extractors (firms). The lease for resource extraction begins at stage 1 and ends at stage 3 for these two firms. Let u_k^i denote the rate of resource extraction of firm i at stage k, for $i \in \{1, 2\}$. Let U^i be the set of admissible extraction rates, and $x_k \in X \subset R^+$ the size of the resource stock at stage k. In particular, we have $U^i \in R^+$ and $u_k^1 + u_k^2 \leq x_k$. The extraction cost for firm $i \in \{1, 2\}$ depends on the quantity of resource extracted u_k^i, the resource stock size x_k, and cost parameters c_1 and c_2. In particular, the extraction cost for firm i at stage k is specified as $c_i(u_k^i)^2/x_k$. The price of the resource is P.

The profits that firms 1 and 2 will obtain at stage k are, respectively,

$$\left[P u_k^1 - \frac{c_1}{x_k}(u_k^1)^2 \right] \quad \text{and} \quad \left[P u_k^2 - \frac{c_2}{x_k}(u_k^2)^2 \right]. \quad (13.16)$$

The growth dynamics of the resource is governed by the stochastic difference equation

$$x_{k+1} = x_k + a - b x_k - \sum_{j=1}^{2} u_k^j + \theta_k, \quad (13.17)$$

for $k \in \{1, 2, 3\}$ and $x_1 = x^0$, where θ_k is a random variable with range $\{\theta_k^1, \theta_k^2, \theta_k^3\}$ and corresponding probabilities $\{\lambda_k^1, \lambda_k^2, \lambda_k^3\}$.

With no human harvesting, the natural growth of the resource stock is $x_{k+1} - x_k = a + \theta_k - b x_k$. The natural death rate is b and the growth parameter $a + \theta_k$ exhibits stochasticity.

The objective of extractor $i \in \{1, 2\}$ is to maximize the present value of the expected stream of future profits

$$E_{\theta_1 \theta_2 \theta_3} \left\{ \sum_{k=1}^{3} \left[P u_k^i - \frac{c_i}{x_k} (u_k^i)^2 \right] \left(\frac{1}{1+r} \right)^{k-1} \right\}, \quad \text{for } i \in \{1, 2\}, \quad (13.18)$$

subject to (13.17).

Invoking Theorem 13.2, one can characterize the noncooperative equilibrium strategies in a feedback solution for the game in (13.17) and (13.18). In particular, a set of strategies $\{\phi_k^i(x), \text{ for } k \in \{1, 2, 3\} \text{ and } i \in \{1, 2\}\}$ provides a Nash equilibrium solution to the game in (13.17) and (13.18) if there exist functions $V^i(k, x)$, for $i \in \{1, 2\}$ and $k \in \{1, 2, 3\}$, such that the following recursive relations are satisfied:

$$V^i(k, x) = \max_{u_k^i} E_{\theta_k} \left\{ \left[P u_k^i - \frac{c_i}{x} (u_k^i)^2 \right] \left(\frac{1}{1+r} \right)^{k-1} \right.$$

$$\left. + V^i \left[k+1, x + a - bx - u_k^i - \phi_k^j(x) + \theta_k \right] \right\}$$

$$= \max_{u_k^i} \left\{ \left[P u_k^i - \frac{c_i}{x} (u_k^i)^2 \right] \left(\frac{1}{1+r} \right)^{k-1} \right. \qquad (13.19)$$

$$\left. + \sum_{y=1}^{3} \lambda_k^y V^i \left[k+1, x + a - bx - u_k^i - \phi_k^j(x) + \theta_k^y \right] \right\};$$

$$V^i(4, x) = 0.$$

Performing the indicated maximization in (13.19) yields

$$\left(P - \frac{2c_i u_k^i}{x} \right) \left(\frac{1}{1+r} \right)^{k-1}$$

$$- \sum_{y=1}^{3} \lambda_k^y V_{x_{k+1}}^i \left[k+1, x + a - bx - u_k^i - \phi_k^j(x) + \theta_k^y \right] = 0, \quad (13.20)$$

for $i \in \{1, 2\}$ and $k \in \{1, 2, 3\}$.

From (13.20), the game equilibrium strategies can be expressed as

$$\phi_k^i(x) = \left(P - \sum_{y=1}^{3} \lambda_k^y V_{x_{k+1}}^i \left[k+1, x + a - bx - \sum_{\ell=1}^{2} \phi_k^\ell(x) + \theta_k^y \right] (1+r)^{k-1} \right)$$

$$\times \frac{x}{2c_i}, \qquad (13.21)$$

for $i \in \{1, 2\}$ and $k \in \{1, 2, 3\}$.

Proposition 13.1 *The value function*

$$V^i(k, x) = \left[A_k^i x + C_k^i\right], \quad \text{for } i \in \{1, 2\} \text{ and } k \in \{1, 2, 3\}, \tag{13.22}$$

where A_k^i and C_k^i, for $i \in \{1, 2\}$ and $k \in \{1, 2, 3\}$, are constants in terms of the parameters of the game in (13.17) and (13.18).

Proof See Appendix 2 of this chapter. □

Substituting the relevant derivatives of the value functions in Proposition 13.1 into the game equilibrium strategies in (13.21) yields a noncooperative feedback equilibrium solution of the game in (13.17) and (13.18).

Now consider the case when the extractors agree to maximize their expected joint profit and share the excess of cooperative gains over their expected noncooperative payoffs equally. To maximize their expected joint payoff, they solve the problem of maximizing

$$E_{\theta_1\theta_2\theta_3}\left\{\sum_{j=1}^{2}\sum_{k=1}^{3}\left[Pu_k^j - \frac{c_j}{x_k}(u_k^j)^2\right]\left(\frac{1}{1+r}\right)^{k-1}\right\}, \tag{13.23}$$

subject to (13.17).

Invoking Theorem 13.4, one can characterize the optimal controls in the stochastic dynamic programming problem in (13.17) and (13.23). In particular, a set of control strategies $\{\psi_k^i(x), \text{ for } k \in \{1, 2, 3\} \text{ and } i \in \{1, 2\}\}$ provides an optimal solution to the problem in (13.17) and (13.23) if there exist functions $W(k, x) : R \to R$, for $k \in \{1, 2, 3\}$, such that the following recursive relations are satisfied:

$$W(k, x) = \max_{u_k^1, u_k^2} E_{\theta_{k+1}}\left\{\sum_{j=1}^{2}\left[Pu_k^j - \frac{c_j}{x}(u_k^j)^2\right]\left(\frac{1}{1+r}\right)^{k-1}\right.$$

$$\left. + W\left[k+1, x+a-bx - \sum_{j=1}^{2}u_k^j + \theta_k\right]\right\}$$

$$= \max_{u_k^1, u_k^2}\left\{\sum_{j=1}^{2}\left[Pu_k^j - \frac{c_j}{x}(u_k^j)^2\right]\left(\frac{1}{1+r}\right)^{k-1}\right. \tag{13.24}$$

$$\left. + \sum_{y=1}^{3}\lambda_k^y W\left[k+1, x+a-bx - \sum_{j=1}^{2}u_k^j + \theta_k^y\right]\right\}, \quad \text{for } k \in \{1, 2, 3\}.$$

$$W(T+1, x) = 0.$$

Performing the indicated maximization in (13.24) yields

$$\left(P - \frac{2c_i u_k^i}{x}\right)\left(\frac{1}{1+r}\right)^{k-1} - \sum_{y=1}^{3}\lambda_k^y W_{x_{k+1}}\left[k+1, x+a-bx-\sum_{j=1}^{2}u_k^j+\theta_k^y\right]$$
$$= 0, \tag{13.25}$$

for $i \in \{1, 2\}$ and $k \in \{1, 2, 3\}$.

In particular, the optimal cooperative strategies can be obtained from (13.25) as

$$u_k^i = \left(P - \sum_{y=1}^{3}\lambda_k^y W_{x_{k+1}}\left[k+1, x+a-bx-\sum_{j=1}^{2}u_k^j+\theta_k^y\right](1+r)^{k-1}\right)\frac{x}{2c_i},$$
$$\tag{13.26}$$

for $i \in \{1, 2\}$ and $k \in \{1, 2, 3\}$.

Proposition 13.2 *The value function*

$$W(k, x) = [A_k x + C_k], \quad for\ k \in \{1, 2, 3\}, \tag{13.27}$$

where A_k and C_k, for $k \in \{1, 2, 3\}$, are constants in terms of the parameters of the problem in (13.23) and (13.17).

Proof See Appendix 3 of this chapter.   □

Using (13.26) and Proposition 13.2, the optimal cooperative strategies of the agents can be expressed as

$$\psi_k^i(x) = [P - A_{k+1}(1+r)^{k-1}]\frac{x}{2c_i}, \quad for\ i \in \{1, 2\}\ and\ k \in \{1, 2, 3\}. \tag{13.28}$$

Substituting $\psi_k^i(x)$ from (13.28) into (13.17) yields the optimal cooperative state trajectory

$$x_{k+1} = x_k + a - bx_k - \sum_{j=1}^{2}[P - A_{k+1}(1+r)^{k-1}]\frac{x_k}{2c_j} + \theta_k, \tag{13.29}$$

for $k \in \{1, 2, 3\}$ and $x_1 = x^0$.

The dynamics in (13.29) is a linear stochastic difference equation readily solvable by standard techniques. Let $\{x_k^*,\ for\ k \in \{1, 2, 3\}\}$ denote the solution to (13.29).

Since the extractors agree to share the excess of cooperative gains over their expected noncooperative payoffs equally, an imputation

$$\xi^i(k, x_k^*) = V^i(k, x_k^*) + \frac{1}{2}\left[W(k, x_k^*) - \sum_{j=1}^{2}V^j(k, x_k^*)\right]$$
$$= (A_k^i x_k^* + C_k^i) + \frac{1}{2}\left[(A_k x_k^* + C_k) - \sum_{j=1}^{2}(A_k^j x_k^* + C_k^j)\right], \tag{13.30}$$

for $k \in \{1, 2, 3\}$ and $i \in \{1, 2\}$ has to be maintained.

Invoking Theorem 13.4, if $x_k^* \in X$ is realized at stage k a payment equaling

$$B_k^i(x_k^*) = (1+r)^{k-1}\big[\xi^i(k, x_k^*) - E_{\theta_{k+1}}\big(\xi^i[k+1, f_k(x_k^*, \psi_k(x_k^*)) + \theta_k]\big)\big]$$

$$= (1+r)^{k-1}\Bigg\{(A_k^i x_k^* + C_k^i) + \frac{1}{2}\Bigg((A_k x_k^* + C_k) - \sum_{j=1}^{2}(A_k^j x_k^* + C_k^j)\Bigg)$$

$$- \sum_{y=1}^{3}\lambda_k^y\Bigg[(A_{k+1}^i x_{k+1}^{*(\theta_k^y)} + C_{k+1}^i)$$

$$+ \frac{1}{2}\Bigg((A_{k+1} x_{k+1}^{*(\theta_k^y)} + C_{k+1}) - \sum_{j=1}^{2}(A_{k+1}^j x_{k+1}^{*(\theta_k^y)} + C_{k+1}^j)\Bigg)\Bigg]\Bigg\},$$

for $i \in \{1, 2\}$, $\qquad\qquad\qquad\qquad\qquad\qquad\qquad\qquad\qquad\qquad$ (13.31)

where $x_{k+1}^{*(\theta_k^y)} = x_k^* + a - bx_k^* - \sum_{j=1}^{2}[P - A_{k+1}(1+r)^{k-1}]\frac{x_k}{2c_j} + \theta_k^y$, for $y \in \{1, 2, 3\}$, given to firm i at stage $k \in \kappa$ will lead to the realization of the imputation in (13.30).

A subgame consistent solution can be readily obtained from (13.28), (13.30), and (13.31).

13.5 Exercises

13.1 Consider an economy endowed with a renewable resource and with two resource extractors (firms). The lease for resource extraction begins at stage 1 and ends at stage 3 for these two firms. Let u_k^i denote the rate of resource extraction of firm i at stage k, for $i \in \{1, 2\}$. Let U^i be the set of admissible extraction rates, and $x_k \in X \subset R^+$ the size of the resource stock at stage k. In particular, we have $U^i \in R^+$ and $u_k^1 + u_k^2 \le x_k$. The extraction cost for firms 1 and 2 are, respectively, $2(u_k^1)^2/x_k$ and $1.5(u_k^i)^2/x_k$. The price of the resource is 25.

The profits that firms 1 and 2 will obtain at stage k are, respectively,

$$\left[25u_k^1 - \frac{2}{x_k}(u_k^1)^2\right] \quad \text{and} \quad \left[25u_k^2 - \frac{1.5}{x_k}(u_k^2)^2\right].$$

The growth dynamics of the resource is governed by the stochastic difference equation

$$x_{k+1} = x_k + 12 - 0.3x_k - \sum_{j=1}^{2}u_k^j + \theta_k,$$

for $k \in \{1, 2, 3\}$ and $x_1 = 55$, where θ_k is a random variable with range $\{0, 0.5, 1\}$ and corresponding probabilities $\{0.2, 0.4, 0.4\}$.

The expected payoffs of extractors 1 and 2 are, respectively,

$$E_{\theta_1\theta_2\theta_3}\left\{\sum_{k=1}^{3}\left[25u_k^1 - \frac{2}{x_k}(u_k^1)^2\right]\left(\frac{1}{1+r}\right)^{k-1}\right\}, \quad \text{and}$$

$$E_{\theta_1\theta_2\theta_3}\left\{\sum_{k=1}^{3}\left[25u_k^2 - \frac{1.5}{x_k}(u_k^2)^2\right]\left(\frac{1}{1+r}\right)^{k-1}\right\}.$$

Characterize a Nash equilibrium solution for the above discrete-time stochastic market game.

13.2 If the extractors agree to cooperate and maximize their expected joint payoff, derive the optimal cooperative strategies.

13.3 Consider the case when the extractors agree to share the excess of the expected cooperative gains over their expected noncooperative payoffs equally. Characterize a subgame consistent solution.

Appendix 1: Proof of Theorem 13.1

The first step is to verify that the statement of Theorem 13.1 yields a stagewise equilibrium solution. In a stagewise Nash equilibrium solution the following inequalities are satisfied:

$$J^i\left(u^{1*}, u^{2*}, \ldots, u^{n*}\right)$$
$$\geq J^i\left(u^{1*}, u^{2*}, \ldots, u^{i-1*}; u_1^{i*}, u_2^{i*}, \ldots, u_{k-1}^{i*}, u_k^i, u_{k+1}^{i*}, \ldots, u_T^{i*}; u^{i+1*}, \ldots, u^{n*}\right),$$

$$(13.32)$$

$\forall u_k^i \in U^i$ for $i \in N$ and $k \in \kappa$, where in the case of Theorem 13.1,

$$J^i\left(u^1, u^2, \ldots, u^n\right) = E_{\theta_1,\theta_2,\ldots,\theta_T}\left\{\sum_{k=1}^{T} g_k^i[x_k, u_k^1, u_k^2, \ldots, u_k^n]\left(\frac{1}{1+r}\right)^{k-1}\right\}$$

$$\equiv E_{\theta_1,\theta_2,\ldots,\theta_T}\left\{L^i\left(u^{1*}, u^{2*}, \ldots, u^{n*}\right)\right\}.$$

Because of the stage-additive nature of $L^i(u^{1*}, u^{2*}, \ldots, u^{n*})$, these inequalities in (13.32) equivalently describe an n-person static game defined by

$$\max_{u_T^i \in U^i} E_{\theta_1,\theta_2,\ldots,\theta_T}\left\{g_T^i\left[x_T^*, \phi_T^1(x_T^*), \phi_T^2(x_T^*), \ldots, \phi_T^{i-1}(x_T^*), u_T^i, \phi_T^{i+1}(x_T^*), \ldots,\right.\right.$$

$$\left.\left.\phi_T^n(x_T^*)\right]\left(\frac{1}{1+r}\right)^{T-1}\right\}$$

$$= E_{\theta_1, \theta_2, \dots, \theta_T} \left\{ g_T^i \left[x_T^*, \phi_T^1(x_T^*), \phi_T^2(x_T^*), \dots, \phi_T^n(x_T^*) \right] \right.$$

$$\left. \times \left(\frac{1}{1+r} \right)^{T-1} \right\}, \tag{13.33}$$

where x_k^* is recursively defined by

$$x_{k+1}^* = f_k^i \left(x_k^*, \gamma_k^1(x_k^*), \gamma_k^2(x_k^*), \dots, \gamma_k^n(x_k^*) \right) + \theta_k, \quad x_1^* = x_1. \tag{13.34}$$

Now since θ_T is statistically independent of $\{\theta_1, \theta_2, \dots, \theta_{T-1}\}$ and initial state x_1, (13.33) can equivalently be written as

$$E_{\theta_1, \theta_2, \dots, \theta_{T-1}} \left\{ \max_{u_T^i} E_{\theta_i} \left[g_T^i \left[x_T^*, \phi_T^1(\cdot), \phi_T^2(\cdot), \dots, \phi_T^{i-1}(\cdot), u_T^i, \phi_T^{i+1}(\cdot), \dots, \phi_T^n(\cdot) \right] \right. \right.$$

$$\left. \left. \times \left(\frac{1}{1+r} \right)^{T-1} \right] \right\}$$

$$= E_{\theta_1, \theta_2, \dots, \theta_{T-1}} \left\{ E_{\theta_T} \left[g_T^i \left[x_T^*, \phi_T^1(\cdot), \phi_T^2(\cdot), \dots, \phi_T^n(\cdot) \right] \left(\frac{1}{1+r} \right)^{T-1} \right] \right\}, \tag{13.35}$$

which implies that the minimizing u_T^i will be a function of x_T^*.

Furthermore (in contrast with the deterministic version), x_T^* cannot be expressed in terms of $x_{T-1}^*, \dots, x_1^*$ since there is a noise term in (13.34) that directly contributes to additional errors in the case of such a substitution (mainly because every strategy has a unique representation when the dynamics are given by (13.34)). Therefore, any stagewise Nash equilibrium strategy for agent i at stage T will only be a function of x_T, that is, a feedback strategy $u_T^{i*} = \phi_T^i(x_T)$, which further implies that the choice of such a strategy is independent of all the optimal (or otherwise) strategies employed by the agents at previous stages. Finally, note that, at stage $k = T$, the maximizing solutions of (13.35) coincide with those of (13.3) since the problems are equivalent.

Let us now consider the inequalities in (13.32) at stage $k = T - 1$, that is,

$$J^i \left(u^{1*}, u^{2*}, \dots, u^{n*} \right)$$

$$\geq J^i \left(u^{1*}, u^{2*}, \dots, u^{i-1*}; u_1^{i*}, u_2^{i*}, \dots, u_{T-2}^{i*}, u_{T-1}^i, u_T^{i*}; u^{i+1*}, \dots, u^{n*} \right)$$

$\forall u_{T-1}^i \in U^i$ for $i \in N$, and because of the stage-additive nature of $L^i(u^{1*}, u^{2*}, \dots, u^{n*})$ and since $\phi_T^i (i \in N)$ have already been determined at stage T as feedback strategies (independent of all the previous strategies), these can further be written as follows:

$$
E_{\theta_1,\theta_2,\ldots,\theta_{T-2}} \Bigg\{ \max_{u^i_{T-1}} E_{\theta_{T-1}} \Big[g^i_{T-1}\big[x^*_{T-1}, \phi^1_{T-1}(\cdot), \phi^2_{T-1}(\cdot), \ldots, \phi^{i-1}_{T-1}(\cdot), u^i_{T-1},
$$

$$
\phi^{i+1}_{T-1}(\cdot), \ldots, \phi^n_{T-1}(\cdot)\Big]
$$

$$
\times \left(\frac{1}{1+r}\right)^{T-2} + V^i\Big[T, \tilde{f}^i_{T-1}\big(x^*_{T-1}, u^i_{T-1}\big) + \theta_{T-1}\Big]\Big]\Bigg\}
$$

$$
= E_{\theta_1,\theta_2,\ldots,\theta_{T-2}} \Bigg\{ E_{\theta_{T-1}} \Big[g^i_{T-1}\big[x^*_{T-1}, \phi^1_{T-1}(\cdot), \phi^2_{T-1}(\cdot), \ldots, \phi^n_{T-1}(\cdot)\big]
$$

$$
\times \left(\frac{1}{1+r}\right)^{T-1} + V^i\Big[T, \tilde{f}^i_{T-1}\big(x^*_{T-1}, \phi^i_{T-1}(\cdot)\big) + \theta_{T-1}\Big]\Big]\Bigg\}. \tag{13.36}
$$

In writing down (13.36), we have also made use of the statistical independence property of $\{\theta_1, \theta_2, \ldots, \theta_n\}, \theta_{T-1}, \theta_T$ and the initial state x_1. Through a reasoning similar to the one employed at stage $k = T$, we readily conclude that any $\phi^i_{T-1}(\cdot)$ that satisfies (13.36) will have to be a feedback strategy, independent of all the past strategies employed and the past values of the state; therefore $u^{i*}_{T-1} = \phi^{i*}_{T-1}(x_{T-1})$ ($i \in N$). This is precisely what (13.3) states for $k = T - 1$.

Proceeding in this manner, the theorem can readily be verified for stagewise equilibrium; that is, every set of strategies $\{\phi^i_k; i \in N \text{ and } k \in \kappa\}$ satisfying (13.3) constitutes a stagewise equilibrium solution for the dynamic game with stochastic disturbances under consideration. Conversely, every stagewise equilibrium solution of the stochastic dynamic game is comprised of feedback strategies that satisfy (13.3).

To complete the last phase of the verification of the theorem we first observe that every stagewise equilibrium solution determined in the foregoing derivation is also a feedback Nash equilibrium solution (satisfying (13.3)) since the construction of stagewise equilibrium strategies at stage $k = l$ did not depend on the strategies employed at stages $k < l$. Under a closed-loop perfect information pattern, the information set available to agent i at stage k includes all the states $\{x_1, x_2, \ldots, x_k\}$, every feedback Nash equilibrium is also a Nash equilibrium, and furthermore, since every Nash equilibrium solution is a stagewise equilibrium solution, it readily follows that the statement of Theorem 13.1 is valid also for the Nash equilibrium solution.

Appendix 2: Proof of Proposition 13.1

Consider first the last stage, that is, stage 3. Remembering that $V^i(3, x) = [A_3^i x + C_3^i]$ from Proposition 13.1 and $V^i(4, x) = 0$, the conditions in (13.19) become

$$V^i(3, x) = [A_3^i x + C_3^i] = \max_{u_3^i}\left\{\left[Pu_3^i - \frac{c_i}{x}(u_3^i)^2\right]\left(\frac{1}{1+r}\right)^2\right\},$$

for $i \in \{1, 2\}$.   (13.37)

Performing the indicated maximization in (13.37) yields

$$\left(P - \frac{2c_i u_3^i}{x}\right)\left(\frac{1}{1+r}\right)^2 = 0, \quad \text{for } i \in \{1, 2\}.$$ (13.38)

The game equilibrium strategies in stage 3 can then be expressed as

$$\phi_3^i(x) = \frac{Px}{2c_i}, \quad \text{for } i \in \{1, 2\}.$$ (13.39)

Substituting (13.39) into (13.37) yields

$$V^i(3, x) = [A_3^i x + C_3^i] = \left[\frac{P^2 x}{2c_i} - \frac{P^2 x}{4c_i}\right]\left(\frac{1}{1+r}\right)^2 = \left(\frac{1}{1+r}\right)^2\frac{P^2}{4c_i}x,$$

for $i \in \{1, 2\}$.   (13.40)

Using (13.40), we obtain

$$V^i(3, x) = [A_3^i x + C_3^i],$$

where $A_3^i = \left(\frac{1}{1+r}\right)^2\frac{P^2}{4c_i}$ and $C_3^i = 0$, for $i \in \{1, 2\}$.   (13.41)

Now we proceed to stage 2, the conditions in (13.19) become

$$V^i(2, x) = [A_2^i x + C_2^i]$$

$$= \max_{u_2^i}\left\{\left[Pu_2^i - \frac{c_i}{x}(u_2^i)^2\right]\left(\frac{1}{1+r}\right)\right.$$

$$\left. + \sum_{y=1}^{3}\lambda_2^y V^i[3, x + a - bx - u_2^i - \phi_2^j(x) + \theta_2^y]\right\},$$

$i, j \in \{1, 2\}$ and $i \neq j$.   (13.42)

Invoking (13.40), the condition in (13.42) can be expressed as

$$
\left[A_2^i x + C_2^i\right] = \max_{u_2^i} \left\{ \left[P u_2^i - \frac{c_i}{x}(u_2^i)^2 \right] \left(\frac{1}{1+r}\right) \right.
$$

$$
\left. + \sum_{y=1}^{3} \lambda_2^y A_3^i \left[x + a - bx - u_2^i - \phi_2^j(x) + \theta_2^y \right] \right\},
$$

for $i, j \in \{1, 2\}$ and $i \neq j$.   (13.43)

Performing the indicated maximization in (13.43) yields

$$
\left(P - \frac{2c_i u_2^i}{x} \right) \left(\frac{1}{1+r} \right) - \sum_{y=1}^{3} \lambda_2^y A_3^i = 0, \quad \text{for } i \in \{1, 2\}.   \tag{13.44}
$$

The game equilibrium strategies in stage 2 can then be expressed as

$$
\phi_2^i(x) = \left[P - (1+r)A_3^i \right] \frac{x}{2c_i}, \quad \text{for } i \in \{1, 2\}.   \tag{13.45}
$$

Substituting (13.45) into (13.43) yields

$$
\left[A_2^i x + C_2^i \right] = \left(\frac{1}{1+r} \right) \left(\left[P - (1+r)A_3^i \right] \frac{P}{2c_i} x - \left[P - (1+r)A_3^i \right]^2 \frac{1}{4c_i} x \right)
$$

$$
+ \sum_{y=1}^{3} \lambda_2^y A_3^i \left(x + a - bx - \left[P - (1+r)A_3^i \right] \frac{1}{2c_i} x \right.
$$

$$
\left. - \left[P - (1+r)A_3^j \right] \frac{1}{2c_j} x + \theta_2^y \right), \quad \text{for } i, j \in \{1, 2\} \text{ and } i \neq j.
$$

(13.46)

Collecting the terms together, (13.46) can be expressed as

$$
V^i(2, x) = \left[A_2^i x + C_2^i \right] = \left\{ \left(\frac{1}{1+r} \right) \left[P - (1+r)A_3^i \right] \frac{P + (1+r)A_3^i}{4c_i} \right.
$$

$$
+ A_3^i(1-b) - \left[P - (1+r)A_3^i \right] \frac{A_3^i}{2c_i} - \left[P - (1+r)A_3^j \right] \frac{A_3^i}{2c_j} \right\} x
$$

$$
+ A_3^i \left(a + \sum_{y=1}^{3} \lambda_2^y \theta_2^y \right), \quad \text{for } i, j \in \{1, 2\} \text{ and } i \neq j.   \tag{13.47}
$$

Substituting $A_3^i = (\frac{1}{1+r})^2 \frac{P^2}{4c_i}$ for $i \in \{1, 2\}$ into (13.47), A_2^i and C_2^i for $i \in \{1, 2\}$ are obtained in explicit terms.

Finally, we proceed to the first stage. The conditions in (13.19) become

$$V^i(1, x) = [A_1^i x + C_1^i]$$

$$= \max_{u_1^i} \left\{ \left[Pu_1^i - \frac{c_i}{x}(u_1^i)^2 \right] \right.$$

$$\left. + \sum_{y=1}^{3} \lambda_1^y V^i [2, x + a - bx - u_1^i - \phi_1^j(x) + \theta_1^y] \right\},$$

for $i, j \in \{1, 2\}$, and $i \neq j$. $\qquad (13.48)$

Invoking (13.47), the condition in (13.48) can be expressed as

$$[A_1^i x + C_1^i] = \max_{u_1^i} \left\{ \left[Pu_1^i - \frac{c_i}{x}(u_1^i)^2 \right] \right.$$

$$\left. + \sum_{y=1}^{3} \lambda_1^y \left(A_2^i [x + a - bx - u_1^i - \phi_1^j(x) + \theta_1^y] + C_2^i \right) \right\},$$

for $i, j \in \{1, 2\}$, and $i \neq j$. $\qquad (13.49)$

Performing the indicated maximization in (13.49) yields

$$\left(P - \frac{2c_i u_1^i}{x} \right) - \sum_{y=1}^{3} \lambda_1^y A_2^i = 0, \quad \text{for } i \in \{1, 2\}. \qquad (13.50)$$

The game equilibrium strategies in stage 1 can then be expressed as

$$\phi_1^i(x) = [P - A_2^i] \frac{x}{2c_i}, \quad \text{for } i \in \{1, 2\}. \qquad (13.51)$$

Substituting (13.51) into (13.49) yields

$$[A_1^i x + C_1^i] = \left((P - A_2^i) \frac{P}{2c_i} x - (P - A_2^i)^2 \frac{1}{4c_i} x \right)$$

$$+ \sum_{y=1}^{3} \lambda_1^y \left[A_2^i \left(x + a - bx - (P - A_2^i) \frac{1}{2c_i} x \right. \right.$$

$$\left. \left. - (P - A_2^j) \frac{1}{2c_j} x + \theta_1^y \right) + C_2^i \right], \qquad (13.52)$$

for $i, j \in \{1, 2\}$, and $i \neq j$.

Collecting the terms together, (13.52) can be expressed as

$$V^i(3, x) = [A_1^i x + C_1^i]$$

$$= \left[(P - A_2^i) \frac{P + A_2^i}{4c_i} + A_2^i(1 - b) - (P - A_2^i) \frac{A_2^i}{2c_i} - (P - A_2^j) \frac{A_2^i}{2c_j} \right] x$$

$$+ A_2^i \left(a + \sum_{y=1}^{3} \lambda_1^y \theta_1^y \right) + C_2^i, \quad \text{for } i, j \in \{1, 2\}, \text{ and } i \neq j. \quad (13.53)$$

Substituting the explicit terms for A_2^i, A_2^j, C_2^i, and C_2^j from (13.47) into (13.53), A_1^i and C_1^i for $i \in \{1, 2\}$ are obtained in explicit terms.

Appendix 3: Proof of Proposition 13.2

Consider first the last stage, that is, stage 3. Remembering that $W(3, x) = [A_3 x + C_3]$ from Proposition 13.2 and $W(4, x) = 0$, the conditions in (13.24) become

$$W(3, x) = [A_3 x + C_3] = \max_{u_3^1, u_3^2} \left\{ \sum_{j=1}^{2} \left[P u_3^j - \frac{c_j}{x} (u_3^j)^2 \right] \left(\frac{1}{1+r} \right)^2 \right\}. \quad (13.54)$$

Performing the indicated maximization in (13.37) yields

$$\left(P - \frac{2c_i u_3^i}{x} \right) \left(\frac{1}{1+r} \right)^2 = 0, \quad \text{for } i \in \{1, 2\}. \quad (13.55)$$

The optimal cooperative strategies in stage 3 can then be expressed as

$$\psi_3^i(x) = \frac{Px}{2c_i}, \quad \text{for } i \in \{1, 2\}. \quad (13.56)$$

Substituting (13.56) into (13.54) yields

$$[A_3 x + C_3] = \left(\frac{1}{1+r} \right)^2 \sum_{j=1}^{2} \frac{P^2}{4c_j} x, \quad \text{for } i \in \{1, 2\}. \quad (13.57)$$

Using (13.57), we obtain

$$W(3, x) = [A_3 x + C_3], \quad \text{where } A_3 = \left(\frac{1}{1+r} \right)^2 \sum_{j=1}^{2} \frac{P^2}{4c_j} \text{ and } C_3 = 0. \quad (13.58)$$

Now we proceed to stage 2. The conditions in (13.24) become

$$W(2, x) = [A_2 x + C_2]$$

$$= \max_{u_2^1, u_2^2} \left\{ \sum_{j=1}^{2} \left[P u_2^j - \frac{c_j}{x} (u_2^j)^2 \right] \left(\frac{1}{1+r} \right) \right.$$

$$\left. + \sum_{y=1}^{3} \lambda_2^y W \left[3, x + a - bx - \sum_{j=1}^{2} u_2^j + \theta_2^y \right] \right\}. \tag{13.59}$$

Invoking (13.57), the condition in (13.59) can be expressed as

$$[A_2 x + C_2] = \max_{u_2^1, u_2^2} \left\{ \sum_{j=1}^{2} \left[P u_2^j - \frac{c_j}{x} (u_2^j)^2 \right] \left(\frac{1}{1+r} \right) \right.$$

$$\left. + \sum_{y=1}^{3} \lambda_2^y A_3 \left[x + a - bx - \sum_{j=1}^{2} u_2^j + \theta_2^y \right] \right\}. \tag{13.60}$$

Performing the indicated maximization in (13.60) yields

$$\left(P - \frac{2 c_i u_2^i}{x} \right) \left(\frac{1}{1+r} \right) - \sum_{y=1}^{3} \lambda_2^y A_3 = 0, \quad \text{for } i \in \{1, 2\}. \tag{13.61}$$

The optimal cooperative strategies in stage 2 can then be expressed as

$$\psi_2^i(x) = [P - (1+r)A_3] \frac{x}{2 c_i}, \quad \text{for } i \in \{1, 2\}. \tag{13.62}$$

Substituting (13.62) into (13.60) yields

$$[A_2 x + C_2] = \left(\frac{1}{1+r} \right) \sum_{j=1}^{2} [P - (1+r)A_3] \frac{P + (1+r)A_3}{4 c_j} x$$

$$+ \sum_{y=1}^{3} \lambda_2^y A_3 \left(x + a - bx - [P - (1+r)A_3] \frac{1}{2 c_1} x \right.$$

$$\left. - [P - (1+r)A_3] \frac{1}{2 c_2} x + \theta_2^y \right). \tag{13.63}$$

Collecting the terms together, (13.63) can be expressed as

$$W(2, x) = [A_2 x + C_2]$$

$$= \left[\left(\frac{1}{1+r} \right) \sum_{j=1}^{2} [P - (1+r)A_3] \frac{P + (1+r)A_3}{4 c_j} \right.$$

$$+ A_3(1 - b) - [P - (1 + r)A_3]\frac{A_3}{2c_1} - [P - (1 + r)A_3]\frac{A_3^i}{2c_2}\bigg] x$$

$$+ A_3\bigg(a + \sum_{y=1}^{3} \lambda_2^y \theta_2^y\bigg).$$

(13.64)

Substituting $A_3 = (\frac{1}{1+r})^2 \sum_{j=1}^{2} \frac{P^2}{4c_j}$ into (13.64), A_2 and C_2 are obtained in explicit terms.

Finally, we proceed to the first stage. The conditions in (13.24) become

$$W(1, x) = [A_1 x + C_1]$$

$$= \max_{u_1^1, u_1^2}\bigg\{ \sum_{j=1}^{2} \bigg[Pu_1^j - \frac{c_j}{x}(u_1^j)^2 \bigg]$$

$$+ \sum_{y=1}^{3} \lambda_1^y W\bigg[2, x + a - bx - \sum_{j=1}^{2} u_1^j + \theta_1^y \bigg]\bigg\}.$$

(13.65)

Invoking (13.64), the condition in (13.65) can be expressed as

$$[A_1 x + C_1] = \max_{u_1^1, u_1^2}\bigg\{ \sum_{j=1}^{2} \bigg[Pu_1^j - \frac{c_j}{x}(u_1^j)^2 \bigg]$$

$$+ \sum_{y=1}^{3} \lambda_1^y \bigg(A_2\bigg[x + a - bx - \sum_{j=1}^{2} u_1^j + \theta_1^y \bigg] + C_2 \bigg)\bigg\}.$$

(13.66)

Performing the indicated maximization in (13.66) yields

$$\bigg(P - \frac{2c_i u_1^i}{x} \bigg) - \sum_{y=1}^{3} \lambda_1^y A_2 = 0, \quad \text{for } i \in \{1, 2\}.$$

(13.67)

The optimal cooperative strategies in stage 1 can then be expressed as

$$\psi_1^i(x) = (P - A_2)\frac{x}{2c_i}, \quad \text{for } i \in \{1, 2\}.$$

(13.68)

Substituting (13.68) into (13.66) yields

$$[A_1 x + C_1] = \sum_{j=1}^{2} (P - A_2)\frac{P + A_2}{4c_j} x$$

$$+ \sum_{y=1}^{3} \lambda_1^y \bigg[A_2\bigg(x + a - bx - (P - A_2)\frac{1}{2c_1} x$$

$$- (P - A_2)\frac{1}{2c_2} x + \theta_1^y \bigg) + C_2 \bigg].$$

(13.69)

Collecting the terms together, (13.69) can be expressed as

$$W(2,x) = [A_1 x + C_1]$$

$$= \left[\sum_{j=1}^{2} (P - A_2) \frac{P + A_2}{4c_j} + A_2(1-b) - (P-A_2)\frac{A_2}{2c_1} - (P-A_2)\frac{A_2}{2c_2} \right] x$$

$$+ A_2 \left(a + \sum_{y=1}^{3} \lambda_1^y \theta_1^y \right) + C_2. \qquad (13.70)$$

Substituting the explicit terms for A_2 and C_2 from (13.64) into (13.70), A_1 and C_1 are obtained in explicit terms.

Technical Appendixes: Dynamic Optimization Techniques

Consider the dynamic optimization problem in which the single decision maker

$$\max_u \left\{ \int_{t_0}^{T} g[s, x(s), u(s)] \exp\left[-\int_{t_0}^{s} r(y)\,dy\right] ds \right.$$

$$\left. + \exp\left[-\int_{t_0}^{T} r(y)\,dy\right] q\big(x(T)\big) \right\}, \tag{A.1}$$

subject to the vector-valued differential equation

$$\dot{x}(s) = f[s, x(s), u(s)]\,ds, \quad x(t_0) = x_0, \tag{A.2}$$

where $x(s) \in X \subset R^m$ denotes the state variables of game and $u \in U$ is the control.

The functions $f[s, x, u]$ and $g[s, x, u]$ are continuously differentiable in x and continuous in u and s. The function $q(x)$ is continuously differentiable in x. Dynamic programming and optimal control are used to identify optimal solutions for the problems in (A.1) and (A.2).

A.1 Dynamic Programming

A frequently adopted approach to dynamic optimization problems is the technique of dynamic programming. The technique was developed by Bellman (1957). The technique is given in Theorem A.1 below.

Theorem A.1 (Bellman's Dynamic Programming) *A set of controls* $u^*(t) = \phi^*(t, x)$ *constitutes an optimal solution to the control problem in* (A.1) *and* (A.2) *if there exists a continuously differentiable function* $V(t, x)$ *defined by* $[t_0, T] \times R^m \to R$ *and satisfying the following Bellman equation*:

$$-V_t(t, x) = \max_u \left\{ g[t, x, u] \exp\left[-\int_{t_0}^{t} r(y)\,dy\right] + V_x(t, x) f[t, x, u] \right\}$$

D.W.K. Yeung, L.A. Petrosyan, *Subgame Consistent Economic Optimization*, 367
Static & Dynamic Game Theory: Foundations & Applications,
DOI 10.1007/978-0-8176-8262-0, © Springer Science+Business Media, LLC 2012

$$= \left\{ g\left[t, x, \phi^*(t, x)\right] \exp\left[-\int_{t_0}^{t} r(y)\,dy\right] + V_x(t, x) f\left[t, x, \phi^*(t, x)\right] \right\},$$

$$V(T, x) = q(x) \exp\left[-\int_{t_0}^{T} r(y)\,dy\right].$$

Proof Define the maximized payoff at time t with current state x as a *value function* in the form

$$V(t, x) = \max_{u} \left\{ \int_{t}^{T} g\left(s, x(s), u(s)\right) \exp\left[-\int_{t_0}^{s} r(y)\,dy\right] ds \right.$$

$$\left. + \exp\left[-\int_{t_0}^{T} r(y)\,dy\right] q\left(x(T)\right) \right\}$$

$$= \int_{t}^{T} g\left[s, x^*(s), \phi^*\left(s, x^*(s)\right)\right] \exp\left[-\int_{t_0}^{s} r(y)\,dy\right] ds$$

$$+ \exp\left[-\int_{t_0}^{T} r(y)\,dy\right] q\left(x^*(T)\right),$$

satisfying the boundary condition

$$V\left(T, x^*(T)\right) = q\left(x^*(T)\right) \exp\left[-\int_{t_0}^{T} r(y)\,dy\right], \quad \text{and}$$

$$\dot{x}^*(s) = f\left[s, x^*(s), \phi^*\left(s, x^*(s)\right)\right], \quad x^*(t_0) = x_0.$$

If, in addition to $u^*(s) \equiv \phi^*(s, x)$, we are given another set of strategies $u(s) \in U$, with the corresponding trajectory $x(s)$, then Theorem A.1 implies

$$g(t, x, u) \exp\left[-\int_{t_0}^{t} r(y)\,dy\right] + V_x(t, x) f(t, x, u) + V_t(t, x) \leq 0, \quad \text{and}$$

$$g\left(t, x^*, u^*\right) \exp\left[-\int_{t_0}^{t} r(y)\,dy\right] + V_{x^*}\left(t, x^*\right) f\left(t, x^*, u^*\right) + V_t\left(t, x^*\right) = 0.$$

Integrating the above expressions from t_0 to T, we obtain

$$\int_{t_0}^{T} g\left(s, x(s), u(s)\right) \exp\left[-\int_{t_0}^{s} r(y)\,dy\right] ds + V\left(T, x(T)\right) - V(t_0, x_0) \leq 0, \quad \text{and}$$

$$\int_{t_0}^{T} g\left(s, x^*(s), u^*(s)\right) \exp\left[-\int_{t_0}^{s} r(y)\,dy\right] ds + V\left(T, x^*(T)\right) - V(t_0, x_0) = 0.$$

The elimination of $V(t_0, x_0)$ yields

$$\int_{t_0}^{T} g\big(s, x(s), u(s)\big) \exp\left[-\int_{t_0}^{s} r(y)\,dy\right] ds + q\big(x(T)\big) \exp\left[-\int_{t_0}^{T} r(y)\,dy\right]$$

$$\leq \int_{t_0}^{T} g\big(s, x^*(s), u^*(s)\big) \exp\left[-\int_{t_0}^{s} r(y)\,dy\right] ds + q\big(x^*(T)\big)$$

$$\times \exp\left[-\int_{t_0}^{T} r(y)\,dy\right],$$

from which it readily follows that u^* is the optimal strategy.   □

Substituting the optimal strategy $\phi^*(t, x)$ into (A.2) yields the dynamics of the optimal state trajectory as

$$\dot{x}(s) = f\big[s, x(s), \phi^*\big(s, x(s)\big)\big]\,ds, \quad x(t_0) = x_0. \tag{A.3}$$

Let $x^*(t)$ denote the solution to (A.3). The optimal trajectory $\{x^*(t)\}_{t=t_0}^{T}$ can be expressed as

$$x^*(t) = x_0 + \int_{t_0}^{t} f\big[s, x^*(s), \phi^*\big(s, x^*(s)\big)\big]\,ds. \tag{A.4}$$

For notational convenience, we use the terms $x^*(t)$ and x_t^* interchangeably.

The value function $V(t, x)$, where $x = x_t^*$, can be expressed as

$$V\big(t, x_t^*\big) = \int_{t}^{T} g\big[s, x^*(s), \phi^*\big(s, x^*(s)\big)\big] \exp\left[-\int_{t_0}^{s} r(y)\,dy\right] ds + q\big(x^*(T)\big).$$

Example A.1 Consider the following dynamic economic optimization problem involving cost management. Let $x(s)$ denote the level of pollution and $u(s)$ the amount of pollution abatement effort at time s. The cost of the pollution abatement effort is $cu(s)^2$ and the damage of pollution is $x(s)$. The planning horizon is $[t_0, T]$, and at time T the terminal damage cost of the pollution stock is $qx(T)$. There is a constant discount rate r. The economic agent involved seeks to minimize the integral of discounted costs by choosing an optimal abatement effort path, that is,

$$\min_{u}\left\{\int_{0}^{T} \exp[-rs]\big[x(s) + cu(s)^2\big]\,ds + \exp[-rT]qx(T)\right\}.$$

The problem can alternatively be expressed as

$$\max_{u}\left\{\int_{0}^{T} \exp[-rs]\big[-x(s) - cu(s)^2\big]\,ds - \exp[-rT]qx(T)\right\}, \tag{A.5}$$

subject to the pollution accumulation dynamics

$$\dot{x}(s) = a - u(s)\big(x(s)\big)^{1/2}, \quad x(0) = x_0, u(s) \geq 0, \tag{A.6}$$

where a, c, x_0, and q are positive parameters.

Invoking Theorem A.1 we have

$$-V_t(t, x) = \max_u \left\{ \left[-x - cu^2 \right] \exp[-rt] + V_x(t, x) \left[a - ux^{1/2} \right] \right\},$$

and

$$V(T, x) = -\exp[-rT]qx. \tag{A.7}$$

Performing the indicated maximization in (A.7) yields

$$\phi(t, x) = \frac{-V_x(t, x)x^{1/2}}{2c} \exp[rt].$$

Substituting $\phi(t, x)$ into (A.7) and solving (A.7), one obtains

$$V(t, x) = \exp[-rt] \left[A(t)x + B(t) \right],$$

where $A(t)$ and $B(t)$ satisfy

$$\dot{A}(t) = rA(t) - \frac{A(t)^2}{4c} + 1,$$

$$\dot{B}(t) = rB(t) - aA(t),$$

$$A(T) = -q, \quad \text{and} \quad B(T) = 0.$$

The optimal control can be solved explicitly as $\phi(t, x) = \frac{-A(t)x^{1/2}}{2c} \exp[rt]$.

Now consider the infinite-horizon dynamic optimization problem with a constant discount rate

$$\max_u \left\{ \int_{t_0}^{\infty} g\left[x(s), u(s)\right] \exp\left[-r(s - t_0)\right] ds \right\}, \tag{A.8}$$

subject to the vector-valued differential equation

$$\dot{x}(s) = f\left[x(s), u(s)\right] ds, \quad x(t_0) = x_0. \tag{A.9}$$

Since s does not appear in $g[x(s), u(s)]$ and the state dynamics explicitly, the problem in (A.8) and (A.9) is an autonomous problem.

Consider the alternative problem that starts at time $t \in [t_0, \infty)$ with initial state $x(t) = x$:

$$\max_u \int_{t}^{\infty} g\left[x(s), u(s)\right] \exp\left[-r(s - t)\right] ds, \tag{A.10}$$

subject to

$$\dot{x}(s) = f\left[x(s), u(s)\right], \quad x(t) = x. \tag{A.11}$$

The infinite-horizon autonomous problem in (A.10) and (A.11) is independent of the choice of t and dependent only upon the state at the starting time, that is, x.

The elimination of $V(t_0, x_0)$ yields

$$\int_{t_0}^{T} g(s, x(s), u(s)) \exp\left[-\int_{t_0}^{s} r(y)\,dy\right]ds + q(x(T))\exp\left[-\int_{t_0}^{T} r(y)\,dy\right]$$

$$\leq \int_{t_0}^{T} g(s, x^*(s), u^*(s)) \exp\left[-\int_{t_0}^{s} r(y)\,dy\right]ds + q(x^*(T))$$

$$\times \exp\left[-\int_{t_0}^{T} r(y)\,dy\right],$$

from which it readily follows that u^* is the optimal strategy.   □

Substituting the optimal strategy $\phi^*(t, x)$ into (A.2) yields the dynamics of the optimal state trajectory as

$$\dot{x}(s) = f[s, x(s), \phi^*(s, x(s))]\,ds, \quad x(t_0) = x_0. \tag{A.3}$$

Let $x^*(t)$ denote the solution to (A.3). The optimal trajectory $\{x^*(t)\}_{t=t_0}^{T}$ can be expressed as

$$x^*(t) = x_0 + \int_{t_0}^{t} f[s, x^*(s), \phi^*(s, x^*(s))]\,ds. \tag{A.4}$$

For notational convenience, we use the terms $x^*(t)$ and x_t^* interchangeably.

The value function $V(t, x)$, where $x = x_t^*$, can be expressed as

$$V(t, x_t^*) = \int_{t}^{T} g[s, x^*(s), \phi^*(s, x^*(s))]\exp\left[-\int_{t_0}^{s} r(y)\,dy\right]ds + q(x^*(T)).$$

Example A.1 Consider the following dynamic economic optimization problem involving cost management. Let $x(s)$ denote the level of pollution and $u(s)$ the amount of pollution abatement effort at time s. The cost of the pollution abatement effort is $cu(s)^2$ and the damage of pollution is $x(s)$. The planning horizon is $[t_0, T]$, and at time T the terminal damage cost of the pollution stock is $qx(T)$. There is a constant discount rate r. The economic agent involved seeks to minimize the integral of discounted costs by choosing an optimal abatement effort path, that is,

$$\min_{u}\left\{\int_{0}^{T} \exp[-rs][x(s) + cu(s)^2]\,ds + \exp[-rT]qx(T)\right\}.$$

The problem can alternatively be expressed as

$$\max_{u}\left\{\int_{0}^{T} \exp[-rs][-x(s) - cu(s)^2]\,ds - \exp[-rT]qx(T)\right\}, \tag{A.5}$$

subject to the pollution accumulation dynamics

$$\dot{x}(s) = a - u(s)(x(s))^{1/2}, \quad x(0) = x_0, u(s) \geq 0, \tag{A.6}$$

where a, c, x_0, and q are positive parameters.

Invoking Theorem A.1 we have

$$-V_t(t, x) = \max_u \{[-x - cu^2] \exp[-rt] + V_x(t, x)[a - ux^{1/2}]\},$$

and

$$V(T, x) = -\exp[-rT]qx. \tag{A.7}$$

Performing the indicated maximization in (A.7) yields

$$\phi(t, x) = \frac{-V_x(t, x)x^{1/2}}{2c} \exp[rt].$$

Substituting $\phi(t, x)$ into (A.7) and solving (A.7), one obtains

$$V(t, x) = \exp[-rt][A(t)x + B(t)],$$

where $A(t)$ and $B(t)$ satisfy

$$\dot{A}(t) = rA(t) - \frac{A(t)^2}{4c} + 1,$$

$$\dot{B}(t) = rB(t) - aA(t),$$

$$A(T) = -q, \quad \text{and} \quad B(T) = 0.$$

The optimal control can be solved explicitly as $\phi(t, x) = \frac{-A(t)x^{1/2}}{2c} \exp[rt]$.

Now consider the infinite-horizon dynamic optimization problem with a constant discount rate

$$\max_u \left\{ \int_{t_0}^{\infty} g[x(s), u(s)] \exp[-r(s - t_0)] \, ds \right\}, \tag{A.8}$$

subject to the vector-valued differential equation

$$\dot{x}(s) = f[x(s), u(s)] \, ds, \quad x(t_0) = x_0. \tag{A.9}$$

Since s does not appear in $g[x(s), u(s)]$ and the state dynamics explicitly, the problem in (A.8) and (A.9) is an autonomous problem.

Consider the alternative problem that starts at time $t \in [t_0, \infty)$ with initial state $x(t) = x$:

$$\max_u \int_t^{\infty} g[x(s), u(s)] \exp[-r(s - t)] \, ds, \tag{A.10}$$

subject to

$$\dot{x}(s) = f[x(s), u(s)], \quad x(t) = x. \tag{A.11}$$

The infinite-horizon autonomous problem in (A.10) and (A.11) is independent of the choice of t and dependent only upon the state at the starting time, that is, x.

Define the value function to the problem in (A.8) and (A.9) by

$$V(t, x) = \max_u \left\{ \int_t^\infty g[x(s), u(s)] \exp[-r(s - t_0)] \, ds \, \middle| \, x(t) = x = x_t^* \right\},$$

where x_t^* is the state at time t along the optimal trajectory. Moreover, we can write

$$V(t, x) = \exp[-r(t - t_0)]$$
$$\times \max_u \left\{ \int_t^\infty g[x(s), u(s)] \exp[-r(s - t)] \, ds \, \middle| \, x(t) = x = x_t^* \right\}.$$

Since the problem

$$\max_u \left\{ \int_t^\infty g[x(s), u(s)] \exp[-r(s - t)] \, ds \, \middle| \, x(t) = x = x_t^* \right\}$$

depends on the current state x only, we can write

$$W(x) = \max_u \left\{ \int_t^\infty g[x(s), u(s)] \exp[-r(s - t)] \, ds \, \middle| \, x(t) = x = x_t^* \right\}.$$

It follows that

$$V(t, x) = \exp[-r(t - t_0)] W(x),$$
$$V_t(t, x) = -r \exp[-r(t - t_0)] W(x), \quad \text{and} \tag{A.12}$$
$$V_x(t, x) = -r \exp[-r(t - t_0)] W_x(x).$$

Substituting the results from (A.12) into Theorem A.1 yields

$$r W(x) = \max_u \{g[x, u] + W_x(x) f[x, u]\}. \tag{A.13}$$

Since time is not explicitly involved in (A.13), the derived control u will be a function of x only. Hence one can obtain the following.

Theorem A.2 *A set of controls $u = \phi^*(x)$ constitutes an optimal solution to the infinite-horizon control problem in (A.10) and (A.11) if there exists a continuously differentiable function $W(x)$ defined by $R^m \to R$ that satisfies the following equation:*

$$r W(x) = \max_u \{g[x, u] + W_x(x) f[x, u]\} = \{g[x, \phi^*(x)] + W_x(x) f[x, \phi^*(x)]\}.$$

Substituting the optimal control in Theorem A.2 into (A.9) yields the dynamics of the optimal state path as

$$\dot{x}(s) = f[x(s), \phi^*(x(s))] \, ds, \quad x(t_0) = x_0.$$

Solving the above dynamics yields the optimal state trajectory $\{x^*(t)\}_{t \geq t_0}$ as

$$x^*(t) = x_0 + \int_{t_0}^{t} f\left[x^*(s), \psi^*\left(x^*(s)\right)\right] ds, \quad \text{for } t \geq t_0.$$

We denote the term $x^*(t)$ by x_t^*. The optimal control to the infinite-horizon problem in (A.8) and (A.9) can be expressed as $\psi^*(x_t^*)$ in the time interval $[t_0, \infty)$.

Example A.2 Consider the infinite-horizon dynamic optimization problem

$$\max_{u} \int_0^{\infty} \exp[-rs]\left[-x(s) - cu(s)^2\right] ds, \tag{A.14}$$

subject to the dynamics in (A.6).

Invoking Theorem A.2 we have

$$rW(x) = \max_{u}\{\left[-x - cu^2\right] + W_x(x)\left[a - ux^{1/2}\right]\}. \tag{A.15}$$

Performing the indicated maximization in (A.15) yields

$$\phi^*(x) = \frac{-V_x(x)x^{1/2}}{2c}.$$

Substituting $\phi(x)$ into (A.15) and solving (A.15), one obtains

$$V(t, x) = \exp[-rt][Ax + B],$$

where A and B satisfy

$$0 = rA - \frac{A^2}{4c} + 1 \quad \text{and} \quad B = \frac{-a}{r}A.$$

Solve A to be $2c[r \pm (r^2 + \frac{1}{c})^{1/2}]$. For a maximum, the negative root of A holds. The optimal control can be obtained as

$$\phi^*(x) = \frac{-Ax^{1/2}}{2c}.$$

Substituting $\phi^*(x) = -Ax^{1/2}/(2c)$ into (A.6) yields the dynamics of the optimal trajectory as

$$\dot{x}(s) = a + \frac{A}{2c}(x(s)), \quad x(0) = x_0.$$

The above dynamical equation yields the optimal trajectory $\{x^*(t)\}_{t \geq t_0}$ as

$$x^*(t) = \left[x_0 + \frac{2ac}{A}\right]\exp\left(\frac{A}{2c}t\right) - \frac{2ac}{A} = x_t^*, \quad \text{for } t \geq t_0.$$

The optimal control of the problem in (A.14) and (A.15) is then

$$\phi^*\left(x_t^*\right) = \frac{-A(x_t^*)^{1/2}}{2c}.$$

A.2 Optimal Control

The maximum principle of optimal control was developed by Pontryagin (details in Pontryagin et al. (1962) and Pontryagin (1966)). Consider again the dynamic optimization problem in (A.1) and (A.2).

Theorem A.3 (Pontryagin's Maximum Principle) *If a set of controls* $u^*(s) = \zeta^*(s, x_0)$ *provides an optimal solution to the control problem in* (A.1) *and* (A.2), *and* $\{x^*(s), t_0 \leq s \leq T\}$ *is the corresponding optimal state trajectory, then there exist costate functions* $\Lambda(s) : [t_0, T] \to R^m$ *such that the following relations are satisfied:*

$$\zeta^*(s, x_0) \equiv u^*(s) = \arg\max_u \left\{ g\left[s, x^*(s), u(s)\right] \exp\left[-\int_{t_0}^s r(y)\,dy\right] \right.$$

$$\left. + \Lambda(s) f\left[s, x^*(s), u(s)\right] \right\},$$

$$\dot{x}^*(s) = f\left[s, x^*(s), u^*(s)\right], \quad x^*(t_0) = x_0,$$

$$\dot{\Lambda}(s) = -\frac{\partial}{\partial x}\left\{g\left[s, x^*(s), u^*(s)\right] + \Lambda(s) f\left[s, x^*(s), u^*(s)\right]\right\},$$

$$\Lambda(T) = \frac{\partial}{\partial x^*} q\left(x^*(T)\right) \exp\left[-\int_{t_0}^T r(y)\,dy\right].$$

Proof First define the function (Hamiltonian)

$$H(t, x, u) = g(t, x, u) \exp\left[-\int_{t_0}^t r(y)\,dy\right] + V_x(t, x) f(t, x, u).$$

From Theorem A.1, we obtain

$$-V_t(t, x) = \max_u H(t, x, u).$$

This yields the first condition of Theorem A.1. Using u^* to denote the payoff maximizing control, we obtain

$$H\left(t, x, u^*\right) + V_t(t, x) \equiv 0,$$

which is an identity in x. If $V(t, x)$ is twice continuously differentiable, differentiating this identity partially with respect to x yields

$$V_{tx}(t, x) + g_x(t, x, u^*) \exp\left[-\int_{t_0}^{t} r(y) \, dy\right] + V_x(t, x) f_x(t, x, u^*)$$

$$+ V_{xx}(t, x) f(t, x, u^*) + \left[g_u(t, x, u^*) \exp\left[-\int_{t_0}^{t} r(y) \, dy\right]\right.$$

$$\left. + V_x(t, x) f_u(t, x, u^*)\right] \frac{\partial u^*}{\partial x} = 0.$$

If u^* is an interior point, then

$$\left[g_u(t, x, u^*) \exp\left[-\int_{t_0}^{t} r(y) \, dy\right] + V_x(t, x) f_u(t, x, u^*)\right] = 0$$

according to the condition $-V_t(t, x) = \max_u H(t, x, u)$. If u^* is not an interior point, then it can be shown that

$$\left[g_u(t, x, u^*) \exp\left[-\int_{t_0}^{t} r(y) \, dy\right] + V_x(t, x) f_u(t, x, u^*)\right] \frac{\partial u^*}{\partial x} = 0$$

(because of optimality, $[g_u(t, x, u^*) \exp[-\int_{t_0}^{t} r(y) \, dy] + V_x(t, x) f_u(t, x, u^*)]$ and $\partial u^*/\partial x$ are orthogonal and for specific problems we may have $\partial u^*/\partial x = 0$). Moreover, the expression $V_{tx}(t, x) + V_{xx}(t, x) f(t, x, u^*) \equiv V_{tx}(t, x) + V_{xx}(t, x) \dot{x}$ can be written as $\frac{dV_x(t,x)}{dt}$. Hence we obtain

$$\frac{dV_x(t, x)}{dt} + g_x(t, x, u^*) \exp\left[-\int_{t_0}^{t} r(y) \, dy\right] + V_x(t, x) f_x(t, x, u^*) = 0.$$

By introducing the *costate vector* $\Lambda(t) = V_{x^*}(t, x^*)$, where x^* denotes the state trajectory corresponding to u^*, we arrive at

$$\frac{dV_x(t, x^*)}{dt} = \dot{\Lambda}(s) = -\frac{\partial}{\partial x} \left\{ g[s, x^*(s), u^*(s)] \exp\left[-\int_{t_0}^{t} r(y) \, dy\right] \right.$$

$$\left. + \Lambda(s) f[s, x^*(s), u^*(s)] \right\}.$$

Finally, the boundary condition for $\Lambda(t)$ is determined from the terminal condition of optimal control in Theorem A.1 as

$$\Lambda(T) = \frac{\partial V(T, x^*)}{\partial x} = \frac{\partial q(x^*)}{\partial x} \exp\left[-\int_{t_0}^{T} r(y) \, dy\right].$$

Hence Theorem A.3 follows.   □

Note that in the proof of Theorem A.3, the assumption of the existence of a twice differentiable $V(t, x)$ is used to expedite the proof, but the Pontryagin Maximum Principle does not require this assumption to hold.

Example A.3 Consider the problem in Example A.1. Invoking Theorem A.3, we first solve the control $u(s)$ that satisfies

$$\arg \max_{u} \{ [-x^*(s) - cu(s)^2] \exp[-rs] + \Lambda(s)[a - u(s)x^*(s)^{1/2}] \}.$$

Performing the indicated maximization

$$u^*(s) = \frac{-\Lambda(s)x^*(s)^{1/2}}{2c} \exp[rs]. \tag{A.16}$$

We also obtain

$$\dot{\Lambda}(s) = \exp[-rs] + \frac{1}{2} \Lambda(s) u^*(s) x^*(s)^{-1/2}. \tag{A.17}$$

Substituting $u^*(s)$ from (A.16) into (A.6) and (A.17) yields a pair of differential equations

$$\dot{x}^*(s) = a + \frac{1}{2c} \Lambda(s) (x^*(s)) \exp[rs],$$

$$\dot{\Lambda}(s) = \exp[-rs] - \frac{1}{4c} \Lambda(s)^2 \exp[rs], \tag{A.18}$$

with boundary conditions

$$x^*(0) = x_0 \quad \text{and} \quad \Lambda(T) = -\exp[-rT]q.$$

Solving (A.18) yields

$$\Lambda(s) = 2c \left(\theta_1 - \theta_2 \frac{q - 2c\theta_1}{q - 2c\theta_2} \exp\left[\frac{\theta_1 - \theta_2}{2} (T - s) \right] \right) \exp(-rs)$$

$$\div \left(1 - \frac{q - 2c\theta_1}{q - 2c\theta_2} \exp\left[\frac{\theta_1 - \theta_2}{2} (T - s) \right] \right), \quad \text{and}$$

$$x^*(s) = \varpi(0, s) \left[x_0 + \int_0^s \varpi^{-1}(0, t) a \, dt \right], \quad \text{for } s \in [0, T],$$

where

$$\theta_1 = r - \sqrt{r^2 + \frac{1}{c}} \quad \text{and} \quad \theta_2 = r + \sqrt{r^2 + \frac{1}{c}};$$

$$\varpi(0, s) = \exp\left[\int_0^s H(\tau) \, d\tau \right], \quad \text{and}$$

$$H(\tau) = \left(\theta_1 - \theta_2 \frac{q - 2c\theta_1}{q - 2c\theta_2} \exp\left[\frac{\theta_1 - \theta_2}{2}(T - \tau)\right]\right)$$
$$\div \left(1 - \frac{q - 2c\theta_1}{q - 2c\theta_2} \exp\left[\frac{\theta_1 - \theta_2}{2}(T - \tau)\right]\right).$$

The substitution of $\Lambda(s)$ and $x^*(s)$ into (A.16) yields $u^*(s) = \zeta^*(s, x_0)$, which is a function of s and x_0.

Now consider the infinite-horizon dynamic optimization problem in (A.10) and (A.11).
The Hamiltonian function can be expressed as

$$H(t, x, u) = g(x, u) \exp\left[-r(t - t_0)\right] + \Lambda(t) f(x, u).$$

Define $\lambda(t) = \Lambda(t) \exp[r(t - t_0)]$ and the current value Hamiltonian

$$\hat{H}(t, x, u) = H(t, x, u) \exp\left[r(t - t_0)\right] = g(x, u) + \lambda(t) f(x, u). \qquad \text{(A.19)}$$

Substituting (A.19) into Theorem A.3 yield the maximum principle for the game equations (A.10) and (A.11).

Theorem A.4 *If a set of controls $u^*(s) = \zeta^*(s, x_t)$ provides an optimal solution to the infinite-horizon control problem in (A.10) and (A.11), and $\{x^*(s), s \geq t\}$ is the corresponding optimal state trajectory, then there exist costate functions $\lambda(s) : [t, \infty) \to R^m$ such that the following relations are satisfied:*

$$\zeta^*(s, x_t) \equiv u^*(s) = \arg\max_u \{g[x^*(s), u(s)] + \lambda(s) f[x^*(s), u(s)]\},$$

$$\dot{x}^*(s) = f[x^*(s), u^*(s)], \quad x^*(t) = x_t = x,$$

$$\dot{\lambda}(s) = r\lambda(s) - \frac{\partial}{\partial x}\{g[x^*(s), u^*(s)] + \lambda(s) f[x^*(s), u^*(s)]\}.$$

Example A.4 Consider the infinite-horizon problem in Example A.2.
Invoking Theorem A.4 we have

$$\zeta^*(s, x_t) \equiv u^*(s) = \arg\max_u \{[-x^*(s) - cu(s)^2] + \lambda(s)[a - u(s)x^*(s)^{1/2}]\},$$

$$\dot{x}^*(s) = a - u^*(s)(x^*(s))^{1/2}, \quad x^*(t) = x_t, \qquad \text{(A.20)}$$

$$\dot{\lambda}(s) = r\lambda(s) + \left[1 + \frac{1}{2}\lambda(s)u^*(s)x^*(s)^{-1/2}\right].$$

Performing the indicated maximization yields

$$u^*(s) = \frac{-\lambda(s)x^*(s)^{1/2}}{2c}.$$

Substituting $u^*(s)$ into (A.20) , one obtains

$$
\dot{x}^*(s) = a + \frac{\lambda(s)}{2c} u^*(s) x^*(s), \quad x^*(t) = x_t,
$$

$$
\dot{\lambda}(s) = r\lambda(s) + \left[1 - \frac{1}{4c} \lambda(s)^2 \right].
$$

(A.21)

Solving (A.21) in a manner similar to that in Example A.3 yields the solutions of $x^*(s)$ and $\lambda(s)$. The substitution of them into $u^*(s)$ gives the optimal control of the problem.

A.3 Stochastic Control

Consider the dynamic optimization problem in which the single decision maker

$$
\max_u E_{t_0} \left\{ \int_{t_0}^T g^i \left[s, x(s), u(s) \right] \exp\left[- \int_{t_0}^s r(y)\, dy \right] ds \right.
$$

$$
\left. + q\big(x(T)\big) \exp\left[- \int_{t_0}^T r(y)\, dy \right] \right\},
$$

(A.22)

subject to the vector-valued stochastic differential equation

$$
dx(s) = f\left[s, x(s), u(s) \right] ds + \sigma\left[s, x(s) \right] dz(s), \quad x(t_0) = x_0,
$$

(A.23)

where E_{t_0} denotes the expectation operator performed at time t_0, $\sigma[s, x(s)]$ is a $m \times \Theta$ matrix, $z(s)$ is a Θ-dimensional Wiener process, and the initial state x_0 is given. Let $\boldsymbol{\Omega}[s, x(s)] = \sigma[s, x(s)]\sigma[s, x(s)]^\mathsf{T}$ denote the covariance matrix with its element in row h and column ζ denoted by $\Omega^{h\zeta}[s, x(s)]$.

The technique of stochastic control developed by Fleming (1969) can be applied to solve the problem.

Theorem A.5 *A set of controls $u^*(t) = \phi^*(t, x)$ constitutes an optimal solution to the problem in (A.22) and (A.23) if there exist continuously twice differentiable functions $V(t, x) : [t_0, T] \times R^m \to R$, satisfying the following partial differential equation*:

$$
-V_t(t, x) - \frac{1}{2} \sum_{h,\zeta=1}^m \Omega^{h\zeta}(t, x) V_{x^h x^\zeta}(t, x)
$$

$$
= \max_u \left\{ g^i[t, x, u] \exp\left[- \int_{t_0}^t r(y)\, dy \right] + V_x(t, x) f[t, x, u] \right\}, \quad \text{and}
$$

$$
V(T, x) = q(x) \exp\left[- \int_{t_0}^T r(y)\, dy \right].
$$

Proof Substitute the optimal control $\phi^*(t, x)$ into (A.23) to obtain the optimal state dynamics as

$$dx(s) = f\big[s, x(s), \phi^*(s, x(s))\big] ds + \sigma\big[s, x(s)\big] dz(s), \quad x(t_0) = x_0. \qquad \text{(A.24)}$$

The solution to (A.24), denoted by $x^*(t)$, can be expressed as

$$x^*(t) = x_0 + \int_{t_0}^t f\big[s, x^*(s), \psi_1^{(t_0)*}(s, x^*(s)), \psi_2^{(t_0)*}(s, x^*(s))\big] ds$$

$$+ \int_{t_0}^t \sigma\big[s, x^*(s)\big] dz(s). \qquad \text{(A.25)}$$

We use X_t^* to denote the set of realizable values of $x^*(t)$ at time t generated by (A.25). The term x_t^* is used to denote an element in the set X_t^*.

Define the maximized payoff at time t with current state x_t^* as a *value function* in the form

$$V(t, x_t^*) = \max_u E_{t_0} \bigg\{ \int_t^T g^i\big[s, x(s), u(s)\big] \exp\bigg[-\int_{t_0}^s r(y)\, dy \bigg] ds$$

$$+ q(x(T)) \exp\bigg[-\int_{t_0}^T r(y)\, dy \bigg] \bigg| x(t) = x_t^* \bigg\}$$

$$= E_{t_0} \bigg\{ \int_t^T g\big[s, x^*(s), \phi^*(s, x^*(s))\big] \exp\bigg[-\int_{t_0}^s r(y)\, dy \bigg] ds$$

$$+ q(x^*(T)) \exp\bigg[-\int_{t_0}^T r(y)\, dy \bigg] \bigg\},$$

satisfying the boundary condition

$$V(T, x^*(T)) = q(x^*(T)) \exp\bigg[-\int_{t_0}^T r(y)\, dy \bigg].$$

One can express $V(t, x_t^*)$ as

$$V(t, x_t^*) = \max_u E_{t_0} \bigg\{ \int_t^T g^i\big[s, x(s), u(s)\big] \exp\bigg[-\int_{t_0}^s r(y)\, dy \bigg] ds$$

$$+ q(x(T)) \exp\bigg[-\int_{t_0}^T r(y)\, dy \bigg] \bigg| x(t) = x_t^* \bigg\}$$

$$= \max_u E_{t_0} \bigg\{ \int_t^{t+\Delta t} g^i\big[s, x(s), u(s)\big] \exp\bigg[-\int_{t_0}^s r(y)\, dy \bigg] ds$$

$$+ V(t + \Delta t, x_t^* + \Delta x_t^*) \bigg| x(t) = x_t^* \bigg\}, \qquad \text{(A.26)}$$

where

$$\Delta x_t^* = f\big[t, x_t^*, \phi^*(t, x_t^*)\big]\Delta t + \sigma\big[t, x_t^*\big]\Delta z_t + o(\Delta t),$$

$$\Delta z_t = z(t + \Delta t) - z(t), \quad \text{and} \quad E_t\big[o(\Delta t)\big]/\Delta t \to 0 \quad \text{as } \Delta t \to 0.$$

With $\Delta t \to 0$, applying Ito's lemma, (A.26) can be expressed as

$$
\begin{aligned}
V(t, x_t^*) = \max_u E_{t_0} \bigg\{ & g^i\big[t, x_t^*, u\big]\exp\bigg[-\int_{t_0}^t r(y)\,dy\bigg]\Delta t + V(t, x_t^*) + V_t(t, x_t^*)\Delta t \\
& + V_{x_t}(t, x_t^*)f\big[t, x_t^*, \phi^*(t, x_t^*)\big]\Delta t + V_{x_t}(t, x_t^*)\sigma\big[t, x_t^*\big]\Delta z_t \\
& + \frac{1}{2}\sum_{h,\zeta=1}^m \Omega^{h\zeta}(t, x)V_{x^h x^\zeta}(t, x)\Delta t + o(\Delta t)\bigg\}.
\end{aligned}
\tag{A.27}
$$

Dividing (A.27) throughout by Δt, with $\Delta t \to 0$, and taking the expectation yields

$$
\begin{aligned}
& -V_t(t, x_t^*) - \frac{1}{2}\sum_{h,\zeta=1}^m \Omega^{h\zeta}(t, x)V_{x^h x^\zeta}(t, x) \\
& = \max_u \bigg\{ g^i\big[t, x_t^*, u\big]\exp\bigg[-\int_{t_0}^t r(y)\,dy\bigg] + V_{x_t}(t, x_t^*)f\big[t, x_t^*, \phi^*(t, x_t^*)\big]\bigg\},
\end{aligned}
\tag{A.28}
$$

with the boundary condition

$$V(T, x^*(T)) = q(x^*(T))\exp\bigg[-\int_{t_0}^T r(y)\,dy\bigg].$$

Hence Theorem A.5 is proved.   □

Example A.5 Consider a resource extractor (firm) endowed with a renewable resource. Let $u(s)$ denote the quantity of the resource extracted and $x(s)$ the size of the resource stock at time s. In particular, $u(s) \le x(s)$ for all time s.

The extraction cost depends on the quantity of the resource extracted $u(s)$, the resource stock size $x(s)$, and a parameter c. In particular, the extraction cost can be specified as follows:

$$C = \frac{c}{x(s)^{1/2}}u(s).$$

This specification implies that the cost per unit of resource extraction $cx(s)^{-1/2}$ decreases when $x(s)$ increases. The above cost structure was also adopted by Jørgensen and Yeung (1996). A decreasing unit cost follows from two assumptions: (i) the cost of extraction is proportional to the extraction effort, and (ii) the amount of the resource extracted, seen as the output of a production function of two inputs (effort and stock level), is increasing in both inputs (cf. Clark 1976).

The market price of the resource depends on the total amount extracted and supplied to the market. The price-output relationship at time s is given by the following inverse demand curve:

$$P(s) = u(s)^{-1/2}.$$

The resource extractor is given an extraction license for the period $[t_0, T]$. When the license expires, the extractor will receive a settlement equaling $qx(T)^{\frac{1}{2}}$. The discount rate is r.

The stochastic problem of the economic agent can be expressed as

$$\max_u E_{t_0} \left\{ \int_{t_0}^T \left[u(s)^{1/2} - \frac{c}{x(s)^{1/2}} u(s) \right] \exp\left[-r(s - t_0)\right] ds \right.$$

$$\left. + \exp\left[-r(T - t_0)\right] q x(T)^{\frac{1}{2}} \right\}, \tag{A.29}$$

subject to

$$dx(s) = \left[ax(s)^{1/2} - bx(s) - u(s) \right] ds + \sigma x(s) dz(s), \quad x(t_0) = x_0 \in X, \tag{A.30}$$

where $c, a, b,$ and σ are positive parameters.

Invoking Theorem A.5 we have

$$-V_t(t, x) - \frac{1}{2}\sigma^2 x^2 V_{xx}(t, x) = \max_u \left\{ \left[u^{1/2} - \frac{c}{x^{1/2}} u \right] \exp\left[-r(t - t_0)\right] \right.$$

$$\left. + V_x(t, x)\left[ax^{1/2} - bx - u \right] \right\}, \quad \text{and} \tag{A.31}$$

$$V(T, x) = \exp\left[-r(T - t_0)\right] q x^{\frac{1}{2}}.$$

Performing the indicated maximization in (A.31) yields

$$\phi^*(t, x) = \frac{x}{4[c + V_x \exp[r(t - t_0)]x^{1/2}]^2}. \tag{A.32}$$

Substituting $\phi^*(t, x)$ from (A.32) into (A.31) and solving (A.31) yields the value function

$$V(t, x) = \exp\left[-r(t - t_0)\right]\left[A(t)x^{1/2} + B(t)\right],$$

where $A(t)$ and $B(t)$ satisfy

$$\dot{A}(t) = \left[r + \frac{1}{8}\sigma^2 + \frac{b}{2} \right] A(t) - \frac{1}{2[c + A(t)/2]} + \frac{c}{4[c + A(t)/2]^2}$$

$$+ \frac{A(t)}{8[c + A(t)/2]^2},$$

$$\dot{B}(t) = r B(t) - \frac{a}{2} A(t),$$

$$A(T) = q, \quad \text{and} \quad B(T) = 0.$$

The optimal control for the problem in (A.29)–(A.30) can be obtained as $\phi^*(t, x) = \frac{x}{4[c + \frac{A(t)}{2}]^2}$.

Now consider the infinite-horizon stochastic control problem with a constant discount rate

$$\max_u E_{t_0} \left\{ \int_{t_0}^{\infty} g^i[x(s), u(s)] \exp[-r(s - t_0)] \, ds \right\}, \tag{A.33}$$

subject to the vector-valued stochastic differential equation

$$dx(s) = f[x(s), u(s)] \, ds + \boldsymbol{\sigma}[x(s)] \, dz(s), \quad x(t_0) = x_0. \tag{A.34}$$

Let $\boldsymbol{\Omega}[x(s)] = \boldsymbol{\sigma}[x(s)]\boldsymbol{\sigma}[x(s)]^{\mathsf{T}}$ denote the covariance matrix with its element in row h and column ζ denoted by $\Omega^{h\zeta}[x(s)]$.

Since s does not appear in $g[x(s), u(s)]$ and the state dynamics explicitly, the problem in (A.33) and (A.34) is an autonomous problem.

Consider the alternative problem that starts at time $t \in [t_0, \infty)$ with initial state $x(t) = x$

$$\max_u E_t \left\{ \int_t^{\infty} g^i[x(s), u(s)] \exp[-r(s - t)] \, ds \right\}, \tag{A.35}$$

subject to the vector-valued stochastic differential equation

$$dx(s) = f[x(s), u(s)] \, ds + \boldsymbol{\sigma}[x(s)] \, dz(s), \quad x(t) = x_t. \tag{A.36}$$

The infinite-horizon autonomous problem in (A.35)–(A.36) is independent of the choice of t and dependent only upon the state at the starting time, that is, x_t.

Define the value function to the problem in (A.35) and (A.36) by

$$V(t, x_t^*) = \max_u E_{t_0} \left\{ \int_t^{\infty} g[x(s), u(s)] \exp[-r(s - t_0)] \, ds \, \bigg| \, x(t) = x_t^* \right\},$$

where x_t^* is an element belonging to the set of realizable values along the optimal state trajectory at time t. Moreover, we can write

$$V(t, x_t^*) = \exp[-r(t - t_0)]$$
$$\times \max_u E_{t_0} \left\{ \int_t^{\infty} g[x(s), u(s)] \exp[-r(s - t)] \, ds \, \bigg| \, x(t) = x_t^* \right\}.$$

Since the problem

$$\max_u E_{t_0} \left\{ \int_t^{\infty} g[x(s), u(s)] \exp[-r(s - t)] \, ds \, \bigg| \, x(t) = x_t^* \right\}$$

depends on the current state x_t^* only, we can write

$$W(x_t^*) = \max_u E_{t_0}\left\{\int_t^\infty g[x(s), u(s)]\exp[-r(s-t)]\,ds \,\Big|\, x(t) = x_t^*\right\}.$$

It follows that

$$
\begin{aligned}
V(t, x_t^*) &= \exp[-r(t-t_0)]W(x_t^*),\\
V_t(t, x_t^*) &= -r\exp[-r(t-t_0)]W(x_t^*),\\
V_{x_t}(t, x_t^*) &= -r\exp[-r(t-t_0)]W_{x_t}(x_t^*), \quad \text{and}\\
V_{x_t x_t}(t, x_t^*) &= -r\exp[-r(t-t_0)]W_{x_t x_t}(x_t^*).
\end{aligned}
\tag{A.37}
$$

Substituting the results from (A.37) into Theorem A.5 yields

$$rW(x) - \frac{1}{2}\sum_{h,\zeta=1}^m \Omega^{h\zeta}(x)W_{x^h x^\zeta}(t, x) = \max_u\{g[x, u] + W_x(x)f[x, u]\}. \tag{A.38}$$

Since time is not explicitly involved in (A.38), the derived control u will be a function of x only. Hence one can obtain the following theorem.

Theorem A.6 *A set of controls $u = \phi^*(x)$ constitutes an optimal solution to the infinite-horizon stochastic control problem in (A.33) and (A.34) if there exists continuously twice differentiable function $W(x)$ defined by $R^m \to R$, which satisfies the following equation:*

$$rW(x) - \frac{1}{2}\sum_{h,\zeta=1}^m \Omega^{h\zeta}(x)W_{x^h x^\zeta}(t, x) = \max_u\{g[x, u] + W_x(x)f[x, u]\}$$

$$= \{g[x, \phi^*(x)] + W_x(x)f[x, \phi^*(x)]\}.$$

Substituting the optimal control in Theorem A.6 into (A.34) yields the dynamics of the optimal state path as

$$dx(s) = f[x(s), \phi^*(x(s))]\,ds + \sigma[x(s)]\,dz(s), \quad x(t_0) = x_0.$$

Solving the above vector-valued stochastic differential equations yields the optimal state trajectory $\{x^*(t)\}_{t\geq t_0}$ as

$$x^*(t) = x_0 + \int_{t_0}^t f[x^*(s), \psi^*(x^*(s))]\,ds + \int_{t_0}^t \sigma[x^*(s)]\,dz(s). \tag{A.39}$$

We use X_t^* to denote the set of realizable values of $x^*(t)$ at time t generated by (A.39). The term x_t^* is used to denote an element in the set X_t^*.

Given that x_t^* is realized at time t, the optimal control to the infinite-horizon problem in (A.33) and (A.34) can be expressed as $\psi^*(x_t^*)$.

Example A.6 Consider the infinite-horizon problem

$$E_{t_0}\left\{\int_{t_0}^{\infty}\left[u(s)^{1/2} - \frac{c}{x(s)^{1/2}}u(s)\right]\exp[-r(s-t_0)]\,ds\right\},\qquad(A.40)$$

subject to

$$dx(s) = \left[ax(s)^{1/2} - bx(s) - u(s)\right]ds + \sigma x(s)\,dz(s),\quad x(t_0) = x_0 \in X,\quad(A.41)$$

where c, a, b, and σ are positive parameters.

Invoking Theorem A.6 we have

$$rW(x) - \frac{1}{2}\sigma^2 x^2 W_{xx}(x) = \max_u\left\{\left[u^{1/2} - \frac{c}{x^{1/2}}u\right] + W_x(x)\left[ax^{1/2} - bx - u\right]\right\}.$$
$$(A.42)$$

Performing the indicated maximization in (A.42) yields the control

$$\phi^*(x) = \frac{x}{4[c + W_x(x)x^{1/2}]^2}.$$

Substituting $\phi^*(t,x)$ above into (A.42) and solving (A.42) yields the value function

$$W(x) = \left[Ax^{1/2} + B\right],$$

where A and B satisfy

$$0 = \left[r + \frac{1}{8}\sigma^2 + \frac{b}{2}\right]A - \frac{1}{2[c + A/2]} + \frac{c}{4[c + A/2]^2} + \frac{A}{8[c + A/2]^2},$$

$$B = \frac{a}{2r}A.$$

The optimal control can then be expressed as

$$\phi^*(x) = \frac{x}{4[c + A/2]^2}.$$

Substituting $\phi^*(x) = x/\{4[c + A/2]^2\}$ into (A.41) yields the dynamics of the optimal trajectory as

$$dx(s) = \left[ax(s)^{1/2} - bx(s) - \frac{x(s)}{4[c + A/2]^2}\right]ds + \sigma x(s)\,dz(s),\quad x(t_0) = x_0 \in X.$$

The above dynamical equation yields the optimal trajectory $\{x^*(t)\}_{t\geq t_0}$ as

$$x^*(t) = \varpi(t_0, t)^2\left[x_0^{1/2} + \int_{t_0}^{t}\varpi^{-1}(t_0, s)H_1\,ds\right]^2,\qquad\text{for } t \geq t_0,\qquad(A.43)$$

where

$$\varpi(t_0, t) = \exp\left[\int_{t_0}^{t}\left[H_2 - \frac{\sigma^2}{8}\right] dv + \int_{t_0}^{t} \frac{\sigma}{2} dz(v)\right],$$

$$H_1 = \frac{1}{2}a, \quad \text{and} \quad H_2 = -\left[\frac{1}{2}b + \frac{1}{4[c + A/2]^2} + \frac{\sigma^2}{8}\right].$$

We use X_t^* to denote the set of realizable values of $x^*(t)$ at time t generated by (A.43). The term x_t^* is used to denote an element in the set X_t^*. Given that x_t^* is realized at time t, the optimal control to the infinite-horizon problem in (A.40) and (A.41) can be expressed as $\psi^*(x_t^*)$.

References

Amir, R., Nannerup, N.: Information structure and the tragedy of the commons in resource extraction. J. Bioecon. **8**, 147–165 (2006)

Basar, T.: A counter example in linear-quadratic games: existence of non-linear Nash solutions. J. Optim. Theory Appl. **14**, 425–430 (1974)

Basar, T.: Nash strategies for M-person differential games with mixed information structures. Automatica **11**, 547–551 (1975)

Basar, T.: On the uniqueness of the Nash solution in linear-quadratic differential games. Int. J. Game Theory **5**, 65–90 (1976)

Basar, T.: Two general properties of the saddle-point solutions of dynamic games. IEEE Trans. Autom. Control **AC-22**, 124–126 (1977a)

Basar, T.: Existence of unique equilibrium solutions in nonzero-sum stochastic differential games. In: Roxin, E.O., Liu, P.T., Sternberg, R. (eds.) Differential Games and Control Theory II, pp. 201–228. Marcel Dekker, New York (1977b)

Basar, T.: Decentralized multicriteria optimization of linear stochastic systems. IEEE Trans. Autom. Control **AC-23**, 233–243 (1978)

Basar, T.: Information structures and equilibria in dynamic games. In: Aoki, M., Marzollo, A. (eds.) New Trends in Dynamic System Theory and Economics, pp. 3–5. Academic Press, New York (1979)

Basar, T.: On the existence and uniqueness of closed-loop sampled-data Nash controls in linear-quadratic stochastic differential games in optimization techniques. In: Iracki, K., et al. (eds.) Lecture Notes in Control and information Sciences, pp. 193–203. Springer, New York (1980). Chap. 22

Basar, T.: Time consistency and robustness of equilibria in noncooperative dynamic games. In: Van der Ploeg, F., de Zeeuw, A. (eds.) Dynamic Policy Games in Economics, pp. 9–54. North-Holland, Amsterdam (1989)

Basar, T., Ho, Y.C.: Informational properties of the Nash solutions of two stochastic nonzero-sum games. J. Econ. Theory **7**, 370–387 (1974)

Basar, T., Mintz, M.: On the existence of linear saddle-point strategies for a two-person zero-sum stochastic game. In: Proceedings of the IEEE 11th Conference on Decision and Control, New Orleans, LA, 1972, pp. 188–192. IEEE Computer Society Press, Los Alamitos (1972)

Basar, T., Mintz, M.: A multistage pursuit-evasion game that admits a Gaussian random process as a maximum control policy. Stochastics **1**, 25–69 (1973)

Basar, T., Olsder, G.J.: Dynamic Noncooperative Game Theory, 2nd edn. Academic Press, London (1995)

Beard, R., McDonald, S.: Time-consistent fair water sharing agreements. Ann. Int. Soc. Dyn. Games **9**, 393–410 (2007)

Bellman, R.: Dynamic Programming. Princeton University Press, Princeton (1957)

D.W.K. Yeung, L.A. Petrosyan, *Subgame Consistent Economic Optimization*, Static & Dynamic Game Theory: Foundations & Applications, DOI 10.1007/978-0-8176-8262-0, © Springer Science+Business Media, LLC 2012

Berkovitz, L.D.: A variational approach to differential games. In: Dresher, M., Shapley, L.S., Tucker, A.W. (eds.) Advances in Game Theory, pp. 127–174. Princeton University Press, Princeton (1964)

Bleeke, J., Ernst, D.: Collaborating to Compete. Wiley, New York (1993)

Blodgett, L.L.: Factors in the instability of international joint ventures: an event history analysis. Strateg. Manag. J. **13**, 475–481 (1992)

Breton, M., Zaccour, G., Zahaf, M.: A differential game of joint implementation of environmental projects. Automatica **41**, 1737–1749 (2005)

Breton, M., Zaccour, G., Zahaf, M.: A game-theoretic formulation of joint implementation of environmental projects. Eur. J. Oper. Res. **168**, 221–239 (2006)

Bylka, S., Ambroszkiewicz, S., Komar, J.: Discrete time dynamic game model for price competition in an oligopoly. Ann. Oper. Res. **97**, 69–89 (2000)

Case, J.H.: Equilibrium points of N-person differential games. PhD Thesis, University of Michigan, Ann Arbor, MI (1967). Department of Industrial Engineering, Tech. Report No. 1967-1

Case, J.H.: Toward a theory of many player differential games. SIAM J. Control Optim. **7**, 179–197 (1969)

Case, J.H.: Economics and the Competitive Process. New York University Press, New York (1979)

Cellini, R., Lambertini, L.: A differential game approach to investment product differentiation. J. Econ. Dyn. Control **27**, 51–62 (2002)

Cellini, R., Lambertini, L.: Private and social incentives towards investment in product differentiation. Int. Game Theory Rev. **6**(4), 493–508 (2004)

Chiarella, C., Kemp, M.C., Long, N.V., Okuguchi, K.: On the economics of international fisheries. Int. Econ. Rev. **25**, 85–92 (1984)

Chintagunta, P.K.: Investigating the equilibrium profits to advertising dynamics and competitive effects. Manag. Sci. **39**, 1146–1162 (1993)

Chintagunta, P.K., Jain, D.: Empirical analysis of a dynamic duopoly model of competition. J. Econ. Manag. Strategy **4**, 109–131 (1995)

Chintagunta, P.K., Vilcassim, N.J.: Marketing investment decision in a dynamic duopoly: a model and empirical analysis. Int. J. Res. Mark. **11**, 287–306 (1994)

Clark, C.W.: Mathematical Bioeconomics: The Optimal Management of Renewable Resources. Wiley, New York (1976)

Clark, C.W.: Mathematical Bioeconomics: The Optimal Management of Renewable Resources, 2nd edn. Wiley, New York (1990)

Clemhout, S., Wan, H.Y. Jr.: Environmental problem as a common-property resource game. In: Ehtamo, H., Hämäläinen, R.P. (eds.) Dynamic Games in Economic Analysis, pp. 132–154. Springer, New York (1985a)

Clemhout, S., Wan, H.Y. Jr.: Dynamic common-property resources and environmental problems. J. Optim. Theory Appl. **46**, 471–481 (1985b)

Clemhout, S., Wan, H.Y. Jr.: Differential games-economic applications. In: Aumann, R.J., Hart, S. (eds.) Handbook of Game Theory with Economic Applications II, pp. 801–825. Amsterdam, Elsevier (1994)

D'Aspremont, C., Jacquemin, A.: Cooperative and noncooperative R&D in duopoly with spillovers. Am. Econ. Rev. **78**(5), 1133–1137 (1988)

Deal, K.: Optimizing advertising expenditure in a dynamic duopoly. Oper. Res. **27**, 682–692 (1979)

Dixit, A.K.: A model of duopoly suggesting a theory of entry barriers. Bell J. Econ. **10**, 20–32 (1979)

Dockner, E.J.: A dynamic theory of conjectural variations. J. Ind. Econ. **40**, 377–395 (1992)

Dockner, E.J., Jørgensen, S.: Cooperative and non-cooperative differential game solutions to an investment and pricing problem. J. Oper. Res. Soc. **35**, 731–739 (1984)

Dockner, E.J., Jørgensen, S.: Optimal pricing strategies for new products in dynamic oligopolies. Mark. Sci. **7**, 315–334 (1988)

Dockner, E.J., Jørgensen, S.: New product advertising in dynamic oligopolies. ZOR, Z. Oper.-Res. **36**, 459–473 (1992)

Dockner, E.J., Kaitala, V.: On efficient equilibrium solutions in dynamic games of resource management. Resour. Energy **11**, 23–34 (1989)

Dockner, E.J., Leitmann, G.: Coordinate transformation and derivation of open-loop Nash equilibria. J. Econ. Dyn. Control **110**, 1–15 (2001)

Dockner, E.J., Long, N.V.: International pollution control: cooperative versus noncooperative strategies. J. Environ. Econ. Manag. **24**, 13–29 (1993)

Dockner, E.J., Feichtinger, G., Mehlmann, A.: Noncooperative solutions for a differential game model of fishery. J. Econ. Dyn. Control **13**, 1–20 (1989)

Dockner, E.J., Jørgensen, S., Long, N.V., Sorger, G.: Differential Games in Economics and Management Science. Cambridge University Press, Cambridge (2000)

Erickson, G.: A model of advertising competition. J. Mark. Res. **22**, 297–304 (1985)

Erickson, G.: Empirical analysis of closed-loop duopoly advertising strategies. Manag. Sci. **38**, 1732–1749 (1992)

Erickson, G.: Offensive and defensive marketing: closed-loop duopoly strategies. Mark. Lett. **4**, 285–295 (1993)

Erickson, G.: Differential game models of advertising competition. Cent. Eur. J. Oper. Res. **83**, 431–438 (1995)

Erickson, G.: Dynamic conjectural variation in a Lanchester duopoly. Manag. Sci. **43**, 1603–1608 (1997)

Esteban-Bravo, M., Nogales, F.J.: Solving dynamic stochastic economic models by mathematical programming decomposition methods. Comput. Oper. Res. **35**, 226–240 (2008)

Feenstra, T., Kort, P.M., De Zeeuw, A.: Environmental policy instruments in an international duopoly with feedback investment strategies. J. Econ. Dyn. Control **25**, 1665–1687 (2001)

Feichtinger, G., Dockner, E.J.: A note to Jorgensen's logarithmic advertising game. ZOR, Z. Oper.-Res. **28**, 133–153 (1984)

Feichtinger, G., Hartl, R.F., Sethi, S.P.: Dynamic optimal control models in advertising, recent developments. Manag. Sci. **40**, 195–226 (1994)

Fershtman, C.: Goodwill and market shares in oligopoly. Economica **51**, 271–281 (1984)

Fershtman, C., Muller, E.: Capital accumulation games of infinite duration. J. Econ. Theory **33**, 322–339 (1984)

Fershtman, C., Muller, E.: Turnpike properties of capital accumulation games. J. Econ. Theory **38**, 167–177 (1986)

Filar, J.A., Petrosjan, L.A.: Dynamic cooperative games. Int. Game Theory Rev. **2**(1), 47–65 (2000)

Fischer, R., Mirman, L.: Strategic dynamic interaction. J. Econ. Dyn. Control **16**, 267–287 (1992)

Fleming, W.H.: Optimal continuous-parameter stochastic control. SIAM Rev. **11**, 470–509 (1969)

Fornell, C., Robinson, W.T., Wernerfelt, B.: Consumption experience and sales promotion expenditure. Manag. Sci. **31**, 1084–1105 (1985)

Forster, B.A.: Optimal consumption planning in a polluted environment. Econ. Rec. **49**, 534–545 (1973)

Forster, B.A.: Optimal pollution control with a nonconstant exponential rate of decay. J. Environ. Econ. Manag. **2**, 1–6 (1975)

Fredj, K., Martin-Herran, G., Zaccour, G.: Slowing deforestation pace through subsidies: a differential game. Automatica **40**, 301–309 (2004)

Fruchter, G.E.: Optimal advertising strategies with market expansion. Optim. Control Appl. Methods **20**, 199–211 (1999a)

Fruchter, G.E.: The many-player advertising game. Manag. Sci. **45**, 1609–1611 (1999b)

Fruchter, G.E.: Advertising in a competitive product line. Int. Game Theory Rev. **3**(4), 301–314 (2001)

Fruchter, G.E., Kalish, S.: Dynamic promotional budgeting and media allocation. Eur. J. Oper. Res. **111**, 15–27 (1998)

Fruchter, G.E., Erickson, G.M., Kalish, S.: Feedback competitive advertising strategies with a general objective function. J. Optim. Theory Appl. **109**, 601–613 (2001)

Fudenberg, D., Tirole, J.: Capital as a commitment: strategic investment to deter mobility. J. Econ. Theory **31**, 227–250 (1983)

Fudenberg, D., Tirole, J.: Dynamic Models of Oligopoly. Harwood, London (1986)

Fudenberg, D., Tirole, J.: Game Theory. MIT Press, Cambridge (1991)

Gomes-Casseres, B.: Joint venture instability: is it a problem? Columbia J. World Bus. **XXII**(2), 97–102 (1987)

Hamel, G., Doz, Y., Prahalad, C.K.: Collaborate with your competitors—and win. Harv. Bus. Rev. **67**(1), 133–139 (1989)

Hansen, L.P., Sargent, T.J.: Robustness. Princeton University Press, Princeton (2008)

Haurie, A.: A note on nonzero-sum differential games with bargaining solutions. J. Optim. Theory Appl. **18**, 31–39 (1976)

Haurie, A., Zaccour, G.: A differential game model of power exchange between interconnected utilizes. In: Proceedings of the 25th IEEE Conference on Decision and Control, Athens, Greece, pp. 262–266 (1986)

Haurie, A., Zaccour, G.: A game programming approach to efficient management of interconnected power networks. In: Hämäläinen, R.P., Ehtamo, H. (eds.) Differential Game-Developments in Modeling and Computation. Springer, Berlin (1991)

Haurie, J., Krawczyk, J.B., Roche, M.: Monitoring cooperative equilibria in a stochastic differential game. J. Optim. Theory Appl. **81**, 73–95 (1994)

Ho, Y.C., Bryson, A.E. Jr., Baron, S.: Differential games and optimal pursuit evasion strategies. IEEE Trans. Autom. Control **AC-10**, 385–389 (1965)

Horsky, D., Simon, L.S.: Advertising and the diffusion of new products. Mark. Sci. **2**, 1–18 (1983)

Hurwicz, L.: The design of mechanisms for resource allocation. Am. Econ. Rev. **63**, 1–30 (1973)

Jørgensen, S.: A differential games solution to a logarithmic advertising model. J. Oper. Res. Soc. **33**, 425–432 (1982)

Jørgensen, S., Sorger, G.: Feedback Nash equilibria in a problem of optimal fishery management. J. Optim. Theory Appl. **64**, 293–310 (1990)

Jørgensen, S., Yeung, D.W.K.: Stochastic differential game model of a common property fishery. J. Optim. Theory Appl. **90**, 381–403 (1996)

Jørgensen, S., Yeung, D.W.K.: Inter-and intragenerational renewable resource extraction. Ann. Oper. Res. **88**, 275–289 (1999)

Jørgensen, S., Zaccour, G.: Time consistent side payments in a dynamic game of downstream pollution. J. Econ. Dyn. Control **25**, 1973–1987 (2001)

Kaitala, V.: Equilibria in a stochastic resource management game under imperfect information. Eur. J. Oper. Res. **71**, 439–453 (1993)

Kaitala, V., Pohjola, M.: Optimal recovery of a shared resource stock: a differential game with efficient memory equilibria. Nat. Resour. Model. **3**, 91–118 (1988)

Kaitala, V., Pohjola, M.: Economic development and agreeable redistribution in capitalism: efficient game equilibria in a two-class neoclassical growth model. Int. Econ. Rev. **31**, 421–427 (1990)

Kaitala, V., Pohjola, M.: Sustainable international agreements on greenhouse warming: a game theory study. Ann. Int. Soc. Dyn. Games **2**, 67–87 (1995)

Kaitala, V., Mäler, K.G., Tulkens, H.: The acid rain game as a resource allocation process with an application to the international cooperation among Finland, Russia and Estonia. Scand. J. Econ. **97**, 325–343 (1995)

Kamien, M.I., Muller, E., Zang, I.: Research joint ventures and R&D cartels. Am. Econ. Rev. **82**(5), 1293–1306 (1992)

Little, J.D.C.: Aggregate advertising models: the state of art. Oper. Res. **27**, 629–667 (1979)

Long, N.V.: A Survey of Dynamic Games in Economics. World Scientific, Singapore (2010)

Mäler, K.-G., de Zeeuw, A.: The acid rain differential game. Environ. Resour. Econ. **12**, 167–184 (1998)

Maskin, E.: Nash equilibrium and welfare optimality. Rev. Econ. Stud. **66**, 23–38 (1999)

Mesak, H.I., Calloway, J.A.: A pulsing model of advertising competition: a game theoretical. Part A—theoretical foundation. Eur. J. Oper. Res. **86**, 231–248 (1995)

Mesak, H.I., Darrat, A.F.: A competitive advertising model: some theoretical and empirical results. J. Oper. Res. Soc. **44**, 491–502 (1993)

Mukundan, R., Elsner, W.B.: Linear feedback strategies in non-zero-sum differential games. Int. J. Syst. Sci. **6**, 513–532 (1975)

Murray, A.I., Siehl, C.: Joint Venture and Other Alliances: Creating a Successful Cooperative Linkage. Financial Executive Research Foundation, Morristown (1989)

Myerson, R.: Mechanisms design. In: Eatwell, J., Milgate, M., Newman, P. (eds.) The New Palgrave: Allocation, Information and Markets. Norton, New York (1989)

Nash, J.F. Jr.: Equilibrium points in n-person games. Proc. Natl. Acad. Sci. USA **36**, 48–49 (1950)

Olsder, G.J.: On open- and closed-loop bang-bang control in nonzero-sum differential games. SIAM J. Control Optim. **40**(4), 1087–1106 (2001)

Parkhe, A.: Messy research, methodological predispositions and theory development in international joint ventures. Acad. Manag. Rev. **18**(2), 227–268 (1993)

Petrosyan, L.A.: Stability of solutions of nonzero sum differential games with many participants. Vestn. Leningr. Univ. **19**, 6–52 (1977)

Petrosyan, L.A., Danilov, N.N.: Stability of solutions in non-zero sum differential games with transferable payoffs. Vestn. Leningr. Univ. **1**, 52–59 (1979)

Petrosyan, L.A., Yeung, D.W.K.: Dynamically stable solutions in randomly-furcating differential games. Trans. Steklov Inst. Math. **253**(Supplement 1), S208–S220 (2006)

Petrosyan, L.A., Danilov, N.N.: Cooperative differential games and their applications. Izd. Tomskogo University, Tomsk (1982)

Petrosyan, L.A.: The regularization of NB-scheme in differential games. Dyn. Control **5**, 31–35 (1995)

Petrosyan, L.A.: Agreeable solutions in differential games. Int. J. Math. Game Theory Algebr. **7**, 165–177 (1997)

Petrosyan, L.A.: Bargaining in dynamic games. In: Petrosyan, L.A., Yeung, D.W.K. (eds.) ICM Millennium Lectures on Games, pp. 139–143. Springer, Berlin (2003)

Petrosyan, L.A., Yeung, D.W.K.: Subgame-consistent cooperative solutions in randomly-furcating stochastic differential games. Int. J. Math. Comput. Model. **45**, 1294–1307 (2007) (Special Issue on Lyapunov's Methods in Stability and Control)

Petrosyan, L.A., Zaccour, G.: Time-consistent Shapley value allocation of pollution cost reduction. J. Econ. Dyn. Control **27**(3), 381–398 (2003)

Petrosyan, L.A., Zenkevich, N.A.: Game Theory. World Scientific, Singapore (1996)

Plourde, C., Yeung, D.: Harvesting of a transboundary replenishable fish stock: a non-cooperative game solution. Mar. Resour. Econ. **6**, 57–71 (1989)

Pontryagin, L.S.: On the theory of differential games. Usp. Mat. Nauk **21**, 219–274 (1966)

Pontryagin, L.S., Boltyanskii, V.G., Gamkrelidze, R.V., Mishchenko, E.F.: The Mathematical Theory of Optimal Processes. Interscience, New York (1962)

Reinganum, J.F., Stokey, N.L.: Oligopoly extraction of a common property resource: the importance of the period of commitment in dynamic games. Int. Econ. Rev. **26**, 161–173 (1985)

Reynolds, S.S.: Capacity investment, preemption and commitment in an infinite horizon model. Int. Econ. Rev. **28**, 69–88 (1987)

Reynolds, S.S.: Dynamic oligopoly with capacity adjustment costs. J. Econ. Dyn. Control **15**, 491–514 (1991)

Rubio, S.J., Casino, B.: A note on cooperative versus non-cooperative strategies in international pollution control. Resour. Energy Econ. **24**, 251–261 (2002)

Smith, A.E., Zenou, Y.: A discrete-time stochastic model of job matching. Rev. Econ. Dyn. **6**(1), 54–79 (2003)

Sethi, S.P.: Optimal control of the Vidale–Wolfe advertising model. Oper. Res. **21**, 998–1013 (1973)

Sethi, S.P., Thompson, G.L.: Optimal Control Theory: Applications to Management Science and Economics, 2nd edn. Kluwer Academic, Boston (2000)

Shapley, L.S.: A value for N-person games. In: Kuhn, H.W., Tucker, A.W. (eds.) Contributions to the Theory of Games, pp. 307–317. Princeton University Press, Princeton (1953)

Singh, N., Vives, X.: Price and quantity competition in a differentiated duopoly. Rand J. Econ. **15**, 546–554 (1984)

Sorger, G.: Competitive dynamic advertising: a modification of the case game. J. Econ. Dyn. Control **13**, 55–80 (1989)

Soros, G.: The Crisis of Global Capitalism: Open Society Endangered. BBS/PublicAffairs, New York (1998)

Soros, G.: The Crash of 2008 and What it Means: The New Paradigm for Financial Markets. PublicAffairs, New York (2009). Revised edition

Spence, A.M.: Investment strategy and growth in a new market. Bell J. Econ. **10**, 1–19 (1979)

Starr, A.W., Ho, Y.C.: Further properties of nonzero-sum differential games. J. Optim. Theory Appl. **3**, 207–219 (1969a)

Starr, A.W., Ho, Y.C.: Nonzero-sum differential games. J. Optim. Theory Appl. **3**, 184–206 (1969b)

Stimming, M.: Capital accumulation subject to pollution control: open-loop versus feedback investment strategies. Ann. Oper. Res. **88**, 309–336 (1999)

Suzumura, K.: Cooperative and noncooperative R&D in an oligopoly with spillovers. Am. Econ. Rev. **82**(5), 1307–1320 (1992)

Tahvonen, O.: Carbon dioxide abatement as a differential game. Eur. J. Polit. Econ. **10**, 685–705 (1994)

Tapiero, C.S.: A generalization of the Nerlove–Arrow model to multi-firms advertising under uncertainty. Manag. Sci. **25**, 907–915 (1979)

Tolwinski, B., Haurie, A., Leitmann, G.: Cooperative equilibria in differential games. J. Math. Anal. Appl. **119**, 182–202 (1986)

Tsutsui, S., Mino, K.: Nonlinear strategies in dynamic duopolistic competition with sticky prices. J. Econ. Theory **52**, 136–161 (1990)

von Neumann, J., Morgenstern, O.: Theory of Games and Economic Behavior. Princeton University Press, Princeton (1944)

Wang, Q., Wu, Z.: A duopolistic model of dynamic competitive advertising. Eur. J. Oper. Res. **128**, 213–226 (2001)

Wie, B.W., Choi, K.: The computation of dynamic Cournot–Nask traffic network equilibria in discrete time. KSCE J. Civ. Eng. **4**(4), 239–248 (2000)

Yeung, D.W.K.: A differential game of industrial pollution management. Ann. Oper. Res. **37**, 297–311 (1992)

Yeung, D.W.K.: A class of differential games which admits a feedback solution with linear value functions. Eur. J. Oper. Res. **107**, 737–754 (1998)

Yeung, D.W.K.: A stochastic differential game model of institutional investor speculation. J. Optim. Theory Appl. **102**, 463–477 (1999)

Yeung, D.W.K.: Infinite-horizon stochastic differential games with branching payoffs. J. Optim. Theory Appl. **111**, 445–460 (2001)

Yeung, D.W.K.: Subgame consistent dormant-firm cartel. In: Haurie, A., Zaccour, G. (eds.) Dynamic Games and Applications, pp. 255–271. Springer, Berlin (2005)

Yeung, D.W.K.: Solution mechanisms for cooperative stochastic differential games. Int. Game Theory Rev. **8**(2), 309–326 (2006)

Yeung, D.W.K.: Dynamically consistent cooperative solution in a differential game of transboundary industrial pollution. J. Optim. Theory Appl. **134**, 143–160 (2007)

Yeung, D.W.K.: Dynamically consistent solution for a pollution management game in collaborative abatement with uncertain future payoffs. Int. Game Theory Rev. **10**(4), 517–538 (2008). In: Yeung, D.W.K. and Petrosyan, L.A. (Guest Eds.) Special Issue on Frontiers in Game Theory: In Honour of John F. Nash

Yeung, D.W.K.: Time consistent Shapley value imputation for cost-saving joint ventures. Math. Game Theory Appl. **2**(3), 137–149 (2010)

Yeung, D.W.K.: Dynamically consistent cooperative solutions in differential games with asynchronous players' horizons. Ann. Int. Soc. Dyn. Games **11**, 375–395 (2011a)

Yeung, D.W.K.: Dynamically consistent collaborative environmental management with production technique choices. Ann. Oper. Res. (2011b). doi:10.1007/s10479-011-0844-0

Yeung, D.W.K., Petrosyan, L.A.: Proportional time-consistent solution in differential games. In: Yanovskaya, E.B. (ed.) International Conference on Logic, Game Theory and Social Choice, pp. 254–256. St Petersburg State University, St Petersburg (2001)

Yeung, D.W.K., Petrosyan, L.A.: Subgame consistent cooperative solutions in stochastic differential games. J. Optim. Theory Appl. **120**, 651–666 (2004)

Yeung, D.W.K., Petrosyan, L.A.: Subgame consistent solution of a cooperative stochastic differential game with nontransferable payoffs. J. Optim. Theory Appl. **124**(3), 701–724 (2005)

Yeung, D.W.K., Petrosyan, L.A.: Cooperative Stochastic Differential Games. Springer, New York (2006a)

Yeung, D.W.K., Petrosyan, L.A.: Dynamically stable corporate joint ventures. Automatica **42**, 365–370 (2006b)

Yeung, D.W.K., Petrosyan, L.A.: The crux of dynamic economic cooperation: subgame consistency and equilibrating transitory compensation. Game Theory Appl. **11**, 207–221 (2007a)

Yeung, D.W.K., Petrosyan, L.A.: The tenet of transitory compensation in dynamically stable cooperation. Int. J. Tomogr. Stat. **7**, 60–65 (2007b)

Yeung, D.W.K., Petrosyan, L.A.: Managing catastrophe-bound industrial pollution with game-theoretic algorithm: the St Petersburg initiative. In: Petrosyan, L.A., Zenkevich, N.A. (eds.) Contributions to Game Theory and Management, pp. 524–538. St Petersburg State University, St Petersburg (2007c)

Yeung, D.W.K., Petrosyan, L.A.: A cooperative stochastic differential game of transboundary industrial pollution. Automatica **44**(6), 1532–1544 (2008)

Yeung, D.W.K., Petrosyan, L.A.: Subgame-consistent solutions for cooperative stochastic dynamic games. J. Optim. Theory Appl. **145**(3), 579–596 (2010)

Yeung, D.W.K., Petrosyan, L.A.: Subgame consistent cooperative solution of dynamic games with random horizon. J. Optim. Theory Appl. **150**, 78–97 (2011)

Yeung, D.W.K., Petrosyan, L.A., Yeung, P.M.: Subgame consistent solutions for a class of cooperative stochastic differential games with nontransferable payoffs. Ann. Int. Soc. Dyn. Games **9**, 153–170 (2007)

Zaccour, G.: Computation of characteristic function values for linear-state differential games. J. Optim. Theory Appl. **117**(1), 183–194 (2003)